All about TELESCOPES

Contents

section

by Sam Brown

Library of Congress Catalog Card Number: 67-31540

Edmund Scientific Co., Barrington 08007

First Edition published 1967
Second Edition 1975
Third Edition 1976
Fourth Edition 1979
Fifth Edition 1981
Sixth Edition 1985
Seventh Edition 1987
Eighth Edition 1989
Ninth Edition 1991
Tenth Edition 1993
Eleventh Edition 1994

Printed in the United States of America
ISBN: 0-933346-20-4

Edmund Scientific Company
101 East Gloucester Pike
Barrington, New Jersey 08007

it began in 1608

GALILEO (1564-1642) WAS FIRST TO VIEW THE MANY WONDERS OF THE NIGHT SKY WITH A TELESCOPE. NAMED FOR HIM, THE GALILEAN TELESCOPE IS ACTUALLY THE INVENTION (IN 1608) OF JAN LIPPERSHEY.

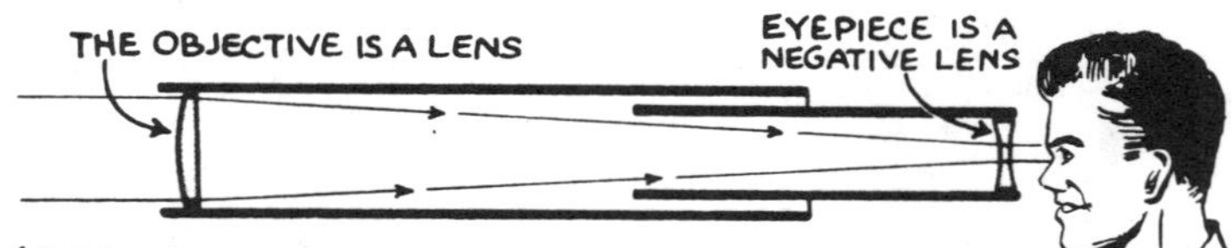

1608 - GALILEAN TELESCOPE - Jan Lippershey - *HOLLAND*

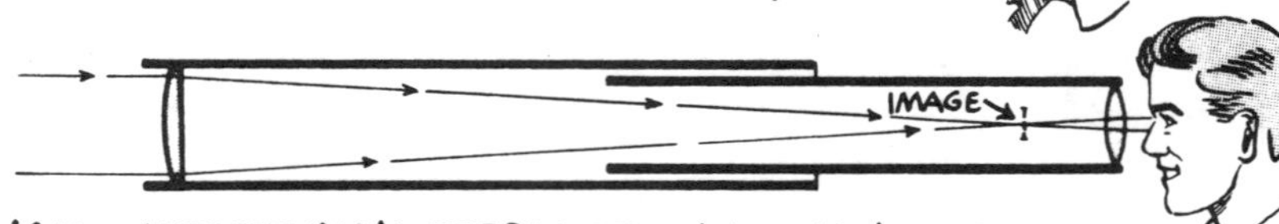

1611 - ASTRONOMICAL REFRACTOR - Johann Kepler - *GERMANY*

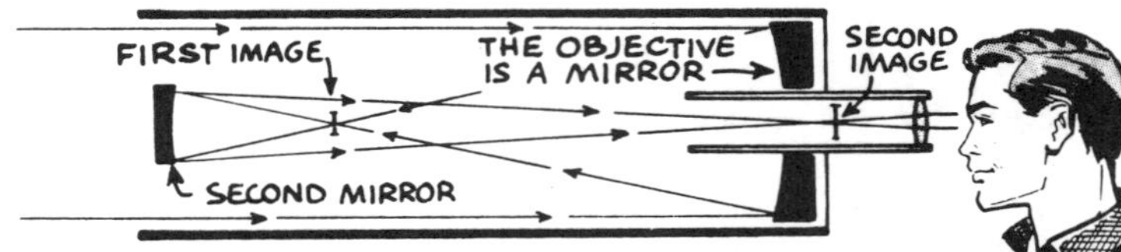

1663 - GREGORIAN REFLECTOR - James Gregory - *SCOTLAND*

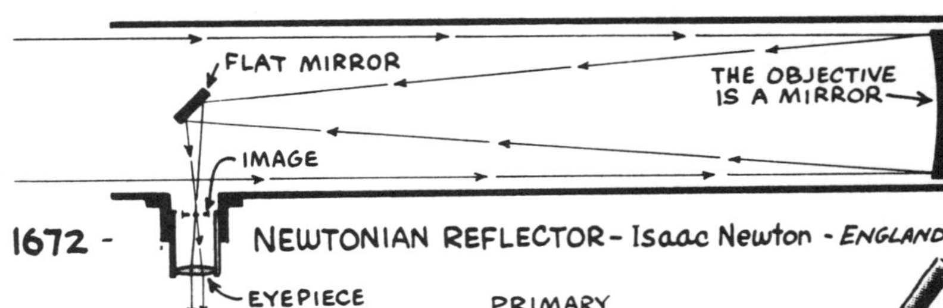

1672 - NEWTONIAN REFLECTOR - Isaac Newton - *ENGLAND*

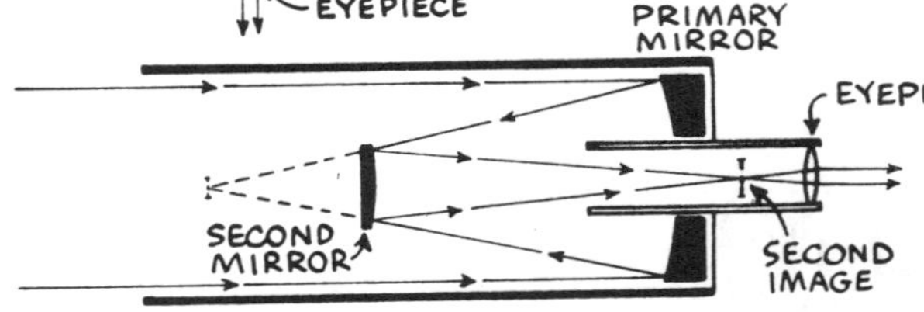

1672 - CASSEGRAIN REFLECTOR - Guillaume Cassegrain - *FRANCE*

THE GIANTS IN MODERN TELESCOPES ARE THE 200-inch APERTURE REFLECTOR ON MT. PALOMAR, Calif., WITH FOCAL LENGTH OF 55 FT., and THE 40-inch YERKES REFRACTOR AT WILLIAMS BAY, Wis., WITH FOCAL LENGTH OF 63 FT.

AMATEURS TODAY BUILD AND USE TELESCOPES TO 12½" APERTURE. TOP CHOICE WITH BEGINNERS IS THE NEWTONIAN REFLECTOR WITH 6 INCH MIRROR, WHICH ITSELF IS OFTEN HOMEMADE

DRAWINGS ADAPTED from "ASTRONOMY and YOU" EDMUND SCIENTIFIC CO.

Getting Acquainted with the Telescope

THE STORY of the telescope begins in Middelburg, Holland, where spectacle maker Jan Lippershey was amazed to discover that a positive lens placed some distance from the eye and a negative lens placed right at the eye, and the two looked through together, brought the distant church steeple so near as to appear right within his shop. The news of the magic glass spread, eventually in 1609 reaching the ear of Galileo Galilei of Padua (Padova), Italy, then 45 years of age, teacher, astronomer and scientist. While Lippershey saw the telescope as being of aid to the military, Galileo was fascinated with the idea of putting it to use in revealing the secrets of the night sky.

"Beyond the stars of the sixth magnitude you will behold through the telescope a host of other stars, so numerous as to be almost beyond belief."

Even in his own time, Galileo was given credit for the instrument which now bears his name--the Galilean telescope. It had the advantage of an erect image, but the grave fault of a very small field of view. Galileo's largest instrument of 1-3/4 inch aperture and 32-power showed less than one quarter of the moon's diameter. A needed step forward was made by Johann Kepler who set forth in 1611 the principles of the astronomical refractor, using a positive eye lens. That the star images were now upside down made no great difference; the big improvement was in the field of view, now expanded some four times.

Even in those long gone days, the big attraction in telescopes was power -- and more power. Even as today, the direct road to power was obvious: the magnification of any telescope is the focal length of the front or objective lens, divided by the focal length of the rear or eye lens. All you have to do is make the objective long and the eyepiece short. And so there soon blossomed aerial rigs of 200 ft. and more in length, all quite shaky and so afflicted with spherical and chromatic aberrations as to fall far short of producing results commensurate with their size.

The famous Isaac Newton (1643-1727) saw a possible solution. Color faults in a refracting telescope result when light is bent or refracted in passing through a glass lens, some colors being refracted much more than others. On the other hand, a reflecting objective reflects all colors the same -- it has no chromatic aberration at all. Of course the spherical aberration is still there, and Newton's small model of his reflector with spherical mirror, built in 1672, was not of convincing optical quality. The reflector languished for fifty years until 1723 when Englishman John Hadley presented to the Royal Astronomical Society the first parabolized reflector. Its performance was excellent, made more dramatic by direct comparison with a 123-ft. focal length refractor. Hadley's instrument was but 5 ft. focal length and a scant 6 in. aperture, yet it showed as much as the larger refractor, and showed it sharper.

While Hadley's demonstration put a stop to the aerial circus of refractors, the reflector too had growing pains, capped by Sir William Herschel's 40 - footer of 48 inches clear aperture erected in 1789. This would be ranked big even today, the first of a great line of modern reflecting telescopes of which the 240-inch in Russia is the current "Mr. Big." The talked-of 300-inch is not yet in the story.

Selecting Your TELESCOPE

A TELESCOPE is the kind of product where the biggest is the best. What you have to do is weigh size against money--and always with an eye on quality. Also you must keep in mind that while the big scope will let you see more, it is a bother to lug around and set up, and is more affected by atmospheric conditions. It is assumed that you want your telescope to be portable. In such case, 6 inches is the accepted maximum size for a reflector; 4 inches for a refractor. Larger telescopes up to 10 inches for reflectors and 6 inches for refractors work well semi-portable, that is you have a permanent mount outdoors and transport only the telescope. For serious star-gazing, a 3-inch refractor or a 4-1/4-inch reflector are generally considered minimum sizes. However, interesting sky objects can be seen with smaller telescopes.

REFRACTOR VS. REFLECTOR. You can read page after page comparing refractors and reflectors, but unless you have actual first-hand experience in using telescopes, such information is of little actual value. The general situation is simply that refractors and reflectors are both good telescopes. Inch-for-inch of aperture, the refractor is generally rated a shade better because (1) it is more likely to be perfectly collimated,

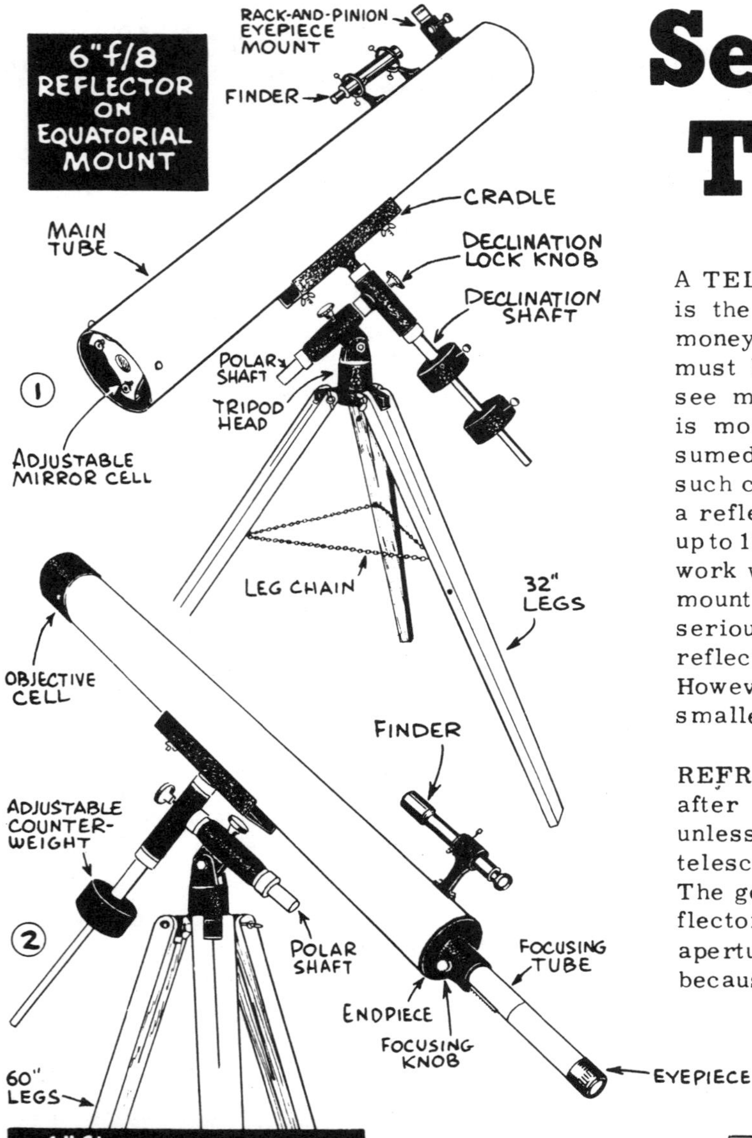

6"f/8 REFLECTOR ON EQUATORIAL MOUNT

4"f/15 REFRACTOR

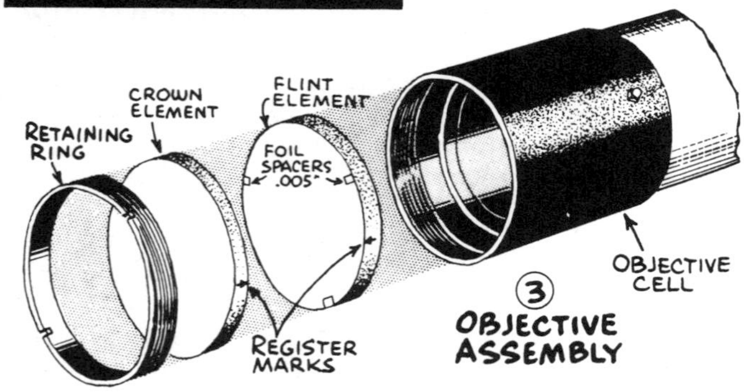

OBJECTIVE ASSEMBLY

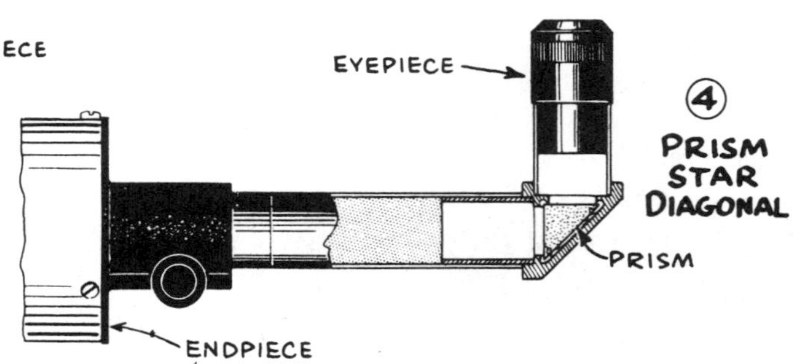

PRISM STAR DIAGONAL

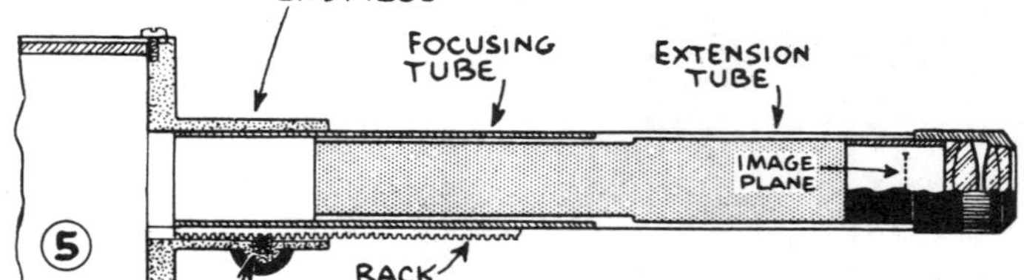

EYE END OF A REFRACTOR

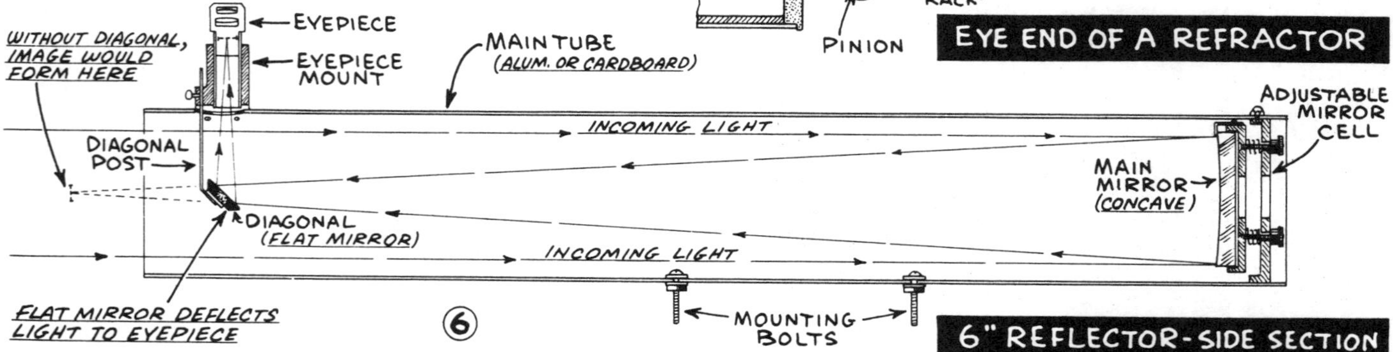

6" REFLECTOR-SIDE SECTION

(2) its closed tube means cleaner optics and less air disturbance within the tube, (3) it can be made nearly 100% glare-proof.

The reflector has two big talking points: (1) it costs only about one-third as much as a refractor of similar size, (2) it is 100% achromatic. Dollar-for-dollar, reflectors are considerably superior to refractors.

Convenience of operation should always be considered. For the most part, a refractor is a neck-breaker. This trouble can be eliminated by the use of a star diagonal, Fig. 4, but the introduction of this extra element causes a slight light loss and may cause serious image deterioration if not perfectly ground and polished. For south sky objects, the reflector is very comfortable to use, the eyepiece being at an approximate 45 degree angle and near eye level for a person standing. It is less convenient for north sky objects unless the tube is rotatable to maintain a horizontal position of the eyepiece tube.

THE MOUNT. If you contemplate serious star-gazing, there is only one type of mount and that is the equatorial. There are dozens of variations of the equatorial mount but all feature two axes at right angles, the commonest type being the German style shown in Figs. 1 and 2. The equatorial is easy to use for south sky objects where the movements are pretty much straight up-and-down in declination and east-and-west in right ascension. The action is more complex in the north sky, as, for example, you sometimes have to move the tube in an east-west direction in order to move it up or down. Because of this slight difficulty, many casual star-gazers prefer the simpler altazimuth mount where the movements are up-and-down and right-left for any and all sky or land objects.

The advantage of the equatorial mount is that it can follow a star by movement around the polar axis only. Also, its movements correspond to the grid of co-ordinate lines used in plotting star atlas maps, making it somewhat easier to find sky objects once you become acquainted with the system. Refinements such as setting circles and manual or clock slow motion are not essential. However, it is a good idea to purchase a mount for which these accessory items are available--once you become an expert observer, you will find circles and slow motion of considerable value. Slow motion with a clock drive is a "must" if you want to do astro photography.

OPTICAL QUALITY. This is where you have to depend on the reputation of the manufacturer unless you can personally inspect and test the telescope. It is best not to believe everything you read because scope makers like soap makers can make even the commonplace sound wonderful. The usual tolerance for precision telescope optics is 1/4-wave, which means the surface must not depart more than 5-1/2 millionths-of-an-inch from the required shape. The required shape is a sphere for lenses; a parabola for concave mirrors. If the f/number of a mirror is f/10 or higher, a spherical shape will not depart from a parabola more than the 1/4-wave tolerance. In other words, a parabola can be a sphere and a sphere can be a parabola with no distinction or difference in image quality. Since the spherical mirror can be machine-polished, it eliminates the expensive hand-figuring necessary to obtain a perfect parabola; hence, it is less expensive. At f/8 and lower f/numbers the sphere departs more than 1/4-wave from the parabola and the mirror must be parabolized for best results.

THE BEST TELESCOPE. All things considered, the 6-inch f/6 reflector on sturdy, simple equatorial mount is the best telescope for the amateur star-gazer. The f/6 means that the focal length is 6 times the mirror diameter, or 36 inches. The mirror is parabolized. A more economical and sensible option would be a 4-1/4 inch f/4.1 reflector at about one half the cost. The f/4.1 parabolic mirror has a focal length of about 17.5 inches, giving the scope approximately half the magnification range as the 6-inch but only half the light grasp. The cheapest reflector you can buy with any claim to quality is the 3-inch size at about half the 4-1/4 inch. You can have a lot of fun with this small instrument but it is not a telescope for serious star study. Good quality refractors of 3-inch or more aperture are excellent but their higher price range usually puts them beyond the beginner's consideration.

In all cases, light grasp is the key feature--objective diameter is what you pay for in any telescope, and is also the yardstick used to measure telescope performance. If you wish merely to view the moon and bright planets, then the smallest refractor or reflector will give nice views at 50 to 100x. Most other sky objects are dim, and what you need to see them is not magnifying power but light power. The situation is sometimes compared to hunting for a lost object in a dark basement, where it is obvious that a light is what you need rather than a magnifying glass.

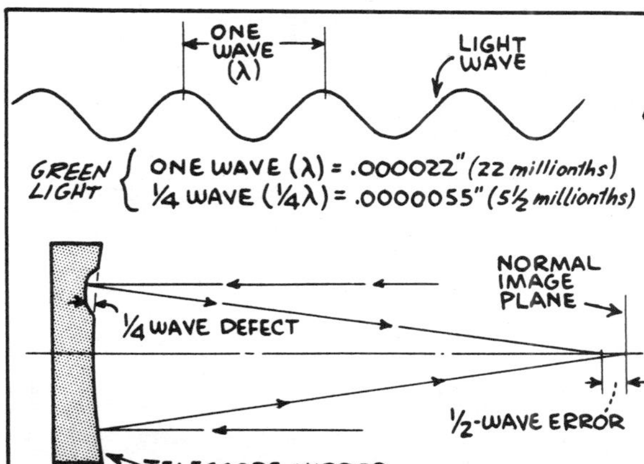

1/4-Wave OPTICS

The usual tolerance for high-precision optics is one-quarter of the wavelength of light--no part of the glass surface must depart more than 1/4-wave or 5-1/2 millionths-of-an-inch from the specified shape. Compare this with a sheet of paper which has a thickness of about 200 waves! In the case of a telescope mirror where light traverses the distance twice, a 1/4-wave defect on the mirror will result in a 1/2-wave error at the image plane as shown. This gives nearly perfect imagery and further narrowing of the tolerance to 1/8-wave, 1/10-wave, etc., is more in the nature of advertising than any appreciable gain in definition.

TELESCOPE PERFORMANCE

OBJECTIVE DIA.		MAGNIFICATION								LIGHT		RESOLUTION	
INCH =	MM	3½x PER INCH	6x PER INCH	10x PER INCH	20x PER INCH	30x PER INCH	40x PER INCH	50x PER INCH	60x PER INCH	COMPARATIVE SCALE	FAINTEST STAR	DAWES' LIMIT	WORKING VALUE
1"	25MM	3½x	6x	10x	20x	30x	40x	50x	60x	9 EYES	8.8MAG.	4.5 SEC.	8.0 SEC.
1¼"	32	4½x	7½x	12x	25x	38x	50x	62x	75x	14	9.3	3.6	6.4
1½"	38	5x	9x	15x	30x	45x	60x	75x	90x	20	9.7	3.0	5.3
1¾"	44	6x	10½x	17x	35x	52x	70x	87x	105x	28	10.0	2.6	4.6
2"	51	7x	12x	20x	40x	60x	80x	100x	120x	36	10.3	2.3	4.0
2½"	64	9x	15x	25x	50x	75x	100x	125x	150x	56	10.8	1.8	3.2
3"	76	10½x	18x	30x	60x	90x	120x	150x	180x	81	11.2	1.5	2.7
4"	101	14x	24x	40x	80x	120x	160x	200x	240x	144	11.8	1.1	2.0
4¼"	108	15x	26x	43x	85x	128x	170x	212x	255x	162	11.9	1.0	1.9
5"	127	18x	30x	50x	100x	150x	200x	250x	300x	225	12.3	.9	1.6
6"	152	21x	36x	60x	120x	180x	240x	300x	360x	324	12.7	.8	1.3
8"	203	28x	48x	80x	160x	240x	320x	400x	480x	576	13.3	.6	1.0
10"	254	35x	60x	100x	200x	300x	400x	500x	600x	900	13.8	.5	.8
12"	305	42x	72x	120x	240x	360x	480x	600x	720x	1296	14.2	.4	.7
REMARKS		LOWEST USEFUL POWER ...GIVES 7MM EXIT PUPIL	BEST VISUAL ACUITY ...4MM EXIT PUPIL	IDEAL FOR LAND OBJECTS AND WIDE VIEWS OF SKY	GOOD FOR PLANETS, M-OBJECTS AND GENERAL VIEWING	30 TO 40x PER INCH IS NORMAL HI-POWER. USE FOR PLANET DETAIL, DOUBLE STARS, ETC.		MAX. USEFUL HI-POWER ...½MM EXIT PUPIL	USEFUL ONLY FOR CLOSE DOUBLE STARS	BASED ON DARK-ADAPTED UNAIDED EYE = 1	BASED ON ABILITY OF NAKED EYE TO SEE MAG. 6.2 STAR	NEEDS "GOOD SEEING" AND POWER OF 50x PER INCH OR MORE	BASED ON AVERAGE SEEING. NEEDS 30x PER INCH OR MORE

MAGNIFICATION. The top useful power is about 60x per inch of objective aperture. No additional detail in an extended object (moon, planets, etc.) can be obtained with higher powers; the diffraction pattern itself begins to show at about 50x per inch and further magnification tends to destroy definition rather than improve it.

LIGHT. The comparative scale can be compared with the eye; a 1 inch objective equals 9 eyes. Also one value can be compared with another, as, for example, a 3 inch objective picks up four times as much light as a 1-1/2 inch objective (81 to 20). Faintest star figures can be increased by one magnitude if "seeing" is excellent, especially with averted vision.

RESOLUTION. A star is a mathematical point, subtending an angle of 1/20 of 1 second of arc or less. No telescope can show such a tiny object as it really is but instead expands the angle to form a small disk of light, known as a diffraction disk or pattern. The smaller the diffraction pattern, the better the resolution. Resolution means simply the ability to show fine detail, and an exact measure of this is offered by close double stars. It should be noted that the common standard--Dawes Limit--does not call for actual separation; the values in right column above are approximate minimums for complete separation. Dawes Limit recognizes only the bright center of the star image, and the angular and linear diameters given in table below are for this part of the diffraction pattern.

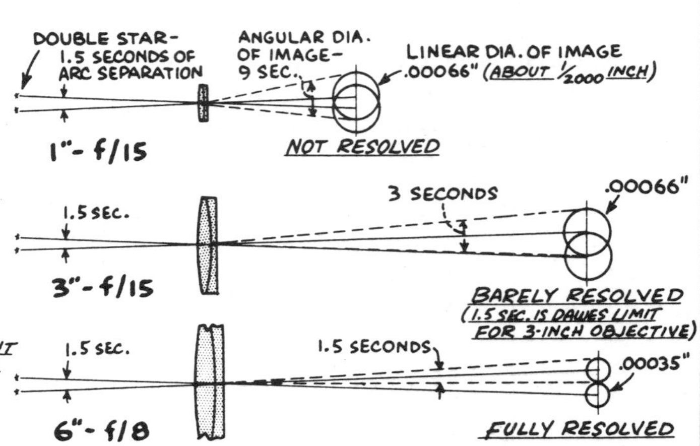

RESOLUTION APPEARANCE OF TYPICAL DOUBLE STAR AT IMAGE PLANE OF THREE DIFFERENT TELESCOPES

SIZE OF STAR IMAGE

ANGULAR DIA.		LINEAR DIA.	
DIA. OF OBJECTIVE	SECONDS OF ARC	f/VALUE	INCHES
1"	9.0	f/4	.00017
2"	4.5	f/6	.00026
3"	3.0	f/8	.00035
4"	2.2	f/10	.00044
5"	1.8	f/15	.00066
6"	1.5	f/20	.00088
8"	1.1	f/25	.0011

ANGULAR DIAMETER DEPENDS ONLY ON THE DIAMETER OF OBJECTIVE. DAWES' LIMIT IS ½ ANGLE OF STAR IMAGE

LINEAR DIAMETER DEPENDS ONLY ON f/VALUE AND IS THE SAME FOR ANY TELESCOPE OF A SPECIFIC f/VALUE

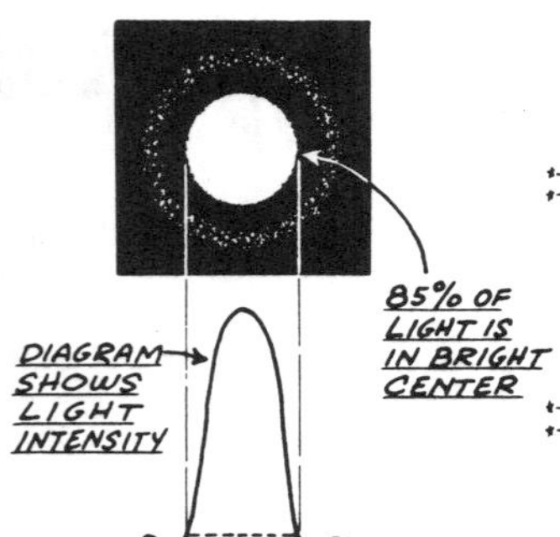

THE STAR IMAGE IS A DIFFRACTION DISK

What to Expect in TELESCOPE PERFORMANCE

IF YOUR telescope has good optics in good alignment and if seeing conditions are good, then you can expect performance to the values given in the table on opposite page, the critical item being resolving power. So far as light grasp and magnification are concerned, even the cheapest, poorest telescope will measure up to standard. The catch, of course, is that light power and magnifying power mean nothing if the telescope image lacks clarity and sharpness.

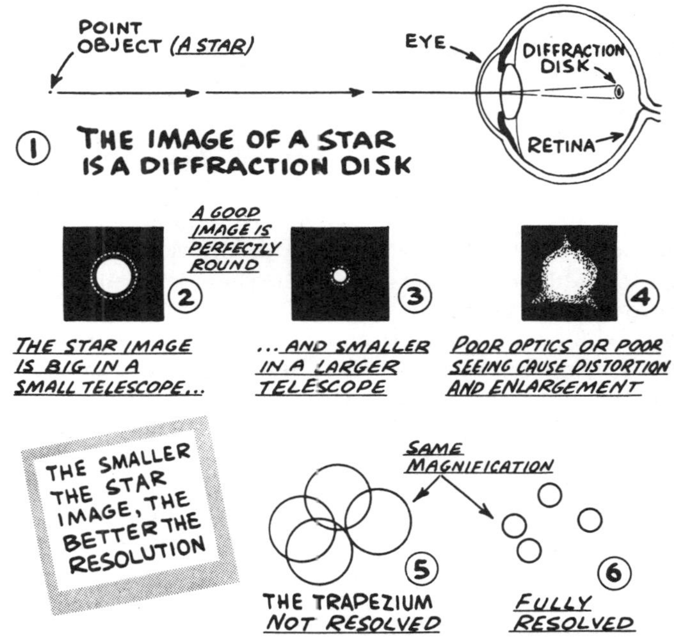

LIGHT POWER. Objective diameter alone determines the light power of your telescope--the bigger the lens or mirror, the more light it will pick up. If the diameter of the eye is taken as about one-third inch and given a value of 1, the comparative light power of objectives will be as given in the table. The base for the "faintest star" is the magnitude the eye can see unaided, which is generally taken as 6.2 magnitude. Any departure from this base should be added or subtracted. For example, if you see stars to only 5th magnitude naked-eye, you are 1.2 magnitude under the base figure and must deduct this amount from the values given.

The telescope lets you see all stars brighter. With a 3-inch telescope, you will see 11th magnitude stars as bright as 6th magnitude viewed with the eyes alone; 6th magnitude will look like 1st magnitude, and a 1st magnitude star will be a real sparkler at an apparent minus 4 magnitude. Since stars look big in direct proportion to their brightness, a considerable amount of apparent magnification is gained in this manner. You get increased light only with point objects (stars) where all of the extra light picked up by the telescope goes into a retinal image about the same size as seen naked-eye. All extended objects--moon, planets, nebulae--are seen less bright in the telescope than with eyes alone. You have the same amount of light as before but now it is diluted by being spread over the much greater area of the magnified image.

STARS ARE NOT MAGNIFIED. A star is very tiny in angular diameter. Even the giants and supergiants subtend at most a mere .05 second, which is 1/20 of one second of arc. Just how small an angle this is may be realized from the fact that the type you are reading subtends an angle of about 1600 seconds of arc at your eye--try to imagine 32,000 stars piled one on top of the other to make a stack as high as this type! As a matter of fact, such tiny objects are invisible.

Although stars are too small in angular size to be seen, they are also too bright to be ignored. The light-receiving cones and rods in your eyes are actuated by any light beam, even though the beam itself may be so small as to light only a small portion of one cone. Hence, one star can "trigger" a light cone, and the brain gets the same impression as if the cone were fully illuminated. But the telescope has to magnify a star 13x per inch of objective diameter to fully illuminate one cone. Hence, even at the top magnification of 50x per inch, stars are magnified only about four times the naked-eye view.

THE DIFFRACTION PATTERN. Diffraction is an optical effect caused by the interference of light waves in passing around or through any opening, such as a lens or your eye. Why it happens is not explained here, but what it does can be seen in Fig. 1--a point object is seen as a tiny disk, surrounded by one or more faint rings of light. About 85% of the light is in the central disk and this is the part you see. The diffraction disk is still very small but it is substantially bigger than the angular diameter of the point object itself. By optical laws, the angular diameter of the diffraction disk becomes smaller as the size of the lens is increased. The smaller the diffraction disk, the better the resolving power. If you have a 1-inch telescope with just fair optics, you may not be able to split the Trapezium, Fig. 5. The smaller diffraction pattern of a 3-inch telescope allows a clean split, Fig. 6.

DAWES LIMIT. The common way of rating the resolving power of a telescope is by giving the minimum separation between two stars which yet allows the two stars to be seen as separate points. This is Dawes Limit. It does not mean that you will see the stars cleanly separated but only that you can tell there are two stars. The general theory is that you can tell there are two stars if the edge of one diffraction disk does not extend beyond the center of the other. In other words, Dawes limit is one-half the diameter of the diffraction disk; or, the diffraction disk is two times Dawes limit. Dawes limit is definitely a "limit", and you will get more fun out of double stars by making your minimum double star twice Dawes limit. This will show the two stars just touching.

To test the resolution of your telescope, select a double star with a separation of twice Dawes limit. Both stars must be about the same magnitude and not too bright (mag. 5 or 6), the stars to be located near the zenith. Further, seeing conditions must be good. You will need at least 30x per inch magnification and you can use as much additional power as you like. Then, if you can see the two stars just touching, the resolution of your scope is equal to Dawes limit, and this in turn means the optics are excellent. If the stars overlap but yet are recognizable as a pair, you are doing very well although not to Dawes limit. If you see only one star, then the seeing is bad or your optics are under par.

DETAIL IN EXTENDED OBJECTS. When you look at any extended object, the telescope image is made up of an arrangement of many diffraction disks of a size and spacing determined by the diameter of the objective. The smaller the disks, the more detail you can see, just as more detail is visible in a fine-screen magazine halftone than in a coarse newspaper halftone. When the telescope image is magnified 13x per inch of objective diameter, the diffraction pattern becomes equal to the resolving power of the eye. This is equivalent to a 150-line halftone screen. Such a pattern or screen readily allows about 2-1/2x magnification in order to produce a larger picture and yet not make the screen pattern too prominent. The pattern or screen is then 60-line, the same as used in newspaper halftones. This is the approximate effect you get at 32x per inch magnification. This is ideal high power. Definition remains good to about 50x per inch and then deteriorates sharply; at 60x per inch, the diffraction pattern has a structure of about 25 lines per inch. Being a picture painted with light disks, you can't view a telescope image and actually count off the lines-per-inch structure. However, the effect of the too-coarse screen is readily apparent, the picture becoming soft and woolly like a photograph with too much enlargement. Remember: You can use all the power you like when looking at single stars or double stars or open clusters, because what you are looking at is a single diffraction disk or a pair of disks or an open cluster of disks. However, when you look at an extended object, you want to see the picture as a whole without making the disks of light which comprise it too prominent.

GOOD SEEING. High power magnifies--but everything! You have already seen how it magnifies the diffraction pattern to the point of producing a fuzzy image. It also magnifies heat waves, dust, clouds and air currents. All of these things cause poor imagery, and air currents most of all. You have to slice your way through ten miles or more of swirls, eddies, updrafts and downdrafts, cold air and hot spots. When the air is quiet and steady, then "seeing is good" and star images shrink to tiny points. A steady breeze is often helpful; good seeing often occurs when the sky is dull and hazy. When you pick up star images three or four times normal size and maybe wandering all over the field of the eyepiece, then you know seeing is bad. Since the average backyard astronomer uses modest equipment at 50x to 150x, good seeing is not quite the problem it is to the professional observer, although there are times when even a 50x image gets the shakes and shivers. You will not be long at observing the moon until you notice this. Like the weather, nothing can be done about it although reducing power and aperture makes the commotion less disturbing. To distinguish between "seeing" and optical quality, test your scope on a near object in quiet air in daylight. A clear, sharp image proves you have good optics--the "good seeing" is something you have to wait for.

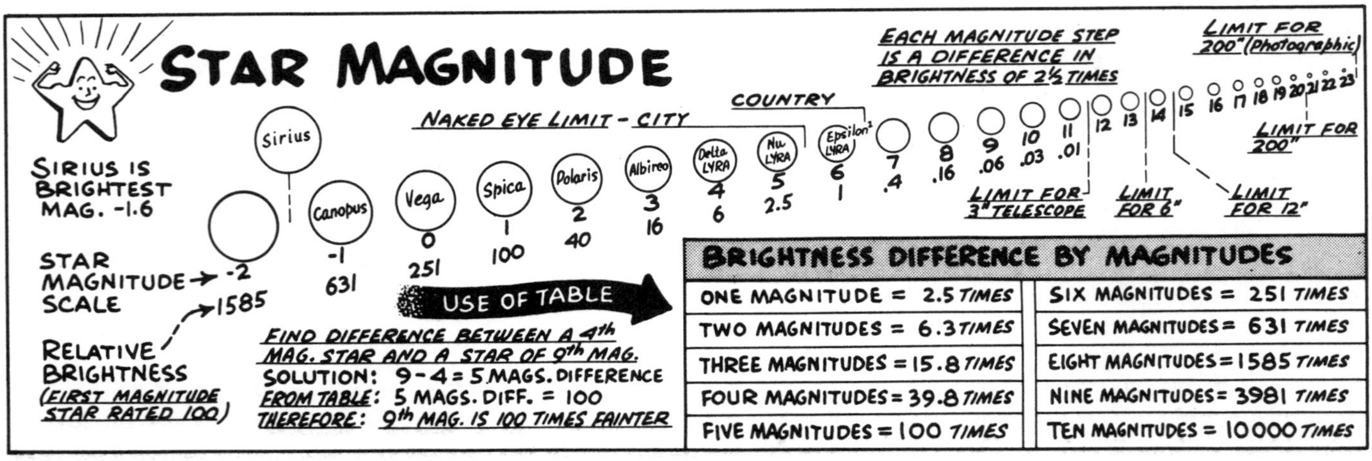

USE OF TABLE

FIND DIFFERENCE BETWEEN A 4th MAG. STAR AND A STAR OF 9th MAG.
SOLUTION: 9 - 4 = 5 MAGS. DIFFERENCE
FROM TABLE: 5 MAGS. DIFF. = 100
THEREFORE: 9th MAG. IS 100 TIMES FAINTER

BRIGHTNESS DIFFERENCE BY MAGNITUDES

ONE MAGNITUDE = 2.5 TIMES	SIX MAGNITUDES = 251 TIMES
TWO MAGNITUDES = 6.3 TIMES	SEVEN MAGNITUDES = 631 TIMES
THREE MAGNITUDES = 15.8 TIMES	EIGHT MAGNITUDES = 1585 TIMES
FOUR MAGNITUDES = 39.8 TIMES	NINE MAGNITUDES = 3981 TIMES
FIVE MAGNITUDES = 100 TIMES	TEN MAGNITUDES = 10000 TIMES

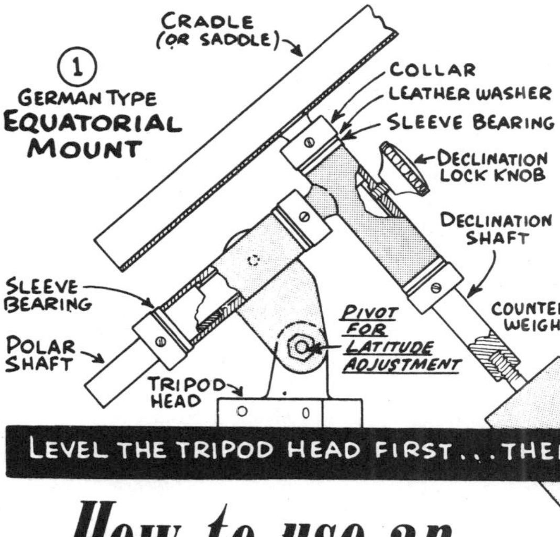

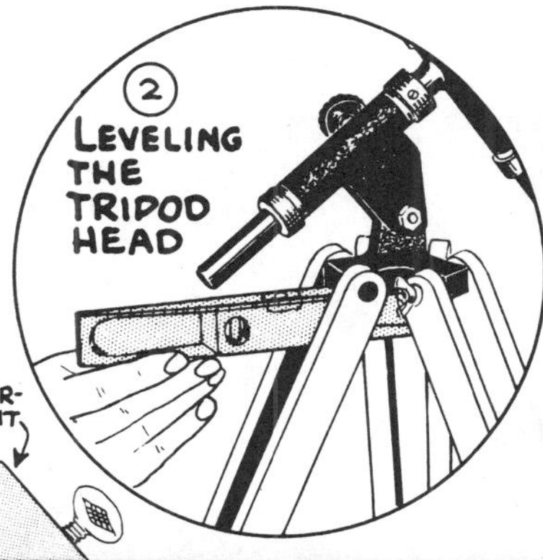

LEVEL THE TRIPOD HEAD FIRST . . . THEN SET POLAR AXIS TO YOUR LATITUDE

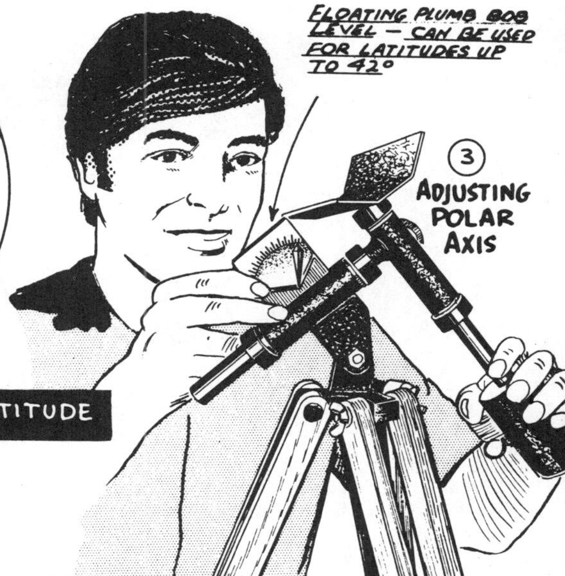

How to use an EQUATORIAL MOUNT

THE TELESCOPE MOUNTING most familiar to amateur stargazers is the German-type equatorial shown in Fig. 1. In its simple form it consists merely of two axes at right angles, with the polar axis adjusted to the same angle as the latitude of the observer's location. Refinements include setting circles, slow motion and clock drive--all features which can be added at any time to the original mount.

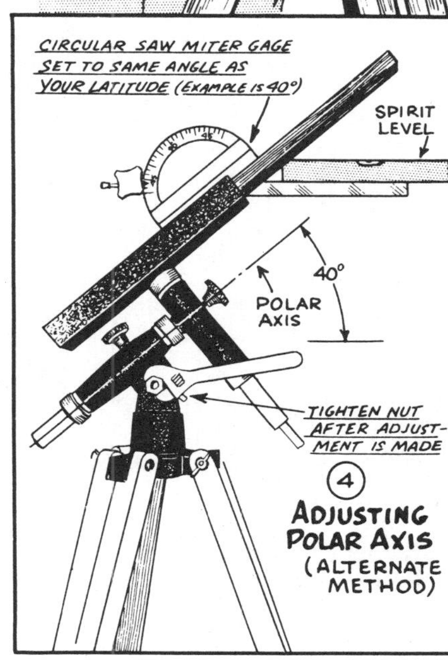

ADJUSTING POLAR AXIS. The first thing you do with an equatorial mount is to adjust the polar axis to your latitude. This is done indoors. First, level the tripod head, Fig. 2. Then loosen the nut at top of tripod head so that the polar axis can be tilted to an angle equal to your latitude. Fig. 3 shows the adjustment being checked with the use of a floating plumb bob level. Another common tool used for jobs like this is the level protractor head of a combination square. Still another way of checking the angle is shown in Fig. 4, which makes use of a circular saw miter gage in combination with a spirit level. With the miter gage in place as shown, the polar axis

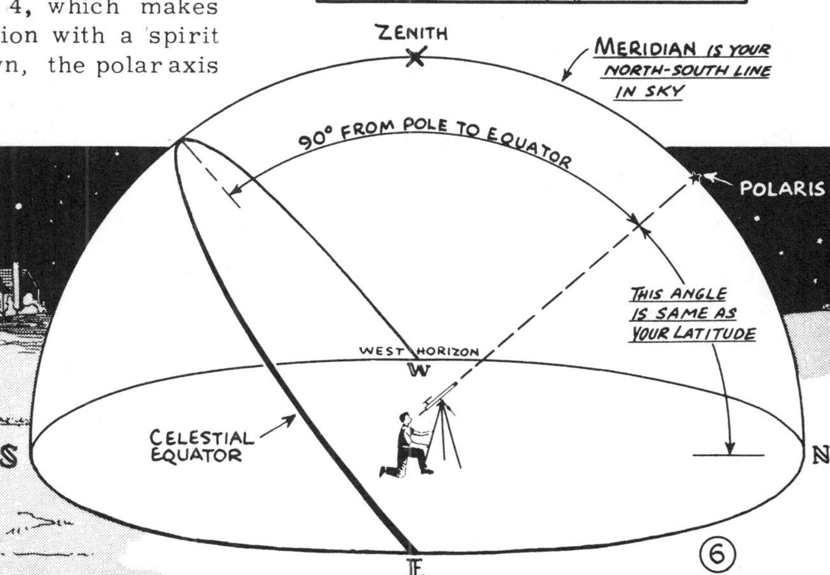

SETTING UP IS SIMPLY A MATTER OF POINTING THE TELESCOPE AT POLARIS . . . EXACT ACCURACY IS NOT NEEDED EXCEPT FOR EQUATORIALS FITTED WITH SETTING CIRCLES

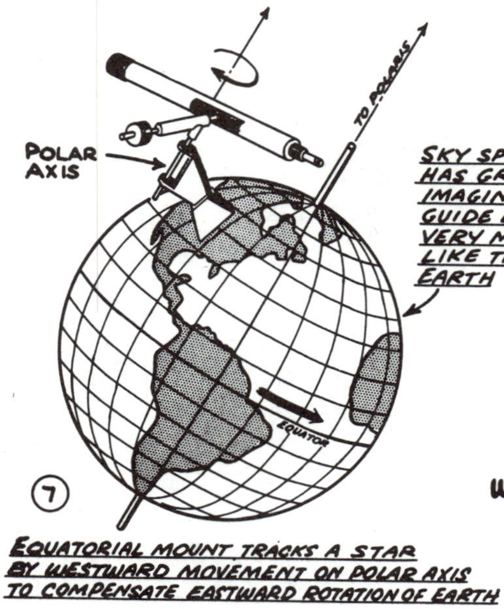

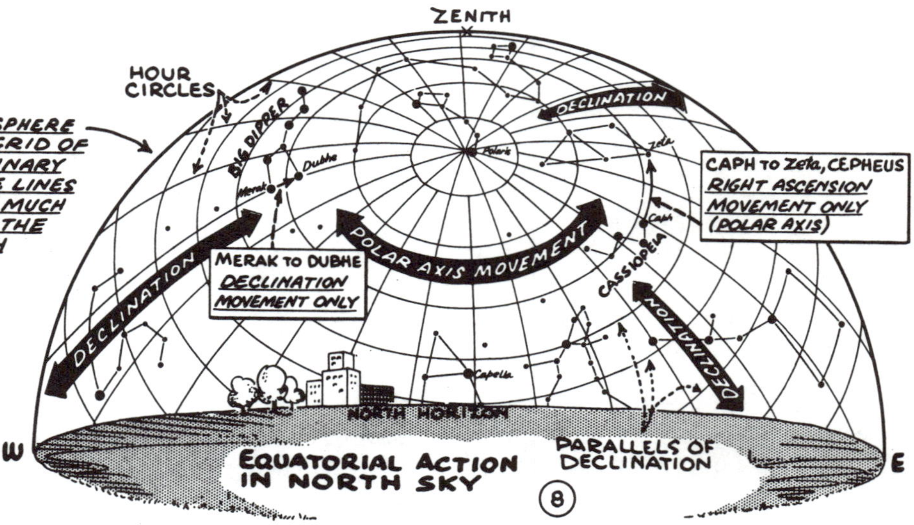

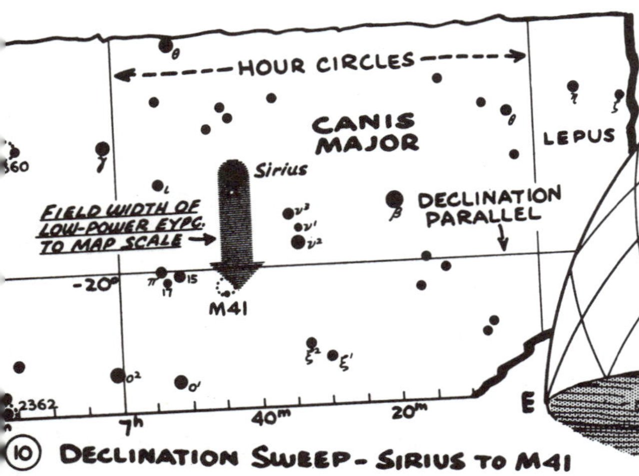

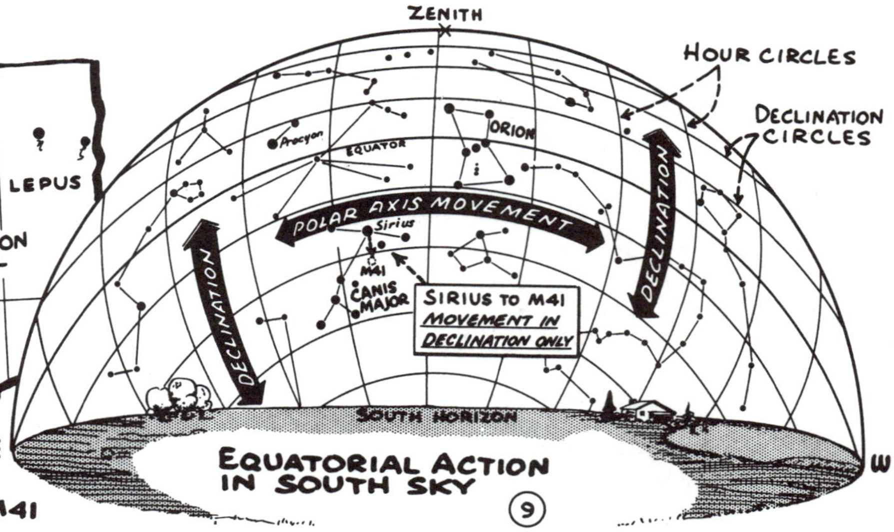

is tilted until a spirit level applied to miter gage guide bar shows level. Accuracy to the nearest degree is good enough.

SETTING UP OUTDOORS. If you spread the tripod legs to make the tripod head approximately level, the telescope will point to the same altitude as Polaris, Fig. 5. The other part of the setup calls for rotation of the whole tripod a little east or west as needed in order to center on Polaris in a cross-wise or east-west position. Rough-sighting along the telescope tube can be followed by a peek in the finder, which should show Polaris near the center of the field. No great accuracy is needed--even a rough setup will put you within five degrees of the pole and this is all the accuracy needed for the short movements normal to either following or finding.

EQUATORIAL MOVEMENTS. By sighting on Polaris, you make the polar axis parallel with the imaginary shaft on which the earth turns, Fig. 7. Hence, as the earth turns to the east, you make a corresponding movement to the west, and in this way you can track a star across the sky with a movement of the polar axis only. An equally important feature is that any movement on either axis corresponds to the grid of plotting lines used on all star atlas maps.

Any movement on the declination axis is a movement in declination. This means that your telescope will move directly toward or away from the North Pole, as shown in Figs. 8 and 9, approximately parallel to the nearest hour circle. Any movement on the polar axis is a movement in right ascension, commonly referred to as R. A. A movement in R. A. is always a circle around the pole, exactly parallel to the parallels of declination. Once you become familiar with this system, you can use it to advantage in finding sky objects. Three examples of "finding" are shown in the drawings (Figs. 8, 9 and 10) and the subject is covered in more detail on other pages.

It will be obvious that the south sky is by far the easiest to work. Here, despite the tilted position of the mount, the action is pretty much a plain up-down, east-west movement very nearly like a simple altazimuth mount. The north sky is a radical departure; both declination and R. A. can run in any direction. However, if you keep the basic movements in mind, you will not find the north sky too difficult.

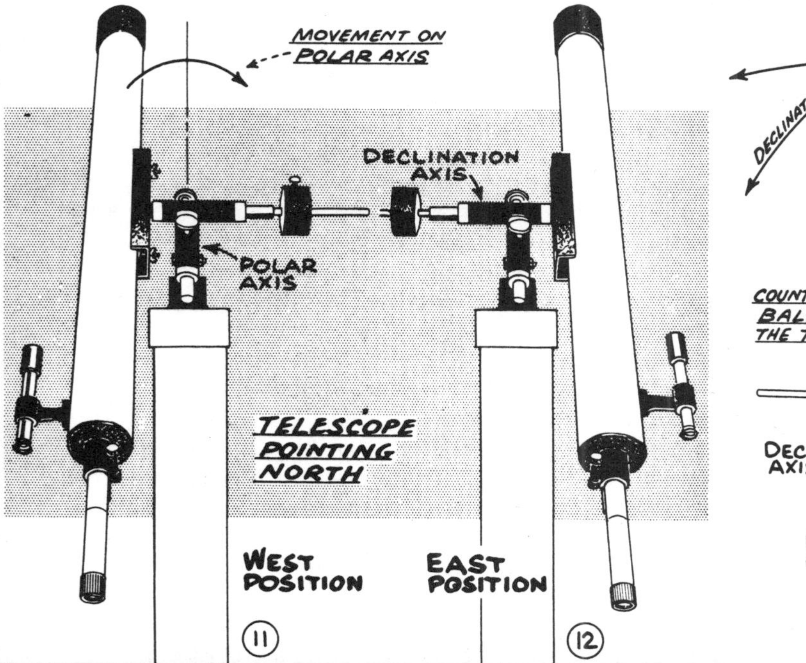

TELESCOPE CAN BE SWUNG TO EITHER SIDE OF TRIPOD

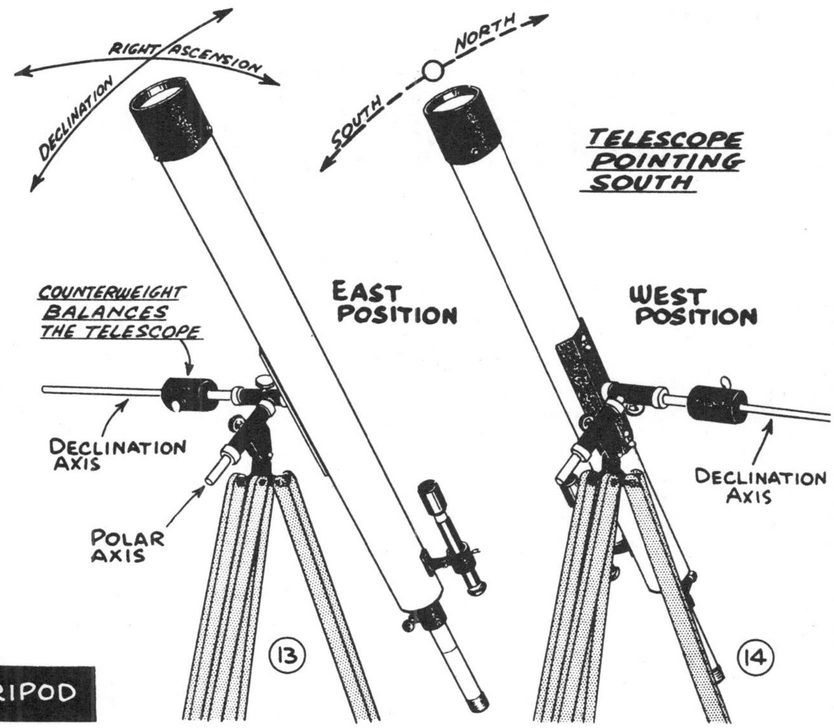

MECHANICAL MOVEMENTS. The only mechanical movement not immediately apparent is the rollover from one side of the tripod head to the other. The rollover is necessary because in most positions the German equatorial can't swing through the meridian--you can't sweep from east to west or west to east. The general rule is that if you want to look at a star in the eastern sky, you use the telescope on west side of tripod; if you want to look west, you roll over to the east side. These rules are reversed if you are observing in the north sky below the pole. Also, in the south sky below the equator, a fair amount of movement through the meridian is possible. It does not mean much to talk about these movements, but if you go through a practice session indoors of pointing to various points in the sky, the whole thing will become clear.

The rollover itself can be done in two ways. With telescope pointing north, the movement from one side to the other is a simple movement around the polar axis, as can be seen in Figs. 11 and 12. From either of these positions you can move in declination into the south sky, Figs. 13 and 14. Alternately, if you are already observing in the south sky, the fastest way to make the rollover is by pointing the front of telescope down, as shown in Fig. 15. Indoor practice should be carried out until all of the movements of the telescope can be made quickly and automatically.

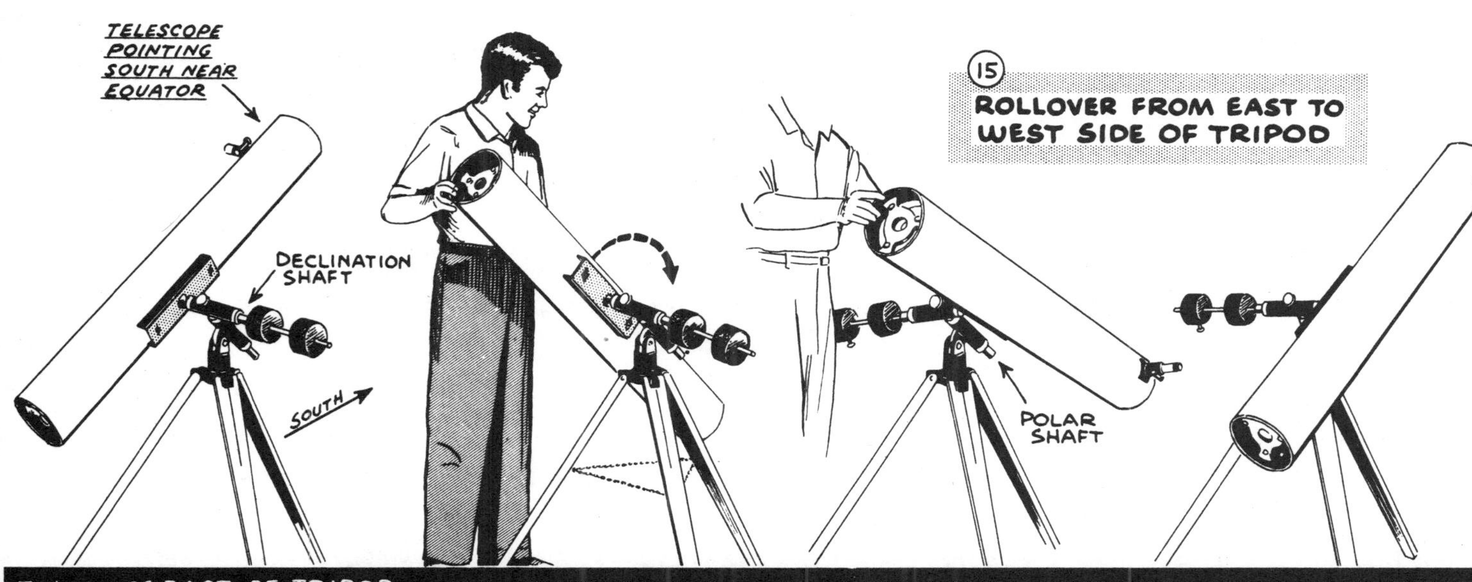

TELESCOPE EAST OF TRIPOD ...TILT FRONT OF TUBE DOWN ...ROLL OVER TO WEST SIDE ...ELEVATE THE TUBE

Telescope ARITHMETIC

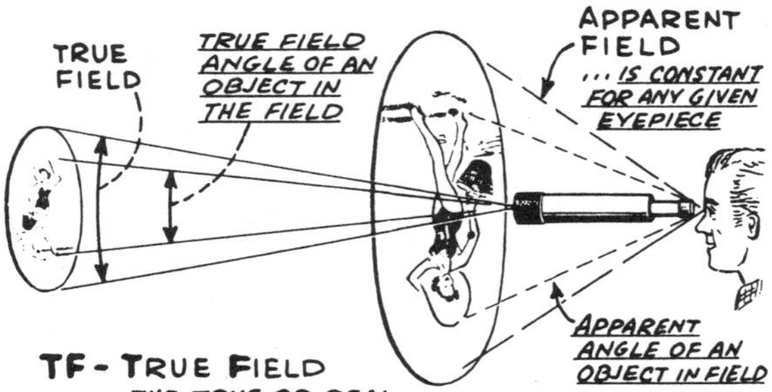

TF - True Field *THE TRUE OR REAL ANGULAR FIELD-OF-VIEW SEEN IN TELESCOPE. ANY OBJECT IN FIELD ALSO HAS A TRUE FIELD ANGLE... IS SAME ANGLE SEEN WITH EYE ALONE*

AF - Apparent Field *IS THE APPARENT ANGLE OF FIELD SEEN THROUGH THE TELESCOPE*

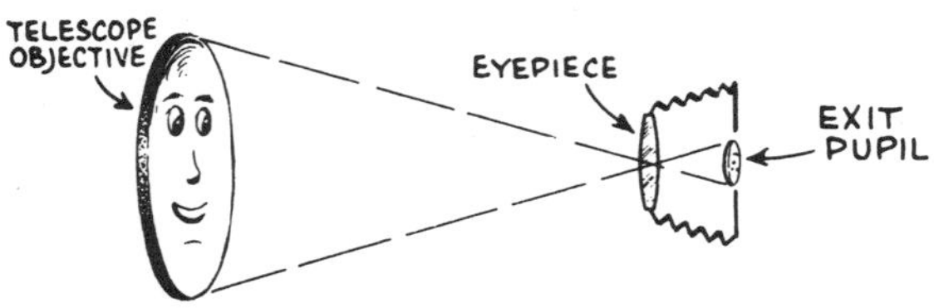

EP - Exit Pupil. *THE EYEPIECE ACTS LIKE A CAMERA AND "TAKES" A LITTLE PICTURE OF THE OBJECTIVE. THIS IS THE EXIT PUPIL*

M - Magnification
DO - Diameter of Objective
FO - Focal Length of Objective
FE - Focal Length of Eyepiece
f/obj. - *THE f-VALUE OF OBJECTIVE (NUMBER OF TIMES FOCAL LENGTH OF OBJECTIVE IS GREATER THAN DIA.)*

ONE OF THE first things you learn about telescopes is that the magnification is equal to the focal length of the objective divided by the focal length of the eyepiece. This basic calculation is No. 1 in the chart below. Like all equations, it can be transposed, as shown by formulas 2 and 3. No. 2 determines the f.l. of eyepiece (FE) needed to obtain a certain magnification; No. 3 gives the f.l. of objective if M and FE are known.

Calculations using the exit pupil are useful for finding the power of a telescope when you know nothing about the instrument. All you have to do is measure the clear diameter of objective. Then on a piece of tracing paper you can see and measure the exit pupil behind the eyepiece. Formula 4 gives the power. A direct-reading magnifier (pocket comparator) is a handy instrument for finding the exact diameter of the exit pupil.

Calculations involving the true field and apparent field, Nos. 7, 8 and 9, are the ones you will use most in actual observing. The apparent field of any eyepiece is a fixed angle. For example, a certain Kellner eyepiece may have an apparent field of 50 degrees--this is a fixed value just the same as the focal length is a fixed value. The true field of the telescope equals AF divided by M (formula 8). Formula 7 gives the magnification when AF and TF are known. Suppose you want to look at the moon (angular dia. 1/2 degree) using the highest power which will show the full disk. If your eyepiece is 50 degrees apparent field, M equals 50 divided by .5 equals 100x magnification. The same calculation can also be applied to any part of the whole field: Assume you want to look at a double star with separation of 15 seconds of arc. You will learn from experience that if this true field angle can be increased to 10 minutes apparent field, the double will be nicely separated. Formula 7 solves the problem, but you must first convert 10 minutes to seconds, equals 600 seconds. Then, 600 seconds (AF) divided by 15 seconds (TF) equals 40x, which is the power needed.

TELESCOPE CALCULATIONS

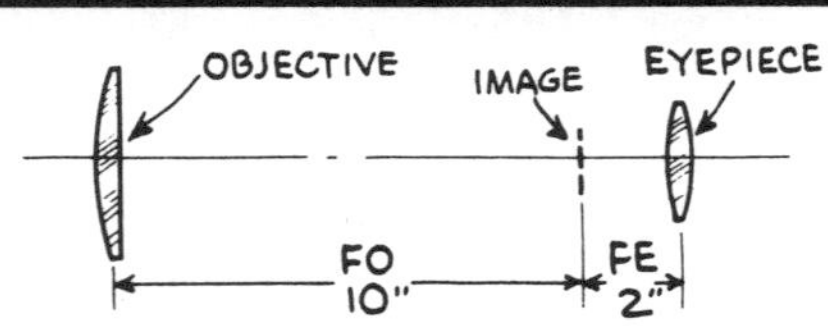

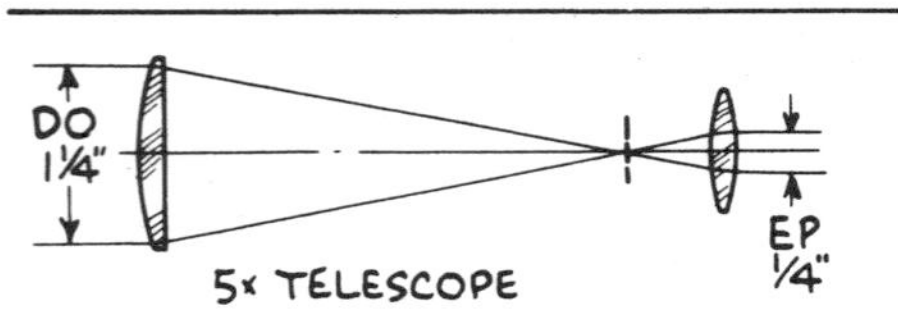

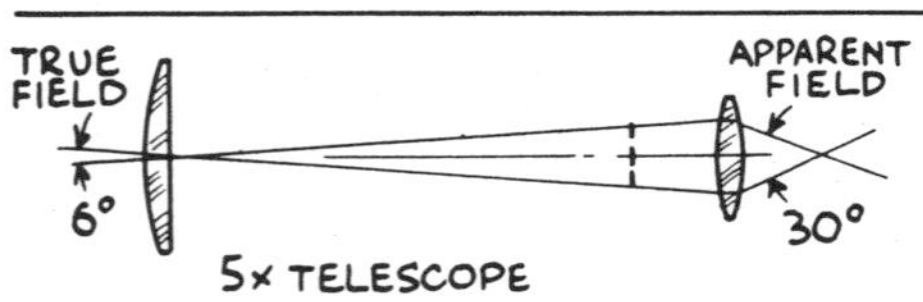

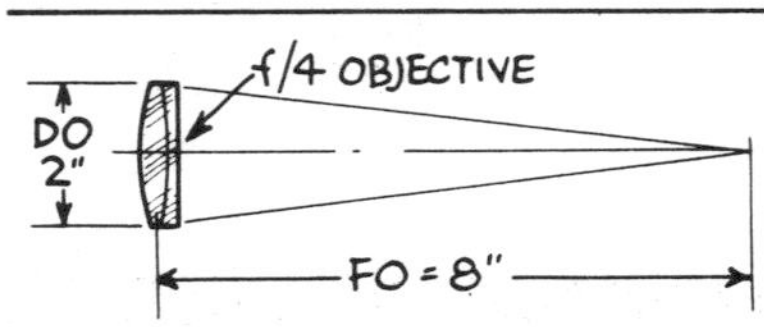

M = FO ÷ FE [1] *EXAMPLE:* M = 10 ÷ 2 = 5x	FE = FO ÷ M [2] FE = 10 ÷ 5 = 2"	FO = M x FE [3] FO = 5 x 2 = 10"
M = DO ÷ EP [4] *EXAMPLE:* M = 1.25 ÷ .25 = 5x	EP = DO ÷ M [5] EP = 1.25 ÷ 5 = .25"	DO = M x EP [6] DO = 5 x .25 = 1.25"
M = AF ÷ TF [7] *EXAMPLE:* M = 30 ÷ 6 = 5x	TF = AF ÷ M [8] TF = 30 ÷ 5 = 6°	AF = M x TF [9] AF = 5 x 6 = 30°
f/obj. = FO ÷ DO [10] *EXAMPLE:* f/obj. = 8 ÷ 2 = 4	DO = FO ÷ f/obj. [11] DO = 8 ÷ 4 = 2"	FO = f/obj. x DO [12] FO = 4 x 2 = 8"

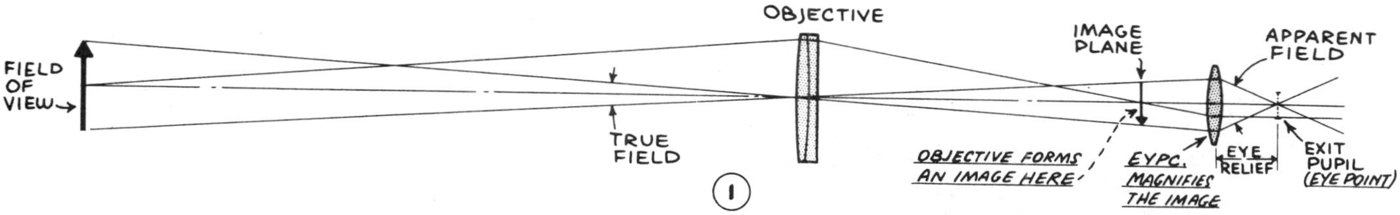

OPTICAL SYSTEM OF THE TELESCOPE

Most Popular Eyepieces

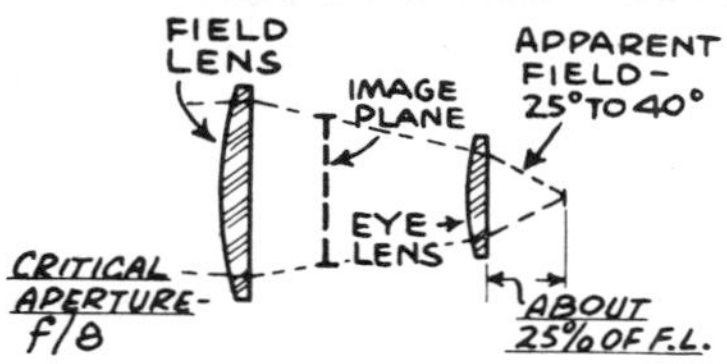

HUYGENS - *FIRST TWO LENS EYEPIECE. FULLY CORRECTED FOR LATERAL COLOR. DEFINITION IS GOOD, ESPECIALLY AT CENTER*

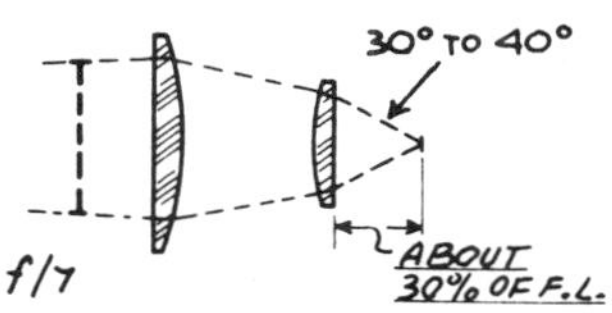

RAMSDEN - *THE ECONOMY EYEPIECE. BETTER CORRECTED THAN HUYGENS IN ALL RESPECTS EXCEPT LATERAL COLOR*

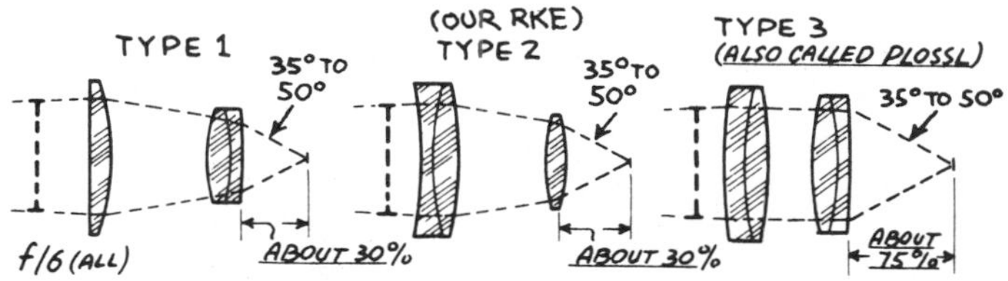

KELLNER - *BASIC FORM (TYPE 1) HAS ACHROMATIC EYE LENS FOR BETTER COLOR CORRECTION THAN RAMSDEN. TYPE 3 (PLOSSL) RATES WITH THE FINEST TELESCOPE EYEPIECES... IT OFFERS WIDE FIELD AND COMFORTABLE EYE RELIEF*

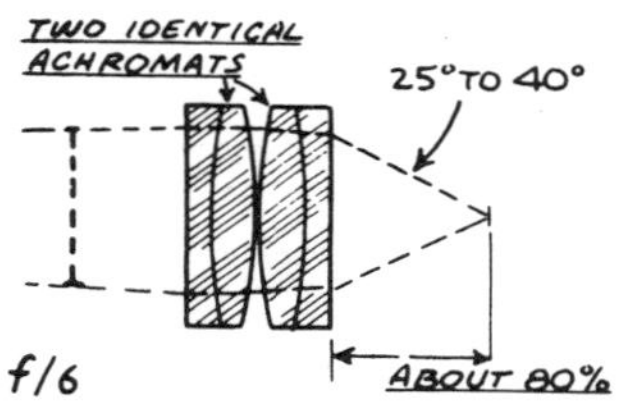

SYMMETRICAL - *A POPULAR HOMEMADE STYLE. FOCAL LENGTH IS ABOUT ONE-HALF F.L. OF LENSES USED... IS GOOD ASTRO EYEPIECE*

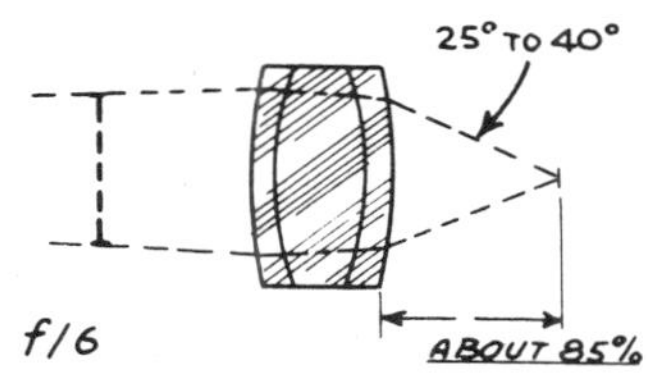

TRIPLET - *ESSENTIALLY THE SAME AS A SYMMETRICAL EXCEPT CEMENTING ELIMINATES ONE ELEMENT... EXPENSIVE*

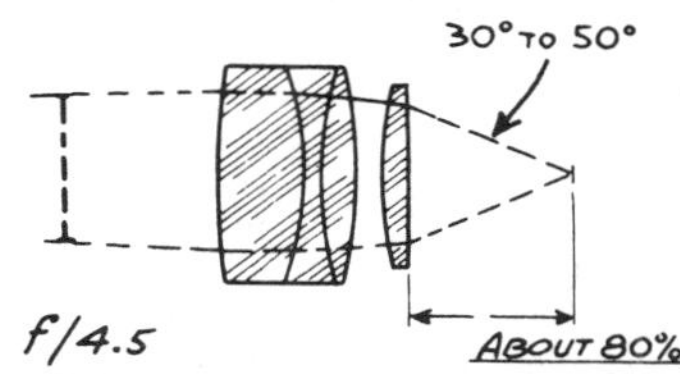

ORTHOSCOPIC - *FULLY CORRECTED FOR DISTORTION. WHEN MADE PROPERLY IT RATES AS BEST ASTRO EYEPIECE FOR HIGH POWER*

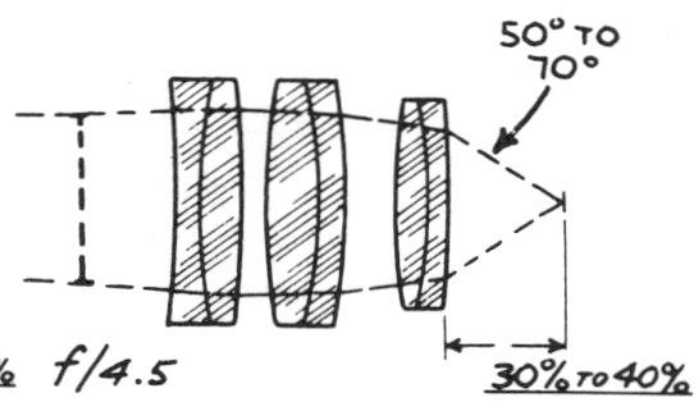

ERFLE - *DESIGNED ESPECIALLY FOR A WIDE FLAT FIELD. GOOD ASTRO EYEPIECE ALTHO MUCH OF THE DESIGN IS WASTED AT HI-POWER*

What Eyepiece is Best?

BY FAR, the objective is the most important part of the telescope. On the other hand, the eyepiece offers a variable in terms of performance and magnification, and as such gets the major share of attention. However, it should be obvious that unless the objective produces a clean, sharp image, no super-duper eyepiece is going to magnify it into a nice sharp picture.

The eyepiece is a simple magnifier. All types except the Huygens have an image or focal plane a short distance outside the front or field lens. You can easily locate the image plane by experimenting with a slip of tracing or wax paper, Fig. 2, having ink or pencil marks on the turned-over end. If you can't see the marks at any position, it is likely you have the Huygens eyepiece, sometimes called a "negative" eyepiece because it can't be used as a simple magnifier.

EYE POSITION. The eyepiece faces the objective and all the while is "taking a picture" of the objective. This image of the objective is known as the exit pupil, and can be located by holding a piece of tracing or wax paper behind the telescope, Fig. 4. If you have a reflector, the exit pupil is a miniature picture of the mirror, even showing the holding clips as well as the black silhouette of the diagonal.

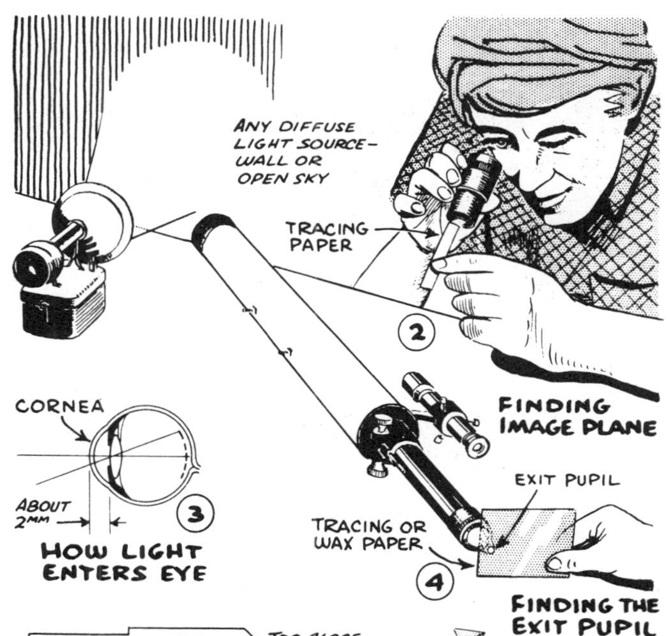

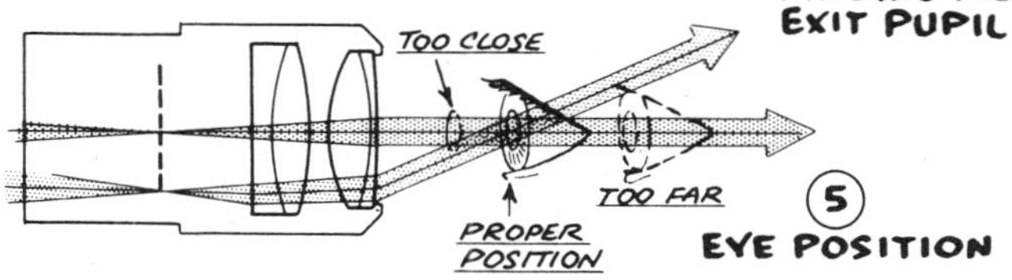

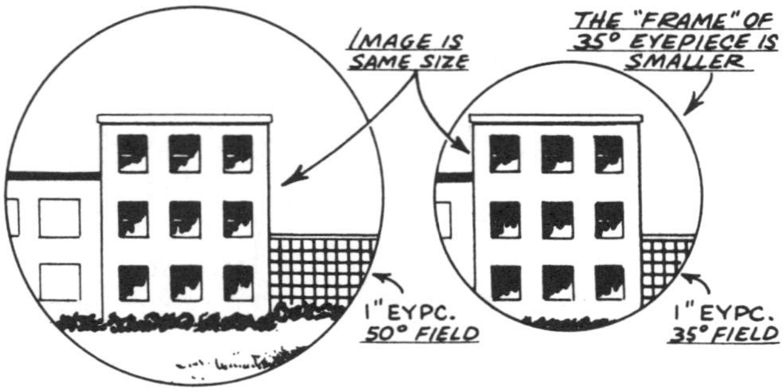

⑥ DIFFERENCE IN APPARENT FIELD

⑦ DIFFERENCE IN MAGNIFICATION

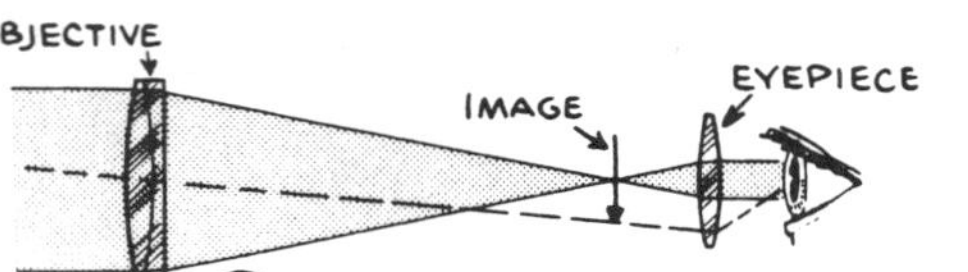

⑧ NORMAL TELESCOPE SYSTEM

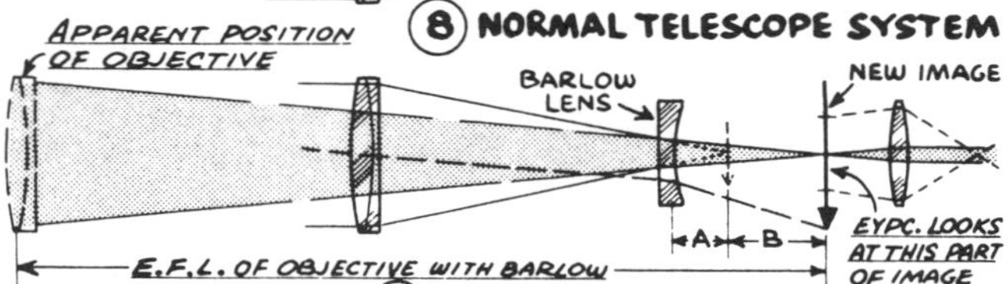

⑨ ...BARLOW LENS ADDED

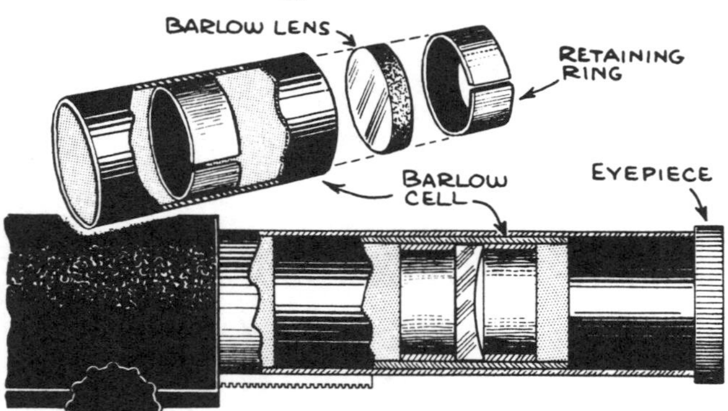

⑩ BARLOW CELL IN REFRACTOR

The distance from the eyepiece to the exit pupil is the eye relief. Ideally this should be about 1/2 inch for normal eyes, or 3/4 inch if you wear glasses. The eye relief decreases with focal length and becomes too short with Huygens and Ramsden eyepieces less than about 1/2 inch focal length.

When you use a telescope, your eye should be at or near the exit pupil to see the widest field and to capture all of the light entering the objective, Fig. 5. If you get inside the exit pupil you will note a "blinker" effect as important light rays will alternately strike and miss your eye with the slightest head movement. Outside of the exit pupil, edge-of-field rays miss the eye cleanly and completely, resulting in a loss of field. The long eye position is sometimes necessary, as, for example, a 1/4 inch Ramsden will have only about 1/16 inch eye relief. Since the entrance to the eye is about 1/12 inch inside the eyeball itself, Fig. 3, it is plain you can't get in close enough. With the long eye position the center of field remains fully illuminated so that no harm is done other than the loss of field.

APPARENT FIELD. The extent of field you can see through any astronomical telescope depends solely on the diameter of the lenses in the eyepiece. Therefore, if you want a wide field, all you need is an eyepiece with big lenses. However, a big lens shows increased faults or aberrations. To reduce aberrations to a practical permissible amount, the lenses must not exceed a certain diameter for a certain focal length of eyepiece. Hence, all eyepieces have a limit as regards extent of field--you can make a Ramsden with a wide 65 degree field but it will not perform well at all, whereas the same lenses in a smaller size or reduced by field stop to cover about 35 degrees will show a clean, sharp picture.

The apparent field of a telescope is the field of the eyepiece alone. It is normally some 30 to 50 degrees in angular extent. This is a fixed and constant angle for any specific eyepiece, and can also be maintained in a series of different focal length eyepieces of the same type. The apparent field is very much like the frame around your television screen--it is always the same size. The apparent field of a 50 degree eyepiece is bigger than the field of a 35 degree eyepiece, Fig. 6. When the power

"Piggy-back" POWER BOOSTERS

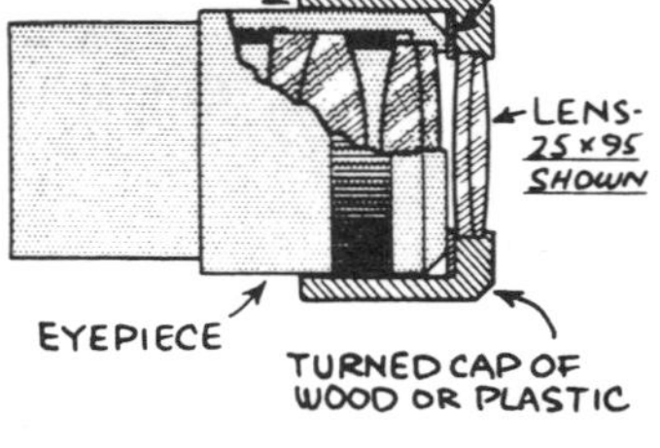

If you have an eyepiece with a good amount of eye relief, it is easy to boost the power as much as 40% by adding an extra lens, which is used as a cap over the eyepiece, as shown. The extra lens should be kept as close to the eyepiece as possible, and can face curved side in or out as desired. The shortened eye relief is still enough for comfortable use.

TYPICAL EXAMPLES USING EDMUND No. 30,787 EYEPIECE

	F.L.	EYE RELIEF	FOCUS*	POWER 3" REFR.	POWER 6" REFL.
No. 30,787 EYPC. ALONE	28mm	7/8"	—	41x	43x
32×132mm ACHROMAT	24mm	1/2"	1/4" IN	48x	51x
29×99mm PLANO-CONVEX	23mm	5/8"	1/4" IN	50x	53x
25×95mm ACHROMAT	22mm	1/2"	1/4" IN	52x	55x
23×48mm ACHROMAT	20mm	1/2"	1/2" IN	56x	60x

*FOCUSING MOVEMENT FROM NORMAL POSITION OF EYEPIECE USED ALONE

EYEPIECE Data (11)

	Focal Length: Inch	= MM.	Name Power [1]	Field Dia. [2]	3" f/15 Refractor: Mag.	E.P.	True Field	Moons	Stars [4]	6" f/8 Reflector: Mag.	E.P.	True Field	Moons	Stars [4]
LOW POWER All types will work well if properly made. Wide field and long eye relief are desirable features obtained with Erfle, Kellner type 3, symmetrical	2"	50.8	5x	1.36" [3]	**23x**	1/8"	1°47'*	3 1/2*	62*	**24x**	1/4"	1°39'*	3 1/3*	194*
	1 3/4"	44.5	5.7x	1.19"	**26x**	1/9"	1°32'	3	46	**27x**	1/4"-	1°28'	2 3/4	150
	1 1/2"	38.1	6.6x	1.02"	**30x**	1/10"	1°20'	2 2/3	35	**32x**	1/5"	1°15'	2 1/2	111
	1 1/4"	31.7	8x	.85"	**36x**	1/12"	1°6'	2	24	**38x**	1/6"	1°3'	2	76
	1 1/8"	28.6	8.9x	.76"	**40x**	1/13"	1°0'	2	20	**43x**	1/7"	56'	1 3/4	61
MEDIUM POWER Ramsden and Huygens are economy choice and have satisfactory eye relief down to about 1/2" F.L. Kellner type 1 is often used, also symmetrical doublets	1"	25.4	10x	.68"	**45x**	1/15"	53'	1 3/4	15	**48x**	1/8"	50'	1 1/2	49
	7/8"	22.2	11.4x	.60"	**52x**	1/17"	48'	1 1/2	13	**55x**	1/9"	44'	1 1/2	38
	3/4"	19.1	13.3x	.51"	**60x**	1/20"	40'	1 1/3	9	**64x**	1/11"	37'	1 1/4	27
	5/8"	15.9	16x	.43"	**72x**	1/24"	33'	1	6	**77x**	1/13"	32'	1	20
	1/2"	12.7	20x	.34"	**90x**	1/30"	27'	1	4	**96x**	1/16"	25'	3/4	12
HIGH POWER Below 1/2" F.L. it is best to use orthos or others with max. eye relief. Simple single lenses sometimes used	3/8"	9.5	26.7x	.26"	**120x**	1/40"	20'	2/3	2	**128x**	1/21"	19'	1/2	7
	5/16"	7.9	32x	.22"	**145x**	1/48"	16'	1/2	1	**154x**	1/26"	16'	1/2	5
	1/4"	6.4	40x	.17"	**180x**	1/59"	13'	1/3	1	**192x**	1/32"	13'	1/3	3
	1/6"	4.2	60x	.11"	**268x**	1/91"	9'	1/4	0	**289x**	1/48"	8'	1/4	2

[1] *Focal length divided into 10 inches. "Name power" is used more for microscope eyepieces than telescope eyepieces*

[2] *Linear diameter of field at focal plane of eyepiece (or objective). Based on 40° apparent field*

[3] *1.36" is larger than standard 1 1/4" focusing tube... image this size needs bigger tube*

[4] *Average view of Milky Way. Stars to 11th mag. for 3" — 12.5 mag. for 6"*

* *Not actually obtained as per note [3]*

of the telescope is increased, the apparent field remains the same, Fig. 7, and only the objects in the field show magnification. The true field is the angular field covered by the telescope and is equal to the apparent field of the eyepiece divided by the magnification. If the apparent field is large, the true field also is comparatively larger than an eyepiece of less apparent field.

WHAT EYEPIECE IS BEST? There are no perfect eyepieces--and no poor ones either if properly made. Some of the more popular types and their main characteristics are shown under Fig. 1 drawing. The "critical aperture" of an eyepiece indicates the maximum light beam it will handle. The Huygens, for example, is just fair in fielding a wide f/8 light cone; it works much better with the narrow f/15 beam of a refractor. The more-expensive eyepieces give wider field and better eye relief but only a little improvement in optical excellence. Apart from type, the focal length is the important feature, Fig. 11. For the average portable telescope, an eyepiece of around 1 inch f.l. or a little more is first choice. If in addition it offers wide field and comfortable eye relief, you have the perfect No. 1 eyepiece. A second eyepiece should approximately double the power--little is gained with mild increases in magnification. The ideal battery could be 1-1/8 inch, 5/8 inch and 1/3 inch, although 1, 1/2 and 1/4-inch sets are more common. It is convenient to have all the same apparent field and all parfocal. Parfocal means that the shoulder or flange distances are such that the eyepieces occupy the same position--focus one and you focus all. Stock eyepieces can sometimes be parfocalized by the addition of spacer rings.

THE BARLOW. A popular way to get extra power (if you must) is to use a Barlow lens with an eyepiece of medium focal length. A Barlow lens is a short f.l. negative lens used a little inside the focal plane of an objective. What happens can be seen in Figs. 8 and 9--with just a small increase in the physical length of the system, the equivalent or effective focal length can be doubled or tripled. This doubling and tripling applies also to the f/value; if you have an f/8 mirror and use 2x Barlow magnification, the system works at f/16. This narrower cone of light will often work wonders in improving the performance of a Huygens or Ramsden eyepiece. Any amount of power can be obtained by simply locating the Barlow at a greater distance from the primary image plane. In Fig. 9, this is distance A. Distance B is how much the focusing tube must be extended from its normal position with eyepiece alone to regain focus with the eyepiece plus Barlow. The combined distance, A plus B, is actually all you need to know since this sets the power; distances A and B adjust automatically when the telescope is focused. Spacing information is supplied with the lens. The usual setup for a Barlow is a metal cell with adjustable spacer rings, Fig. 10. This fits inside the focusing tube of the telescope and the eyepiece is then used in a normal manner except a little extra "out" focusing travel is needed.

THE SUN

... INTERESTING TELESCOPE OBJECT AT 40x TO 70x BUT YOU MUST USE A SUN FILTER TO AVOID SERIOUS INJURY TO YOUR EYE. THE SUN SPOTS ARE EASY TO SEE

THE MOON

MAGNITUDE -12 WHEN FULL IS 190,000 TIMES BRIGHTER THAN FIRST MAGNITUDE STAR. CRATER TYCHO *(TIE-CO)* IS ON SOUTH SIDE - MOST PHOTOS ARE SHOWN INVERTED

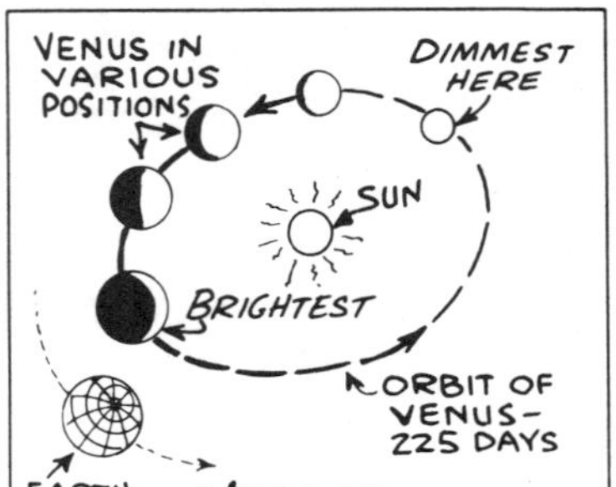

VENUS

LIKE ALL OF THE PLANETS, VENUS ORBITS AROUND THE SUN AND IS LIGHTED BY THE SUN. ON HER NEAR APPROACHES TO THE EARTH SHE IS BRILLIANT AT -4 MAGNITUDE

JUPITER

BIG JUPE IS THE EASIEST PLANET TO SEE--ALWAYS BRIGHTER THAN -1½ MAG. HIS FOUR BRIGHTEST MOONS OF MAG. 6 SHUTTLE BACK AND FORTH, CHANGING NIGHTLY

SATURN

SATURN IS THE PRETTIEST PLANET. THE RINGS ARE SEEN PLAINLY AT 40x ALTHOUGH INVISIBLE WITH 7x BINOCULAR. WITH HIGHER POWER YOU MAY BE ABLE TO SEE CASSINI'S DIVISION

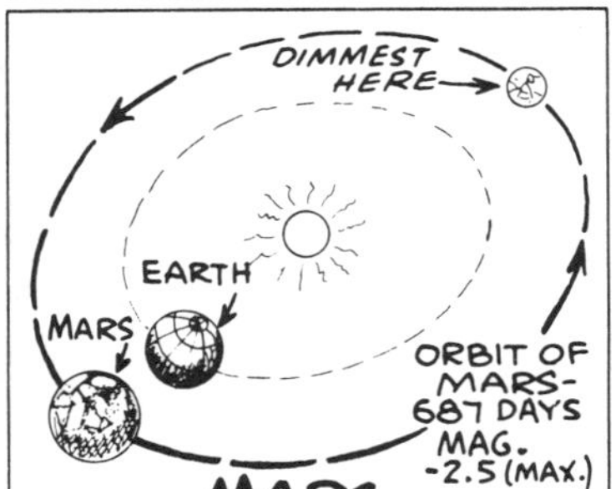

MARS

RED MARS MAKES A NEAR APPROACH TO THE EARTH EVERY OTHER YEAR, AND AT SUCH TIMES SOME SURFACE DETAIL CAN BE SEEN WITH TELESCOPES AT 200-300x

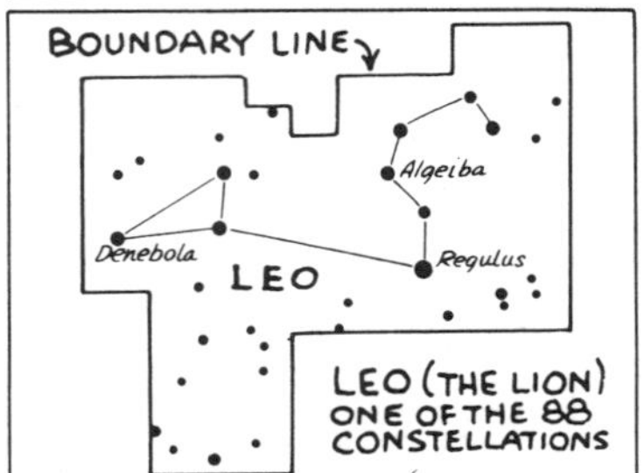

CONSTELLATIONS

A CONSTELLATION IS A GROUP OF STARS, USUALLY FORMING SOME KIND OF PATTERN OR "PICTURE." PROPERLY, A CONSTELLATION IS A SPECIFIC AREA OF THE SKY

DOUBLE STARS

ONE OUT OF 15 STARS IS A DOUBLE OR MULTIPLE STAR AND ABOUT 500 OF THESE FROM 2 SECONDS TO 1 MINUTE OF ARC SEPARATION CAN BE "SPLIT" WITH SMALL TELESCOPES

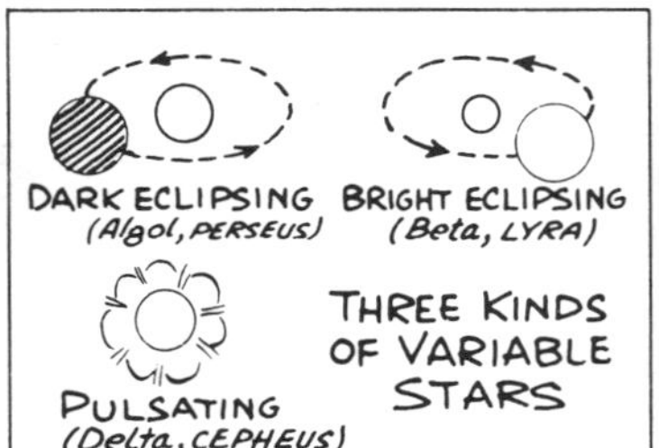

VARIABLE STARS

A VARIABLE STAR VARIES IN BRIGHTNESS. THE CHANGE TAKES 2 DAYS *(AVERAGE)*, MAKING THE V.S. A POOR "SHOW" OBJECT ALTHOUGH IDEAL FOR SYSTEMATIC STUDY

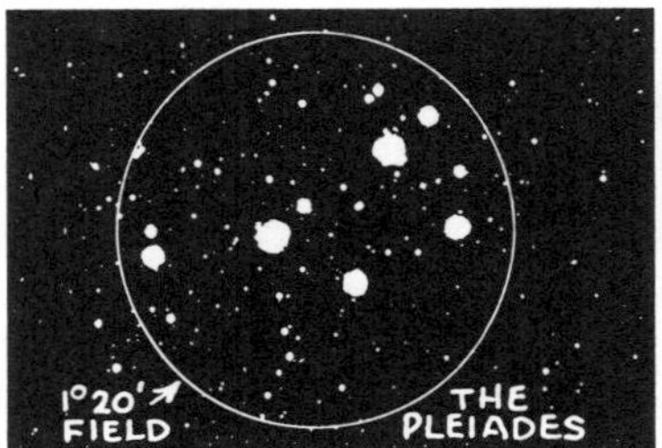

OPEN CLUSTERS

OPEN CLUSTERS OF STARS ARE A FAVORITE TARGET FOR THE TELESCOPE. 40 TO 60x IS ENOUGH FOR MOST GROUPS. POPULAR PLEIADES CLUSTER IS A FINE BINOCULAR OBJECT

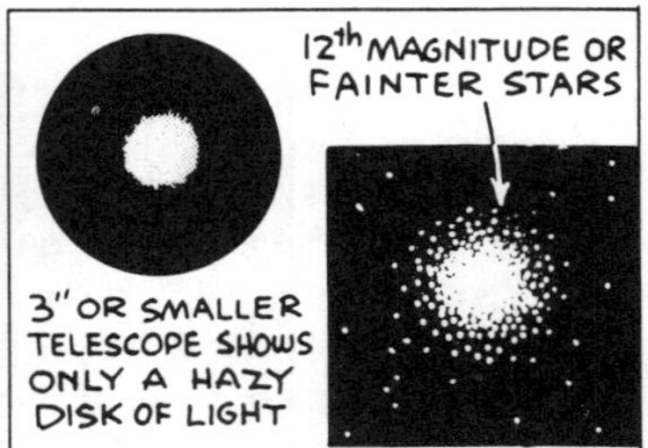

GLOBULAR CLUSTERS

A GLOBULAR CLUSTER IS A BALL OF STARS. INDIVIDUAL STARS ARE FAINT AND NEED 6" OR MORE APERTURE FOR RESOLUTION. M13 AND M22 ARE TWO BRIGHTEST

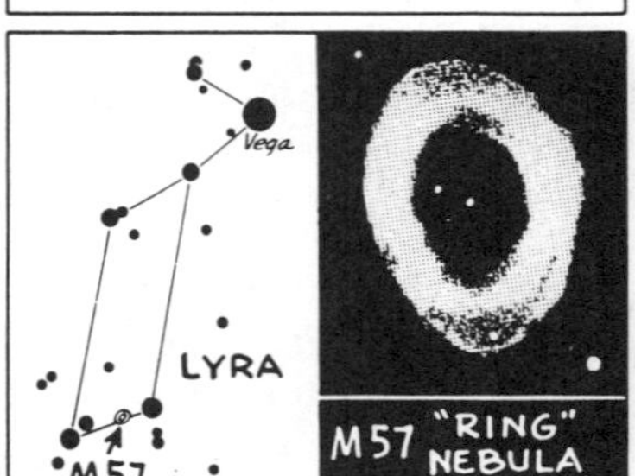

PLANETARY NEBULAE

PLANETARY NEBULAE ARE SO NAMED ONLY BECAUSE THEY ARE ROUND LIKE PLANETS. THEY ARE LUMINOUS GAS CLOUDS AND ARE A PART OF OUR GALAXY

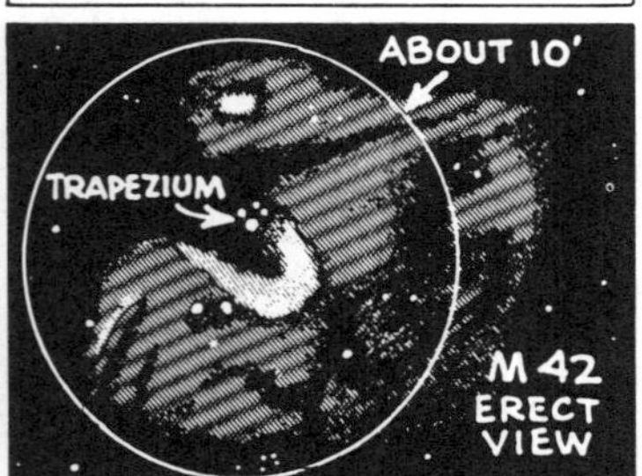

DIFFUSE NEBULAE

A LARGE DIFFUSE GAS CLOUD LIGHTED BY THE STARS IN ITS VICINITY IS KNOWN AS A BRIGHT DIFFUSE NEBULA. M42 IN ORION IS IMPRESSIVE, EASILY SEEN WITH ANY TELESCOPE

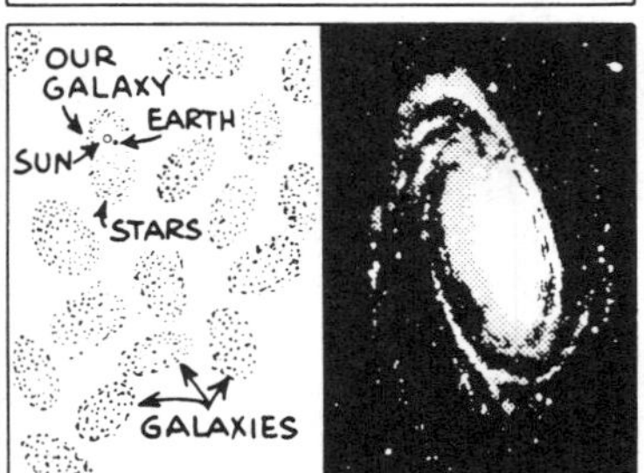

EXTERNAL GALAXIES

GALAXIES ARE COMPLETE STAR SYSTEMS LIKE OUR OWN GALAXY. ALL ARE VERY DISTANT. M81 SHOWN IS ABOUT AS BRIGHT AS A STAR OF 9th MAGNITUDE

No.	TYPE	CONS.	M.
M 44	OPEN CL.	CANCER	3.7
M 41	OPEN CL.	CANIS MAJ.	4.6
M 24	OPEN CL.	SAGR.	4.6
M 31	GALAXY	ANDR.	4.8
M 35	OPEN CL.	GEMINI	5.3
M 13	GLOBULAR	HERCULES	5.7
M 22	GLOBULAR	SAGR.	5.9
M 8	DIFFUSE NEB.	SAGR.	—
M 42	DIFFUSE NEB.	ORION	—
M 57	PLANETARY	LYRA	9.3

MESSIER OBJECTS

FRENCH ASTRONOMER, CHARLES MESSIER, MADE UP THE FIRST LIST OF SKY OBJECTS OTHER THAN STARS (1784). ALL OF THE 103 M-OBJECTS CAN BE SEEN WITH SMALL TELESCOPES

Observing the SKY SHOW

LIKE a lot of other hobbies, you can go for star-gazing a little or a lot, just as you like. Even with the smallest telescope you have equipment far better than that used by Galileo some 350 years ago when he discovered that Jupiter had four bright moons. There is much to see; also much to learn. It takes the beginner about a year to become an expert star-gazer. In this time he gets over the idea that he is going to see huge fireballs and other fantastic wonders, finding instead an increasing enjoyment in his ability to use a telescope. Without leaving his backyard, he becomes a sky explorer with the skill to guide the big eye of his telescope to the most remote corners of the sky.

For a starter, you will want to look at "show" objects. Naturally, the moon and bright planets come first. Then, in any star book or atlas, you will find other showpieces of the sky: the Lagoon in Sagittarius; the double cluster in Perseus; M42, with the Trapezium set in its greenish glow; the blue-and-gold double star, Albireo; the stardust glitter of M11; the double-double in Lyra; the ever-charming Seven Sisters; distant Andromeda, the farthest you can see.

However, many sky objects can't be seen, and many others are not seen as clearly as the beginner anticipates. Most beginners get a "first impression" of the sky show from photographs, without realizing that these pictures are time exposures. When a light strikes the camera film it makes a bright spot; the longer the film is exposed, the bigger and brighter the spot becomes. Sky photos represent a sincere effort to show sky objects as they really are, but some of the effects have a strong element of trick photography. Don't expect to see most sky objects like they are shown in photographs. There are some exceptions--you can see the moon as big and clear as any photo; Saturn, Jupiter and Venus all look better than their pictures.

You may find the scaled sketches below informative. Look at them with one eye from a distance of about 10 inches and you get the same size effect obtained with a telescope at 100x. Maybe you are surprised that Saturn and Mars are so small and the moon so big. Even scaled drawings are not entirely realistic. M57, for example, looks like a big easy target but is actually quite difficult--close your eye nearly shut to reduce the light and you will get the idea. Saturn for all its apparent small size is bright and clear in any telescope at 40x or more and stands magnification beautifully. Mars is much more difficult. Big objects are not lacking with nearly all of the open clusters and diffuse nebulae ranging from Jupiter to moon size or larger. The easiest type of object to see is the open cluster; globulars are easily visible although you don't get size and detail like photographs. Planetaries and external galaxies are dim, difficult. 100x is enough power for most objects and better seeing is a matter of a bigger diameter objective rather than mere magnification.

Join a club or star-gazing group so that you can exchange ideas and talk shop about the big sky show. Other than star-gazing, you can use your astro telescope for land-gazing to see brilliant daytime views that are amazing in clarity and detail. Photography--land or sky--is another popular hobby where the telescope adds new thrills.

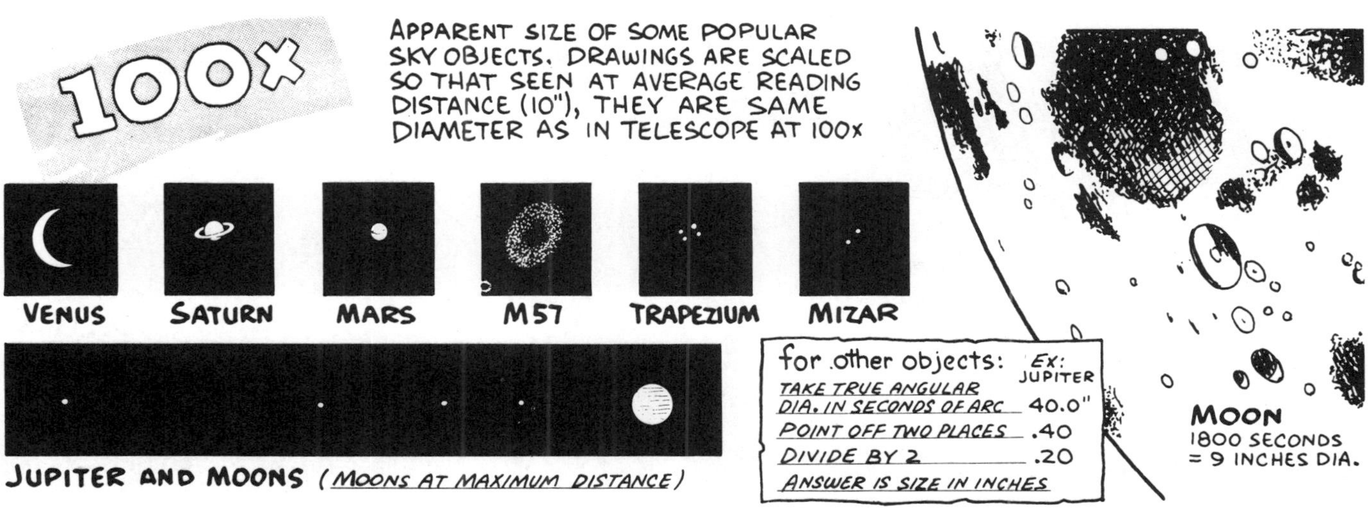

The POSITION of a STAR

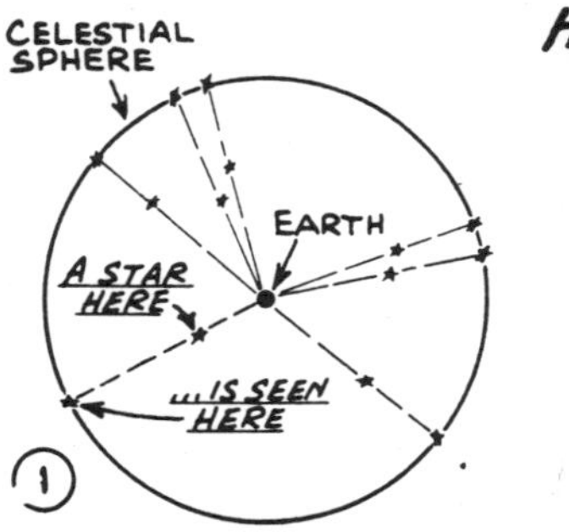

THE CELESTIAL SPHERE IS A BIG IMAGINARY BALL WITH THE EARTH AT ITS CENTER

How it seems...

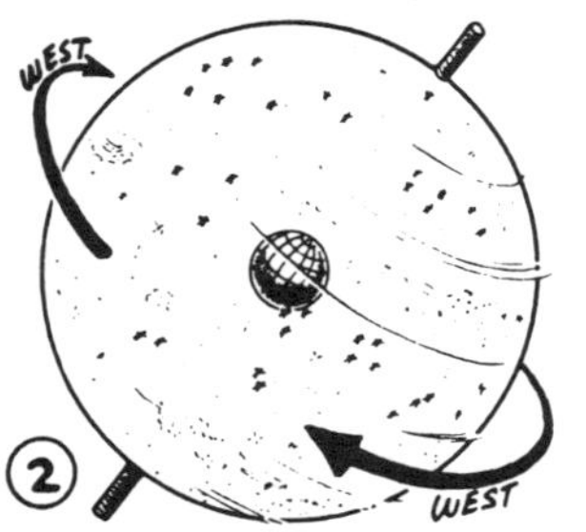

... THE EARTH STANDS STILL WHILE THE CELESTIAL SPHERE ROTATES AROUND IT

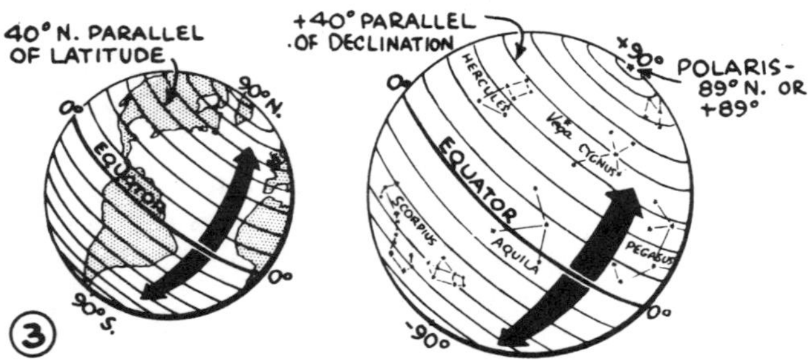

LATITUDE ON EARTH IS DECLINATION IN SKY

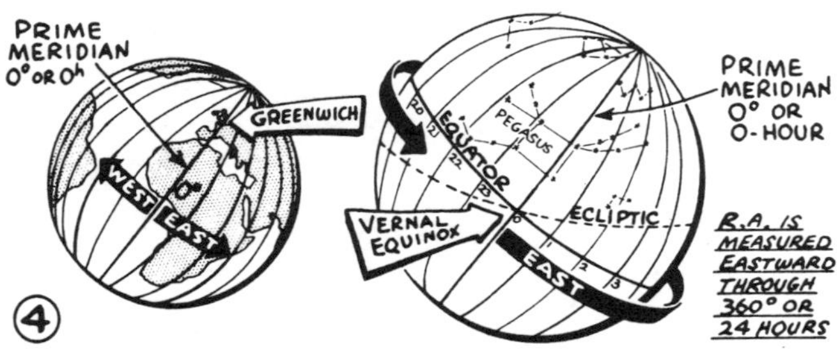

LONGITUDE BECOMES RIGHT ASCENSION

EVERY STAR has a permanent address in the sky--if at times you can't find a certain sky object you can be sure it is not off wandering around in another neighborhood. The star's address is given in terms of right ascension hours and minutes and declination degrees and minutes. A knowledge of star positions by R. A. and Dec. becomes absolutely necessary if you use setting circles and is otherwise an informative study even if you do your star-gazing with nothing more than binoculars.

THE CELESTIAL SPHERE. The celestial sphere is a big imaginary hollow ball with the earth at its center. It is so large that if drawn to the same scale as the earth in Fig. 1, it would take a sheet of paper bigger than the whole United States to chart even the nearest star. Some stars are nearer the earth than others, but all are very distant and all are imagined as being projected to the inside surface of the celestial sphere. We see the celestial sphere from the inside, a giant dome of blue by day and spangled by stars at night, rotating westward. You don't have to be a whiz at astronomy to know that this apparent westward movement is actually caused by the earth rotating to the east.

The surface of the celestial sphere is plotted with a system of imaginary guide lines in about the same manner as latitude and longitude lines

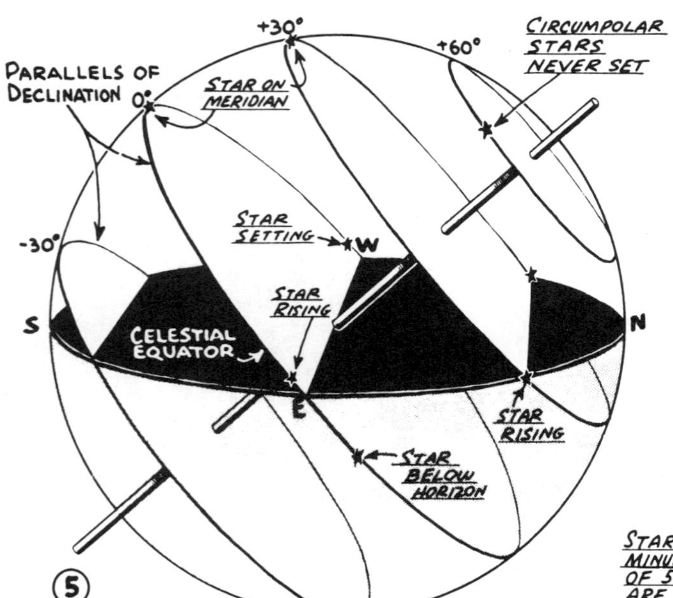

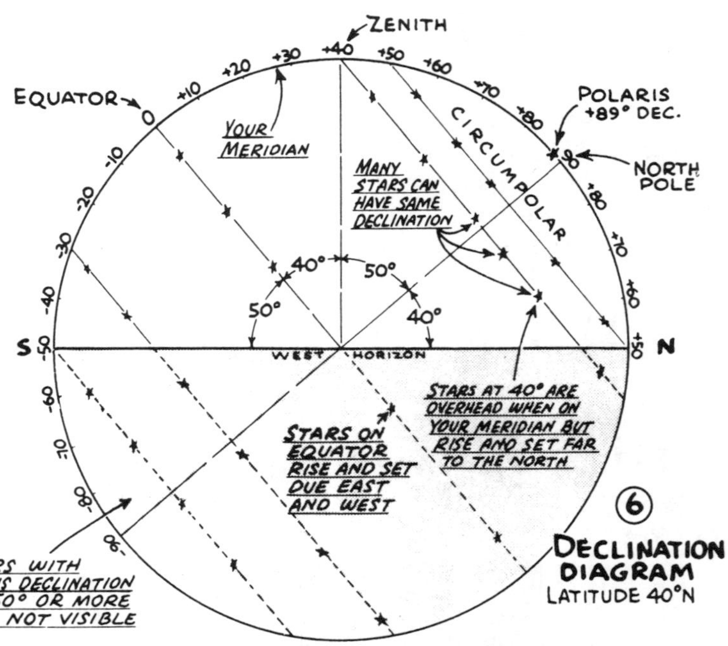

DECLINATION DIAGRAM
LATITUDE 40°N

DECLINATION LINE SHOWS NORTH-SOUTH POSITION OF A STAR AND ALSO INDICATES ITS PATH ACROSS SKY

Hour Circles make a GIANT Sky Clock

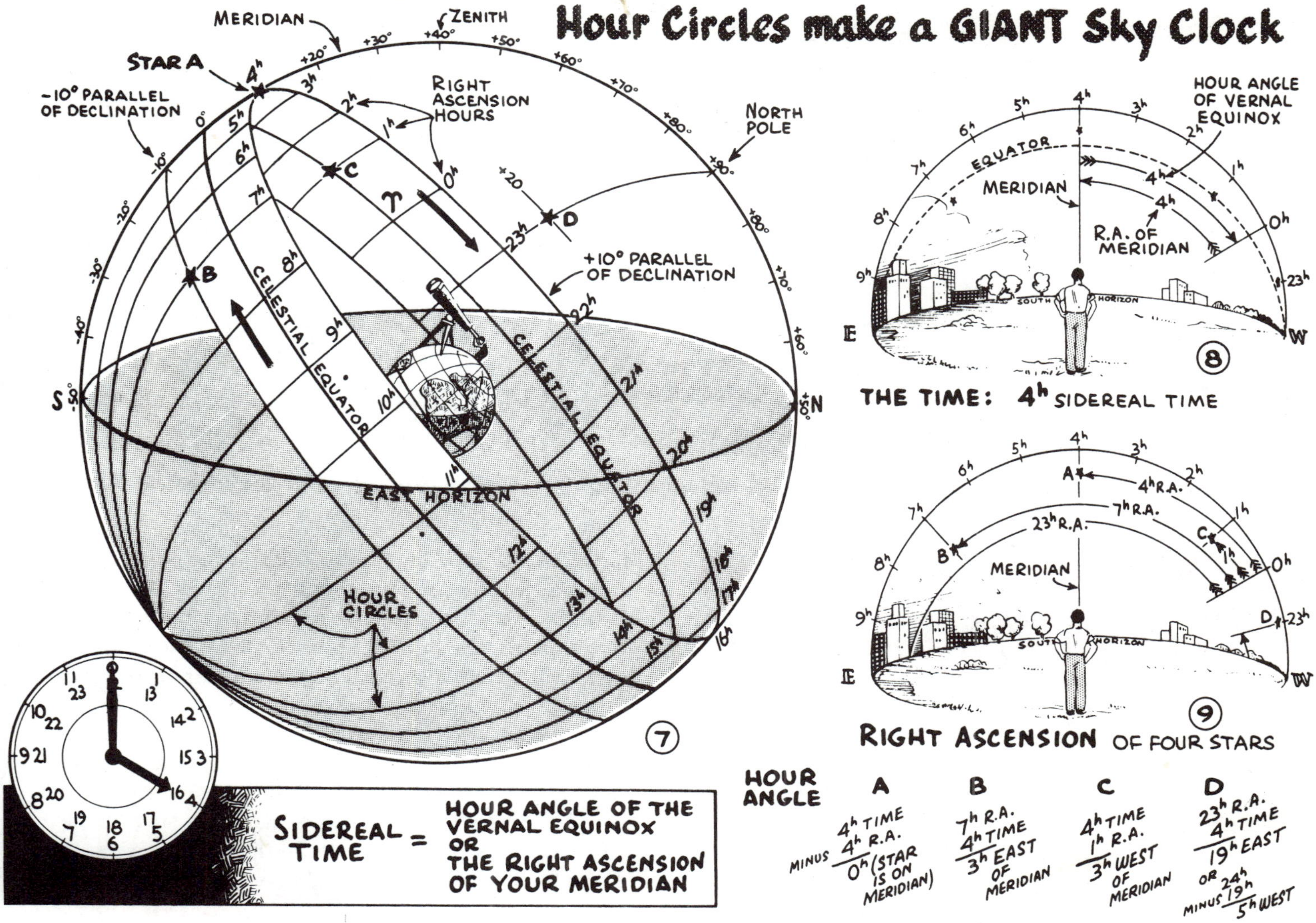

are used in plotting the surface of the earth. Like the earth, the celestial sphere has an equator and north and south poles. Latitude on earth is the angular location of a place north or south of the equator, and, similarly, declination is the angular position of a star north or south of the celestial equator, Fig. 3. Longitude in earth geography gives the position of a place at so-many hours of solar time from the prime meridian; similarly, right ascension is used in sky geography to indicate position at so-many sidereal hours from the prime meridian. The prime meridian for right ascension is the vernal equinox, which is the point on the celestial equator which the sun crosses at the beginning of spring. On this network of declination and right ascension lines, every star has its own fixed position, different from all others except for certain double and multiple stars which reside at the same sky address. Note that in Figs. 3 and 4 you are looking at the celestial sphere from the outside, and the familiar figures of the constellations are thereby reversed left-to-right.

DECLINATION. The declination of a star is its angular distance north or south of the celestial equator. A circle through this point parallel to the equator is a parallel of declination. Such a parallel not only marks the star's position north or south of the equator, but also shows the path it takes in its east to west journey across the sky, Fig. 5. It is worth noting that stars on the celestial equator rise and set due east and west. This can be seen in Fig. 5 and also in the diagram, Fig. 6. If your location is at latitude 40 degrees north, stars at 40 degrees north declination will be exactly overhead when on your meridian, but will rise and set far to the north; stars at 50° N and more declination become circumpolar and never rise or set but wheel forever around the celestial pole. Going the other way from the equator, you can see that stars below the equator make shorter and shorter journeys across the night sky, and a star at 49°S declination will barely lift its eye above the southern horizon.

RIGHT ASCENSION. The right ascension of a star

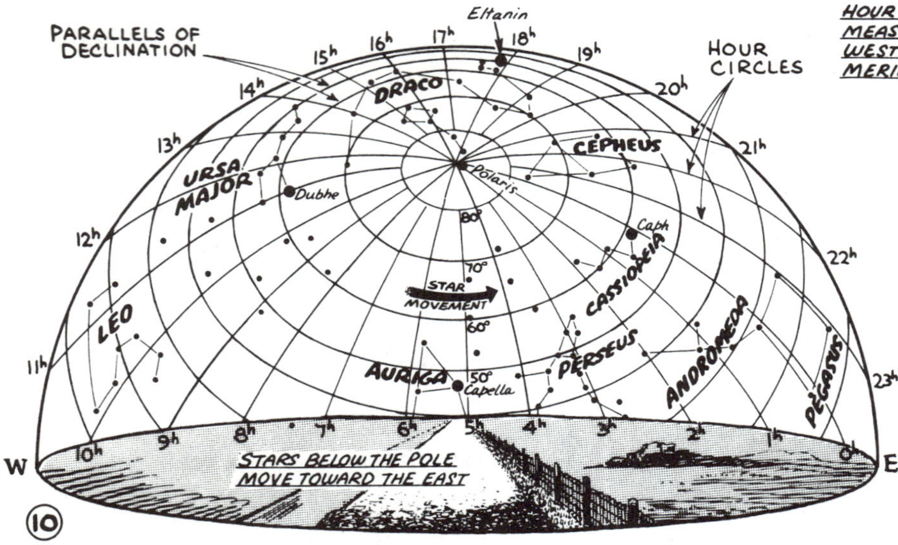

(10)

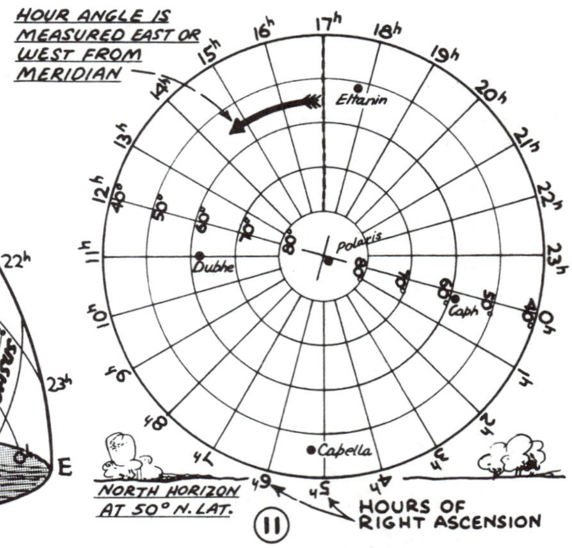

(11)

DECLINATION AND HOUR CIRCLES IN NORTH SKY..... *HOW IT LOOKS ON A CIRCUMPOLAR CHART*

is its distance from the vernal equinox, measured eastward. On sky maps, the grid of hour circles is spaced at convenient 1-hour intervals, marked 0-hour through 23-hour to 0-hour.

Hours of right ascension increase toward the east. This is a fundamental in positional astronomy and should be memorized. It is worth noting again that the grid of declination and right ascension lines on the sky sphere are fixed guide lines which rotate with the sphere. The hours of R.A. make a giant sky clock. Your meridian is the hour hand, but unlike an ordinary clock, it remains fixed while the sky dial rotates to the west, Fig. 7. You learn quickly that it is 0-hour sidereal time when Pegasus is on your meridian, 5:30 when Orion stalks across mid-sky, 18:30 when bright Vega is overhead--and eventually you may master the whole sky clock.

Fig. 7 shows the 4-hour circle of right ascension on the meridian--it is 4:00 sidereal time. Notice how the hours of right ascension increase toward the eastern horizon. Since 4-hour R.A. is on your meridian, the vernal equinox is four hours to the west--it has an hour angle of 4 hours. This illustrates the basic definition of sidereal time: Sidereal Time is the hour angle of the vernal equinox. The position of the vernal equinox is also known as the First Point of Aries. Aries is the Ram, and the symbol like a pair of horns is often used to indicate 0-hour R.A.

It will be obvious that if it is 4 hours from your meridian to the vernal equinox, it is also 4 hours from the vernal equinox to your meridian, Fig. 8. Thus, there is a second definition of sidereal time. Sidereal Time is the right ascension of the meridian. The way the whole thing works out, sidereal time is the R.A. hour which is on your meridian at the moment.

HOUR ANGLE. The hour angle of a star is its distance westward from your meridian. Hour angle always means a measurement to the west, but for convenience hour angles to the east are also used, but must be so designated. Since the sky sphere is always moving westward, the hour angle of a star is always changing and can be determined only for a certain instant of sidereal time. To find the hour angle of any star, you subtract its R.A. from sidereal time. If the result of the calculation is positive, the hour angle of the star is west; if negative, east. Most star-gazers simply subtract the smaller from the larger number. Then you must remember if R.A. is greater, the hour angle is east; if sidereal time is greater, the hour angle is west. Examples are shown in the drawing, applying to Figs. 7, 8 and 9.

THE NORTH SKY CLOCK. Fig. 10 shows the north sky clock. It is about 17:10 sidereal time. Above the pole you will find hours of R. A. increasing toward the east, following the basic rule. Below the pole, R.A. increases toward the west and the stars rotate to the east--everything is backwards! Around a sky circle like this there is no actual east or west; if you start out at the top of the drawing and move east, you are presently going south, then west, then north. In the north sky, directions and hour angles should be taken from that part of your meridian which is above the pole.

SKY MAPS. The sky maps on opposite page and following are sufficiently detailed to give you a start in sky geography. Star positions change a little from year to year, but the change is so slow that maps up to 50 years old are substantially correct.

CHART OF NORTHERN SKY WITH DATE SCALE FOR 8:00 MEAN TIME

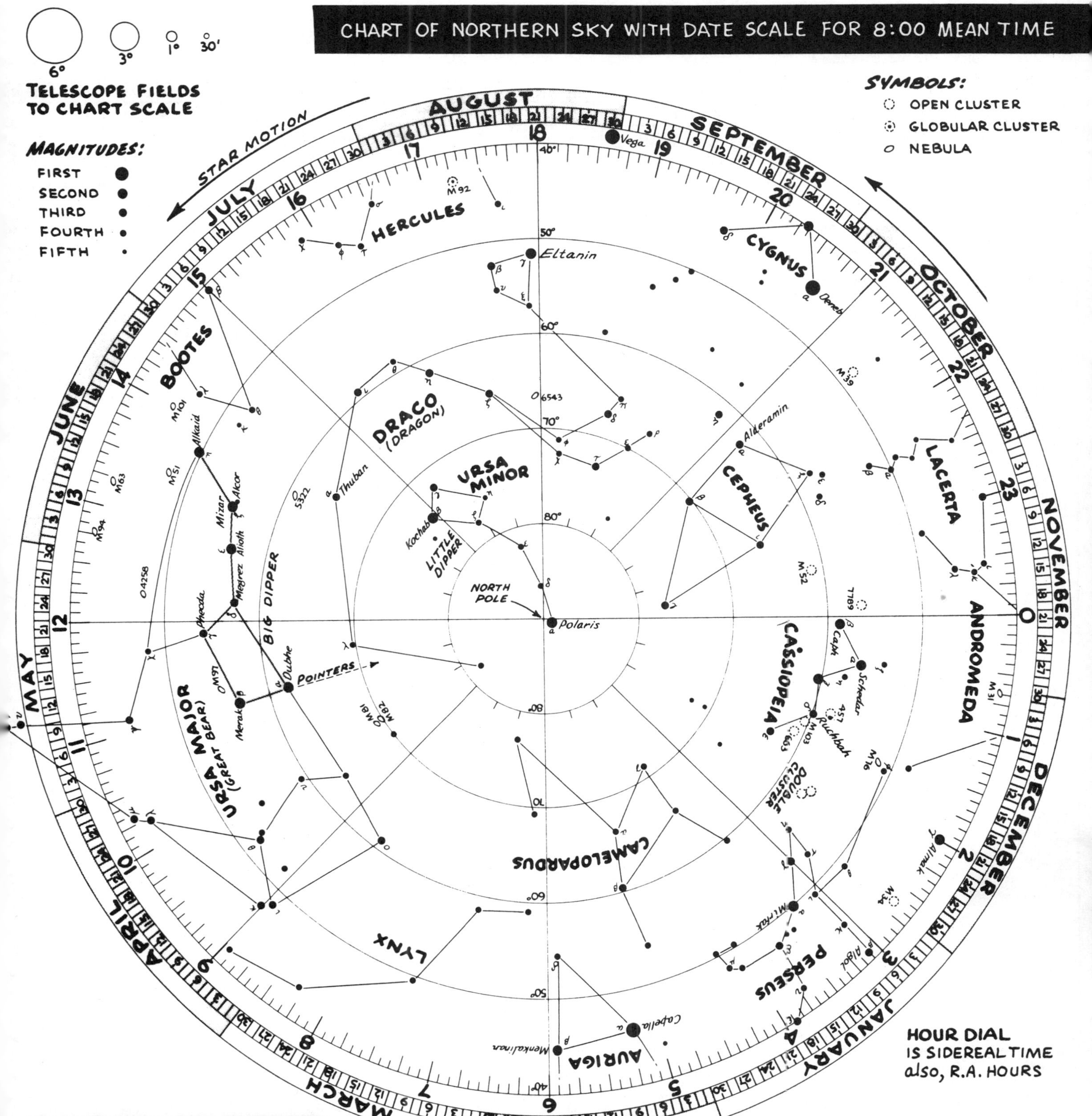

HOW TO USE THIS CHART

FACE NORTH. TURN PAGE TO PUT CURRENT DATE AT TOP. THE STARS WILL THEN BE IN PROPER POSITION FOR 8:00 LOCAL MEAN TIME. <u>THE BOTTOM EDGE OF CHART WILL BE YOUR NORTH HORIZON – TOP PART WILL SHOW STARS SOUTH OF POLARIS TO YOUR ZENITH</u>

IF AFTER 8:00 L.M.T., TURN CHART COUNTERCLOCKWISE THE TIME INTERVAL YOU ARE PAST 8:00 L.M.T.

Example: OCT. 6

FACE NORTH. TURN PAGE TO PUT OCT. 6 AT TOP. <u>IF DESIRED YOU CAN HOLD THE CHART ERECT OR EVEN OVERHEAD TO PUT IT IN SAME PLANE AS SKY</u>

NOTE AT OCT. 6 SETTING, 21^h ON HOUR DIAL IS ALSO AT THE TOP. THIS IS THE SIDEREAL TIME AT 8:00 L.M.T. AN HOUR LATER, THE SIDEREAL TIME WILL BE 22^h AND THAT HOUR SHOULD BE PUT AT THE TOP

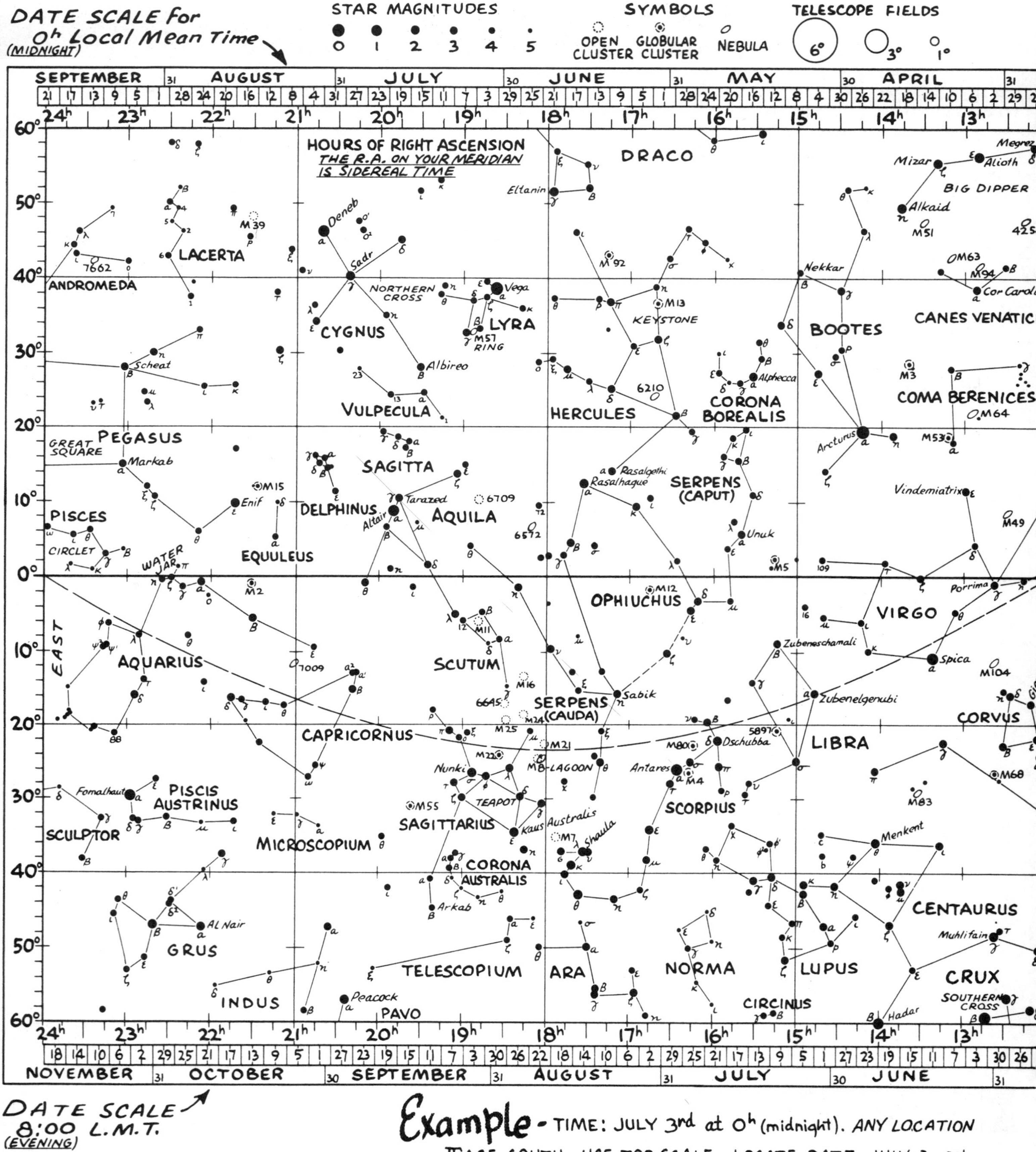

Example - TIME: JULY 3rd at 0h (midnight). ANY LOCATION

FACE SOUTH. USE TOP SCALE. LOCATE DATE, JULY 3, ON SCALE. A VERTICAL LINE THRU DATE SHOWS BRIGHT VEGA NEARLY ON YOUR MERIDIAN. THE TEAPOT (SAGITTARIUS) IS DUE SOUTH. RED ANTARES IS ABOUT 2 HOURS WEST.

SIDEREAL TIME IS 18h 44m APPROX. (from scale)
(COMPARE WITH 18h 43m (TABLE 2))

FACE SOUTH. LOCATE CURRENT DATE ON DATE SCALE. USE BOTTOM SCALE IF TIME IS 8:00 p.m. USE TOP SCALE IF TIME IS MIDNIGHT. A VERTICAL LINE THRU DATE WILL BE YOUR MERIDIAN AT THE MOMENT

EQUATORIAL STAR CHART with DATE SCALES for 0^h and 8:00 p.m. LOCAL MEAN TIME

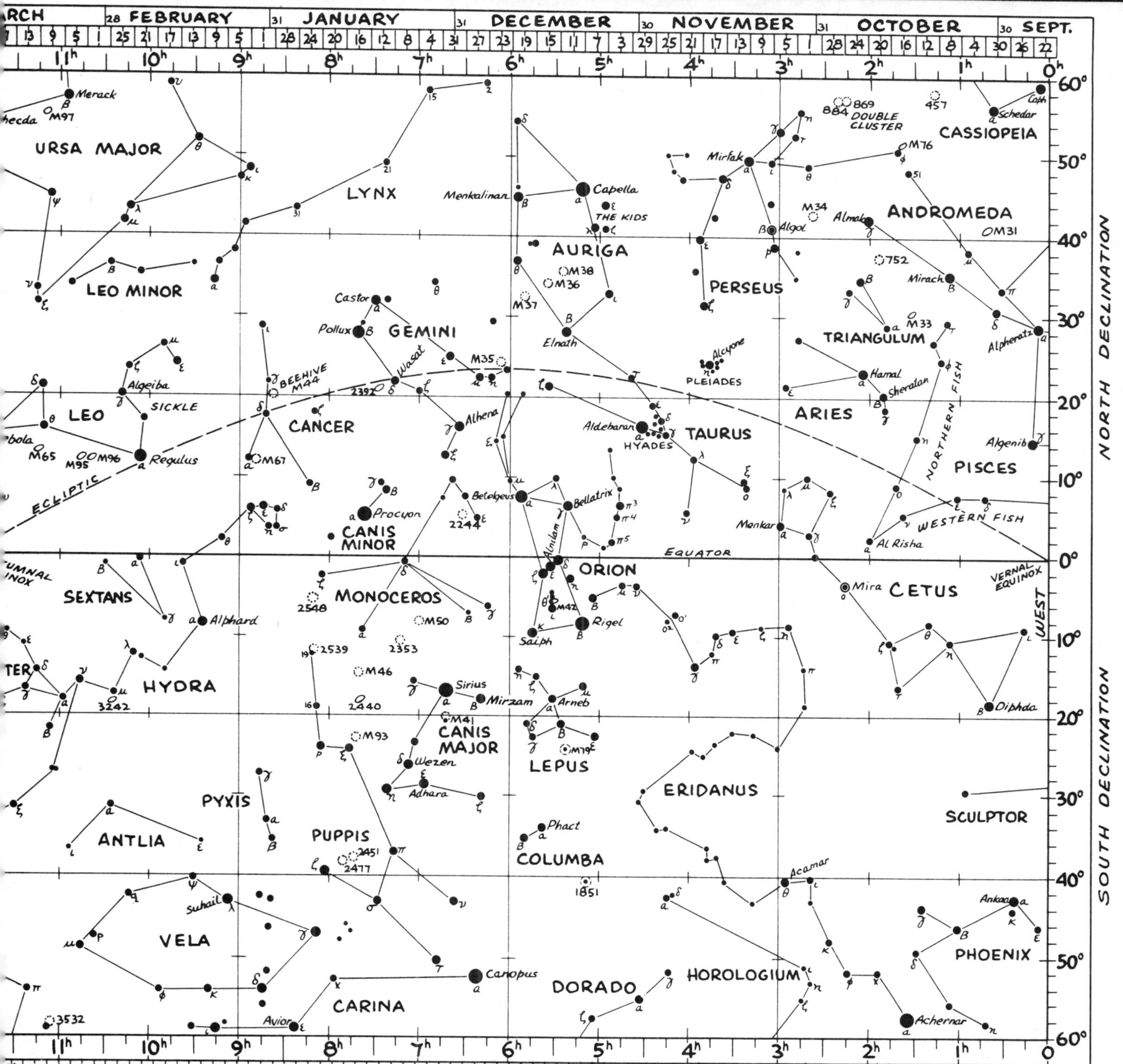

AT OTHER TIMES:

LOCATE YOUR MERIDIAN FOR 8:00 O'CLOCK OR 12:00 O'CLOCK IN THE MANNER DESCRIBED ON OPPOSITE PAGE.

NOTE SIDEREAL TIME ON SCALE.

SHIFT YOUR MERIDIAN EAST THE SAME INTERVAL YOU ARE PAST 8:00 OR MIDNIGHT

Example - TIME: JULY 3rd at 9:15 p.m. L.M.T.

USE BOTTOM SCALE. OPPOSITE JULY 3 DATE LINE, READ $14^h 45^m$ SIDEREAL TIME *(AT 8:00 L.M.T.)*

YOU ARE $1^h 15^m$ PAST 8:00, SO SHIFT YOUR MERIDIAN THIS SAME TIME INTERVAL TO THE EAST. *IT WILL LOCATE YOUR MERIDIAN AT $16^h 00^m$. SIDEREAL TIME IS 16^h -- ANY STAR NEAR R.A. 16^h WILL BE ON YOUR MERIDIAN*

The TIME SCHEDULE is L.M.T.

THE YARDSTICK used to measure time is the period it takes the earth to make one complete rotation. This period is very uniform--it will not vary by as much as a fraction of a second in your lifetime. The period it takes the earth to rotate once is one day. Some kind of "index" mark is needed to show the beginning and end of one complete rotation. When the index mark is the sun, the resulting time is solar or sun time; when the index mark is a star, the time is sidereal or star time.

APPARENT SOLAR TIME. This is true sun time, governed by the passage of the real sun across the sky. This is the time you read on a sundial, Fig. 1. It is also the kind of time you are using when you guess-estimate the time by the position of the apparent (real) sun. If the real sun is right on your meridian, it is exactly 12 o'clock apparent solar time.

MEAN SOLAR TIME. Uniform solar time is obtained when all of the time in a year is divided into 365 days of equal length. This averaged type of time is paced by an imaginary sun known as the mean sun. The daily difference between the apparent (real) sun and the mean (average) sun ranges from zero to about 16 minutes.

STANDARD TIME. Both apparent and mean sun time are local times, either being about 4 minutes different for two locations about 50 miles apart, Fig. 3. Mean time is used in astronomy, but it is not practical for everyday use because every location has a different time. It was not until 1884 that the people of the world got together and agreed on a world-wide standard of time. This system divides the world into 24 standard time zones, each comprising an area of some 15 degrees of longitude. All places in a specific time zone keep the same standard time as the mean solar time of the central meridian of the time zone, Fig. 4. The Zero zone is centered at Greenwich, England. Zones to the west are plus 1 to 12; zones to the east are minus 1 to 12. The mainland of the United States has four zones, Plus 5, 6, 7 and 8, better known as Eastern, Central, Mountain and Pacific standard time, as shown in Fig. 6.

At the central meridian of a zone, standard time coincides exactly with mean solar time. What this amounts to is simply that when it is noon the sun is on the meridian (nearly) at the central meridian of any time zone, and elsewhere in the zone the position of the sun is no more than about 30 minutes off the meridian. Thus, Standard Time keeps the hours of daylight and darkness about where they belong, and yet offers a uniform time system over a large area.

LOCAL MEAN TIME. This is mean solar time applied to your own location. It is also known as Local Civil Time. In general, it is called mean time, or local time or local mean time. It is the time used in astronomy. Standard Time is a useful man-made kind of time, but it is useless for the exact location of sky objects. If you want to locate the sun exactly, you must use apparent solar time; for all other sky objects,

Three Kinds of Solar Time

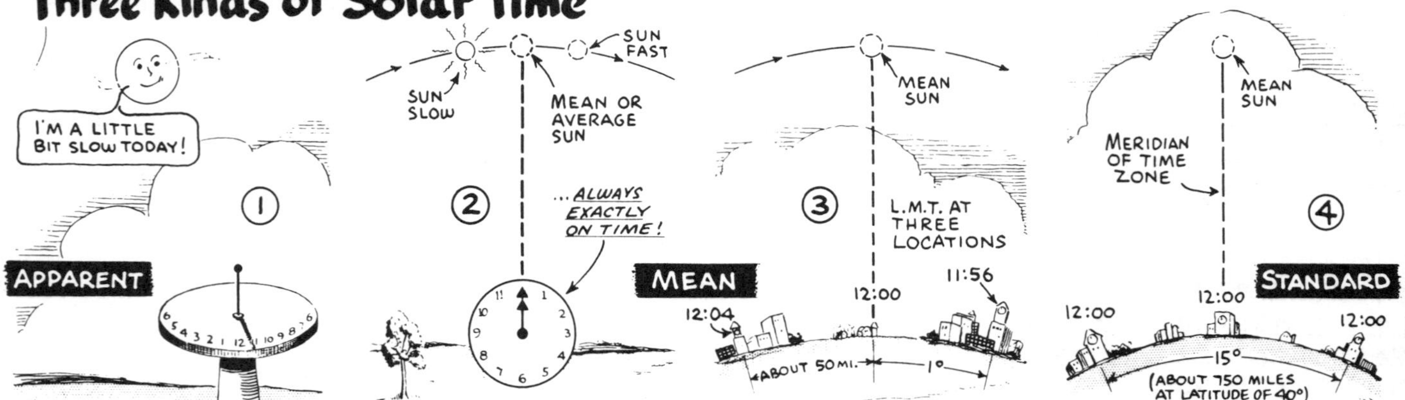

APPARENT TIME IS TIME BY THE REAL SUN. SUN'S MOVEMENT IS NOT UNIFORM - ACCUMULATED ERROR CAN BE AS MUCH AS 16 MINUTES

MEAN TIME MAKES USE OF AN IMAGINARY SUN TRAVELING AT A UNIFORM SPEED EQUAL TO THE MEAN OR AVERAGE SPEED OF THE REAL SUN

LOCAL MEAN TIME IS THE SPECIFIC USE OF MEAN TIME AT ANY PARTICULAR PLACE. IT VARIES FOUR MINUTES PER DEGREE OF LONGITUDE

STANDARD TIME IS THE MEAN TIME OF A SPECIFIC LOCATION USED AS THE STANDARD CLOCK TIME OVER AN AREA OF ABOUT 15°

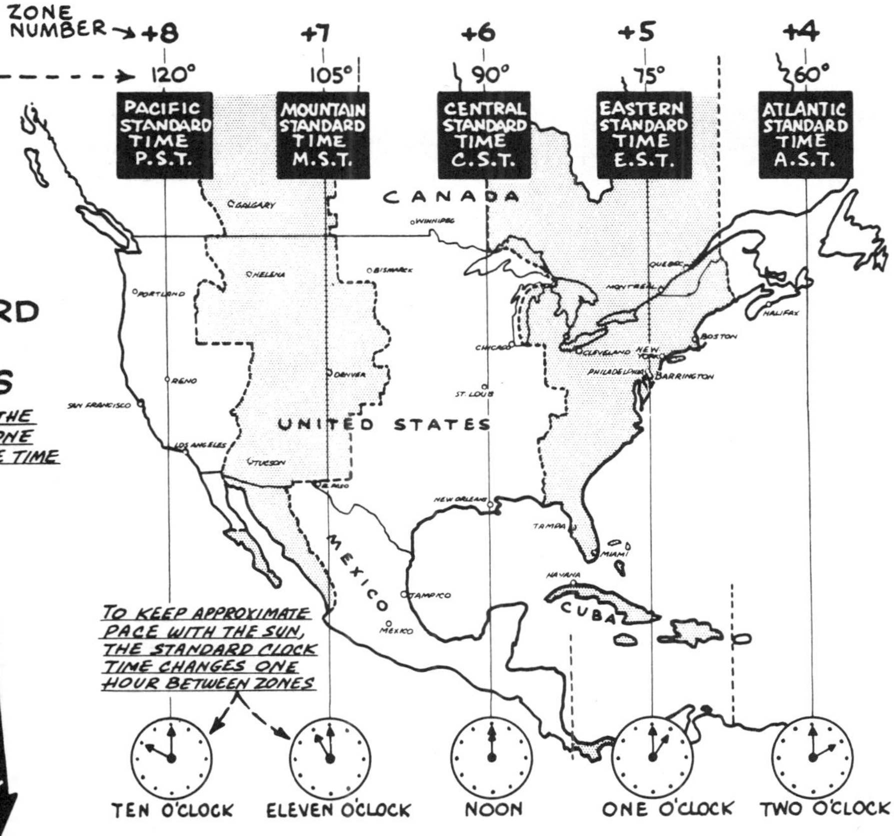

⑥

STANDARD TIME ZONES

ALL PLACES IN THE SAME TIME ZONE HAVE THE SAME TIME

MEAN TO STANDARD

THIS COLUMN GIVES YOUR STANDARD TIME EQUAL TO 8:00 LOCAL MEAN TIME

A TIME CORRECTION IS NEEDED TO CHANGE STANDARD TIME TO MEAN TIME and VICE VERSA ⑤

The CORRECTION FOR ANY LOCATION IS 4 MINUTES FOR EACH DEGREE YOU ARE AWAY FROM THE CENTRAL MERIDIAN OF YOUR TIME ZONE

Example: AKRON, OHIO AT 81°31' WEST LONGITUDE IS 6°31' (6.5°) FROM CENTRAL MERIDIAN OF ZONE 5

CORRECTION = 6.5 × 4 = 26 MIN.

IF YOUR CITY IS IN LIST BELOW, THE TIME CORRECTION IS THE AMOUNT BY WHICH THE TIME GIVEN DIFFERS FROM 8:00

THE FOLLOWING PAGE SHOWS HOW THE CORRECTION IS USED

TABLE 1

PLACE	8:00 L.M.T.	LAT.	LONG.
AKRON, OHIO	8 26 E	41° 05'	81° 31'
ALBANY, N.Y.	7 55 E	42 39	73 45
ALBUQUERQUE	8 07 M	35 05	106 39
ALLENTOWN, PA.	8 02 E	40 36	75 28
ALTOONA, PA.	8 14 E	40 31	78 24
AMARILLO, TEX.	8 47 C	35 12	101 50
ANN ARBOR, MICH	8 35 E	42 17	83 45
ASHEVILLE, N.C.	8 30 E	35 36	82 33
ASHLAND, KY.	7 31 C	38 29	82 38
ATLANTA, GA.	8 38 E	33 45	82 24
ATLANTIC CITY, N.J.	7 58 E	39 22	74 26
AUGUSTA, GA.	8 28 E	33 28	81 58
AUGUSTA, ME.	7 39 E	44 19	69 46
AUSTIN, TEX.	8 31 C	30 16	97 45
BALTIMORE, MD.	8 06 E	39 17	76 37
BANGOR, ME.	7 35 E	44 48	68 46
BARRINGTON, N.J.	8 00 E	39 52	75 02
BATON ROUGE, LA.	8 07 C	30 27	91 11
BATTLE CREEK, MICH	8 41 E	42 19	85 11
BAY CITY, MICH.	8 36 E	43 36	83 53
BEAUMONT, TEX.	8 16 C	30 05	94 06
BELLINGHAM, WSH	8 10 P	48 45	122 29
BERKELEY, CAL.	8 09 P	37 52	122 16
BETHLEHEM, PA.	8 02 E	40 37	75 23
BILLINGS, MONT.	8 14 M	45 47	108 30
BINGHAMPTON, N.Y.	8 04 E	42 06	75 55
BIRMINGHAM, ALA.	7 47 C	33 21	86 49
BISMARCK, N.D.	8 43 C	46 48	100 47
BOISE, IDAHO	8 45 M	43 37	116 12
BOSTON, MASS.	7 44 E	42 21	71 03
BRIDGEPORT, CONN.	7 53 E	41 11	73 11
BROCKTOWN, MASS.	7 44 E	42 05	71 01
BROWNSVILLE, TEX.	8 30 C	25 54	97 30
BUFFALO, N.Y.	8 15 E	42 53	78 52
BURLINGTON, VT.	7 53 E	44 29	73 13
BUTTE, MONT.	8 30 M	46 01	112 32
CAMBRIDGE, MASS.	7 44 E	42 22	71 06
CAMDEN, N.J.	8 00 E	39 57	75 07
CANTON, OHIO	8 26 E	40 48	81 23
CEDAR RAPIDS, IOWA	8 07 C	41 58	91 40
CHAMPAIGN, ILL.	7 53 C	40 07	88 15
CHARLESTON, S.C.	8 20 E	32 47	79 56
CHARLESTON, W.VA.	8 27 E	38 21	81 38
CHARLOTTE, N.C.	8 23 E	35 14	80 51
CHATTANOOGA	7 41 C	35 03	85 19
CHEYENNE, WYO.	7 59 M	41 08	104 49
CHICAGO, ILL.	7 51 C	41 52	87 38
CINCINNATI, OHIO	8 38 E	39 06	84 31
CLEVELAND, OHIO	8 27 E	41 30	81 42
COLORADO SPRINGS	7 59 M	38 50	104 49
COLUMBIA, MO.	8 09 C	38 57	92 20
COLUMBIA, S.C.	8 24 E	34 00	81 02
COLUMBUS, GA.	8 40 E	32 28	84 59
COLUMBUS, OHIO	8 32 E	39 58	83 00
CONCORD, N.H.	7 46 E	43 12	71 32
CORPUS CHRISTI	8 30 C	27 48	97 24
DALLAS, TEX.	8 27 C	32 47	96 48
DAVENPORT, IOWA	8 02 C	41 31	90 35

PLACE	8:00 L.M.T.	LAT.	LONG.
DAYTON, OHIO	8 37 E	39° 46'	84° 12'
DAYTONA BEACH	8 24 E	29 13	81 01
DECATUR, ILL.	7 56 C	39 51	88 57
DENVER, COL.	8 00 M	39 45	104 59
DES MOINES, IOWA	8 14 C	41 35	93 37
DETROIT, MICH.	8 32 E	42 20	83 03
DODGE CITY, KAN.	7 40 M	37 45	100 01
DULUTH, MINN.	8 08 C	46 47	92 06
DURHAM, N.C.	8 16 E	36 01	78 54
EL PASO, TEX.	9 06 C	31 46	106 29
ELIZABETH, N.J.	7 57 E	40 40	74 13
ERIE, PA.	8 20 E	42 07	80 05
EUGENE, ORE.	8 12 P	44 03	123 06
EVANSVILLE, IND.	7 50 C	37 58	87 34
FALL RIVER, MASS.	7 45 E	41 42	71 09
FARGO, N.D.	8 27 C	46 53	96 47
FLINT, MICH.	8 35 E	43 01	83 42
FORT WAYNE, IND.	7 41 C	41 04	85 08
FORT WORTH, TEX.	8 29 C	32 45	97 20
FRESNO, CAL.	7 59 P	36 44	119 47
GADSDEN, ALA.	7 44 C	34 01	86 01
GALVESTON, TEX.	8 15 C	29 18	94 48
GARY, IND.	7 49 C	41 36	87 21
GRAND RAPIDS, MICH.	8 43 E	42 58	85 40
GREEN BAY, WIS.	7 52 C	44 31	88 01
GREENSBORO, N.C.	8 19 E	36 04	79 47
GREENVILLE, S.C.	8 30 E	34 51	82 24
GULFPORT, MISS.	7 56 C	30 22	89 06
HAMILTON, OHIO	8 38 E	39 24	84 33
HARRISBURG, PA.	8 08 E	40 16	76 53
HARTFORD, CONN.	7 51 E	41 46	72 41
HELENA, MONT.	8 28 M	46 36	112 02
HOLYOKE, MASS.	7 50 E	42 12	72 37
HOUSTON, TEX.	8 21 C	29 45	95 22
HUNTINGTON, W.VA.	8 30 E	38 25	82 27
INDIANAPOLIS	7 45 C	39 46	86 10
IOWA CITY, IOWA	8 06 C	41 40	91 32
JACKSON, MICH.	8 38 E	42 15	84 24
JACKSON, MISS.	8 01 C	32 18	90 11
JACKSONVILLE, FLA.	8 27 E	30 20	81 40
JERSEY CITY	7 56 E	40 44	74 04
JOHNSTOWN, PA.	8 16 E	40 20	78 55
KALAMAZOO, MICH.	8 42 E	42 17	85 35
KANSAS CITY, KAN.	8 18 C	39 07	94 36
KANSAS CITY, MO.	8 18 C	39 05	94 35
KENOSHA, WIS.	7 52 C	42 36	87 50
KEY WEST, FLA.	8 27 E	24 34	81 48
KNOXVILLE, TENN.	8 36 E	35 58	83 55
LAFAYETTE, IND.	7 48 C	40 25	86 54
LANCASTER, PA.	8 05 E	40 02	76 18
LANSING, MICH.	8 38 E	42 44	84 33
LAREDO, TEX.	8 38 C	27 30	99 31
LAS VEGAS, NEV.	7 41 P	36 10	115 09
LAWRENCE, MASS.	7 45 E	42 42	71 10
LEXINGTON, KY.	7 38 C	38 03	84 30
LIMA, OHIO	8 36 E	40 45	84 06
LINCOLN, NEB.	8 27 C	40 49	96 42
LITTLE ROCK, ARK.	8 09 C	34 45	92 17

PLACE	8:00 L.M.T.	LAT.	LONG.
LONG BEACH, CAL.	7 53 P	33° 46'	118° 11'
LORAIN, OHIO	8 29 E	41 28	82 10
LOS ANGELES, CAL.	7 53 P	34 03	118 14
LOUISVILLE, KY.	7 43 C	38 15	85 46
LOWELL MASS.	7 45 E	42 38	71 19
LUBBOCK, TEX.	8 47 C	33 35	101 51
MACON, GA.	8 35 E	32 50	83 38
MADISON, WIS.	7 58 C	43 04	89 23
MANCHESTER, N.H.	7 46 E	42 59	71 28
MARION, OHIO	8 33 E	40 35	83 10
MEMPHIS, TENN.	8 00 C	35 09	90 03
MIAMI, FLA.	8 21 E	25 47	80 12
MILWAUKEE, WIS.	7 52 C	43 02	87 54
MINNEAPOLIS	8 13 C	44 59	93 16
MOBILE, ALA.	7 52 C	30 42	88 03
MOLINE, ILL.	8 02 C	41 31	90 31
MONTGOMERY, ALA.	7 45 C	32 23	86 19
MONTPELIER, VT.	7 50 E	44 16	72 35
MUNCIE, IND.	7 42 C	40 11	85 23
NASHVILLE, TENN.	7 47 C	36 10	86 47
NEWARK, N.J.	7 57 E	40 44	74 10
NEW BEDFORD, MASS.	7 44 E	41 38	70 56
NEW BRITAIN, CONN.	7 51 E	41 40	72 47
NEW HAVEN, CONN.	7 52 E	41 18	72 56
NEW ORLEANS, LA.	8 00 C	29 57	90 04
NEW YORK, N.Y.	7 56 E	40 45	74 00
NIAGARA FALLS	8 16 E	43 06	79 03
NORFOLK, VA.	8 05 E	36 51	76 17
OAKLAND, CAL.	8 09 P	37 48	122 16
OGDEN, UTAH	8 28 M	41 14	111 58
OKLAHOMA CITY	8 30 C	35 28	97 31
OMAHA, NEB.	8 24 C	41 16	95 56
ORLANDO, FLA.	8 26 E	28 33	81 23
PADUCAH, KY.	7 54 C	37 05	88 36
PASADENA, CAL.	7 53 P	34 09	118 09
PATERSON, N.J.	7 57 E	40 55	74 09
PENSACOLA, FLA.	7 49 C	30 25	87 13
PEORIA, ILL.	7 58 C	40 42	89 36
PHILADELPHIA	8 01 E	39 57	75 09
PHOENIX, ARIZ.	8 28 M	33 27	112 04
PITTSBURG, PA.	8 20 E	40 26	80 00
PITTSFIELD, MASS.	7 53 E	42 27	73 15
PORT ARTHUR, TEX.	8 16 C	29 53	93 56
PORTLAND, ME.	7 41 E	43 40	70 15
PORTLAND, ORE.	8 11 P	45 31	122 41
PORTSMOUTH, VA.	8 05 E	36 50	76 18
PROVIDENCE, R.I.	7 46 E	41 50	71 25
PUEBLO, COL.	7 58 M	38 16	104 37
RACINE, WIS.	7 51 C	42 44	87 47
RALEIGH, N.C.	8 15 E	35 47	78 38
READING, PA.	8 04 E	40 20	75 56
RENO, NEV.	7 59 P	39 31	119 49
RICHMOND, VA.	8 10 E	37 32	77 26
ROANOKE, VA.	8 20 E	37 16	79 57
ROCHESTER, MINN.	8 10 C	44 01	92 28
ROCHESTER, N.Y.	8 10 E	43 10	77 36
ROCKFORD, ILL.	7 56 C	42 16	89 06
SACRAMENTO	8 06 P	38 35	121 30

PLACE	8:00 L.M.T.	LAT.	LONG.
SAGINAW, MICH.	8 36 E	43° 26'	83° 56'
SCHENECTADY	7 56 E	42 49	73 56
ST. JOSEPH, MO.	8 19 C	39 46	94 51
ST. LOUIS, MO.	8 01 C	38 38	90 12
ST. PAUL, MINN.	8 12 C	44 57	93 06
ST. PETERSBURG	8 31 E	27 46	82 38
SALT LAKE CITY	8 28 M	40 45	111 53
SAN ANTONIO, TEX.	8 34 C	29 26	98 29
SAN BERNARDINO	7 49 P	34 07	117 17
SAN DIEGO, CAL.	7 49 P	32 43	117 09
SAN FRANCISCO	8 10 P	37 47	122 25
SAN JOSE, CAL.	8 08 P	37 20	121 53
SANTA BARBARA	7 59 P	34 25	119 42
SANTA FE, N.M.	8 04 M	35 41	105 56
SAVANNAH, GA.	8 24 E	32 05	81 06
SCRANTON, PA.	8 03 E	41 25	75 40
SEATTLE, WASH.	8 09 P	47 37	122 20
SHREVEPORT, LA.	8 15 C	32 31	93 45
SIOUX CITY, IOWA	8 26 C	42 30	96 25
SIOUX FALLS, S.D.	8 27 C	43 33	96 44
SOMERVILLE, MASS.	7 44 E	42 23	71 06
SOUTH BEND, IND.	7 45 C	41 41	86 15
SPARTANBURG, S.C.	8 28 E	34 57	81 56
SPOKANE, WASH.	7 50 P	47 40	117 26
SPRINGFIELD, ILL.	7 59 C	39 48	89 39
SPRINGFIELD, MASS.	7 50 E	42 06	72 36
SPRINGFIELD, MO.	8 13 C	37 13	93 18
SPRINGFIELD, OHIO	8 35 E	39 56	83 48
STAMFORD, CONN.	7 54 E	41 03	73 32
STEUBENVILLE, O.	8 22 E	40 22	80 37
STOCKTON, CAL.	8 06 P	37 58	121 17
SUPERIOR, WIS.	8 08 C	46 43	92 06
SYRACUSE, N.Y.	8 05 E	43 03	76 09
TACOMA, WASH.	8 10 P	47 15	122 26
TAMPA, FLA.	8 30 E	27 57	82 27
TERRE HAUTE, IND.	7 50 C	39 28	87 24
TOLEDO, OHIO	8 34 E	41 39	83 33
TOPEKA, KAN.	8 23 C	39 03	95 40
TRENTON, N.J.	7 59 E	40 13	74 46
TROY, N.Y.	7 55 E	42 44	73 41
TUCSON, ARIZ.	8 24 M	32 13	110 58
TULSA, OKLA.	8 24 C	36 09	96 00
UTICA, N.Y.	8 01 E	43 06	75 14
WACO, TEX.	8 29 C	31 33	97 08
WALLA WALLA	7 53 P	46 04	118 20
WASHINGTON, D.C.	8 08 E	38 54	77 01
WATERBURY, CONN.	7 52 E	41 33	73 03
WATERLOO, IOWA	8 09 C	42 30	92 20
WHEELING, W.VA.	8 23 E	40 04	80 43
WICHITA, KAN.	8 29 C	37 42	97 20
WICHITA FALLS, TEX.	8 33 C	33 55	98 29
WILKES-BARRE, PA.	8 04 E	41 15	75 53
WILMINGTON, DEL.	8 02 E	39 45	75 33
WINSTON-SALEM	8 21 E	36 06	80 15
WORCESTER, MASS.	7 47 E	42 16	71 48
YONKERS, N.Y.	7 56 E	40 56	73 54
YORK, PA.	8 07 E	39 58	76 44
YOUNGSTOWN, OHIO	8 23 E	41 06	80 39

TABLE COMPILED FROM THE WORLD ALMANAC

you must use local mean time. L.M.T. is calculated by applying a correction equal to four minutes for each degree you are away from the central meridian of your time zone. If your location is in the list of cities in Table 1, you can determine your time correction by inspection since it is simply the amount by which the time given differs from 8:00. However, it is nearly as easy to calculate your correction, Fig. 5. The time correction is used in converting your standard time to mean time or vice versa. Rules for doing this are given below.

GREENWICH MEAN TIME. G.M.T. is the mean solar time at Greenwich, England, located on the zero meridian, which is the central meridian of Zone 0. From the description already given, it will be apparent that Greenwich mean time is also the Standard or Zone time for all of Zone 0.

UNIVERSAL TIME. Universal time is exactly the same time as Greenwich mean time, that is, it is just another name for the same thing. The designation U.T. is popular in astronomy, while navigators prefer G.M.T.

DAYLIGHT TIME. This is called advanced or fast time, daylight-saving time or Summer time. If your standard summer clock time is Daylight time, you must deduct 1 hr. to get regular Standard time, and then take it from there for any needed conversion.

TIME CONVERSIONS. If you dabble in astronomy or use a telescope, you will have to convert one kind of time to another kind of time. The simple formulas for doing this are given at the bottom of this and other pages. Although the formulas are easy in an arithmetical sense, it takes some study to understand the how-and-why of time conversions.

If a football game on the Pacific coast starts at 2:30 Pacific standard time, what is the time at Cleveland, Ohio? Technically, a problem of this kind is covered by the general rule: Standard time varies the same amount as the difference between zone numbers, the place to the east having the greater time. No doubt you are still lost because you don't know that California is Zone 8 and Cleveland Zone 5. But with this information known, you can apply the rule already stated--the difference between zone numbers is 3 (hours) and the place to the east has the greater time. The time in Cleveland is 5:30. At London, zone 0, it is 10:30 p.m. at the same instant.

The conversion of Greenwich mean time to local mean time or vice versa requires the use of your longitude in time instead of your zone number. The general rule for making this conversion has a familiar ring: The difference in time between two places is the same as their difference in longitude, the more easterly place having the greater time. This is the general rule for any kind of time. This same rule governs local mean time within any time zone. If you are west of the central meridian, it means the central meridian is the more easterly place and has the greater time. How much greater is exactly the same as the difference in longitude.

TIME OF POSITION. All sky objects have some movement of their own, but over a period of several hours, even the near and fast-moving moon practically stands still in relation to the rotation of the earth. Fig. 7 shows planet Saturn on the meridian at Greenwich, 10:00 p.m., G.M.T. In the space of a single night the planet is practically stationary. The only movement that matters is the rotation of the earth. If you live on or near the central meridian of time zone No. 5, it will take 5 hours for the earth to rotate enough to put your position directly under the stationary planet. It will then be 10:00 p.m. local mean time at your location, the same transit time as at Greenwich. The point to remember is: If a transit

TIME CONVERSIONS

MEAN TO STANDARD and STANDARD TO MEAN

IF YOU ARE WEST OF TIME ZONE MERIDIAN		Example: AKRON, Ohio — TIME CORRECTION = 26 MIN. — ASSUME CLOCK TIME IS 7:00 a.m. E.S.T.
STANDARD TO MEAN	L.M.T. = STANDARD TIME *MINUS* TIME CORRECTION (SEE PREVIOUS PAGE)	LOCAL MEAN TIME = 7:00 − 26^m = 6:34 L.M.T.
MEAN TO STANDARD	STANDARD TIME = LOCAL MEAN TIME *PLUS* TIME CORRECTION	STANDARD TIME = 6:34 + 26^m = 7:00 a.m., E.S.T.

IF YOU ARE EAST OF TIME ZONE MERIDIAN		Example: LOS ANGELES, Calif. — TIME CORRECTION = 7 MIN. — ASSUME CLOCK TIME IS 4:00 a.m. P.S.T.
STANDARD TO MEAN	L.M.T. = STANDARD TIME *PLUS* TIME CORRECTION	LOCAL MEAN TIME = 4:00 + 7^m = 4:07 L.M.T
MEAN TO STANDARD	STANDARD TIME = LOCAL MEAN TIME *MINUS* TIME CORRECTION	STANDARD TIME = 4:07 − 7^m = 4:00 a.m., P.S.T.

time of any sky object is given in Greenwich mean time, this "time of position" will be just the same at your location.

APPLICATIONS OF TIME. Some astronomical events are time-at-the-same-instant. This is not concerned with the position of the outer space object, but with something that happens on or to the object itself. Phenomena of this type include eclipses of the sun and moon, configuration of Jupiter's satellites, minima of a variable star, maxima of a meteor shower, and others of similar nature. If the time of such event is G.M.T., you deduct your zone number to get time-at-the-same-instant at your location.

Star maps for use at a certain hour always mean local mean time. If the map is for 9 o'clock, it means 9 o'clock L.M.T., and you must convert 9 o'clock L.M.T. to standard time.

A rotating star map (planisphere) presents somewhat the same problem--but exactly opposite. The clock dial on the planisphere is local mean time. You could change all the numbers, and in doing this you would be converting mean-to-standard. However, the usual circumstance of using a star dial is that you look at your standard clock and note that it is a certain standard time. You then convert this to local mean time, which is then used for direct setting of the clock dial.

SIDEREAL OR STAR TIME

SIDEREAL TIME is based on one rotation of the earth in relation to any star. This is a sidereal day. Like solar time, the sidereal day is divided into 24 hours, but the day itself is about 4 minutes shorter. A simple explanation of the difference is

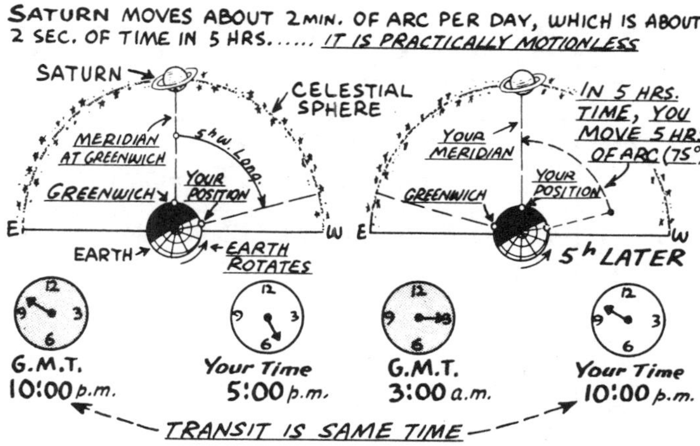

(7) TIME-OF-POSITION IS ALWAYS LOCAL MEAN TIME

shown in Fig. 10.

Some particular point in the sky is needed as a zero index mark for 0-hour sidereal time. For this, astronomers have selected the vernal equinox rather than a star, Fig. 8. However, the eastern edge of the great square in Pegasus is in line with the vernal equinox and supplies a convenient visual guide--whenever you see Pegasus on your meridian, it is about 0-hour sidereal time.

A clock keeping sidereal time will run about four minutes fast per day as compared to a solar clock. More exactly, the daily difference is 3.94 minutes. On about Sept. 23 of each year, the sun crosses the equator going south, the crossover point being the autumnal equinox. When the autumnal equinox is coincident with the sun, solar and sidereal clocks will agree, Fig. 12. The sidereal clock starts to gain immediately, and the 4-min. daily difference soon makes the sidereal clock hopelessly at odds with solar time. The

TIME CONVERSIONS

TO GREENWICH and BACK... TIME AT THE SAME INSTANT

Example: SAME TIME AND PLACE AS OPPOSITE PAGE

PLACE: AKRON, Ohio, $5^h 26^m$ W. LONGITUDE, ZONE 5

TIME: 7:00 a.m. E.S.T. (MORNING)

STANDARD TO G.M.T. and G.M.T. TO STANDARD

STANDARD TO G.M.T.	G.M.T. = STANDARD TIME + ZONE NO.	G.M.T. = 7:00 + 5 = 12:00
G.M.T. TO STANDARD	STANDARD TIME = G.M.T. − ZONE NO.	E.S.T. = 12:00 − 5 = 7:00

SAME EXAMPLE: 7:00 E.S.T. = 6:34 L.M.T. (SEE OPPOSITE PAGE)

LOCAL MEAN TIME TO G.M.T and G.M.T TO L.M.T.

LOCAL MEAN TO GREENWICH MEAN	G.M.T. = L.M.T. + longitude WEST	G.M.T. = 6:34 + 5:26 = $11^h 60^m$ G.M.T. $12^h 00^m$
GREENWICH MEAN TO LOCAL MEAN	L.M.T. = G.M.T. − Longitude WEST (LOCAL MEAN TIME) (GREENWICH MEAN TIME) (YOUR LONGITUDE IN HOURS and MINUTES OF TIME)	G.M.T. $12^h 00^m$ = $11^h 60^m$ MINUS Long. $5^h 26^m$ → $5^h 26^m$ L.M.T. $6^h 34^m$

NOTE: THE TIME FORMULAS ARE FOR ANY LOCATION WEST OF GREENWICH. IF EAST, SWITCH PLUS AND MINUS SIGNS

TABLE 2

SIDEREAL TIME AT 0^h (MIDNIGHT) GREENWICH MEAN TIME (G.M.T.)

DAY	JAN.	FEB.	MARCH	APRIL	MAY	JUNE	JULY	AUG.	SEPT.	OCT.	NOV.	DEC.	DAY
1	$6^h 41^m$	$8^h 44^m$	$10^h 34^m$	$12^h 36^m$	$14^h 35^m$	$16^h 37^m$	$18^h 35^m$	$20^h 37^m$	$22^h 40^m$	$0^h 38^m$	$2^h 40^m$	$4^h 38^m$	1
2	6 45	8 48	10 38	12 40	14 39	16 41	18 39	20 41	22 44	0 42	2 44	4 42	2
3	6 49	8 52	10 42	12 44	14 42	16 45	18 43	20 45	22 47	0 46	2 48	4 46	3
4	6 53	8 56	10 46	12 48	14 46	16 49	18 47	20 49	22 51	0 50	2 52	4 50	4
5	6 57	9 00	10 50	12 52	14 50	16 53	18 51	20 53	22 55	0 54	2 56	4 54	5
6	7 01	9 03	10 54	12 56	14 54	16 57	18 55	20 57	22 59	0 58	3 00	4 58	6
7	7 05	9 07	10 58	13 00	14 58	17 01	18 59	21 01	23 03	1 02	3 04	5 02	7
8	7 09	9 11	11 02	13 04	15 02	17 04	19 03	21 05	23 07	1 05	3 08	5 06	8
9	7 13	9 15	11 06	13 08	15 06	17 08	19 07	21 09	23 11	1 09	3 12	5 10	9
10	7 17	9 19	11 10	13 12	15 10	17 12	19 11	21 13	23 15	1 13	3 16	5 14	10
11	7 21	9 23	11 14	13 16	15 14	17 16	19 15	21 17	23 19	1 17	3 20	5 18	11
12	7 25	9 27	11 18	13 20	15 18	17 20	19 18	21 21	23 23	1 21	3 23	5 22	12
13	7 29	9 31	11 21	13 24	15 22	17 24	19 22	21 25	23 27	1 25	3 27	5 26	13
14	7 33	9 35	11 25	13 28	15 26	17 28	19 26	21 29	23 31	1 29	3 31	5 30	14
15	7 37	9 39	11 29	13 32	15 30	17 32	19 30	21 33	23 35	1 33	3 35	5 34	15
16	7 41	9 43	11 33	13 36	15 34	17 36	19 34	21 36	23 39	1 37	3 39	5 37	16
17	7 45	9 47	11 37	13 39	15 38	17 40	19 38	21 40	23 43	1 41	3 43	5 41	17
18	7 49	9 51	11 41	13 43	15 42	17 44	19 42	21 44	23 47	1 45	3 47	5 45	18
19	7 52	9 55	11 45	13 47	15 46	17 48	19 46	21 48	23 51	1 49	3 51	5 49	19
20	7 56	9 59	11 49	13 51	15 50	17 52	19 50	21 52	23 54	1 53	3 55	5 53	20
21	8 00	10 03	11 53	13 55	15 53	17 56	19 54	21 56	23 58	1 57	3 59	5 57	21
22	8 04	10 07	11 57	13 59	15 57	18 00	19 58	22 00	0 02	2 01	4 03	6 01	22
23	8 08	10 10	12 01	14 03	16 01	18 04	20 02	22 04	0 06	2 05	4 07	6 05	23
24	8 12	10 14	12 05	14 07	16 05	18 08	20 06	22 08	0 10	2 09	4 11	6 09	24
25	8 16	10 18	12 09	14 11	16 09	18 11	20 10	22 12	0 14	2 12	4 15	6 13	25
26	8 20	10 22	12 13	14 15	16 13	18 15	20 14	22 16	0 18	2 16	4 19	6 17	26
27	8 24	10 26	12 17	14 19	16 17	18 19	20 18	22 20	0 22	2 20	4 23	6 21	27
28	8 28	10 30	12 21	14 23	16 21	18 23	20 22	22 24	0 26	2 24	4 27	6 25	28
29	8 32	10 32	12 25	14 27	16 25	18 27	20 26	22 28	0 30	2 28	4 30	6 29	29
30	8 36		12 28	14 31	16 29	18 31	20 29	22 32	0 34	2 32	4 34	6 33	30
31	8 40		12 32		16 33		20 33	22 36		2 36		6 37	31

MEAN TABLE COMPILED FROM THE AMERICAN EPHEMERIS

THIS TABLE IS GREENWICH SIDEREAL TIME (G.S.T.) AT 0^h G.M.T.

TO FIND LOCAL SIDEREAL TIME (L.S.T.) USE THIS FORMULA:

$$\text{L.S.T.} = \text{L.M.T.} + \text{G.S.T.} + \frac{\text{L.M.T. (TO NEAREST HOUR)} \text{ ADD TIME ZONE}}{6}$$

- L.S.T. — LOCAL SIDEREAL TIME
- L.M.T. — LOCAL MEAN TIME
- G.S.T. — GREENWICH SIDEREAL TIME (FROM TABLE)
- MAKE THIS DIVISION AND CALL THE ANSWER MINUTES

Example: YOUR LOCATION IS AKRON, OHIO, IN ZONE 5. IT IS JAN. 1, 20:00 L.M.T. **What** IS SIDEREAL TIME?

SOLUTION:

ADD
- L.M.T. $20^h 00^m$
- G.S.T. $6^h 41^m$
- CORRECTION 04^m ($20 + 5 = 25$; $25 / 6 = 4\frac{1}{6}$ MIN.)

$26^h 45^m$

LESS $24^h 00^m$

L.S.T. IS $2^h 45^m$ AT 20^h L.M.T.

time difference allows the stars to advance westward for 4 min. before the solar day is completed, and this small advance repeated night after night gives us a constantly changing parade of stars.

CONVERTING SIDEREAL TIME. The usual method of doing this is to consult a table which gives the sidereal time at midnight for each day. If, to this tabulated value, you add the number of hours and minutes you are past midnight, you will get the approximate sidereal time at your location.

The complete formula to convert local mean time to sidereal time is given at the bottom of Table 2. The table itself gives the sidereal time for each day when it is 0-hour (midnight) at Greenwich. It should be noted that 0-hour (midnight) is a "station" in time, the same as 12-hour (noon) is a station in time. Your noon will occur later than noon at Greenwich, but when it is noon at your location it is noon, 12 o'clock, and the sun is on the meridian. If your location is 5 hours west of Greenwich, midnight at your location will occur 5 hours after midnight at Greenwhich. But when it comes midnight at your location, the sidereal equivalent of midnight will be the same as at Greenwich.

In brief, sidereal time at Greenwich is the same sidereal time at your location. However, you have a modest correction to make. In practically all cases you will want the sidereal time at some time past midnight. So, to the tabulated time, you must add the hours past midnight. During this interval, the sidereal clock gains on the solar clock at the stated rate of about 4 minutes per day, or 10 seconds per hour. This correction must be added. Also, the tabulated sidereal time at midnight at your location will occur 5 hours (in Zone 5) later, and this means another 50 seconds added. The formula in Table 2 covers everything.

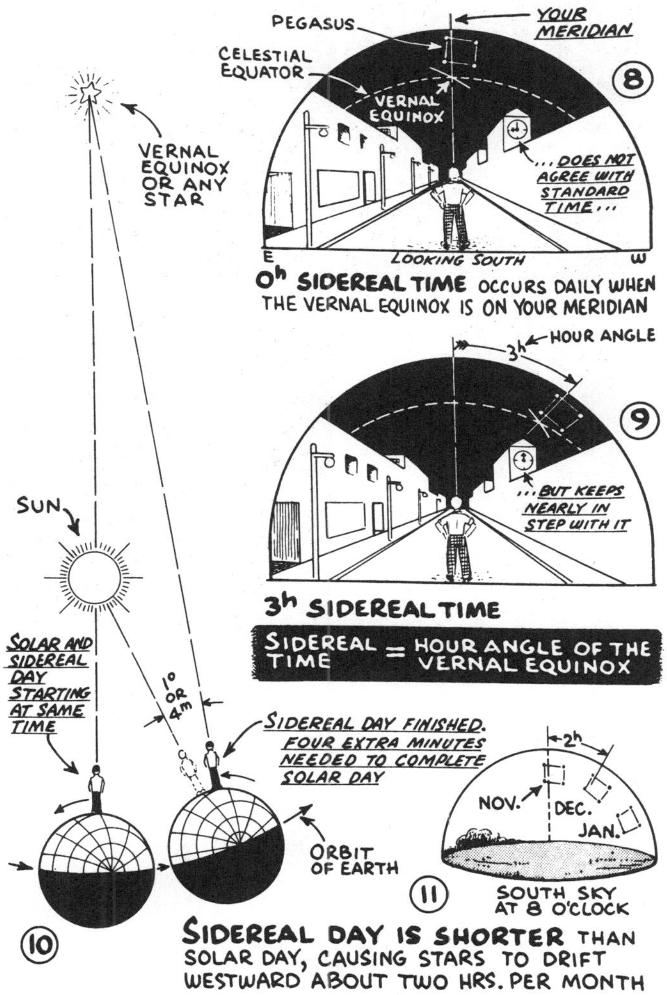

(12) THE SIDEREAL CLOCK GAINS 4 MIN. PER DAY

Two Simple Ways to Find SIDEREAL TIME

METHOD No. 1 - *NO CORRECTION*

L.S.T = L.M.T + G.S.T. at 0^h

Example: SAME AS OPPOSITE PAGE
PLACE: AKRON, Ohio
TIME: 20:00 L.M.T.

SOLUTION:

L.M.T	20^h 00^m
Plus G.S.T. 0^h JAN. 1	6 41 (from TABLE)
	26 41
Less 24h	24 00
L.S.T.	$2^h 41^m$

...Compare with $2^h 45^m$ ...ERROR OF 4 MIN.

METHOD No. 2 - *WITH APPROX. CORRECTION*

L.S.T. = L.M.T + G.S.T. at 0^h NEXT DAY

Example: SAME

SOLUTION:

L.M.T.	20^h 00^m
Plus G.S.T. 0^h JAN. 2	6 45 (from TABLE)
	26 45
Less 24h	24 00
L.S.T.	2^h 45^m

THE NEXT DAY

...NO ERROR! ALTHOUGH NOT ALWAYS SO EXACT, THE ERROR BY THIS METHOD WILL SELDOM EXCEED 1 MIN.

VARIOUS methods are used in locating sky objects with a telescope, ranging from coarse naked-eye sighting to precise pin-pointing with the use of setting circles. All methods require a good mount--it must not vibrate unduly, it must "stay put" at any position and it must work smoothly on both axes. Other requirements are a planisphere and a star atlas. The planisphere is used to determine the general aspect of the sky; the atlas then supplies the detail maps. Don't expect to find sky objects by random sweeping--you must know exactly what you are looking for and how to get there.

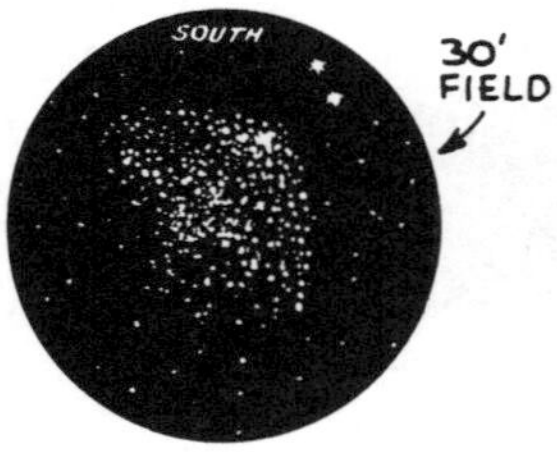

M11 - OPEN CLUSTER
MANY TINY STARS GIVE A GLOWING EFFECT. EASY TO SEE WITH 3 TO 6" AT 40x

STAR HOPPING. This is the finding method most used by beginners. The idea is that you hop from a bright star you know to another star you know, etc., and in this way reach your target, which may be invisible to the naked eye. An important part of this technique is careful plotting of the course on a star atlas. Make a "field plotter" of clear plastic as shown in drawing, scaled to the degree marks which you will find at the edge of all atlas maps. If you don't know the field of your finder, find it by the method shown in boxed drawing below.

Now, let's plot the route to a typical telescope object, such as M 11. Altair will be your pilot star or starting point. Note in drawing that a 6-degree finder field will take in the two guard stars, which will make identification positive when you locate Altair in the sky. Move the field plotter, keeping Altair in field but stretching out to another star along the route. This will be Mu, as shown. Keeping Mu in the field, you can reach Delta. From Delta to Lambda you will have a little bit of blind hop, but it will not be hard to pick up the curved string of stars ending at the top of Scutum (SKYOU-tum), the Shield. Below Eta and Beta in Scutum you will find three faint stars, and about half degree east and south is M 11.

Note that the general direction of your route is west and south. The drawing shows the stars as they appear in a naked-eye view facing south. If your finder is the usual inverting type, all this will be upside down. Hence, turn the drawing (or atlas) upside down and it will then agree with the view you will see later in the eyepiece of the finder. Memorize each step of the route; call out each star by

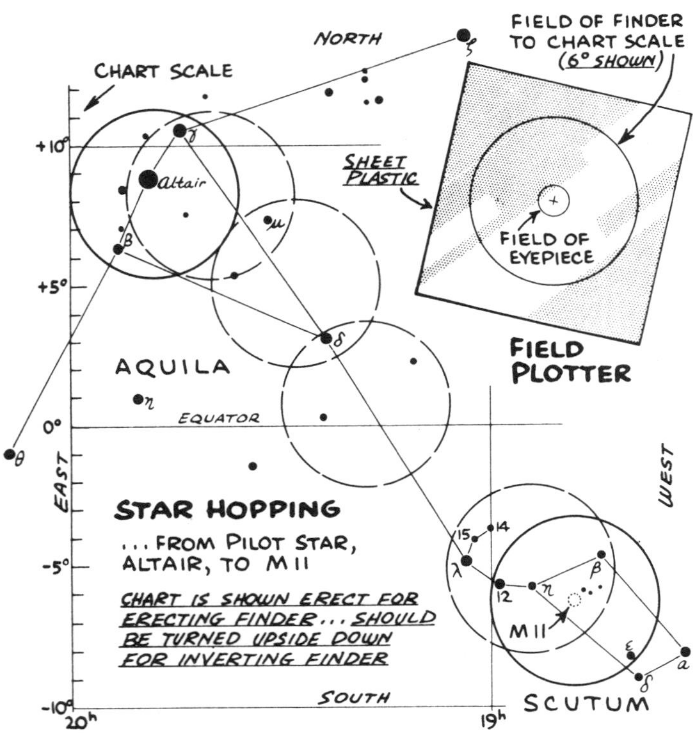

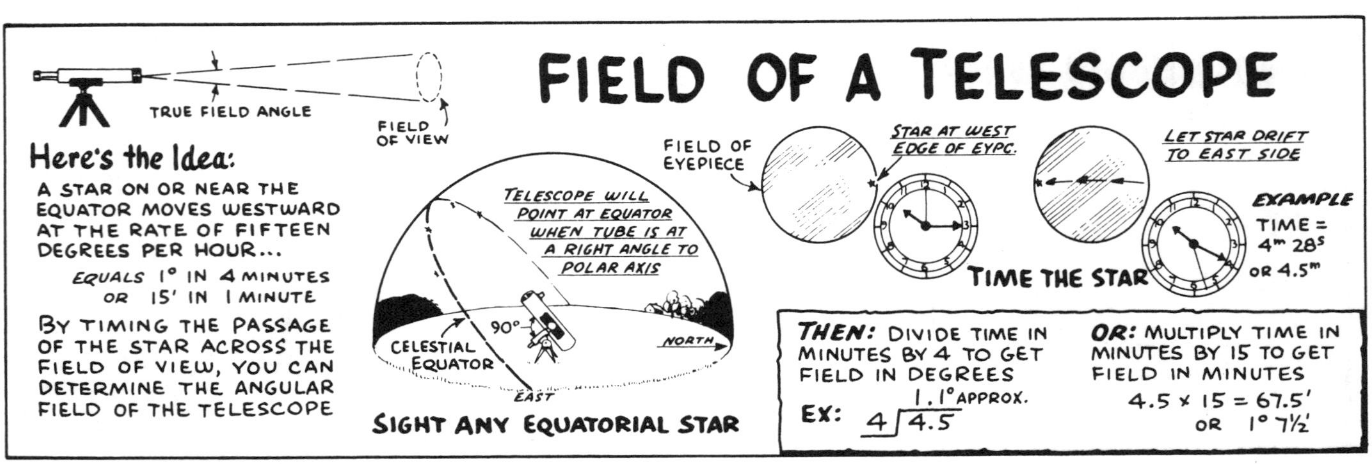

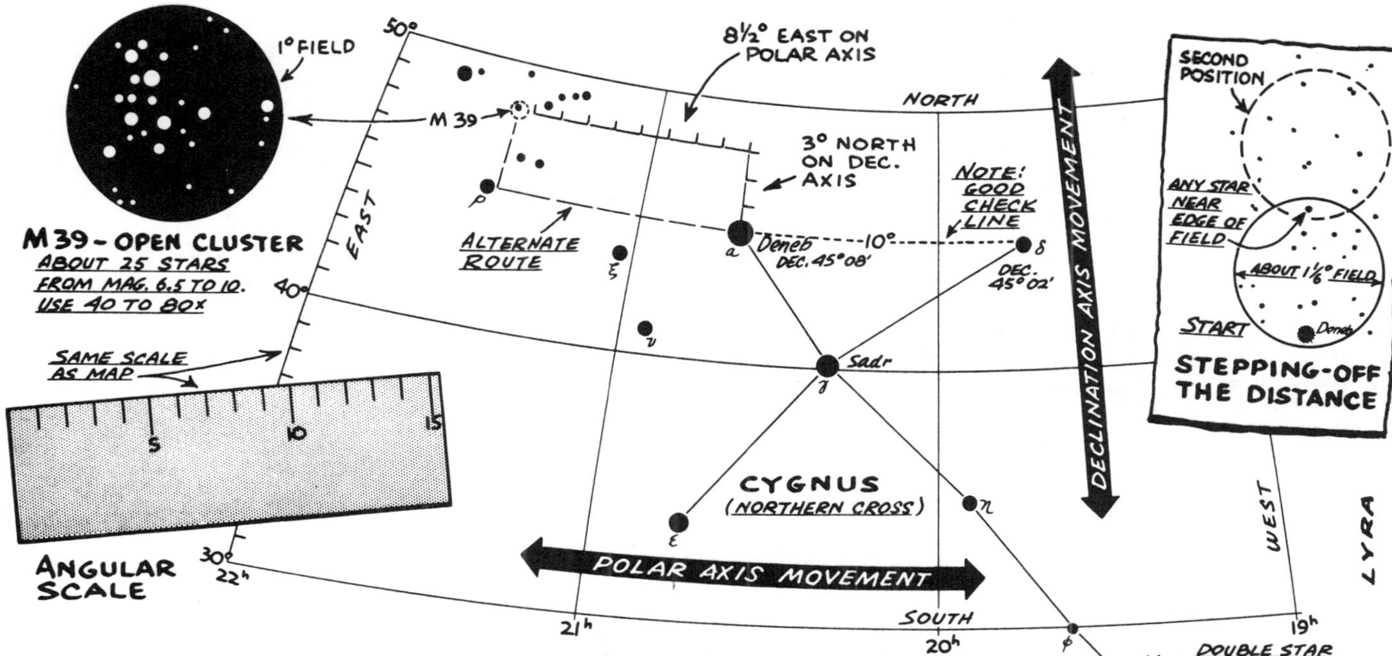

RIGHT ANGLE SWEEP - DENEB TO M 39

name. Careful attention to the atlas plotting will make the actual finding of M 11 a fairly simple matter. It will show as a misty patch of light in the finder, while the telescope will resolve it into a myriad of tiny sparklers.

RIGHT ANGLE SWEEP. With an equatorial mount there are just two possible movements: (1) Any movement on the declination axis will follow a meridian, (2) any polar axis movement describes a circle around the pole. These movements are at right angles and correspond to the grid of lines shown on all atlas maps. In making a right angle sweep you first locate a pilot star and from this step off the required number of degrees in two separate movements, measuring the distance by the field of a low-power eyepiece. The finder is not used.

M 39 is shown as a target object, with Deneb the pilot star. You will need a little cardboard scale, marked to atlas scale. This serves to measure angular distances in both declination and right ascension. The scaling is not exact on most maps, especially for the crosswise R. A. distance, but is accurate enough for the purpose. As shown in the drawing, M 39 is 3 degrees north of Deneb and 8-1/2 degrees east. A low-power eyepiece with a field of about 1-1/6 degrees (the average) is convenient for stepping off the distance, the slightly larger field eliminating the need of measuring exactly to the edge of eyepiece field.

Make the declination sweep first. You will be able to see many faint stars not shown on the atlas and these serve as spacing guides--you pick up any star at one side of the field and then move the telescope to put it at the opposite side--that's one field or 1 degree. After completing the declination step, lock the declination shaft. Step off 8-1/2 fields to the east, moving only on the polar axis. This should bring you to M 39. If not in the field, sweep cautiously in the immediate area--M 39 is just a bright splash of stars and is not outstanding against the rich background of the Milky Way.

In any right-angle sweep, it will be apparent you have a choice of two routes. Sometimes a convenient bright star will simplify the whole operation,

Aa IS FOR ALPHA — THE GREEK ALPHABET (ALPHA) (BETA)

Α	α	ALPHA	AL-fuh	Ι	ι	IOTA	i-OH-tuh	Ρ	ρ	RHO	Row (YOUR BOAT)
Β	β	BETA	BAY-tuh	Κ	κ	KAPPA	CAP-uh	Σ	σ	SIGMA	SIG-muh
Γ	γ	GAMMA	GAM-uh	Λ	λ	LAMBDA	LAM-duh	Τ	τ	TAU	Taw
Δ	δ	DELTA	DELL-tuh	Μ	μ	MU	Mew or Moo	Υ	υ	UPSILON	UP-sih-lon
Ε	ε, ϵ	EPSILON	EPP-sih-lon	Ν	ν	NU	New or Noo	Φ	φ	PHI	Fie
Ζ	ζ	ZETA	ZAY-tuh	Ξ	ξ	XI	Zi (NOTE Z SOUND)	Χ	χ	CHI	Kie
Η	η	ETA	AY-tuh	Ο	ο	OMICRON	⁺OHM-ih-krawn	Ψ	ψ	PSI	Sie (sigh)
Θ	θ	THETA	THAY-tuh	Π	π	PI	Pie	Ω	ω	OMEGA	Oh-ME-guh

⁺LIKE Omelet, Omnibus

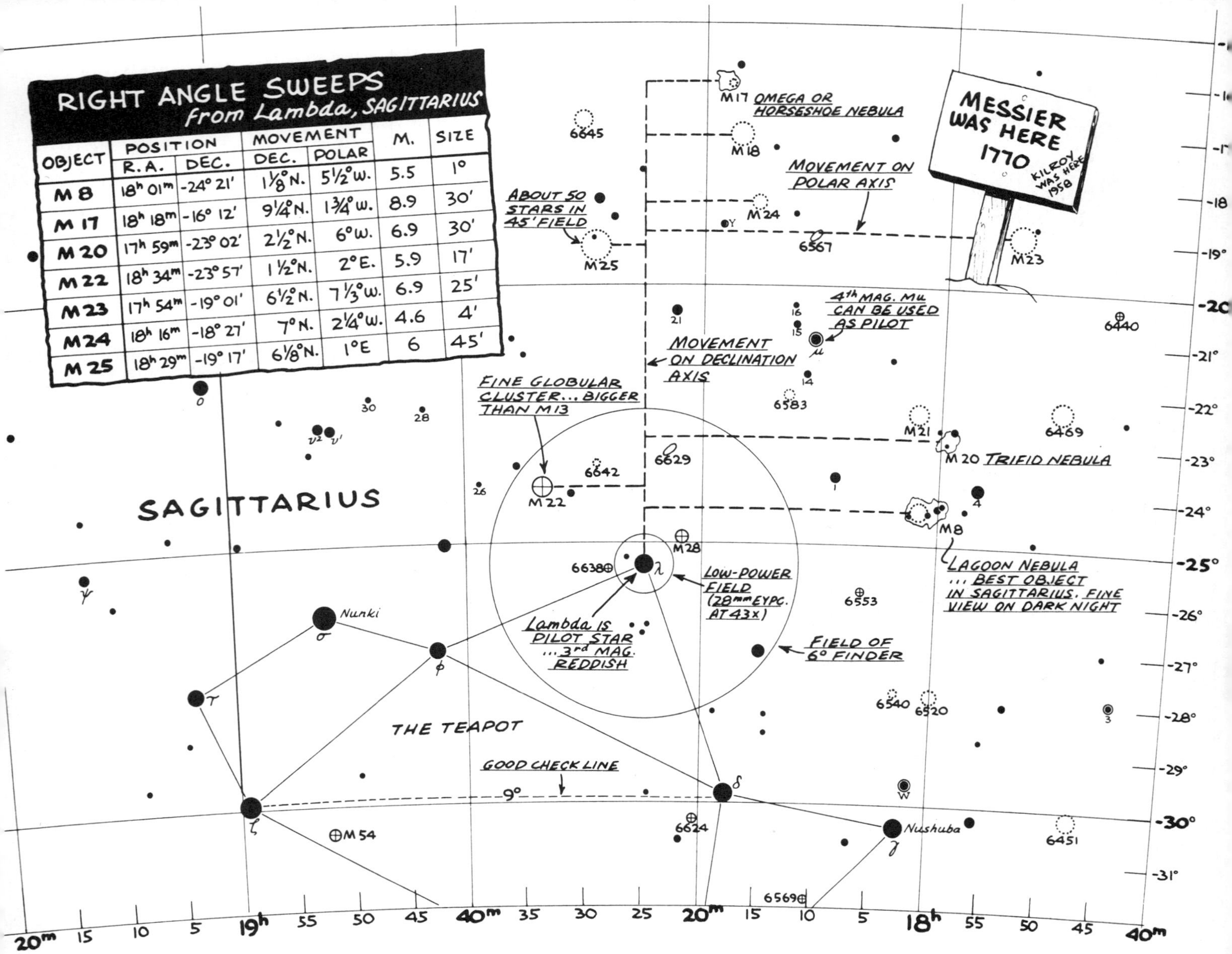

RIGHT ANGLE SWEEPS *from Lambda, SAGITTARIUS*

OBJECT	POSITION R.A.	POSITION DEC.	MOVEMENT DEC.	MOVEMENT POLAR	M.	SIZE
M 8	18h 01m	-24° 21'	1 1/8° N.	5 1/2° W.	5.5	1°
M 17	18h 18m	-16° 12'	9 1/4° N.	1 3/4° W.	8.9	30'
M 20	17h 59m	-23° 02'	2 1/2° N.	6° W.	6.9	30'
M 22	18h 34m	-23° 57'	1 1/2° N.	2° E.	5.9	17'
M 23	17h 54m	-19° 01'	6 1/2° N.	7 1/3° W.	6.9	25'
M 24	18h 16m	-18° 27'	7° N.	2 1/4° W.	4.6	4'
M 25	18h 29m	-19° 17'	6 1/8° N.	1° E	6	45'

as, with M 39, the alternate route shown has a turning point at Rho, eliminating the need of measuring the angular distance. While in the Cygnus area, stars Deneb and Delta provide a convenient check for the alignment of your mounting to the pole--you should be able to sweep from one star to the other with polar axis movement only. The separation of 10 degrees can be used to check atlas scale and also your own ability to step off the distance with eyepiece field.

SWEEPING IN SAGITTARIUS. The Sagittarius-Scorpius region was Messier's favorite hunting ground and more than a quarter of his popular list can be found in this richly-spangled area of the sky. Sagittarius is especially good for measured right-angle sweeps, using reddish, third magnitude Lambda as a pilot star, as shown in map above. A little triangle of stars directly under Lambda will make identification positive. If you put Lambda at the edge of a low-power field you can pick up a faint glow at the opposite edge of field. This is the globular cluster, M 28, of seventh magnitude. A brighter globular is M 22, which tops the famous M 13 cluster in size and is very nearly as bright. If you move 1 degree north from Lambda in declination and then sweep to the west in R. A., you will not fail to pick up M 8, the popular Lagoon nebula. It is unfortunate this splendid object must be viewed under the luminous skies of summer, since even this small amount of light destroys the nebulosity which forms the lagoon, leaving only a fair star cluster. View this on a really dark night about midnight and you will understand how it got its name--it does indeed resemble a misty lake dotted about with lights.

There are many fine open clusters in Sagittarius, all easily located with measured sweeps from Lambda. M 25 and M 23 are popular low-power fields; M 24 packs about fifty stars in a tiny 4-minute field and needs a 3 to 6-inch objective and medium power for good resolution although the bright glow of about fifth magnitude is easily seen with smallest telescope or binoculars. Sagittarius offers a good test sweep from Zeta to Delta of nearly an exact nine degrees separation and on nearly the same parallel of declination. Zeta is a fine double (mags. 3.4 and 3.6) but the separation of less than 1 second puts it beyond the range of portable telescopes.

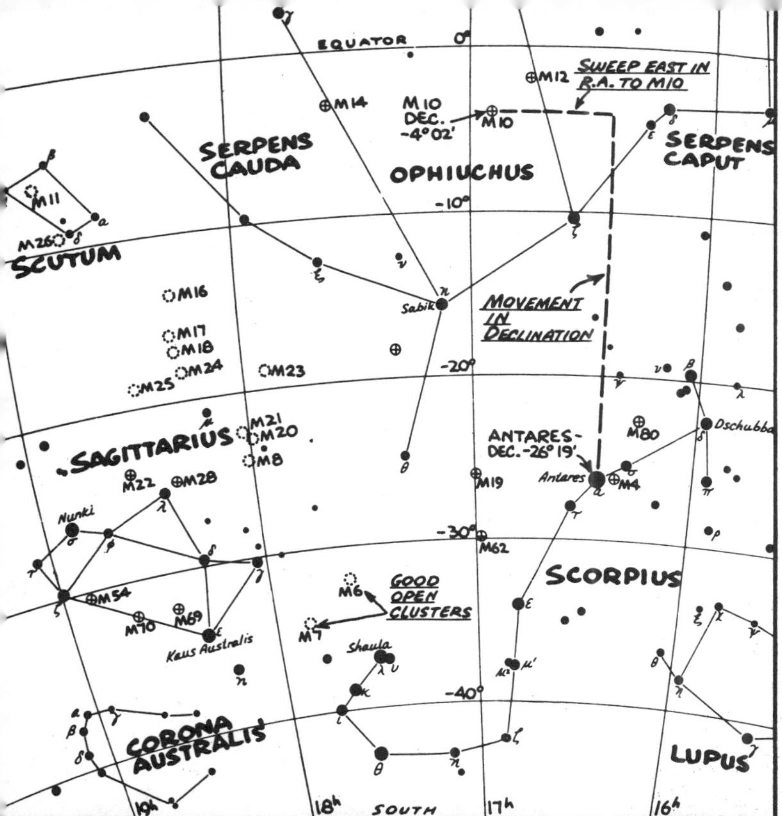

FINDING M 10 – WITH DECLINATION SET BY CIRCLE, IT IS EASY TO SWEEP IN R.A. TO OBJECT

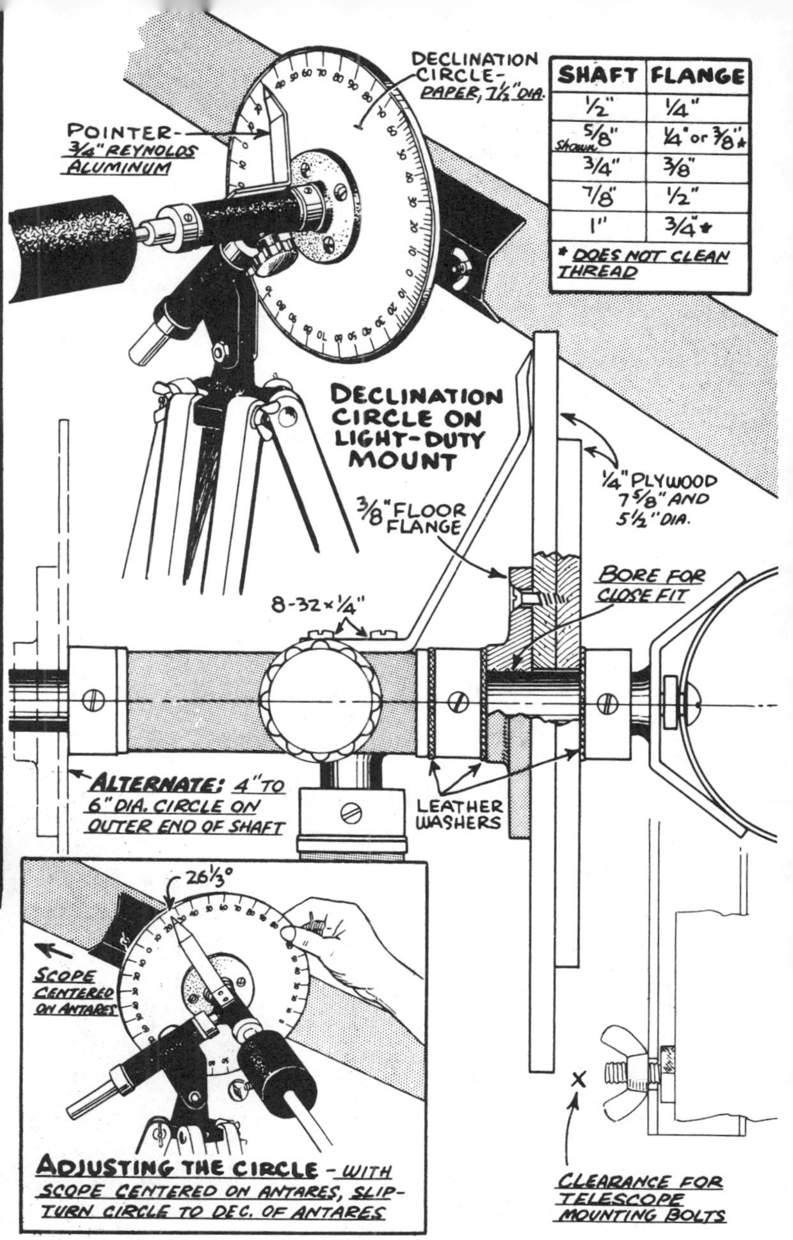

SHAFT	FLANGE
1/2"	1/4"
5/8" Shown	1/4" or 3/8"*
3/4"	3/8"
7/8"	1/2"
1"	3/4"*

* DOES NOT CLEAN THREAD

SETTING CIRCLES. The scientific way to find sky objects is to use setting circles. One circle of the pair is used to set the declination of the object while the other sets off hour angle as determined from right ascension and sidereal time. The hour circle is a bit complicated to use but the declination circle is easy; many beginners find the declination circle alone a big help. A typical setup is shown in the drawing above where a 7-1/2 inch declination circle is used on an Edmund light-duty mount having 5/8 inch shafts. The 3/8 inch floor flange shown is bored out on lathe or drill press for a neat turning fit on the shaft. In this particular size combination, the thread is not entirely removed from the flange but provides enough of a flat for the purpose. Suitable flanges for other sizes of shafts are listed in the table. If the mounted circle is sandwiched between leather washers, it will have enough friction to "stay put", yet at the same time it can be slip-turned by hand.

In use, the declination circle is adjusted by pointing the telescope at any convenient bright star of known declination. In summer skies, a good star for this purpose is Antares at 26° 19' South. Center Antares exactly in the field. Then, lock both shafts to make sure the telescope will not move, and slip-turn the circle to read 26-1/3 degrees, as shown in small drawing.

Now, with this local adjustment in declination, any object in the area can be set off directly in declination. Suppose you want to find M 10, with declination of 4 degrees South. Simply move the telescope until the pointer reads 4 degrees. Then, with declination set and locked, all you have to do is sweep in R. A. until you reach M 10. Other objects in the vicinity can be set off directly. No change in the adjustment need be made unless (1) you switch from one side of the tripod to the other, (2) you move into another area a considerable distance away. By working in this manner, you correct your declination for a star in the immediate vicinity, thereby minimizing position errors. You can use this system perfectly with any portable with only a very rough setting to the pole position.

STAR ATLASES. While the few maps shown enable you to find a few objects, it is plain you need a star atlas to locate objects in other parts of the sky. Among atlases, Norton's Atlas and Telescope Handbook is a long-time favorite and is thoroughly good. Becvar's Atlas of the Heavens is also excellent; it consists of unbound charts, available in two sizes, the smaller field edition (12 by 18 inches) being the best for outdoor use. Webb's Atlas of the Stars is detailed to magnitude 9 and is a useful supplement but less useful as a first atlas. Norton and Webb list sky objects.

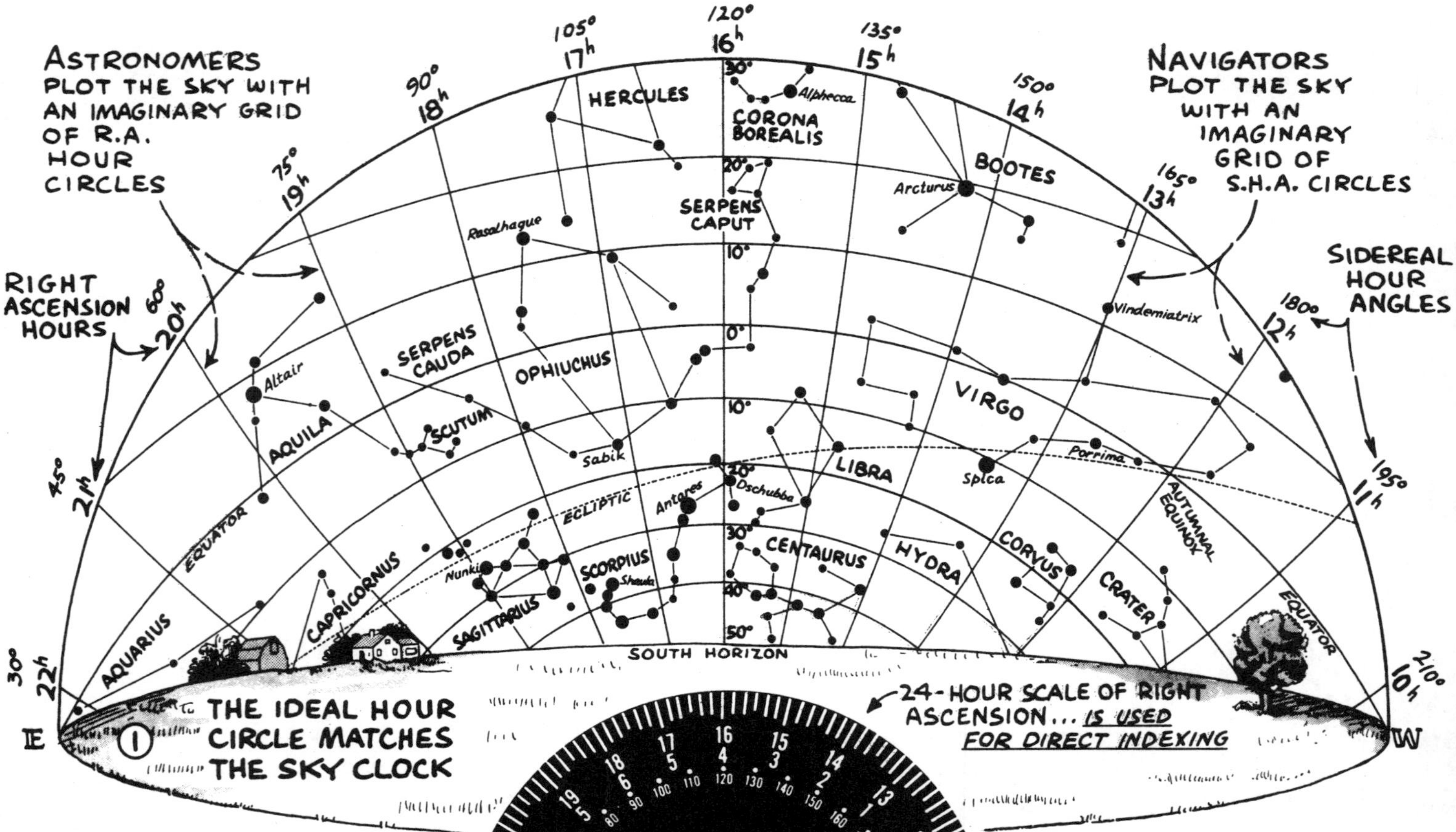

How to use SETTING CIRCLES

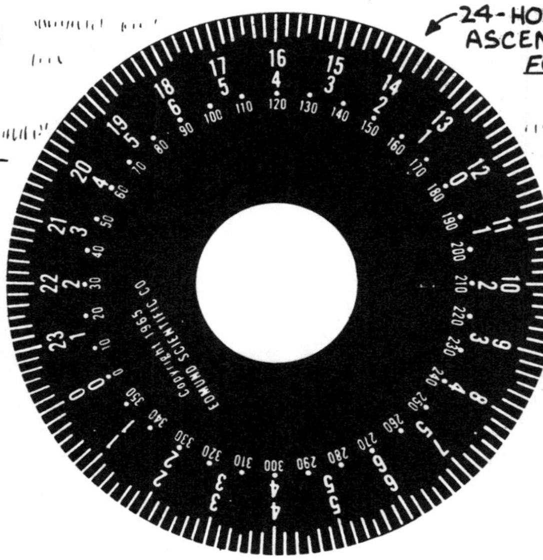

SETTING CIRCLES are fun to use and lend to your star-gazing a certain amount of scientific magic. Even in the daytime sky, with familiar skymarks lost in a field of blue, you can pick up bright stars and planets by simply dialing the proper numbers in right ascension and declination.

Setting circles are made in numerous patterns and various diameters. The size range is about 3 inches to 8 inches diameter for portable telescopes, the 6-inch size being a comfortable medium. Two circles make a set, one for declination being marked in degrees, while the other for right ascension is marked in hours and minutes. The circle measuring R.A. is known generally as the "hour circle." The main scale on an hour circle, **Fig. 1**, is the continuous 24-hour scale, which is comparable to a 24-hour clock dial. This is the only scale actually needed. A second scale with 24 divisions marked 0-6-0-6 will be found on many hour circles. This measures hour angle only and can't be used for direct indexing. Some hour circles include an S.H.A. (Sidereal Hour Angle) scale, which is degrees of arc. This system of sky addresses is explained in another Edmund book, "Time In Astronomy."

FOUR ARRANGEMENTS. There are only four possible arrangements of a 24-hour hour circle. It will be immediately apparent the number sequence can be clockwise, or it can be counterclockwise. The other variable is that the index can be either fixed or moving. A study of the diagrams, Fig. 2, will show that arrangement D is the only completely satisfactory system for direct indexing. Arrangement A would work, but the

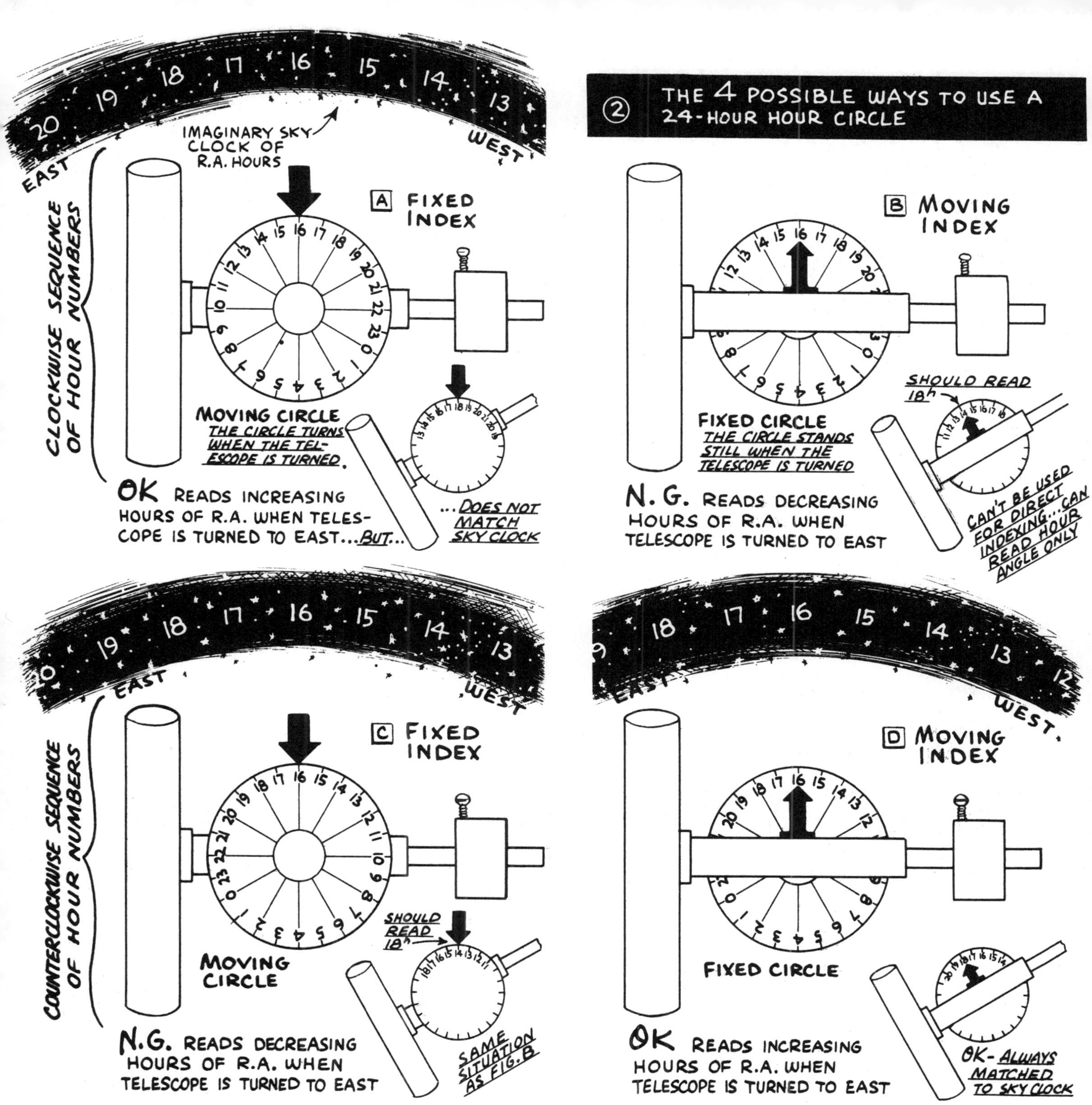

numbers on the hour circle would not match the imaginary clock in the sky. The other two systems do not work at all--they simply do not index or point the telescope the right way. However, even these arrangements can be used to set off an hour angle.

The proper arrangement for direct indexing in R.A is D, Fig. 2. As can be seen, this requires a counter clockwise sequence of hour numbers. The index moves--which means the circle itself is fixed. A "fixed" circle means only that it does not partake of the turning movement of the telescope. The fixed hour circle has movement in that it can be slip-turned as desired. If a clock drive is used, the fixed hour circle is attached to the worm gear to partake of the clock movement, but it is still a "fixed" circle in that it does not turn when the telescope is turned. The clock motor itself should have counterclockwise rotation when viewed from the end of the clock shaft--this will drive the telescope in the required westerly direction.

INSTALLATION OF CIRCLES. Direct indexing

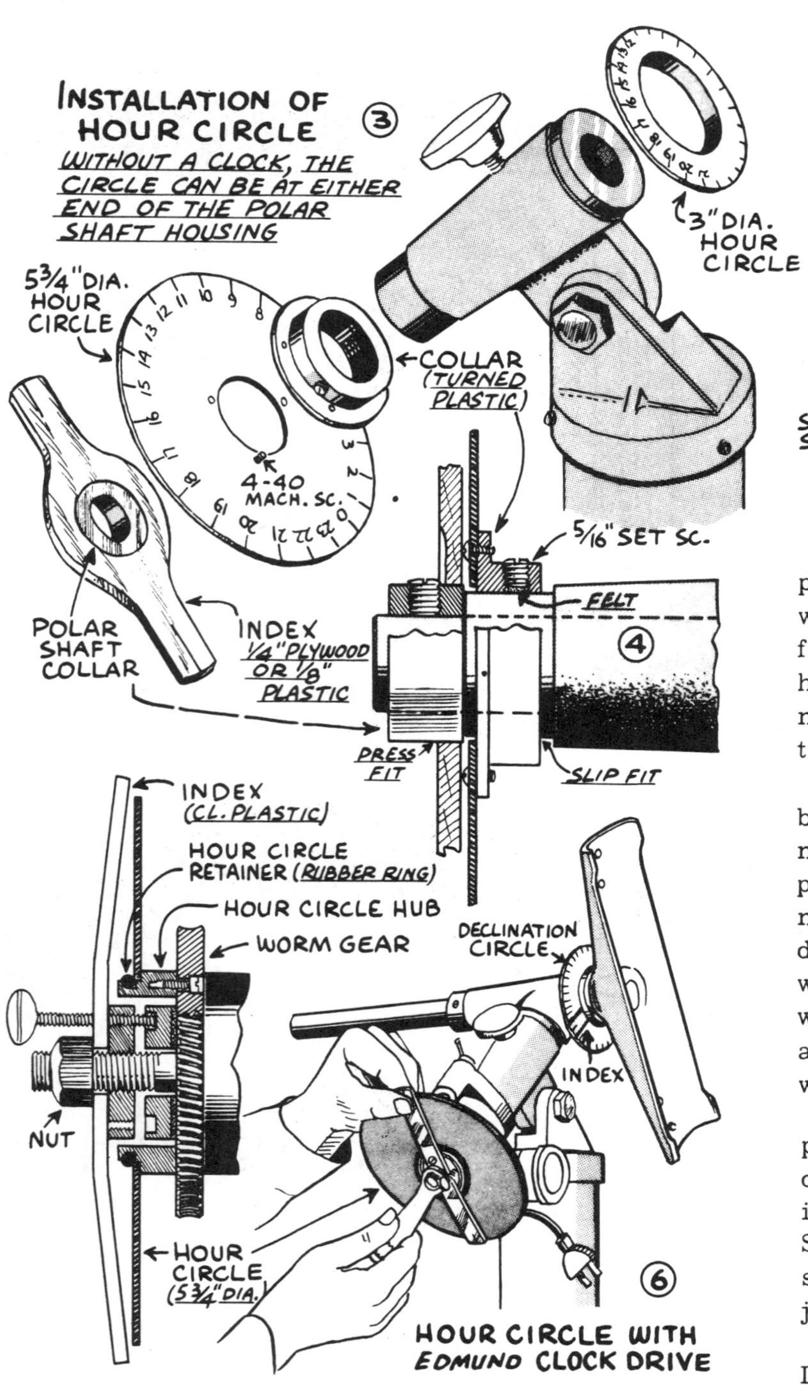

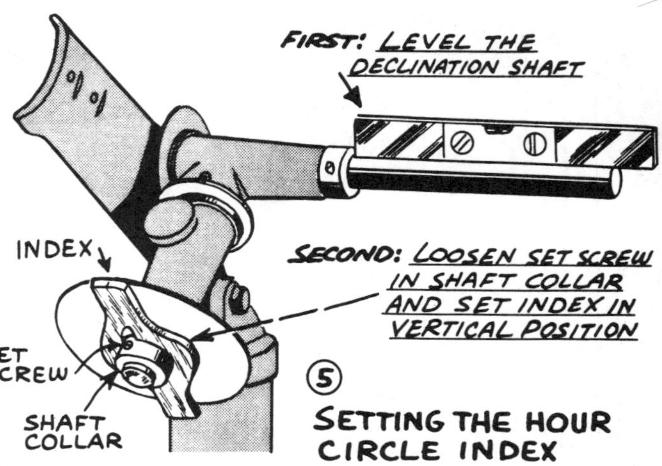

requires a double- ended index in order to read at either of the two common positions of the telescope, i.e., east or west of pedestal. The index is attached to the polar shaft or shaft collar, or to any moving part of the polar shaft. The hour circle is attached to either upper or lower end of the polar shaft housing, Fig. 3. It must allow slip-turning, and the commonest way to obtain this feature is with a set screw, Fig. 4. Some setting circles are supplied with a mounting hub or collar but most are just flat disks and you have to provide the slip-turn mounting yourself. The same situation applies to the index.

The declination circle can be used either fixed or moving. A moving circle means a fixed index, and, other things being equal, the fixed index is preferable because it calls for just one pointer which is always at the same fixed location. If a fixed circle is used, as in Fig. 6, it is best to have an index on both sides of the cradle to permit comfortable viewing at all positions of the telescope.

If you have a clock drive, the hour circle should be mounted as part of the clock. This means it must be attached to the worm gear or to some part of the drive that partakes of the clock movement. A small-diameter circle can be attached directly to the worm gear, but a larger circle will require a position somewhat apart from the worm gear to gain clearance. The usual installation is with a mounting collar or hub, Fig. 6, which is supplied with some units.

With or without clock, the moving index has a permanent position. To obtain this, you set the declination shaft level, Fig. 5, and then turn the index to stand vertical, pointing to the meridian. So adjusted, the moving index will "point" the same way as the telescope itself. The same adjustment is made with clock drive, Fig. 6.

DIRECT INDEXING. Like it sounds, direct indexing means that you index directly to the R.A. and declination of the sky object you want to find. Direct indexing is always used for setting the declination, but direct indexing in R.A. is possible only if you have the proper kind of hour circle properly installed, as already described. The whole procedure of direct indexing in R.A. is shown in Figs. 7 to 10 inclusive. In the first operation, Fig. 8, you obtain sidereal time indirectly by pointing the telescope at a star whose R.A. is known to you. The second operation, Fig. 9, matches the hour circle to the sky clock. With this adjustment made, you can then index directly to any object in the sky. Errors in setting the telescope to the pole position can be minimized if you use a pilot star near your intended target to set the hour circle.

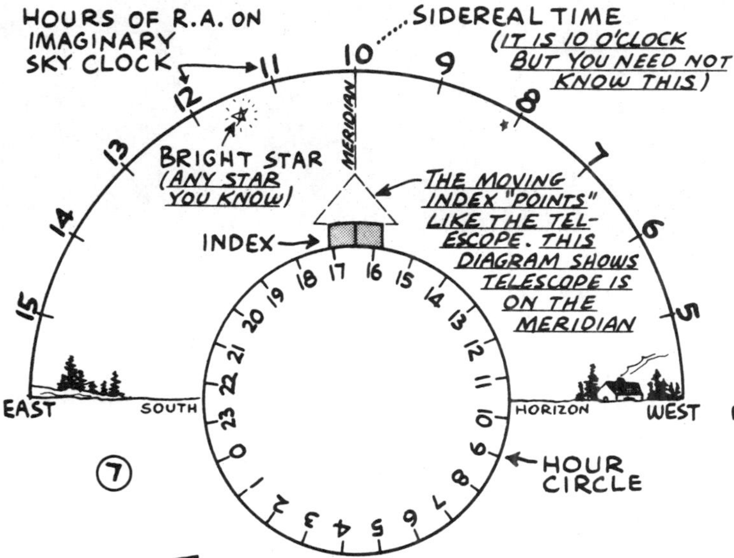

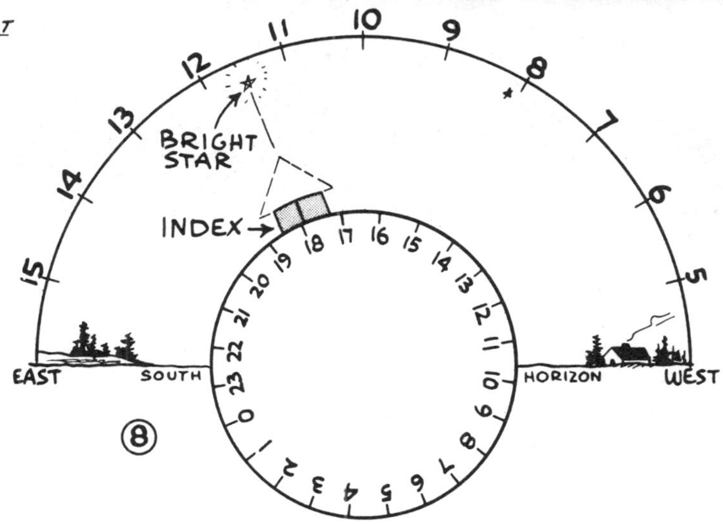

at Start THE HOUR CIRCLE IS IN ANY CHANCE POSITION. THE MOVING INDEX POINTS IN SAME DIRECTION AS TELESCOPE.
THE INDEX SETTING IS A PERMANENT ADJUSTMENT... SEE OPPOSITE PAGE

Operation 1 POINT THE TELESCOPE AT ANY BRIGHT STAR WHOSE R.A. IS KNOWN TO YOU. CENTER THE STAR IN FIELD OF EYEPIECE. R.A. OF STAR SHOWN IS $11^h\ 30^m$

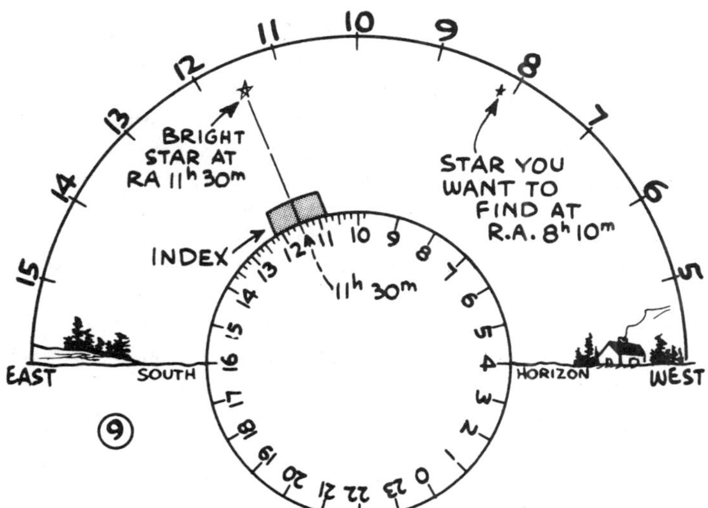

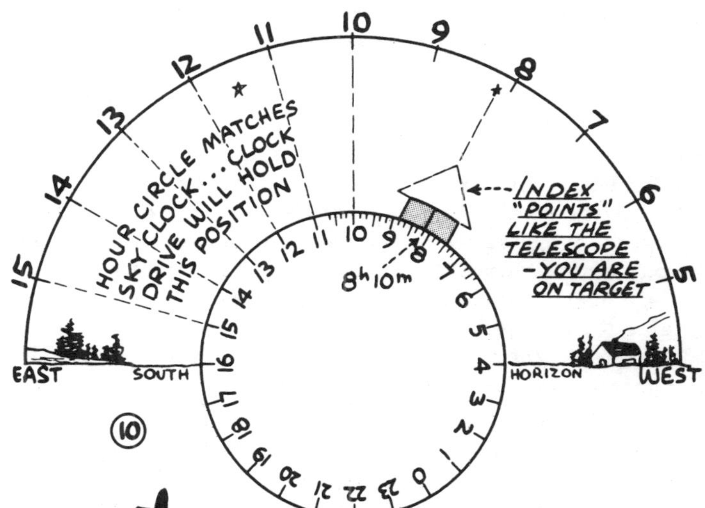

Operation 2 SLIP-TURN THE HOUR CIRCLE TO PUT R.A. OF PILOT STAR ($11^h 30^m$) IN LINE WITH THE INDEX.
NOTE THE HOUR CIRCLE NOW MATCHES THE SKY CLOCK

Operation 3 INDEX DIRECTLY TO THE STAR YOU WANT TO FIND (*$8^h\ 10^m$ IN DRAWING*). *IF ON CLOCK DRIVE, YOU CAN CONTINUE DIRECT INDEXING TO OTHER SKY OBJECTS AS LONG AS YOU LIKE*

INDEXING BY HOUR ANGLE. In this familiar method of star-finding, the R.A. of the star is subtracted from sidereal time at the moment. The result is the hour angle of the star from your meridian. The most convenient scale for setting off an hour angle is the 0-6-0-6 scale, with one of the zero marks set to the meridian. This method is cumbersome and time-consuming--you use it when you have to, but otherwise you are miles ahead with direct indexing.

DIRECT INDEXING WITHOUT A CLOCK. Although direct indexing works best with a clock, it can be used without a clock by frequently resetting the hour circle to any convenient pilot star. Working without a clock drive, the stars will drift to the west at the rate of 1 degree in 4 minutes time. This means that after setting to the R.A. of the target by direct indexing, you must make a small additional westerly movement in R.A., equal to the elapsed time since you set the hour circle. With a little practice, you can get good results with this method by resetting the hour circle at intervals of about 15 minutes.

observing HINTS

THE BEGINNER should put in an hour or so of practice on land objects. Even though the image is upside down, you will gain valuable experience in setting up the telescope, focusing the eyepiece and other basic operations, all of which must be learned by actual practice. Then, in the night sky, the best starting sky is at dusk. This also of course is the only time you can see Mercury or Venus as evening stars.

EYES MUST BE DARK-ADAPTED. It takes at least ten minutes to dark-adapt your eyes and slight improvement can be noted up to half-an-hour. If the weather outdoors is a bit chilly, you can get your night eyes more comfortably by staying indoors with your eyes closed or in a dark room. Meanwhile, you have already setup the telescope and it too is undergoing a slight change in adapting to the weather. If you want to look at maps or notes outdoors, use a lamp or flashlight covered with red or brown paper or a red filter.

EYE POSITION. Your eye must not touch the eyepiece but at the same time it must be centered on the emergent light beam. This is impossible to do when your eyes are not dark-adapted. After you get your night eyes, you will note that the sky as seen in the telescope is not really black but a rather bright, luminous gray. Given this target, your eye will automatically center on the eyepiece. Obviously, a low-power eyepiece is easier to use because it has a bigger exit pupil. If desired, you can cup your hand around the eyepiece to serve as a guide until you get your eye centered on the light beam.

A second feature of proper eye position is that your eye must be at or near the exit pupil point. If you are too close, you will get a hit-and-miss shadow effect; if too far, you will lose valuable area in the field-of-view. High-power eyepieces always require a closer eye position than low-power.

If possible, try to master the trick of keeping both eyes open since this is much less tiring than the usual one-eye squint. Two-eye viewing is easy on daytime objects or any bright night object, such as the moon. For other night targets where the emergent beam is of low luminosity, your only chance of staying "on target" is the one-eye squint. However, practice with two eyes open whenever you can.

IF YOU WEAR GLASSES. Take them off if you are far sighted. Your unaided eyes will then see distant objects clearly, while the removal of the glasses will let you crowd the eyepiece when necessary. Myopes have a different problem: if you remove your glasses you lose your eyes for distant objects. The best practical solution here is to keep your glasses on and use only eyepieces with long eye relief of 1/2 inch or more. Note, however, that even with eyepieces having short eye relief, a long eye position means only that you lose field.

FOCUSING. There is no such thing as exact focusing of a telescope. What happens is that the image forms at a very precise and exact image plane, but you can see the image at various settings of the eyepiece because the eye can adjust for either long or short focus. The best general practice is to focus "long". This is done by extending the eyepiece a little more than necessary and then focusing in just enough to get a sharp image. The "long" focus causes your eye to focus as for a distant object--the most comfortable position. If you focus to the maximum "in" position which yet retains a sharp image, the eye accommodates as for a close object. This position gives slightly greater magnification but is somewhat more tiring. In actual practice, you will use both the long and short focus since frequent changes will allow you to see clearer without eye fatigue. Also, as a matter of fact, objects low in the sky require a slightly different focus than objects at the zenith; a bright object like the moon may require different focusing than a dim nebula. Exact focus on star objects is simply a matter of obtaining the smallest possible star image.

Out-of-focus focusing is sometimes useful. For example, if the finder telescope is set slightly outside focus, the star images will be big and easily seen; you can even make fine crosshairs visible in this manner. Colored doubles are sometimes seen better slightly outside focus although too much of this tends to dilute the colors rather than improve them. If your eyepieces are not parfocal and you use a series from low to high-power for finding and observing, it is sometimes practical to focus only the high-power eyepieces. The others, if not too much out-of-focus, will show large star images which serve quite well for finding and tracking. If you want to see something spectacular, off-focus a bright star near the horizon and then tap the eyepiece tube lightly with your finger--you will see a flaming pinwheel shooting off red and green sparks!

AVERTED VISION. On luminous objects, you can increase visual acuity by one or two magnitudes by using averted vision. The idea is to get the target

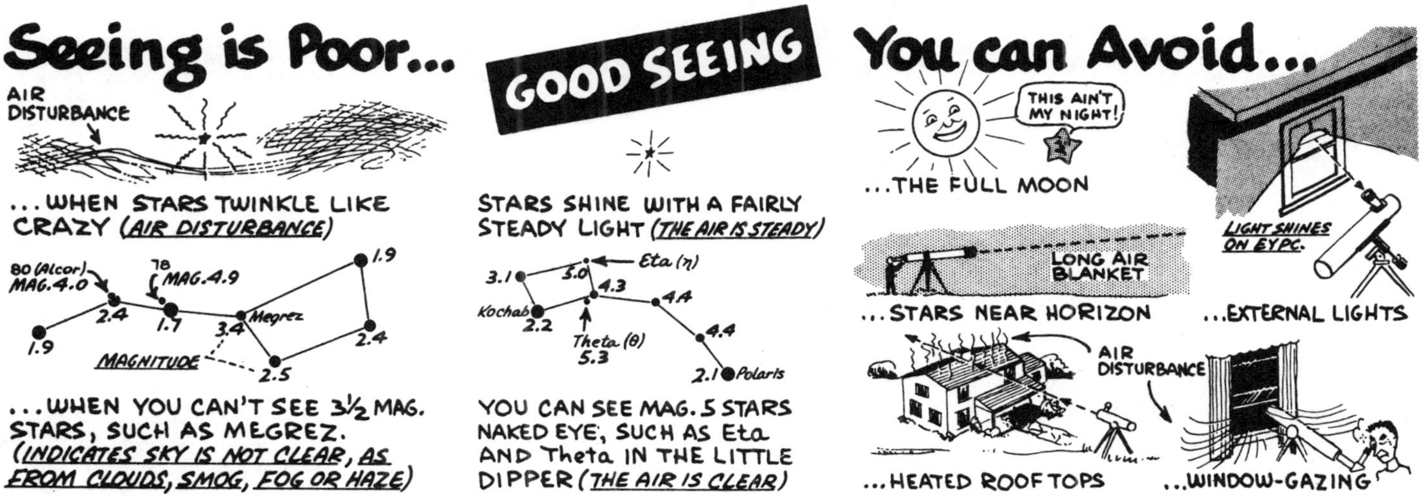

object in the center of the field, and then instead of looking directly at it, direct your gaze a little to one side. This technique is especially useful for star clusters. The center of your eye sees the sharpest, but the outer portion is more sensitive to light and movement. Try looking all around a faint object to determine if a certain part of your eye is more sensitive.

VIBRATION OF MOUNT. At 50x or less, a sturdy mount can be pushed around without disturbing the image: you can do continuous tracking. At higher powers and at all powers with light-duty mounts, it is not practical to do this. Instead, you must hold the scope as steady as possible, focus on the target, and then get your hands off the mount and let it settle down. A vibration period of 6 to 8 seconds is normal; anything over 12 seconds indicates a poor mount. When making the initial sight on the target, allowance should be made for drift. Assuming a south sky object, the object should be positioned at the west side of the eyepiece. Then, you take your hands off the mount, after which it will do its 7-second shimmy and settle down, allowing one to four minutes of viewing time, during which period the object will drift across the field to the east side of eyepiece. The operation can be repeated as often as desired, the telescope being moved around the polar axis only.

GOOD SEEING. The star-gazer's "seeing" depends on many things but specifically it is concerned with the atmosphere or air. The main body of the earth's atmosphere is about ten miles thick, straight up. At 45 degrees, it is about 15 miles thick, and the air blanket increases to over 100 miles for stars near the horizon. Obviously, the best seeing is at the zenith where the air blanket is thin. The atmosphere is constantly in motion--shifting, swirling, boiling--and it is a rare night when you can use powers over 350x regardless of the size or excellence of your telescope. On the other hand, atmospheric disturbances are seldom a problem at 50 to 100x.

You can get a first-hand introduction to air disturbance by trying to look through an open window on a chilly night. The warm indoor air will make a violent rush to get outdoors and will make quite a commotion around the telescope. Turbulence is especially bad with a reflector since the air flows through the tube as well as around it. You don't actually see the air pattern in the telescope, but you do see the image is wavering, wandering, shivering, oversize and distorted--seeing is bad! Open-window viewing is practical only when indoor and outdoor air are about the same temperature. Even so, a power of about 50x represents the maximum for this kind of observing. It is interesting and instructive to note the effect of atmosphere on a daytime object. Use 100x or more and look along the eaves of a house; you will be convinced that "seeing" is more than just a matter of good optics.

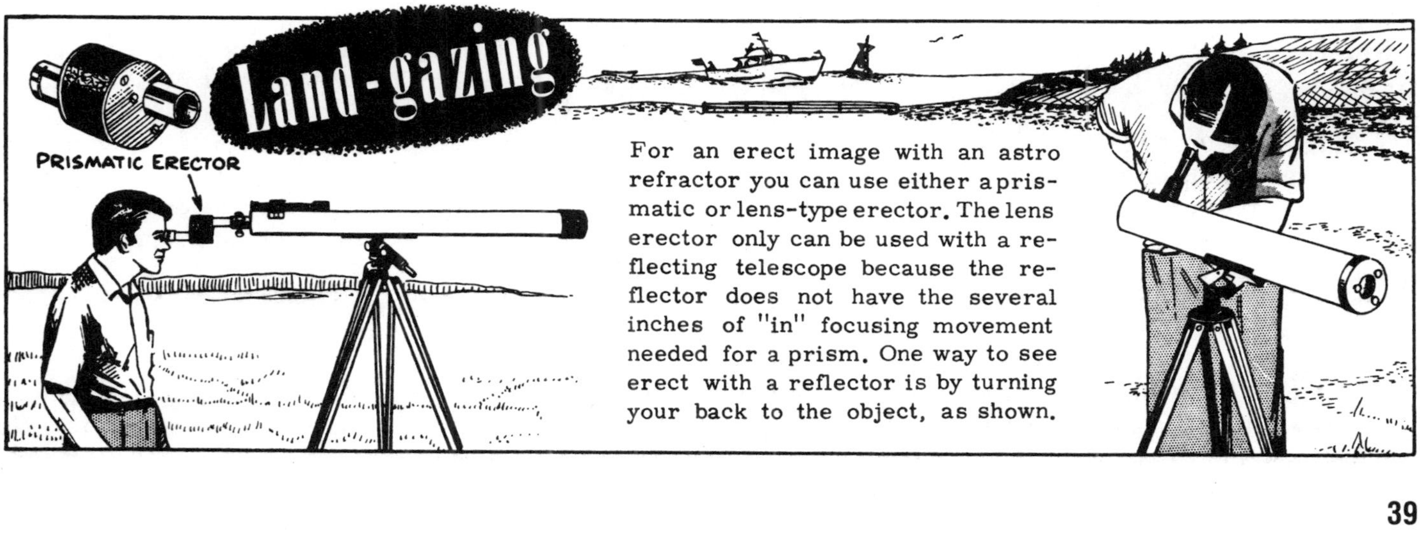

For an erect image with an astro refractor you can use either a prismatic or lens-type erector. The lens erector only can be used with a reflecting telescope because the reflector does not have the several inches of "in" focusing movement needed for a prism. One way to see erect with a reflector is by turning your back to the object, as shown.

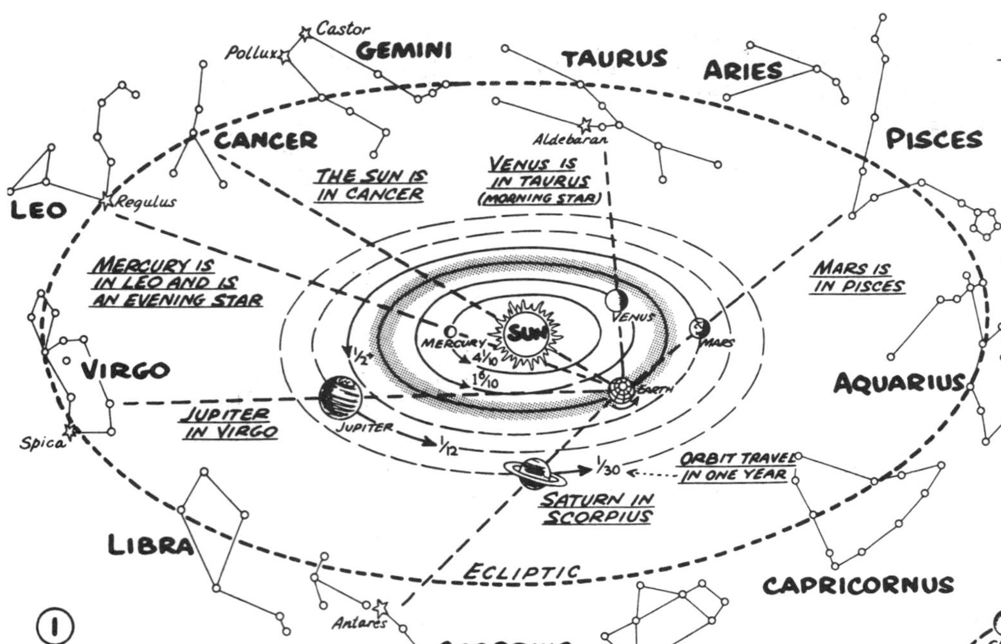

① THE SOLAR SYSTEM AND CONSTELLATIONS OF THE ZODIAC

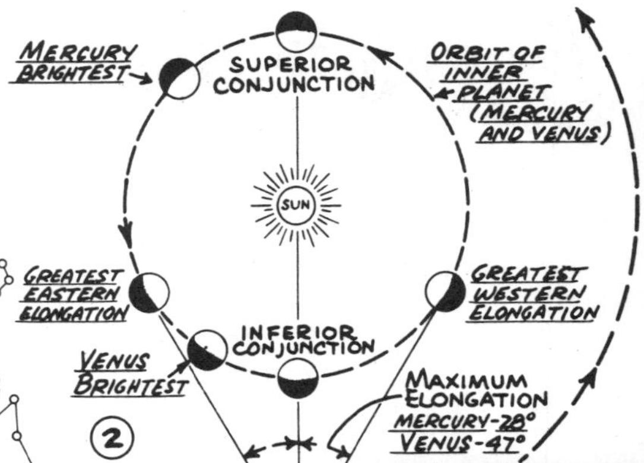

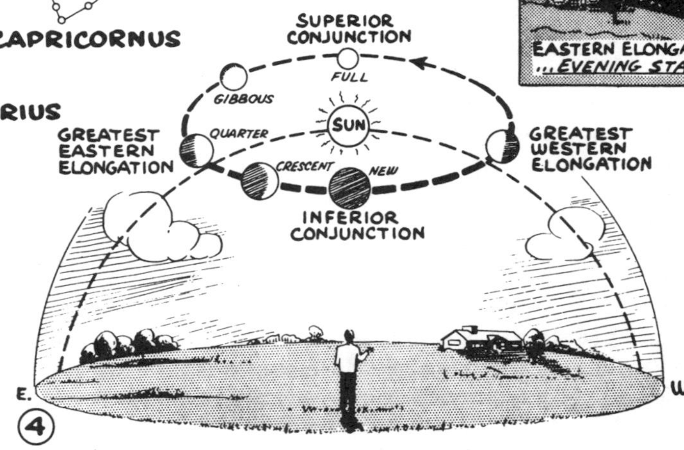

ASPECTS AND PHASES OF AN INNER PLANET

Observing the PLANETS

NINE PLANETS make up the solar system but only Venus, Mars, Jupiter and Saturn rate as ideal telescope objects. Fig. 1 shows all of the bright planets as they might appear in the sky. If you are good at visualizing, try this: Face south. Turn the drawing upside down and put a straightedge (ruler) connecting earth and sun. This is your horizon--turn the page as needed to make it level. Now, keeping the imaginary horizon level, rotate the drawing slowly in a clockwise direction, with the earth as pivot. You will notice that little Mercury is seen briefly after the sun sets; Jupiter and Saturn are prominent in the night sky; Mars will rise in the east and Venus will be seen as a morning star before the rotating diagram shows the sun coming over the horizon.

THE INNER PLANETS. Mercury and Venus are conveniently classified as inner planets because they are inside the orbit of the earth. Both are in the daytime sky every day of the year because Mercury can never get more than 28 degrees or two hours from the sun; Venus, three hours. In the circular race around the sun, the planets are now abreast, now lagging behind or pulling ahead in relation to the earth. Certain of these positions or aspects are named and should be learned. Fig. 2 shows the aspects of an inner planet; the cycle follows the order of orbit travel, that is, from superior conjunction to eastern elongation to inferior conjunction to western elongation. Any angle east or west of the sun is an elongation. You can see from Fig. 4 that if the planet is east of the sun, it will appear as an evening star after the sun has set, Fig. 3.

Like the moon, the inner planets show varying amounts of illuminated surface. Venus is most brilliant at the crescent phase, being then nearly six times larger in angular diameter than when at superior conjunction. Mercury is brightest between greatest elongation and superior conjunction. Nearer, larger and brighter, Venus far outclasses Mercury as a telescope object. Thirty-six days either way from new, she is at her brightest, a glowing crescent outshining Sirius a dozen times and of an apparent size when viewed at 40x equal to the moon as seen with naked eye. Mercury should be viewed at greatest elongation in order to obtain a "high" sky position. Even so, he is hard to see, being lost behind trees and rooftops surrounding the average backyard telescope.

THE OUTER PLANETS. The aspects of an outer planet are shown in Fig. 5. Here, opposition is the aspect of greatest interest since it is at oppo-

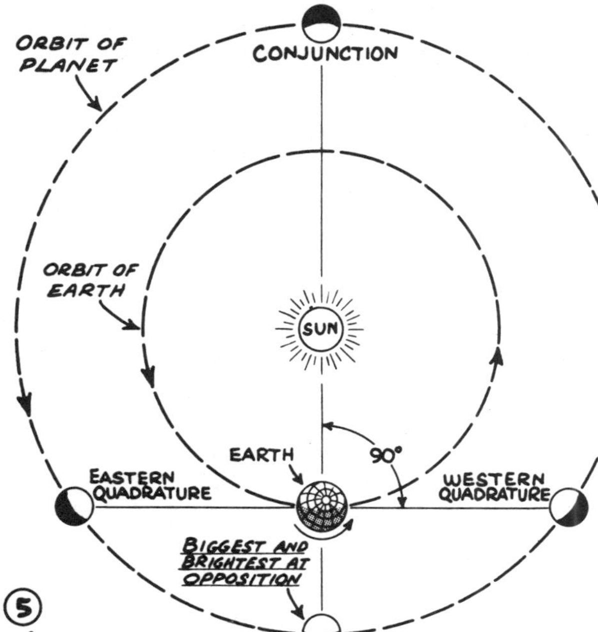

⑤ ASPECTS OF AN OUTER PLANET

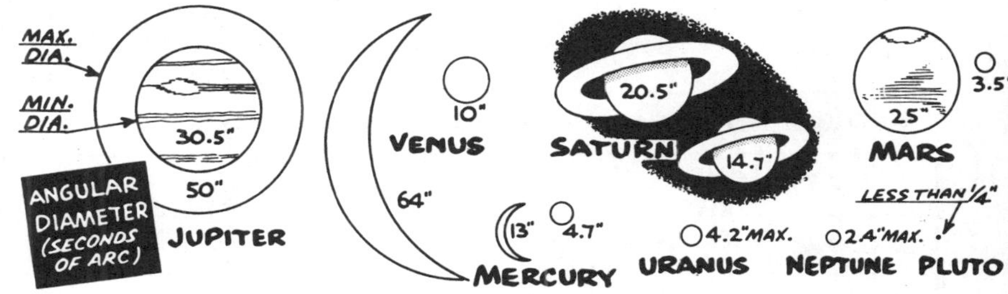

⑦ MAXIMUM-MINIMUM DIAMETERS OF PLANETS to scale

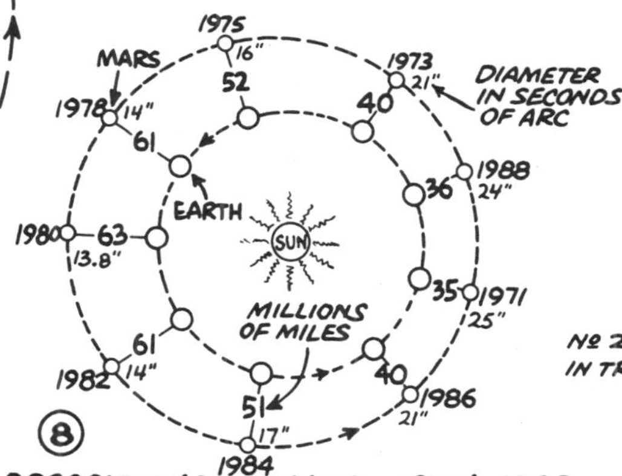

⑧ OPPOSITIONS OF MARS - 1971 to 1988

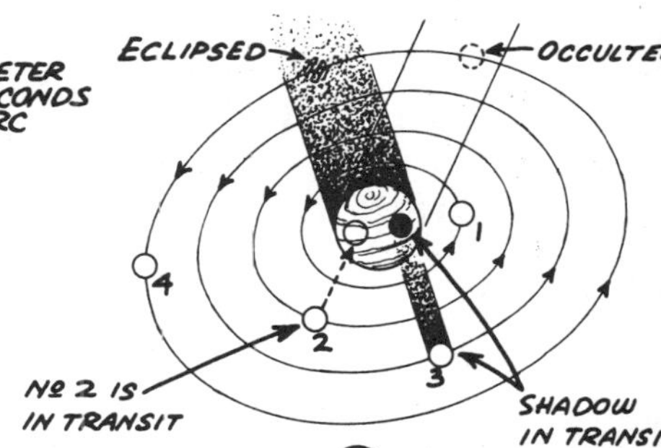

⑨ SATELLITES OF JUPITER

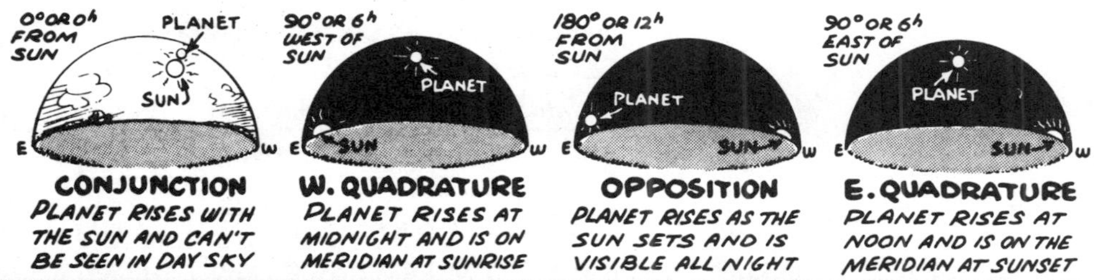

⑥ CYCLE OF ASPECTS OF AN OUTER PLANET (MARS, JUPITER, SATURN)

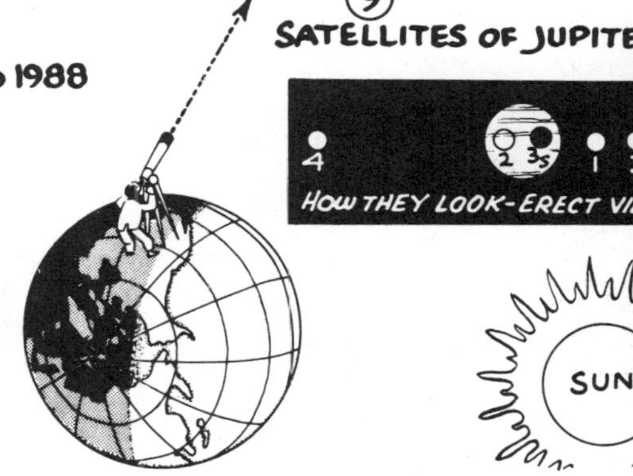

sition that the planet is nearest the earth and at its biggest and brightest best. Opposition means simply that the planet is opposite the sun. When you have an outer planet rising in the east just as the sun sets, you know that the planet is in opposition and ideally placed for observation. The other aspects are readily determined by the location of the planet in regard to the sun, Fig. 6.

MARS. Opposition distances are not uniform, notably in the case of Mars, Fig. 8, where it can be seen that 1971 was the last favorable opposition and 1988 will be the next. This best-to-best cycle runs 15 or 17 years. At a favorable opposition, Mars is both big and bright at about 24 second of arc and -2.5 magnitude. He does not fade a whole lot in the oppositions on either side of a favorable one. The minimum planet disk of about 14 seconds at a poor opposition is still big enough for a nice view. At conjunction, Mars fades to 2nd magnitude; with a long synodic period of over two years, he is bright one year, dim the next. Unlike cloudy Venus, Mars has a fairly clear atmosphere and shows a maze of surface detail, most of which is beyond the range of earth-based telescopes, large or small. In 1965 the Mariner flyby produced photos taken at 6200 miles. These reveal detail never seen before--even the experts were surprised to find Mars heavily cratered very much like the moon. Needless to say, you won't see craters with a small telescope. But Mars is always a nice object at opposition just on the basis of color and brilliance. Dark areas can be detected with even a 2-inch at 50x but it takes at least 5 in. aperture and 200 to 300x to define these with any measure of clarity.

JUPITER. Big Jupiter is the most consistent performer of the bright planets; he is never under 30 seconds in angular diameter, Fig. 7, which is bigger than Mars at his best. Even the four bright satellites are 5th and 6th magnitude, easily seen with binoculars. Fig. 9 is a sketch view of the satellites in orbit, but our actual view is nearly edge-on so that the moons merely shuttle back and forth. A satellite in transit is difficult to see because it is a bright object seen against a bright surface. It is much easier to see a shadow transit, which is a black dot on a bright surface. A random view will usually show all four bright moons. From the data on satellites you can see that the maximum true

Comparing the Planets

PLANET	DIAMETER IN MILES	DISTANCE-MILLIONS OF MILES from SUN MAX.	from SUN MIN.	from EARTH MAX.	from EARTH MIN.	PERIOD OF REVOLUTION SIDEREAL	SYNODIC	ORBITAL SPEED MILES PER SEC.	MAGNITUDE MAX.	MEAN	OTHER	ANGULAR DIAMETER SECONDS OF ARC MAX.	MEAN	MIN.	SUITABLE TELESCOPE POWER
MERCURY ☿	2,900	43.3	28.6	136	50	88 DAYS	116 DAYS	29.7	-1.9 NEAR S.C.	-1.7 NEAR S.C.	+0.2 AV. G.E.	12.9"	6.7"	4.7"	40-120
VENUS ♀	7,600	67.6	66.7	161	25	225 DAYS	584 DAYS	21.7	-4.4 CRESCENT	-4 AV. G.E.	-3.3 NEAR S.C.	64.0	16.0	9.9	20-120
EARTH ⊕	7,913	94.4	91.3	—	—	365 DAYS	—	18.5	—	—	—	—	—	—	—
MARS ♂	4,200	154.7	128.3	248	35	1.9 YRS.	780 DAYS	15.0	-2.8	-1.8 AV. OPP.	+2 CONJ.	25.1	6.1 [1]	3.5	100-300
JUPITER ♃	86,800	506.7	459.9	600	367	11.9 YRS.	399 DAYS	8.1	-2.5	-2.2 AV. OPP.	-1.4 CONJ.	49.8	37.9	30.5	20-300
SATURN ♄	[B] 71,500 [R] 170,000	936	837	1028	744	29.5 YRS.	378 DAYS	6.0	-0.4	+0.0 AV. OPP.	+0.9 RINGLESS	[B] 20.5 [R] 49.2	17.3 41.5	14.7 35.2	40-300
URANUS ⛢	29,400	1867	1699	1960	1606	84 YRS.	370 DAYS	4.2	+5.7	LITTLE CHANGE		4.2	3.8	3.4	ANY
NEPTUNE ♆	28,000	2817	2770	2910	2677	165 YRS.	367 DAYS	3.4	+7.6	LITTLE CHANGE		2.4	2.3	2.2	ANY
PLUTO ♇	3,600	4600	2760	4700	2670	248 YRS.	367 DAYS	3.0	+14	LITTLE CHANGE		0.28	0.2	0.16	NEEDS 10-INCH

[B] BALL [R] RING

SIDEREAL PERIOD: *TIME TO MAKE ONE REVOLUTION AROUND SUN... PLANET'S "YEAR" IN TERMS OF EARTH TIME*

SYNODIC PERIOD: *TIME BETWEEN SUCCESSIVE SIMILAR ASPECTS... EQUAL TO ONE "LAP" IN RACE WITH EARTH AROUND SUN*

[1] *SEE FIG. 8 FOR OPPOSITION DIAMETERS*

S.C. ...... *SUPERIOR CONJUNCTION*

AV. G.E. .. *AVERAGE GREATEST ELONGATION*

AV. OPP. *AVERAGE OR MEAN OPPOSITION*

CONJ. ... *CONJUNCTION*

SATURN'S RINGS ARE SEEN AT VARIOUS ANGLES

(10)

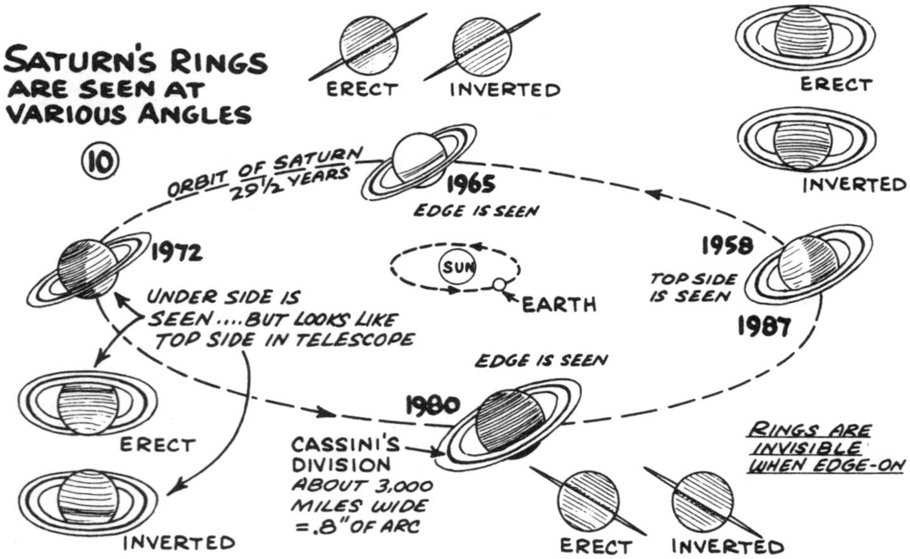

BRIGHT SATELLITES

EARTH	DIA. MILES	PERIOD	ANGULAR DIA.	MAG.
Moon	2160	$27^d\ 8^h$	31'	-12
JUPITER	**DIA. MILES**	**PERIOD**	**SEP. MINUTES**	**MAG.**
1 (Io) EYE-oh	2100	$1^d\ 18^h$	2.3'	5.5
2 (Europa)	1850	$3^d\ 13^h$	3.7	6.1
3 (Ganymede)	3200	$7^d\ 4^h$	5.9	5.1
4 (Callisto)	3100	$16^d\ 18^h$	10.3	6.2
SATURN				
3 (Tethys)	800	$1^d\ 21^h$	.7'	10.6
4 (Dione)	700	$2^d\ 18^h$	1.0	10.7
5 (Rhea)	1150	$4^d\ 12^h$	1.3	10.0
6 (Titan)	3000	16^d	3.1	8.3
8 (Iapetus) eye-APE-uh-tus	1000	80^d	9.0	10.8

SEP.-*SEPARATION FROM PLANET IN MINUTES OF ARC*

field of the moons cannot exceed 16.2 minutes (Ganymede plus Callisto), making it possible to go to 180-200x with a 50-degree eyepiece and still capture all four moons in one view. The planet itself is banded with tan, yellow and brown belts readily visible at 100x, while spots and smaller detail show at higher power. These markings are atmospheric and changeable; new cloud bands often form overnight and few spots last over a month. Jupiter remains bright right through conjunction and if visible at all is always a good telescope object.

SATURN. The ringed planet tops them all as a "show" object. You can't see the ring naked-eye nor can you see it with binoculars, making it all the more delightful to have it flash into view at 30x or more. This planet has a clean-cut appearance and can stand a lot of magnification if you want to blow it up real big. Cassini's division can be detected with a 3-inch when the seeing is good and the rings well-displayed in open position.

Saturn has nine satellites, of which five are brighter than 11th magnitude and can be seen with a 6-inch reflector. The brightest, Titan, of 8th magnitude, can be seen with any small telescope when well-separated from the planet. Saturn takes 29-1/2 years to wheel once around the sun and during this time maintains a constant angle. For this reason, the rings show a top view, edge view, bottom view and edge view in succession, Fig. 10. When edge-on, the ring is practically invisible for a period of about a year. Also, since Saturn gets more than half of its light from the ring system, the magnitude fades as the ring closes.

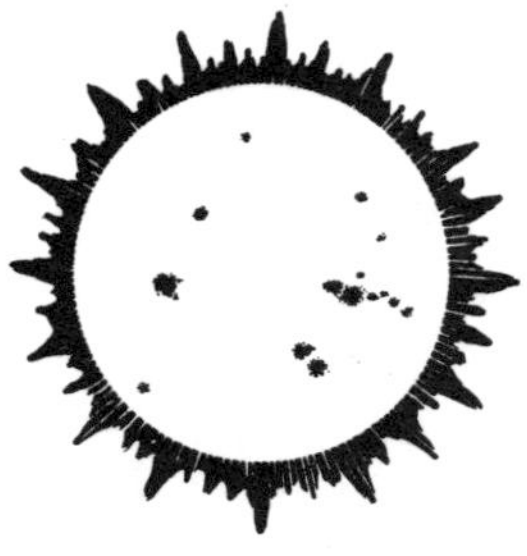

OBSERVING THE SUN

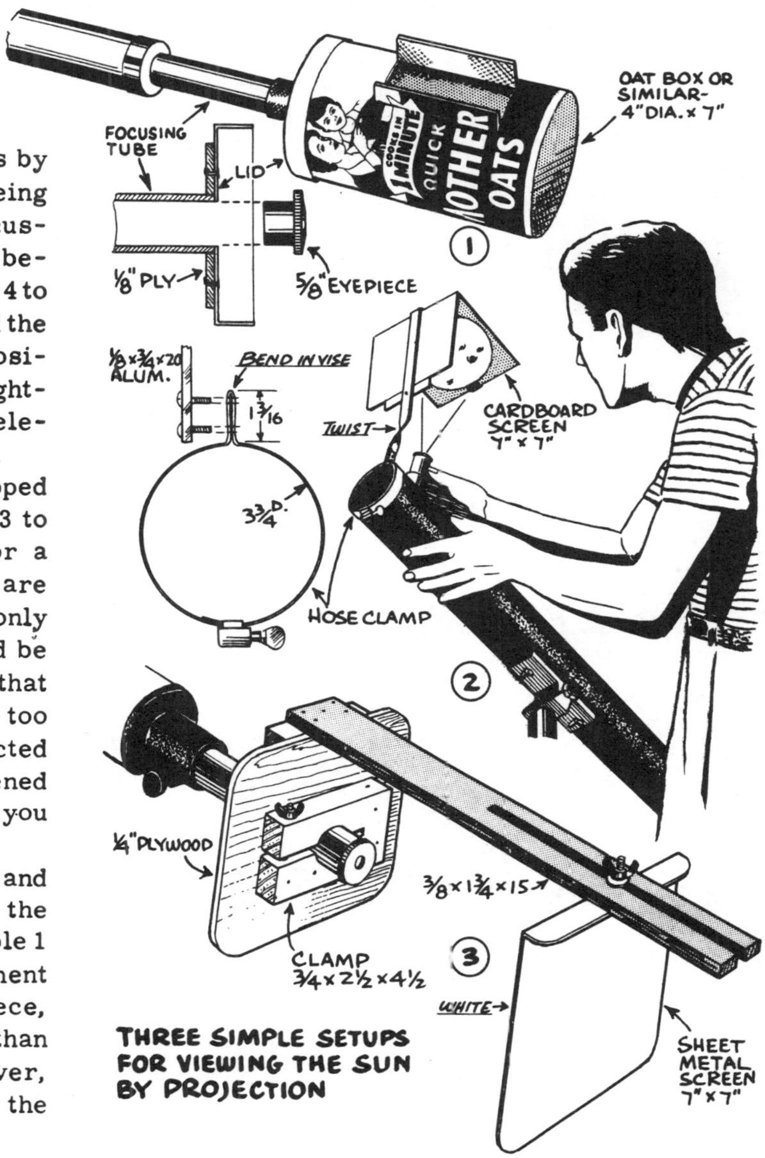

THREE SIMPLE SETUPS FOR VIEWING THE SUN BY PROJECTION

THE SAFE and sane way to observe the sun is by projection. Equipment for this is simple, being merely a cardboard shade slipped over the focusing tube and a piece of white cardboard held behind the eyepiece. Hold the cardboard screen 4 to 6 inches behind the eyepiece and then extend the eyepiece just a little from normal infinity position to focus the sun's image on the screen. Sighting is done by watching the shadow of the telescope tube on the ground or on the sunshade.

A simple setup is a round cereal box slipped over the focusing tube, Fig. 1. This allows a 3 to 4-in. sun image, which is about normal for a small telescope. Other equipment ideas are shown in Figs. 2 and 3. With any setup using only a simple sunshade, the enlargement should be between 10x and 20x. The situation here is that you are in open daylight, and if you enlarge too much the daylight will wash out the projected image. With a closed box, or inside a darkened room or with a cloth thrown over your head, you can go up to 50x enlargement.

Assume for example 30 in. f.l. objective and desired enlargement of 15x. Table 2 shows the image will be 4-1/16 in. diameter. Then, Table 1 shows the "throw" needed for 15x enlargement using various eyepieces. With 5/8 in. eyepiece, the throw is 9-7/8 inches. This is a bit more than provided by the oat box setup, Fig. 1. However, you can get 10x easily (6-7/8 in. throw), and the

SUN PROJECTION DATA

CAUTION: INTENSE HEAT CAN DAMAGE CEMENTED EYEPIECE LENSES — USE RAMSDEN OR HUYGENS, ESPECIALLY WHEN (1) SUN IS BRIGHT, OR (2) OBJECTIVE IS OVER 3" DIAMETER

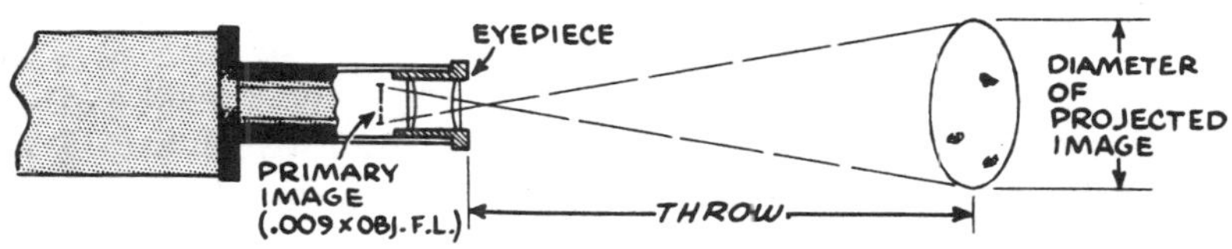

TABLE 1 - EYEPIECE TO SCREEN DISTANCE (THROW)

PROJECTION MAG.	EYEPIECE FOCAL LENGTH						
	1/4"*	1/2"	5/8"	3/4"	7/8"	1"	1 1/4"
5x	1 1/2"	3"	3 3/4"	4 1/2"	5 1/4"	6"	7 1/2"
10x	2 3/4	5 1/2	6 7/8	8 1/4	9 5/8	11	13 3/4
15x	4	8	9 7/8	12	13 7/8	16	20
20x	5 1/4	10 1/2	13	15 3/4	18 1/4	21	26 1/4
30x	7 3/4	15 1/2	19 1/4	23 1/4	27	31	38 3/4
40x	10 1/4	20 1/2	25 3/8	30 3/4	35 3/4	41	51 1/4
50x	12 3/4	25 1/2	31 5/8	38 1/4	44 3/8	51	63 3/4

*WILL NOT COVER FULL DIAMETER OF SUN

TABLE 2 - DIAMETER OF SUN IMAGE

OBJECTIVE F.L.	PRIMARY IMAGE	PROJECTED IMAGE						
		5x	10x	15x	20x	30x	40x	50x
20"	.180"	7/8"	1 13/16"	2 11/16"	3 5/8"	5 7/16"	7 3/16"	9"
30"	.271	1 3/8	2 3/4	4 1/16	5 7/16	8 1/8	10 7/8	13 9/16
40"	.361	1 13/16	3 5/8	5 7/16	7 1/4	10 7/8	14 1/2	18 1/16
45"	.405	2	4 1/16	6 1/16	8 1/8	12 3/16	16 3/16	20 1/4
48"	.432	2 1/8	4 5/16	6 1/2	8 5/8	13	17 1/4	21 5/8
50"	.451	2 1/4	4 1/2	6 3/4	9	13 1/2	18	22 1/2
60"	.541	2 11/16	5 7/16	8 1/8	10 13/16	16 1/4	21 5/8	27
70"	.631	3 1/8	6 5/16	9 1/2	12 5/8	19	25 1/4	31 1/2

TABLE 3 TRANSMISSION OF VARIOUS SETUPS FOR DIRECT VIEWING

TYPE OF LIGHT REDUCTION	TRANS-MISSION	LIGHT AT EYE =	DENSITY NO. =	XD	GEN. RATING
ONE REFLECTION FROM GLASS OR ONE 5% FILTER	5%	5%	1.3	20	MUCH TOO BRIGHT
TWO REFLECTIONS OR 5% FILTERS	5%×5%	.25%	2.6	400	TOO BRIGHT
PENTA PRISM (NOT SILVERED)	96%×5% ×4%×96%	.16%	2.8	600	TOO BRIGHT
ANY COMB. OF THREE REFLECTIONS OR 5% FILTERS	5%×5% ×5%	.0125%	3.9	8000	OK FOR 2" APERTURE
*ONE SUN FILTER (NO. 4 DENSITY)	.01%	.01%	4.0	10,000	OK TO 2" APERTURE
ANY COMB. OF FOUR REFLECTIONS OR 5% FILTERS	5%×5% ×5%×5%	.0006%	5.2	167,000	OK TO 3" APERTURE
ONE REFLECTION OR 5% FILTER and ONE SUN FILTER*	5%×1% OR 1%×5%	.0005%	5.3	200,000	OK TO 3" APERTURE
POLAROID (SEE FIG. 7): PARALLEL	35%	35%	.46	3	NEVER USED ALONE
POLAROID (SEE FIG. 7): AT 45°	20%	20%	.7	5	
POLAROID (SEE FIG. 7): CROSSED	3%	3%	1.52	33	

DENSITY IS STANDARD LOG SCALE USED FOR FILTERS.

$$\text{DENSITY} = \text{LOG}\ \frac{1}{\text{TRANSMISSION}}$$

XD MEANS TIMES DIMINISHED

*NO. 4 DENSITY IS FAIRLY STANDARD AS A "SUN FILTER," ALTHO IT IS SOMETIMES TOO BRIGHT FOR COMFORT

TABLE 4 DENSITY NO. TO TRANSMISSION %

NEUTRAL DENSITY FILTERS: DENSITY NO.	WELDING* FILTERS: DENSITY NO. =	WELDING* FILTERS: SHADE NO.	TRANS-MISSION PER CENT (AVERAGE)	XD (TIMES DIMINISHED)	GEN. RATING
2.5			.32%	300	ALL TOO BRIGHT DENSITY NOS. CAN BE ADDED Ex. #2.5 Plus 3.0 DENSITY 5.5
	2.96	8	.11%	900	
3			.10%	1000	
	3.40	9	.04%	2500	
3.5			.03%	3300	
	3.82	10	.015%	6,600	OK UP TO 2" APERTURE
4			.010%	10,000	
	4.22	11	.006%	17,000	
4.5			.003%	33,000	
	4.70	12	.002%	50,000	
5			.001%	100,000	OK 2" TO 3"
	5.10	13	.0008%	125,000	
5.5			.0003%	330,000	OK SAFE and COMFORTABLE FOR 3"
	5.52	14	.0003%	333,000	
6			.0001%	1,000,000	TOO DIM FOR 3"

*WILLSON-WELD OR EQUAL. THE EQUIVALENT DENSITY and TRANSMISSION ARE AVERAGE VALUES

image at 10x (from Table 2) will be 2-3/4 inches. While small, this size will show sunspots clearly. A little experimenting will show you that the whole procedure is quite simple. If you smoke, it is instructive to blow smoke at the eyepiece to reveal the cone of light concentrating at the exit pupil and then fanning out to form the image. Put your finger at the exit pupil point and you will pull it away--fast; the heat here will char newspaper in an instant.

Huygens and Ramsden eyepieces are the types commonly used for sun projection, with a slight preferance for the Huygens. A Ramsden will perform somewhat better if the spacing is opened up a little. Cemented lenses can be used if desired, but there is always the chance that heat will damage the cement. Eyepieces must be clean since any dirt on the field lens will show more or less in focus on the projected image. So before you get excited about a "new discovery" on the sun, try rotating the eyepiece--a real-for-sure sun spot will stand still.

DIRECT VIEWING. The direct view through the telescope is a trifle sharper than by projection. However, there is an element of danger involved--just one momentary flash of the intensified sunlight can scar the retina of your eye causing partial or total permanent blindness. So far as "sun filters" are concerned, there are good ones and poor ones, but practically all are safe in that they limit the sunlight to an intensity which the eye can endure. The danger is only that the heat of the concentrated sunlight may crack the filter. Then faster than a bullet, the white-hot needle of light will pierce your eye and strike the retina. The least damage you can hope for is a small, permanent black area in the center of your vision.

Perversely, the average sun cap for use over an eyepiece is located very neatly at the hottest part of the light beam, Fig. 4. The coolest location is in front of the objective, Fig. 5, where the filter is exposed to only normal sun heat and is no more likely to crack than your eyeglasses. A similar setup is popular with reflecting telescopes, Fig. 6. The filter should be a free, shake fit to eliminate possible warping pressure which might break the glass. A minor disadvantage of the filter-in-front is that defects in the glass or any departure from a plane surface will cause greater deterioration of the image than the same filter used nearer the focal plane. However, you will invariably want to look at the full diameter of the sun, which automatically limits the magnification to about 50 or 60x. At this comparatively low power, ordinary welding filters are practical for front mounting. These are available in 50mm

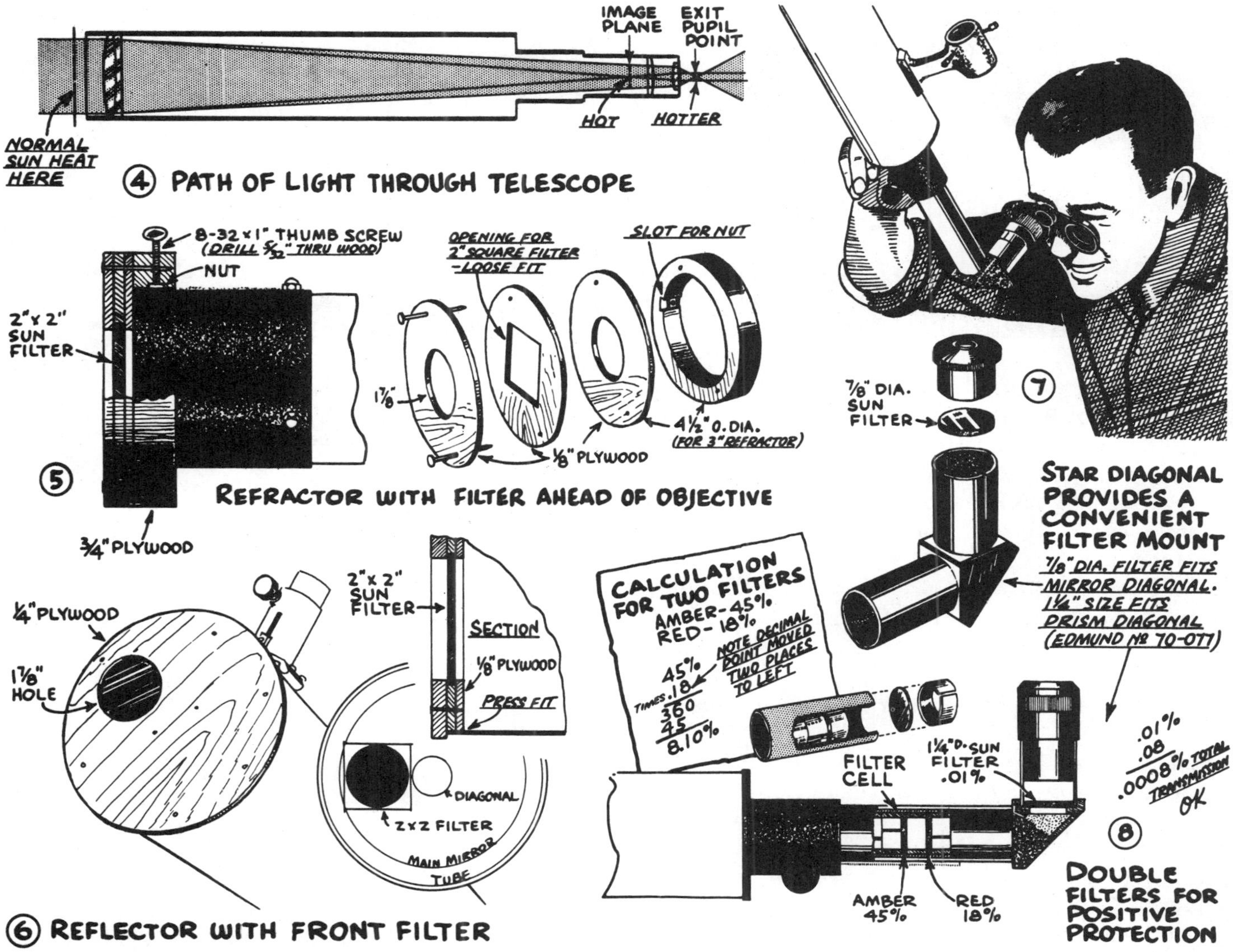

rounds and 2 x 4-in. rectangles in fifteen different shades, Nos. 10 through 14 being the darker shades most useful for sun viewing although lighter shades can be combined for the needed density. Nos. 12, 13 and 14 welding filters are dark enough to use alone; lighter shades may require additional filtering at the eyepiece, often obtained with an inexpensive polarizing filter, Fig. 7, which allows a fair range of density control. A star diagonal provides a convenient mount for a sun or other filter, Fig. 8, and you can increase the safety factor by using three or more filters, the general idea being that all three sure ain't going to pop in unison. As a further precaution when using a filter at or near the eyepiece, the simple action of taking the telescope off the target for a short cooling-off period does much to prevent a dangerous build-up of heat in the glass. Always swing the telescope a little away from the sun when it is not in actual use.

Seen with the usual sun filter, the sun is a greenish yellow. The intensity should be about the same as the full moon viewed without a filter. With a 3-in. refractor, this takes a light reduction of about 200,000 times. Tables 3 and 4 provide useful data. Filter tolerances are very generous. No. 11 welding glass, for example, may run from 13,000 to 33,000 XD, with the average transmission at about 17,000 XD, as listed in Table 4. From this and other variables, the "safe" range for viewing the sun has a considerable spread. A comparatively weak filter may be safe but uncomfortable to use; a "long" eye position will reduce the light intensity.

REFLECTION FROM GLASS. The classic method of observing the sun calls for one reflection from bare glass, plus a filter of about No. 4 density. Unlike the filter alone, this system is practically foolproof; the initial reflection from bare glass reduces the heat 20 times; the reduced heat will rarely crack the filter. Even if the filter cracks, you have enough protection from the bare glass reflection alone to avoid any permanent eye damage.

Oldest and probably the best of the bare glass

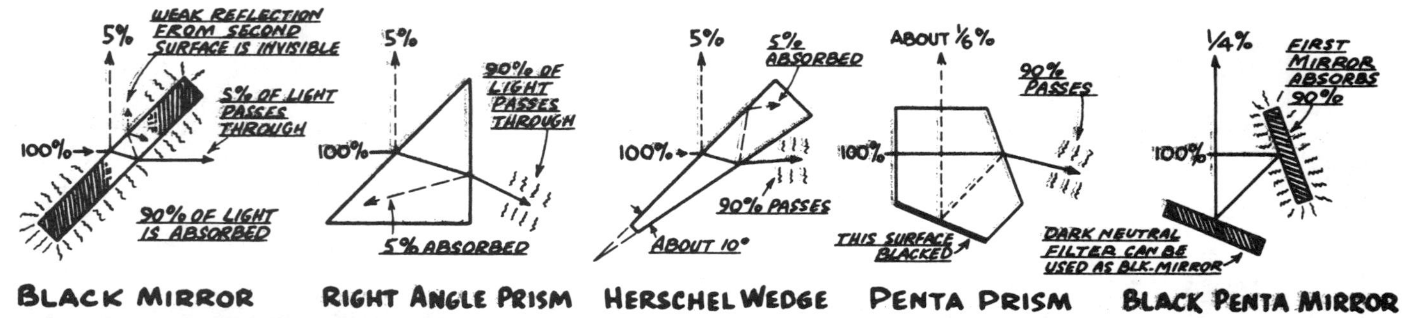

(9) REFLECTION FROM BARE GLASS ELIMINATES HEAT IN SAFE MANNER

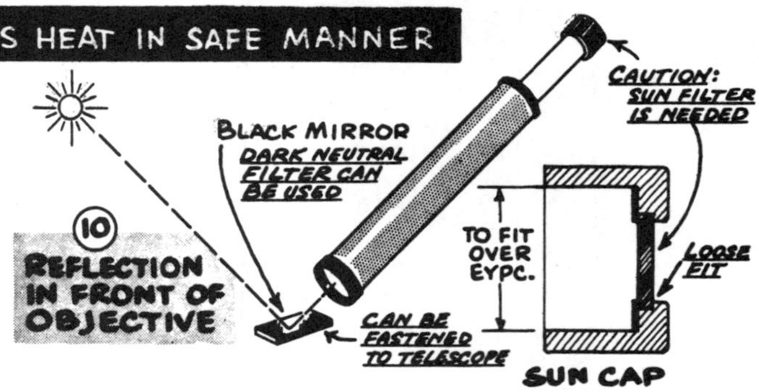

(10) REFLECTION IN FRONT OF OBJECTIVE

reflectors is the Herschel wedge, shown at center, Fig. 9. The usual equipment for combining this with a sun filter for use with a refractor is shown in Fig. 12. Like the more familiar star diagonal, the sun diagonal requires about 3 in. of "in" focusing travel. This is available with most refractors, but not with reflectors. The usual setup for a reflector calls for the substitution of the wedge for the usual flat mirror, Fig. 11. The sun filter can be either in the eyepiece tube or at the front of the main tube in the manner shown in Fig. 6. Even if a front filter is not used, the reflector is preferably masked to 2 or 3 inches aperture in a like manner but without the filter. There will be no loss of resolution from the lesser aperture because the atmosphere in sunlight is always turbulent; practically all the resolution you can get in a sun image can be obtained with 2 or 3 inches of aperture.

In common equipment, one reflection from glass plus a suitable filter is the only method of direct viewing which is entirely safe. You should never but never depend on a single sun filter located behind the eyepiece.

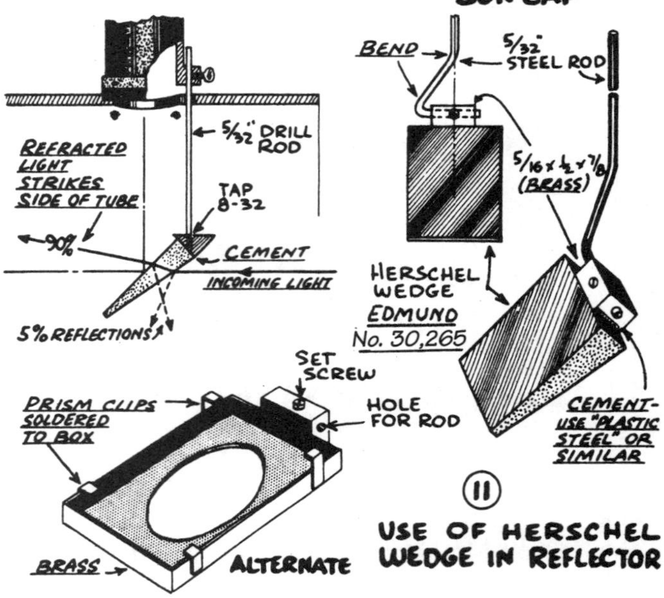

(11) USE OF HERSCHEL WEDGE IN REFLECTOR

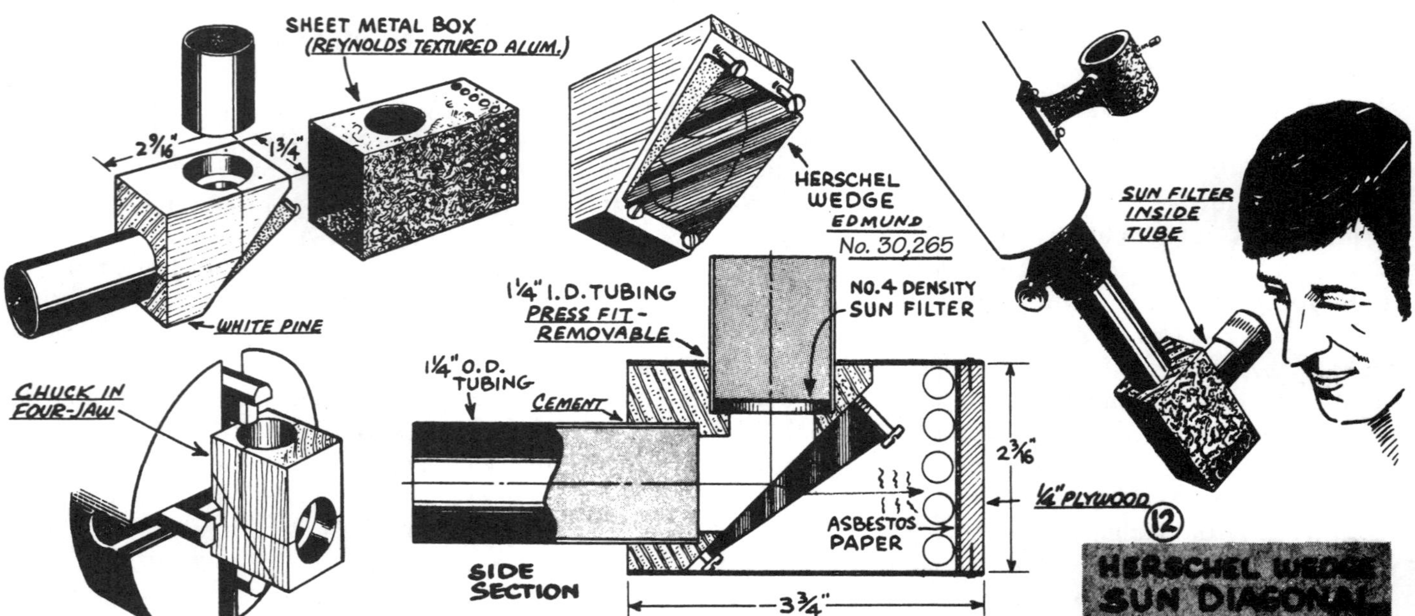

(12) HERSCHEL WEDGE SUN DIAGONAL

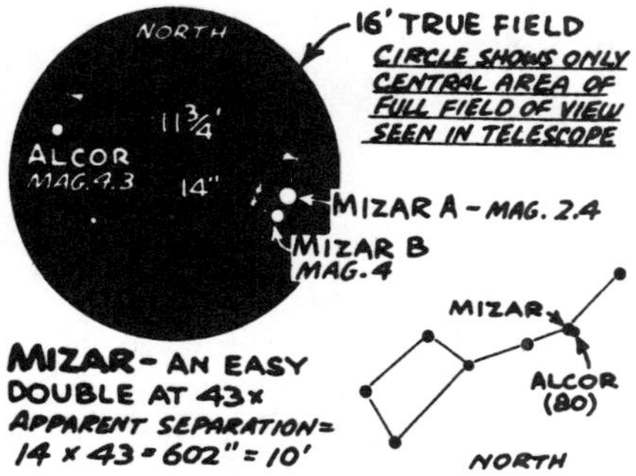

Splitting the DOUBLES

OBSERVING double stars is an interesting phase of star-gazing which can be done with any telescope. The beginner should start with easy doubles, say, 10 seconds of arc or more separation with companion star no fainter than 8th magnitude. It should be noted that close doubles near the limit of resolution for a specific telescope can be split only when seeing conditions are excellent.

How much power? A fundamental rule in telescope optics states that the true angular field of view multiplied by the magnification equals the apparent angular field of view. This rule can also be applied to any part of the field. For example: The popular double star, Mizar, is separated by 14 seconds of arc. This is the real or true field angle, the same angular separation as you see the stars with unaided eye. The human eye can't "split" 14 seconds of arc, so you have to boost it a bit. A magnification of 43x will increase the angle to about 10 minutes of arc (see drawing). This you can see quite easily. About the closest star separation which the eye can distinguish is 4 minutes of arc. Twice this distance or 8' apparent field angle is a more practical value for comfortable viewing, and in some cases you may want a separation of 20 or 25 minutes of arc. The powers needed for these various apparent angles are given in handy tabular form below.

Make Mizar your first telescope double. It is easy to find, with Alcor alongside supplying positive identification. An 8th magnitude star can be seen south from Alcor. Star distances are always amazing: The tiny space you see between A and B Mizar amounts to five thousandths of a light year or about 30 billion miles.

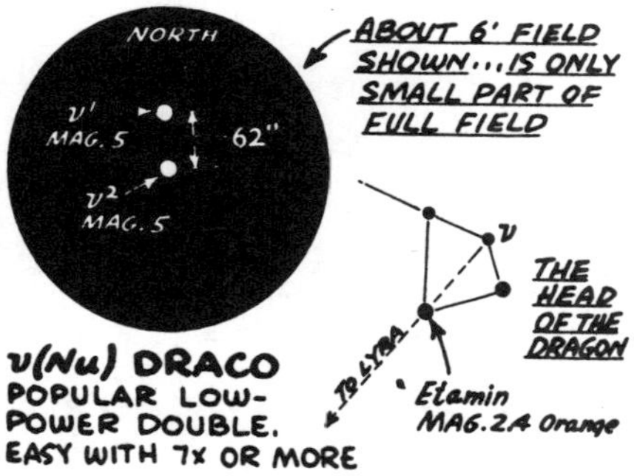

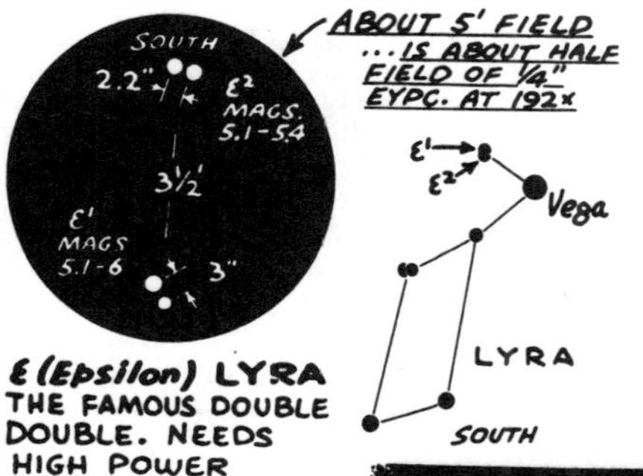

POWER NEEDED TO SPLIT DOUBLE STARS

ANGULAR SEPARATION "	FOR APPARENT SEPARATION OF				
	4'	6'	8'*	20'	25'
1 35 Coma	240	360	480	1200	1500
1.2 ζ Canc	200	300	400	1000	1250
1.5 μ Cygn	160	240	320	800	1000
1.7	140	210	280	700	882
2 ξ U Maj	120	180	240	600	750
2.5 Castor	96	144	192	480	600
3 ζ Aqar	80	120	160	400	500
3.5	68	102	136	343	428
4 γ Leo	60	90	120	300	375
4.5	54	80	108	266	332
5 γ Virgo	48	72	96	240	300
5.5	44	66	88	218	272
6 π Boot	40	60	80	200	250
6.5	36	54	72	184	230
7 32 Erid	34	51	68	171	214
7.5	32	48	64	160	200
8	30	45	60	150	188

ANGULAR SEPARATION "	FOR APPARENT SEPARATION OF				
	4'	6'	8'*	20'	25'
8.5	28	42	56	141	176
9	27	40	54	133	166
9.5	26	39	52	126	156
10 γ Andr	24	36	48	120	150
11 η Cass	22	33	44	109	136
12	20	30	40	100	125
13 8 Mono	18	28	36	91	115
14 Mizar	17	26	34	86	107
15	16	24	32	80	100
16	15	22	30	75	94
17	14	21	28	70	88
18 Polaris	13	20	26	67	83
19	12	19	24	63	78
20 α C Ven	12	18	24	60	75
21	12	17	24	57	71
22 8 Lacr	11	16	22	55	68
23	10	16	20	52	65

ANGULAR SEPARATION "	FOR APPARENT SEPARATION OF				
	4'	6'	8'*	20'	25'
24 ζ Pisc	10	15	20	50	62
25 61 Cygn	10	14	20	48	60
30 ι Canc	8	12	16	40	50
35 Albireo	7	10	14	34	43
40 16 Cygn	6	9	12	30	37
45 ζ Lyra	6	8	12	27	33
50	5	8	10	24	30
55 67 Ophi	5	7	10	22	27
60 ν Drac	4	6	8	20	25
65 τ Taur	4	6	8	19	23
70	4	6	8	17	21
75	4	5	8	16	20
80	3	5	6	15	18
85	3	5	6	14	18
90 ε Sgte	3	4	6	14	17
95 γ Leps	3	4	6	13	16
100 δ Boot	3	4	6	12	15

EQUIVALENT SEPARATIONS

LINE DIAGRAMS SHOW ANGULAR SEPARATION AS SEEN AT 10 INCHES

*8 MINUTES APPARENT SEPARATION IS RECOMMENDED FOR EASY VIEWING. NOTE IN DIAGRAM AT LEFT THIS ANGLE IS CLEARLY VISIBLE, ALTHO LUMINOUS POINTS (STARS) ARE MORE DIFFICULT TARGETS. WIDER SEPARATION IS NEEDED IF THE COMPANION STAR IS MORE THAN FIVE MAGNITUDES FAINTER THAN PRIMARY. NOTE 25' SEPARATION IS NEARLY WIDTH OF MOON SEEN WITH NAKED EYE

selected SKY OBJECTS

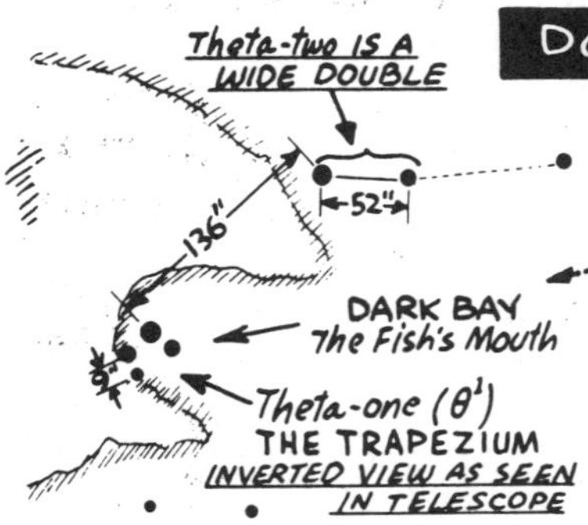

THE TRAPEZIUM IS SEEN PLAINLY IN 6" AT 50X. DIAGRAM SHOWS ABOUT 1/10 THE AREA YOU SEE AT 50X

DOUBLE STARS	R.A.	DEC.	MAGN.	SEPARATION	POWER* NEEDED	REMARKS
γ (Gamma) ARIES *Mesartim*	$1^h\ 51^m$	+19° 03′	4.8 4.8	8.4"	30-140	WHITE AND YEL.-WHT. EASY
Σ401 - TAURUS *W. Struve 401*	$3^h\ 28^m$	+27° 24′	6.5 6.8	11"	25-110	BOTH WHITE. N.W. of PLEIADES
θ^2 (Theta-one) ORION *The Trapezium*	$5^h\ 33^m$	-5° 25′	A 6.8 C 5.4 B 8 D 6.8	A-B 8.7" A-C 13" A-D 21.6"	30-140	IN BRIGHT AREA OF M42. SEE DIAGRAM
a^1, a^2 - CAPRICORNUS	$20^h\ 15^m$	-12° 40′	3.6 4.2	6′ 16"	7x and up	PRETTY YELLOW PAIR. BOTH HAVE FAINT COMPANIONS
γ (Gamma) DELPHINUS	$20^h\ 44^m$	+15° 57′	4.3 5.1	10"	25-120	YELLOW AND WHT. EASY AT 40x
61 - CYGNUS	$21^h\ 05^m$	+38° 28′	5.2 6.0	27.4" (1957)	10-50	BINARY. SEPARATION INCREASING FROM 16" IN 1780
μ (Mu) CYGNUS	$21^h\ 42^m$	+28° 31′	4.7 6.1	1.5"	160-800	TEST FOR 4-INCH
ζ (Zeta) AQUARIUS	$22^h\ 26^m$	-0° 17′	4.4 4.6	2.0"	120-600	BOTH WHITE. TEST FOR 3"

*MIN-MAX POWER IS FOR 4′ AND 20′ APPARENT SEPARATION

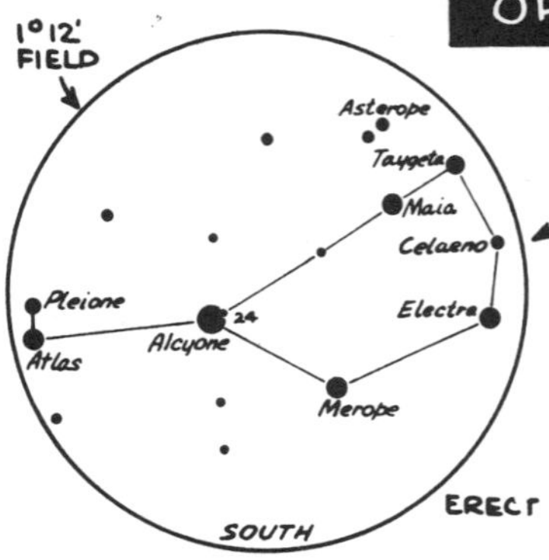

THE PLEIADES - WELL-KNOWN OPEN CLUSTER MAKES A PRETTY PICTURE IN TELESCOPE AT 40x

OPEN CLUSTERS	R.A.	DEC.	MAGN.	DIA.	REMARKS
NGC 752 - ANDROMEDA	$1^h\ 55^m$	+37° 25′	5.7	1°	NICE GROUP OF ABOUT 70 STARS
M 34 - PERSEUS	$2^h\ 39^m$	+42° 34′	5.2	30′	ABOUT 80 8th to 13th MAG. STARS
M45 - TAURUS *The Pleiades*	$3^h\ 44^m$	+24° 00′	1.2	1° 30′	PRETTY OBJECT... USE LOW POWER
M 38 - AURIGA	$5^h\ 26^m$	+35° 48′	6.4	20′	"A MASS OF STARS OF A SQUARE FORM" -Messier, 1764
M 50 - MONOCEROS	$7^h\ 01^m$	-8° 16′	5.9	10′	IRREGULAR GROUP OF ABOUT 100 STARS
M46 - PUPPIS	$7^h\ 40^m$	-14° 42′	6.1	24′	MANY TINY SPARKLERS... FINE FIELD AT 96x
M25 - SAGITTARIUS	$18^h\ 29^m$	-19° 17′	4.6	45′	LARGE AND SMALL STARS
M 11 - SCUTUM	$18^h\ 48^m$	-6° 20′	5.8	15′	NICE OBJECT... VISIBLE WITH BINOC

THE PLEIADES (PLEE-uh-deez)

SEVEN SISTERS		
ALCYONE	al-SIGH-oh-nee	3.0
ASTEROPE	as-TAIR-oh-pee	5.9
CELAENO	seh-LEE-noe	5.4
ELECTRA	eh-LEK-tra	3.8
MAIA	MY-uh	4.0
MEROPE	MARE-oh-pee	4.3
TAYGETA	tah-IJ-ah-tah	4.4
PLEIONE	plee-OH-nee	5.1
ATLAS	AT-lus	3.8

PLEIONE IS MAMA AND ATLAS IS PAPA. STAR № 24 (MAG. 8) MAKES A WIDE DOUBLE WITH ALCYONE AT 117 SECONDS

GLOBULAR CLUSTERS	R.A.	DEC.	MAGN.	DIA.	REMARKS
ω (Omega) CENTAURUS	$13^h\ 24^m$	-47° 03′	3.6	23′	FINEST GLOBULAR IN SKY BUT TOO FAR SOUTH FOR MANY U.S. OBSERVERS
M4 - SCORPIUS	$16^h\ 21^m$	-26° 24′	5.9	14′	A LITTLE OVER 1° WEST OF ANTARES
M 13 - HERCULES	$16^h\ 40^m$	+36° 32′	5.9	10′	RESOLVABLE WITH 6-INCH
M12 - OPHIUCHUS	$16^h\ 45^m$	-1° 52′	6.6	9′	BLAZE AT CENTER ABOUT 2′ DIA.
M 10 - OPHIUCHUS	$16^h\ 55^m$	-4° 02′	6.6	8′	RESOLVED WITH 6-INCH. BRIGHT
M92 - HERCULES	$17^h\ 16^m$	+43° 12′	6.5	8′	LUMINOUS CENTER... RESOLVABLE
M9 - OPHIUCHUS	$17^h\ 16^m$	-18° 28′	7.9	2.5′	SMALL BUT BRIGHT BALL OF STARS
M22 - SAGITTARIUS	$18^h\ 34^m$	-23° 57′	5.1	17′	BIG AND BRIGHT FOR A GLOBULAR. EASY TO SEE WITH 7x BINOCULAR

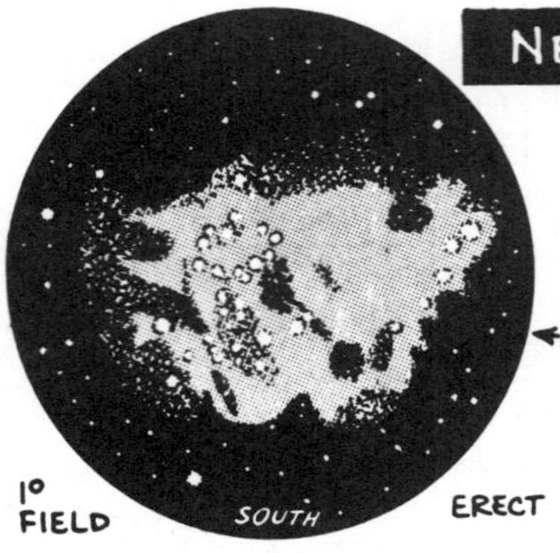

THE LAGOON - FINE OBJECT AT 40x. NEEDS BLACK NIGHT TO SHOW MISTY NEBULA AROUND STAR CLUSTER. VISIBLE WITH BINOC

NEBULAE	R.A.	DEC.	MAGN.*	SIZE	REMARKS
M31 - ANDROMEDA	$0^h\ 41^m$	+41° 03′	3.5	40′×160′	SPIRAL GALAXY. ONLY ABOUT 20′ BRIGHT CENTER IS SEEN. EASY WITH BINOC
M42 - ORION	$5^h\ 33^m$	-5° 25′	2.9	40′	DIFFUSE NEBULA. GOOD OBJECT. INCLUDES θ^1 (THE TRAPEZIUM)
M81 - URSA MAJOR	$9^h\ 52^m$	+69° 18′	6.9	10′×16′	SPIRAL GALAXY. LOW-POWER FIELD SHOWS ALSO M82 ABOUT 3/4° NORTH
M8 - SAGITTARIUS *The Lagoon*	$18^h\ 01^m$	-24° 21′	5.8	35′×60′	DIFFUSE NEB WITH CLUSTER... GOOD
NGC 6572 - OPHIUCHUS	$18^h\ 10^m$	+6° 50′	9.5	10"	PLANETARY. "SMALL BUT BRIGHT" -Norton
M17 - SAGITTARIUS *Horse-shoe NEB*	$18^h\ 18^m$	-16° 12′	9	30′	DIFFUSE NEB AROUND A SMALL CLUSTER OF 9th TO 12th MAG. STARS
M 57 - LYRA *Ring NEBULA*	$18^h\ 52^m$	+32° 58′	9	65"	PLANETARY NEB... NOT BRIGHT... LIKE HAZY 9th MAG STAR. USE HI-POWER
NGC 7662 - ANDROMEDA	$23^h\ 24^m$	+42° 14′	8.9	30"	PLANETARY... 13th MAG CENTER STAR

*APPROX. VISUAL MAGNITUDE

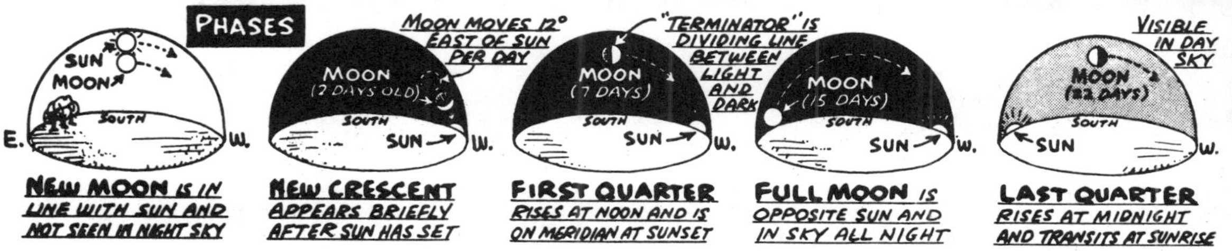

Target MOON

PRIME target for telescope observers everywhere, the moon at a comparatively near 240,000 miles shows amazing detail in even the smallest telescope. Over 100,000 objects are shown on maps and photos, and the beauty of the whole thing is that with a good 3-inch telescope you can see everything as clear as the very finest photos.

Any and all powers can be used. Since the moon is about 30 minutes of arc in angular diameter, the maximum power which will show the whole surface is about 90x. The smallest crater you can hope to see is two times Dawes Limit. Thus, a 3-inch telescope can show craters as small as 3 miles diameter. Twice this (6 miles) is a more practical working value. Much smaller detail can be detected when in the form of a line; you will have no trouble seeing The Straight Wall.

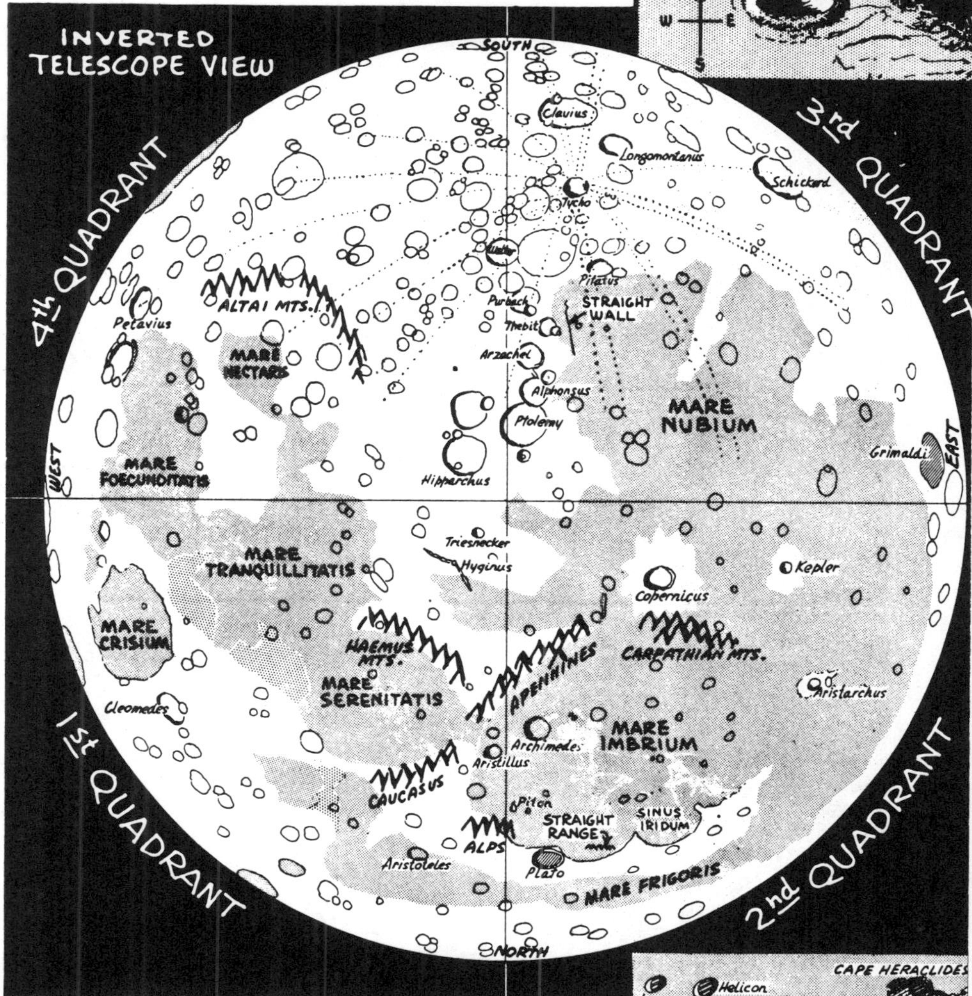

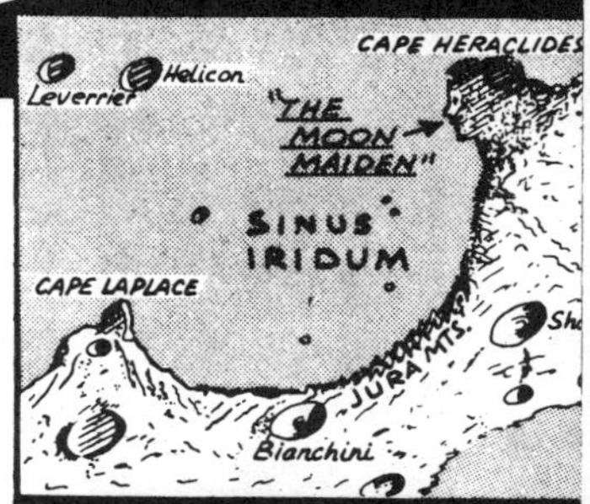

BAY OF RAINBOWS

SELECTED MOON OBJECTS

QUADRANT	NAME OF OBJECT	TYPE	SIZE (MILES)	REMARKS
2	ARISTARCHUS (air-is-TAR-kus)	CRATER	29 DIA.	CRATER ON A ROCKY PLATEAU IS BRIGHTEST OBJECT ON THE MOON
1	ARISTILLUS (air-is-TILL-us)	CRATER	35 DIA.	ONE OF THE BEST-FORMED CRATERS IN A MAGNIFICENT MOUNTAIN SETTING
2	COPERNICUS (ko-PURR-nick-us)	RINGED PLAIN	56 DIA.	FINE EXAMPLE OF A RINGED PLAIN CRATER. MANY BRIGHT RAYS IN IMMEDIATE VICINITY
3	GRIMALDI (gri-MALL-di)	WALLED PLAIN	20 DIA.	DARKEST SPOT ON THE MOON
1	HYGINUS (hih-JINE-us)	CRATER PIT	4 DIA.	SMALL CRATER SPLIT BY A GREAT CLEFT (CRACK). CLEFT CAN BE SEEN IN SMALL TELESCOPE
4	PETAVIUS (peh-TAVE-ih-us)	WALLED PLAIN	100 DIA.	EASY OBJECT WHEN MOON IS NEW CRESCENT. MOUNTAIN AT CENTER WITH CLEFT EXTENDING S.E.
2	PITON (PIE-tun)	SOLITARY MOUNTAIN	1½ HIGH	BRIGHT HIGH PEAK WHICH CASTS A LONG SHADOW WHEN ON TERMINATOR
2	PLATO (PLAY-toe)	WALLED PLAIN	60 DIA.	"THE GREAT BLACK LAKE." ONE OF THE DARKEST SPOTS ON MOON
3	PTOLEMY (TOL-uh-me)	WALLED PLAIN	90 DIA.	WITH ALPHONSUS (70 MILES) AND ARZACHEL (60 MILES), A PROMINENT LANDMARK
2	SINUS IRIDUM (BAY OF RAINBOWS)	PLAIN	135 (BETWEEN CAPES)	PRETTY LUNAR LANDSCAPE. HERACLIDES (hair-uh-KLIE-dees) FORMS "MOON MAIDEN"
2	STRAIGHT RANGE	MOUNTAIN PEAK	40 LONG	STRAIGHT LINE OF MOUNTAIN PEAKS
3	STRAIGHT WALL	CLIFF	70 LONG	SURFACE FAULT WITH EAST SIDE 800 FT. LOWER THAN WEST. BEST SEEN AT LAST QUARTER
3	TYCHO (TIE-co)	RINGED PLAIN	56 DIA.	WHEN MOON IS FULL, BRIGHT RAYS MAKE TYCHO MOST PROMINENT LUNAR OBJECT

MOST OBJECTS ARE SEEN BEST WHEN ON OR NEAR TERMINATOR. ANGULAR SIZE IN SECONDS OF ARC IS APPROX. SAME AS LINEAR MILES - Example: 14 MILES = 14 SECONDS

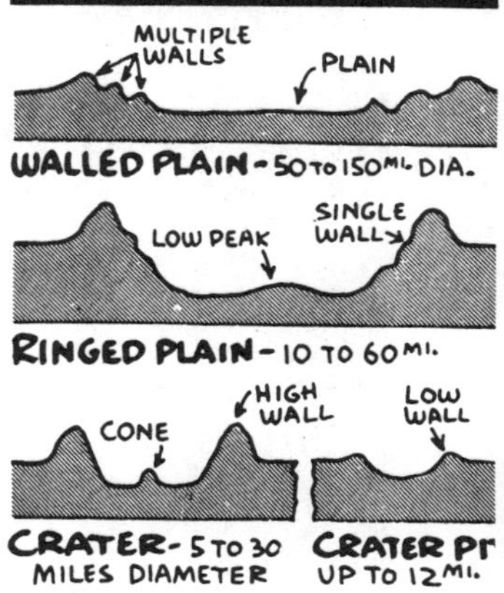

QUICK GUIDE to sky shooting

What to Shoot		EQUIPMENT	OPTICAL SYSTEM	METHOD – REMARKS
	STAR TRAILS	ANY CAMERA FROM 35MM SIZE UP TO 8×10 INCH. CAMERA MUST BE ON TRIPOD OR OTHER SUPPORT	ANY LENS, 2" TO ABOUT 12" F.L. CLEAR APERTURE SHOULD BE AT LEAST 5/8" AND PREFERABLY LARGER	SET CAMERA LENS AT INFINITY AND POINT AT TARGET. MAKE A TIME EXPOSURE OF 10 MIN. OR MORE. *SKY MUST BE DARK– DO NOT SHOOT WITH MOON IN SKY*
	MOON	REFLECTING OR REFRACTING TELESCOPE WITH ADAPTER FOR CAMERA BODY. CLOCK DRIVE USEFUL BUT NOT ESSENTIAL	45" TO 80" E.F.L. APERTURE 1" OR MORE, f/8 TO f/80 MOON IMAGE DIA. = .009 × E.F.L.	WITH FIXED MOUNT, USE FAST FILM TO PERMIT SHORT EXPOSURE OF 1/100 SEC. OR LESS. WITH CLOCK DRIVE USE SLOWER, FINE-GRAIN FILM AND LONGER EXPOSURES
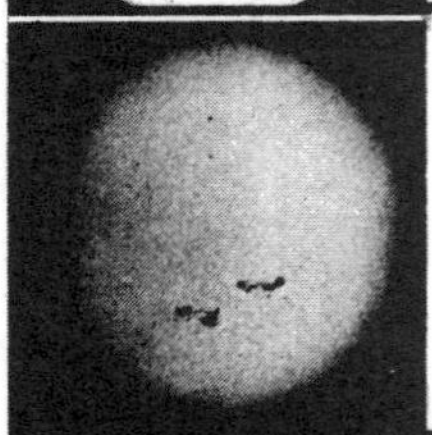	SUN	TELESCOPE AND CAMERA. BY FILTER OR OTHER MEANS, THE LIGHT MUST BE REDUCED TO FULL MOON INTENSITY	SAME AS FOR MOON. USUALLY NOT LESS THAN 2" APERTURE, f/8 TO f/80. USUALLY A PROJECTION SYSTEM	WITH IMAGE AT FULL MOON BRIGHTNESS, FAST EXPOSURES OF 1/100 TO 1/500 SECOND CAN BE MADE ON SLOW FILM, SUCH AS HIGH CONTRAST COPY
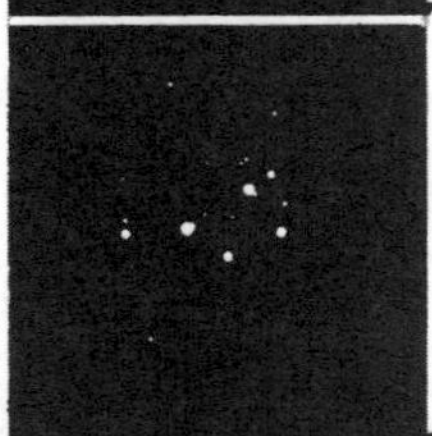	STAR FIELDS	ANY CAMERA. THE TELESCOPE IS USED FOR GUIDING ONLY. A CLOCK DRIVE WITH SLOW-MOTION CONTROL IS ESSENTIAL	PREFERABLY NOT LESS THAN 1" CLEAR APERTURE. 3" TO 12" F.L., DIRECT OBJECTIVE. FIELD SHOULD BE FLAT, COMA-FREE	USE FAST FILM. GUIDING CONSTANTLY WITH TELESCOPE, MAKE TIME EXPOSURE OF 10 MINUTES OR MORE. *GOOD STARTER... EASIEST TARGET FOR GUIDED TELESCOPE*
	OPEN STAR CLUSTERS	CAMERA, 20" OR MORE F.L. TELESCOPE USED FOR GUIDING BUT MAY BE THE CAMERA, IN WHICH CASE A LONG F.L. GUIDESCOPE IS NEEDED	NOT LESS THAN 2" APERTURE -- *THE LARGER THE APERTURE THE MORE STARS YOU CAN PHOTOGRAPH*. F.L. TO SUIT TARGET SIZE	GUIDED TELESCOPE, 15 MIN. OR MORE TIME EXPOSURE. THE LONG F.L. OF CAMERA MAKES GUIDING MORE DIFFICULT – A SLOW-MOTION MAY BE NEEDED ON DECLINATION SHAFT
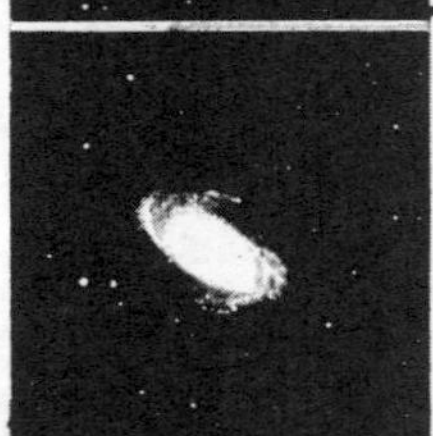	NEBULAE and GALAXIES	CAMERA IS OFTEN THE TELESCOPE ITSELF. BIG APERTURE, LONG F.L. NEEDED. CLOCK DRIVE WITH SLOW-MOTION	NOT LESS THAN 3" APERTURE. 20" TO 100" F.L. USUALLY DIRECT OBJECTIVE FOR LOW f/NUMBER BUT PROJECTION SYSTEM IS USEFUL	USE FAST FILM. SOME NEBS ARE BRIGHT AND CAN BE CAPTURED ON TRI-X FILM IN AS LITTLE AS 10 MINUTES. *MOST TARGETS ARE DIM -- BIG APERTURE AND FAST FILM NEEDED*
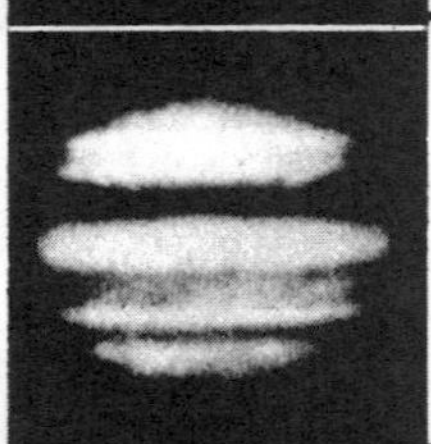	PLANETS	TELESCOPE WITH AMPLIFYING SYSTEM TO GET E.F.L. OF 100" TO 1000." TELESCOPE ON EQUATORIAL MOUNT WITH CLOCK DRIVE	LONG E.F.L. IS MAIN NEED. f/16 TO f/100. POSITIVE OR NEGATIVE PROJECTION OF 3× TO 8× IS USUAL SETUP	USE FINE-GRAIN FILM TO PERMIT ENLARGEMENT. PLANETS ARE BRIGHT BUT HIGH f/NUMBER CALLS FOR LONG EXPOSURE OF 1 TO 10 SEC. *THIS IS DIFFICULT PHOTOGRAPHY*
	COMETS	LENS 8" TO 18" F.L. MANUAL SLOW-MOTION ON BOTH AXES	APERTURE NOT LESS THAN 2", f/3 TO f/8, USED WIDE OPEN	TIME EXPOSURE, 10 SEC. TO 10 MIN. YOU MUST GUIDE ON COMET HEAD DURING EXPOSURE
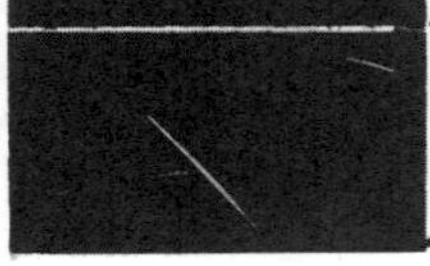	METEORS	ANY CAMERA. WIDE FIELD IS NEEDED -- NOT LESS THAN 40°	WITH 35MM CAMERA, 2" F.L. IS PREFERRED FOR WIDEST FIELD	KNOWING TIME AND GENERAL AREA OF A METEOR SHOWER, MAKE A STAR TRAIL – *AND HOPE!*

Photography with your Telescope

MOST of the pictures taken with a telescope are of an astronomical nature, being pictures of the sun and moon, stars and planets. However, you can also use your "long glass" to advantage for certain types of daytime photography, particularly those views which are too distant to be captured by the short focal length objectives normally used on small cameras.

Whatever the subject, a camera using small-size film is preferable, with the popular 35mm film size being first choice for most work. Photography on 35mm film is inexpensive. There is also the plain fact that the average telescope set-up will not cover larger film sizes. Reflex focusing is highly desirable because you can see to focus and see to position your target object the way you want it. In brief, the 35mm single lens reflex is the camera you can use to best advantage. You can purchase a camera of this kind for as little as $100 or as much as $600; secondhand much less. Much of the cost of a single lens reflex is in the original objective. It is not necessary that the camera have a top-quality fast lens because you will rarely use the lens anyhow for the type of photography we are considering, the general situation being that the telescope itself is the objective, while the camera is mainly a film transport and viewer. Some useful camera equipment can be homemade, and with this in view the original purchase must permit the use of interchangeable objectives, preferably with a screw-in thread.

Four optical systems are generally recognized in photography with the telescope, as shown in drawing below. Actually these boil down to just two basic types: (1) the direct objective, (2) projection system. With the direct objective, light goes direct from the lens or mirror objective to the film; the picture is taken at the primary or first focus. In all other systems, the primary image is projected to form a second image, and the picture is taken at this second image plane.

4 Systems

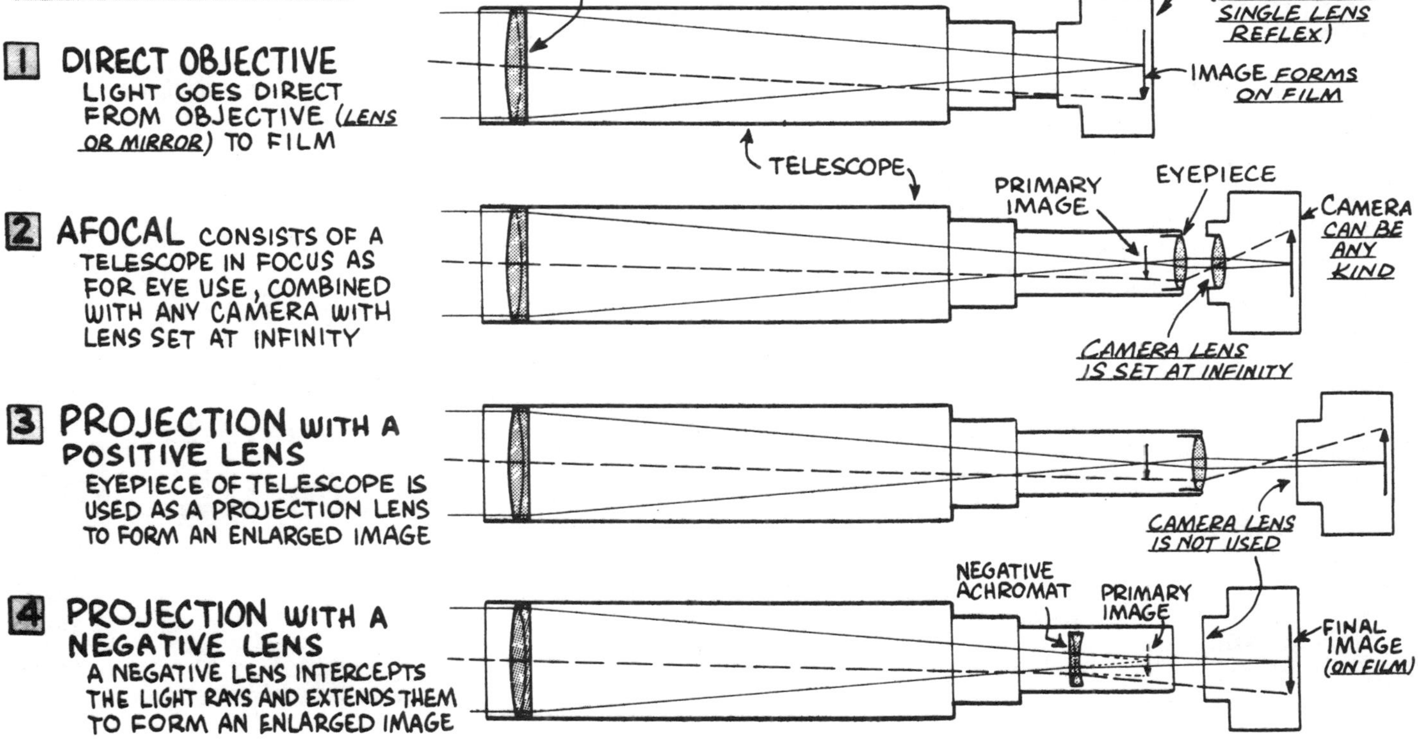

the simplest system...

The DIRECT OBJECTIVE

① 3-INCH REFRACTOR AS A DIRECT OBJECTIVE

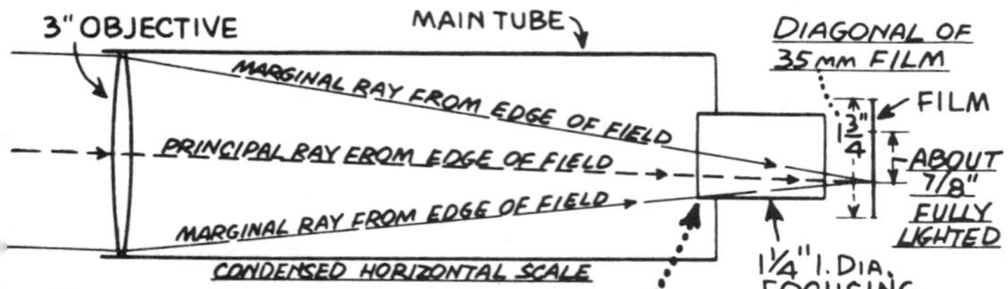

② FOCUSING TUBE IS BOTTLENECK

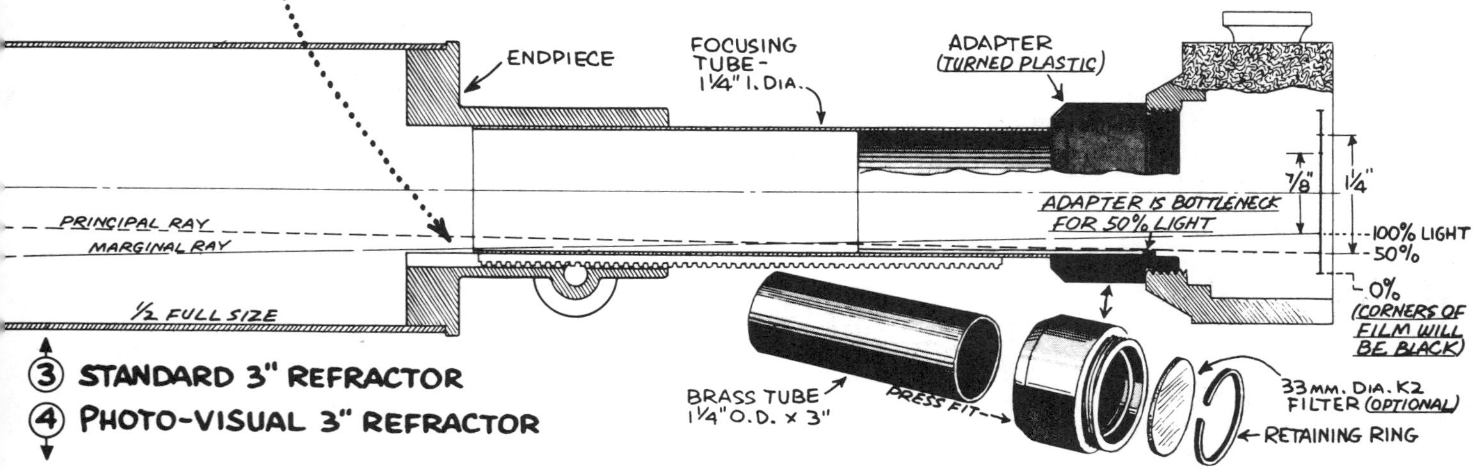

③ STANDARD 3" REFRACTOR

④ PHOTO-VISUAL 3" REFRACTOR

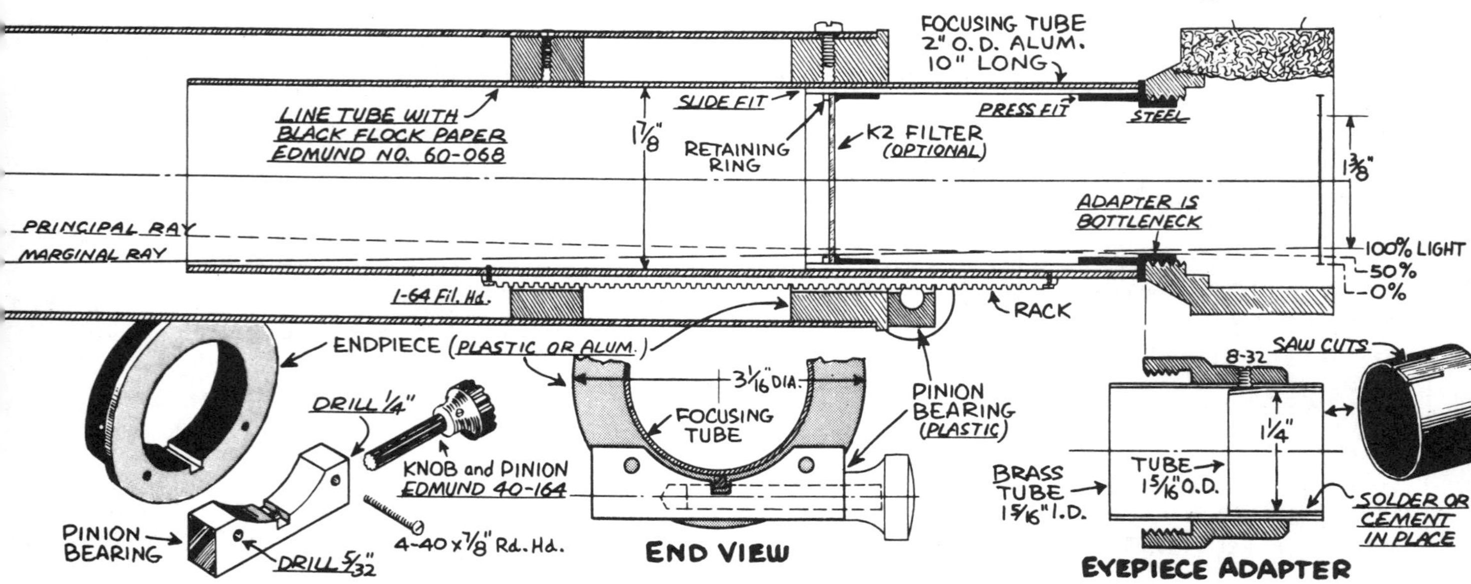

CONTAINING the least possible number of optical elements, the direct objective is the simplest and most-used picture-taking system--99% of all cameras use a direct objective. Normally it is also the simplest system mechanically, although this can't be said of a telescope conversion where the switch may require more work than afocal and projection systems. Advantages of the direct objective are highest light transmission and best definition. In other words, this is the fastest and sharpest optical system for shooting pictures--only once in a blue moon will you run into a compensating optical system where the performance of the primary is improved by the addition of other optical elements. The only

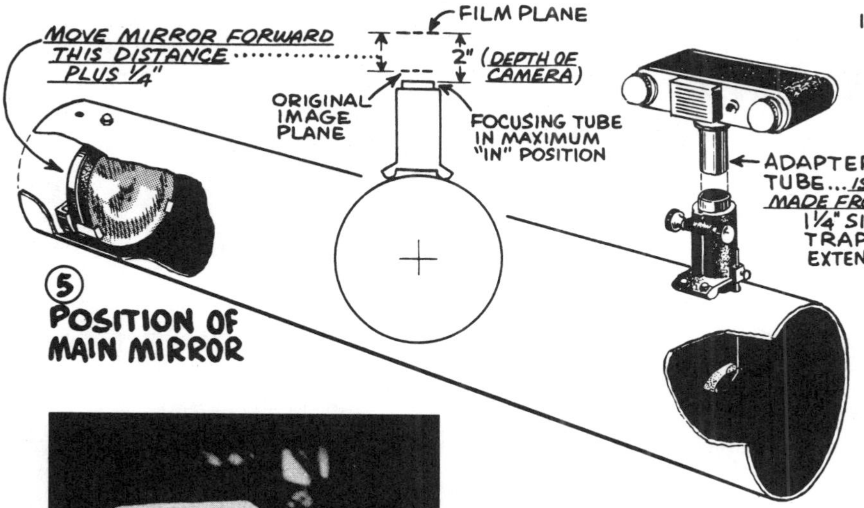

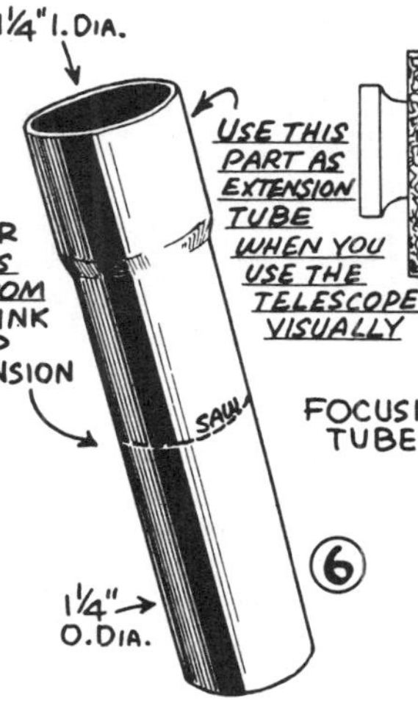

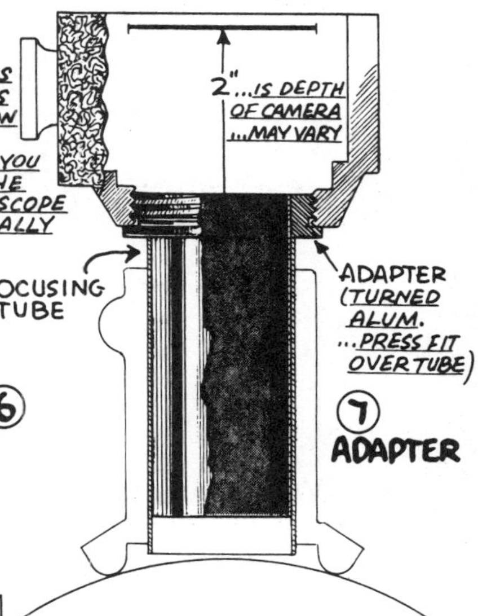

VIGNETTING CAUSED BY SMALL DIAMETER OF FOCUSING TUBE
TEST PIC WITH 4 1/4" REFLECTOR FROM INDOORS POINTING THRU OPEN WINDOW TO OBJECT AT 100 YDS. TRI-X FILM, 1/200 SEC.

(8) DIAMETER OF ILLUMINATED FIELD

6" REFLECTOR	100%	75%	50%	25%	0%
WITH 1 1/4" DIAGONAL	5/16" D	7/8" D	1 1/4" T	1 3/8" T	1 1/2" T
WITH 1 1/2" DIAGONAL*	5/8 D	1 1/8 D	1 1/4 T	1 3/8 T	1 1/2 T
WITH 1 7/8" DIAGONAL	3/4 T	1 1/8 T	1 1/4 T	1 3/8 T	1 1/2 T
4 1/4" REFLECTOR	T - FOCUSING TUBE IS BOTTLENECK D - DIAGONAL IS BOTTLENECK				
WITH 1 1/16" DIAGONAL	1/2" D	7/8" D	1 1/4" T-D	1 3/8" T	1 1/2" T
WITH 1 1/4" DIAGONAL*	3/4 D	1 1/8 T-D	1 1/4 T	1 3/8 T	1 1/2 T
WITH 1 1/2" DIAGONAL	7/8 T	1 1/8 T	1 1/4 T	1 3/8 T	1 1/2 T

*BEST CHOICE

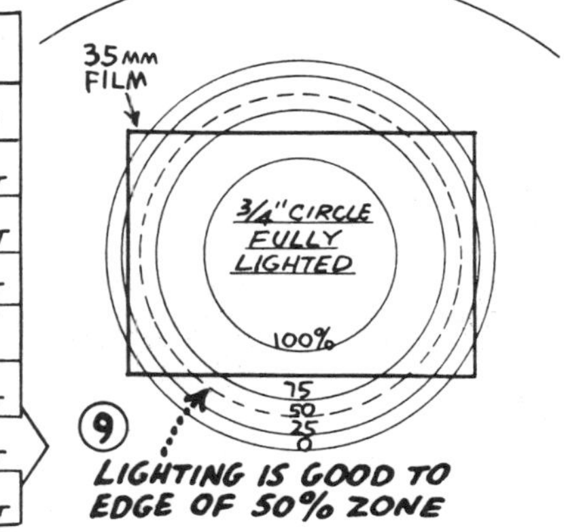

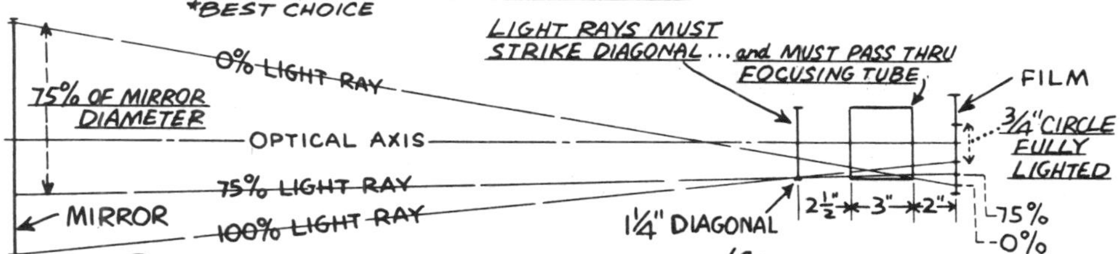

(10) LIGHT DIAGRAM - 4 1/4" REFLECTOR (CONDENSED HORIZONTAL SCALE)

poor feature of the direct objective is lack of compactness, which becomes an item of considerable importance as the focal length is increased.

THE REFRACTOR AS A CAMERA. You can make a good telecamera from a refractor by simply mounting a 35mm single lens reflex camera by means of an adapter tube, as shown in Figs. 1 and 3. You can get nice pictures--land or sky--with such an outfit, but they will show a black vignette about as shown in photo above. This comes from the small diameter of the telescope focusing tube, which restricts the fully-lighted image field to less than an inch diameter. To increase the illuminated field, it is obvious you must enlarge the focusing tube, and this means practically building a new instrument. A simple type of construction is shown in Fig. 4; a new set of glare stops inside the main tube will also be needed. Such a conversion will show only a touch of black at the extreme corners of 35mm film. A visual adapter as shown in Fig. 4 detail makes the conversion back to telescope for use with standard eyepieces.

THE REFLECTOR AS A CAMERA. The reflecting telescope poses the same problem of small-diameter focusing tube. Another snag is that the image plane is not even accessible. What you have to do first of all is move the main mirror forward a short distance, as shown in Fig. 5. This will advance the image plane to a position where it can be made to coincide with the film plane of the camera. Even so, space is at a premium and the adapter, Fig. 7, should be made as shallow as practical. The extended position of the image plane will not interfere with the use

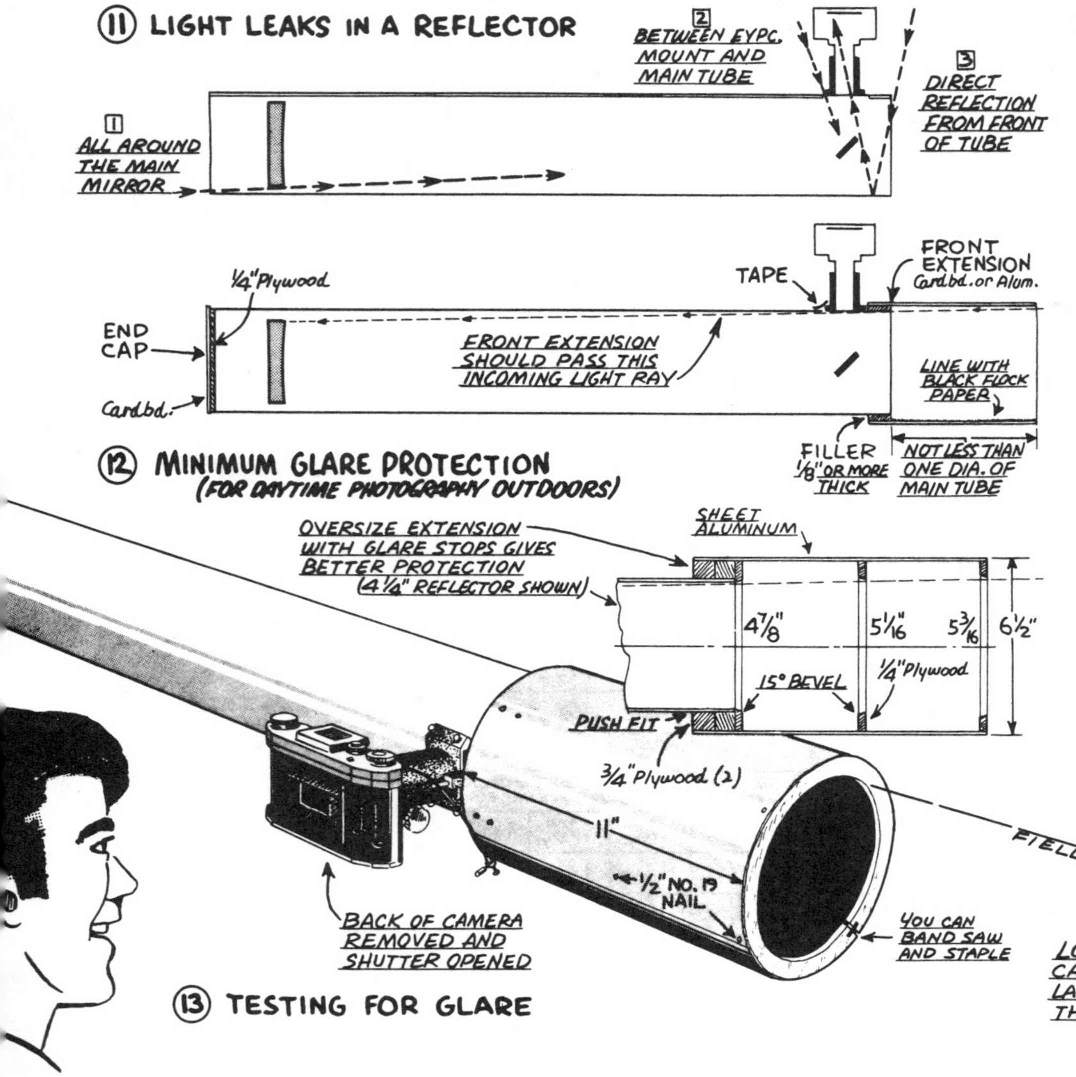

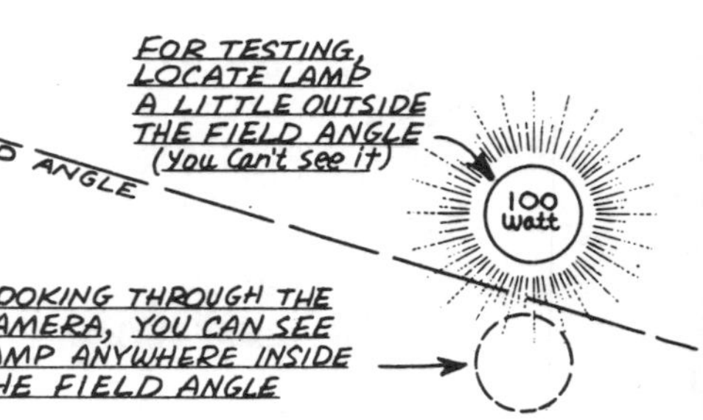

LIGHT-FOGGING OCCURS WHEN STANDARD 6-INCH REFLECTOR IS USED OUTDOORS WITHOUT PROTECTION. *Cloudy-bright, no sun. 1/500 Sec. at f/8 on Tri-X Film*

of the telescope visually, but you will have to use a short extension tube, easily made from a chrome sink trap extension, Fig. 6.

One more obstacle now appears: the standard-size diagonal is too small. Mainly, this results because the linear size of 35mm film is quite a bit larger than what you ordinarily look at with an eyepiece. Contributing is the fact the image plane is at a greater distance from the diagonal. You can correct by using the next larger standard size diagonal, but you still have the considerable vignette caused by the small-diameter focusing tube. Fig. 8 table shows that some vignetting will occur regardless of how big you make the diagonal; the focusing tube is the bottleneck. However, enough light gets through to make useful pictures. A 4-1/4-inch reflector will vignette a little more than a 6-inch, the villain in this case being the main tube itself, which is too small to admit the full incoming light cone. A few things about vignetting in general are worth noting: (1) Practically all camera lenses vignette 25% or more at corners of film when used wide open, (2) a 50% vignette is hardly detectable on the ground glass, and often does not show at all on prints unless negatives are very thin, (3) if you plan only moon shots and similar astro targets, the vignette will have little or no effect because the area outside the moon is black anyhow.

GLARE PROTECTION. The refractor is well-protected against unwanted glare light, and you can shoot pics outdoors in the sun just like any camera. The reflector is another story. Even though no direct glare light can reach the film, you will pick up enough one and two-bounce reflections to fog the picture, Fig. 11. The worst offender is the one-bounce reflection from the front end of the main tube. A simple cure is a sun shade, Fig. 12; a long, over-size extension fitted with glare stops is even more effective, Fig. 13.

Check your setup for glare light outdoors, or, do the job indoors with a lamp in the manner shown in Fig. 13. This little bit of look-and-see should be a routine test for all telecamera systems. You will be able to see immediately any areas which are unduly bright. By sighting from the corners of the film opening, you can also check the extent of illumination. The setup is also a good check for collimation--all of the various apertures should appear concentric with the film opening of camera.

AUXILIARY LENSES. You can shoot a lot of pictures with a telescope, but more than half of all pictures taken with a telescope are not taken with a telescope. The situation in astrophotography is that while you need the equatorial movement and clock drive of a telescope, the actual picture-taking is often done with an ordinary camera mounted on the telescope tube. For short focal length camera objectives up to about 10 in. f.l., some kind of a camera bracket,

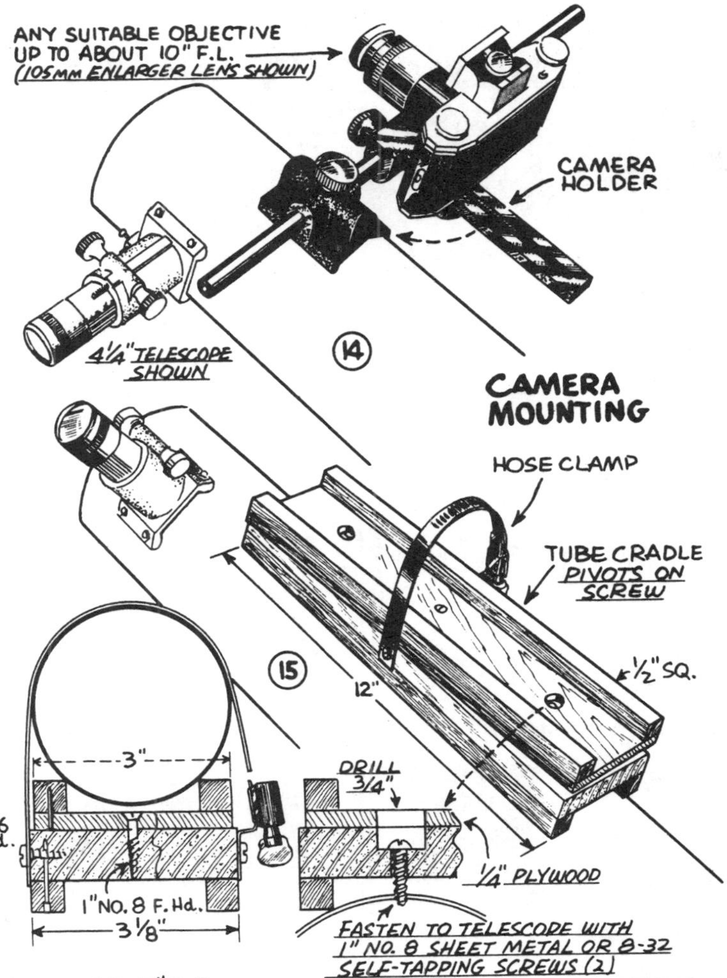

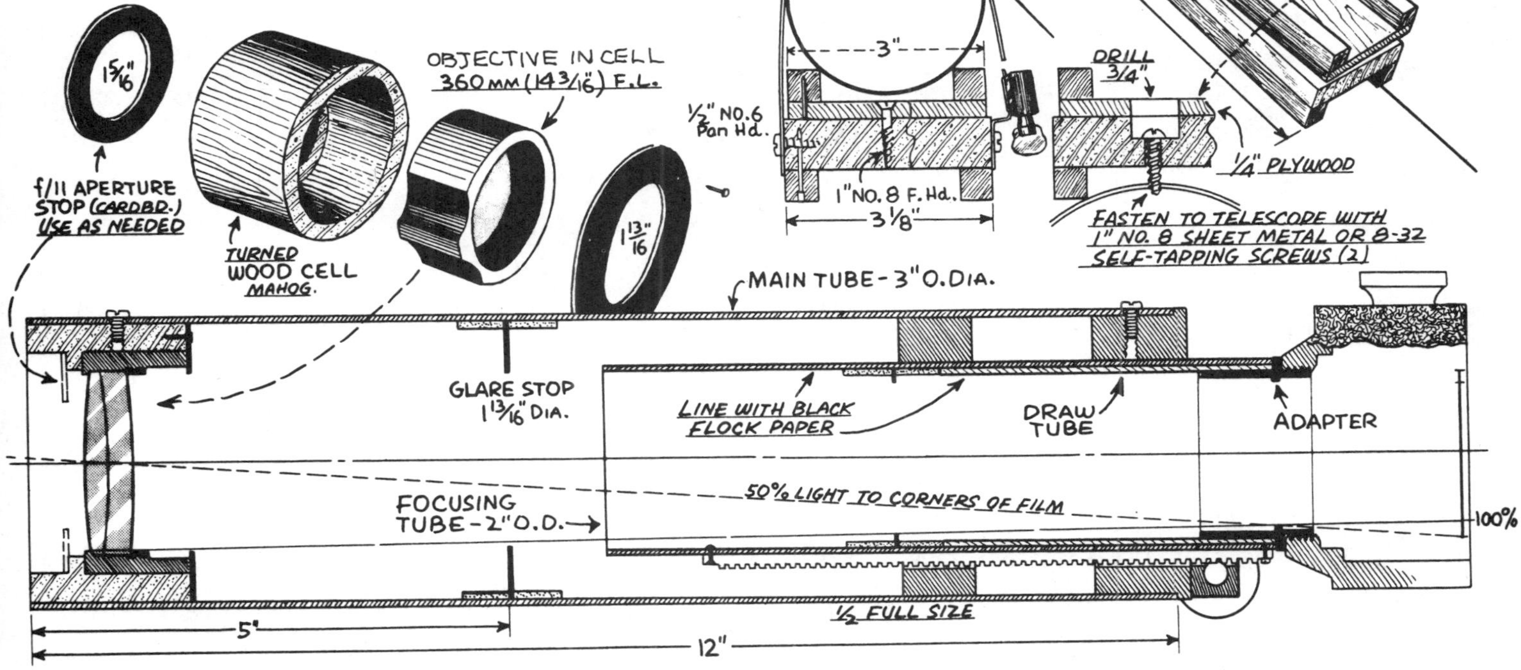

(16) **14-INCH F.L. DIRECT OBJECTIVE**
(SAME CONSTRUCTION AS FIG. 4)

Test Pics: ALL TRI-X PAN FILM, NO FILTER, BRIGHT SUN. DISTANCE ABOUT 100 Yds.

105 MM ENLARGING LENS DIRECT OBJECTIVE. 1/500 SEC., f/11. HAND-HELD

14-INCH ACHROMAT AS SHOWN ABOVE. 1/500 SEC. AT f/8. CAMERA HELD BY HAND

3x BARLOW AND 14" ACHROMAT. COMBO IS 42" F.L., f/24. 1/200 SEC. ON TRIPOD

TABLE 1

Angular and Linear Field Covered by 35mm Film

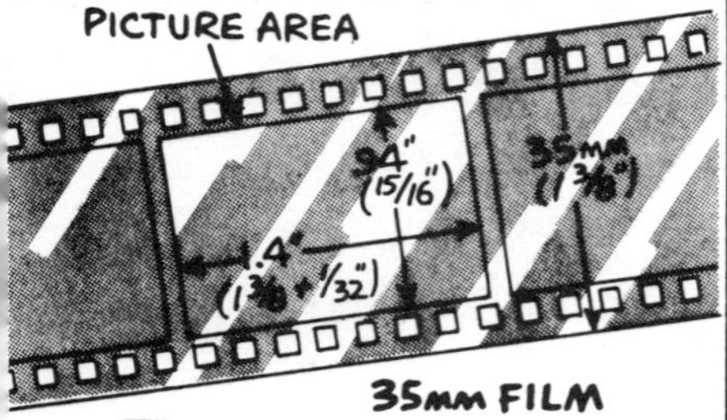

USE THIS TABLE TO DETERMINE IF F.L. YOU ARE USING WILL COVER THE OBJECT YOU PLAN TO SHOOT

Example 1 PLEIADES

THE ANGULAR SIZE CAN BE OBTAINED FROM STAR MAPS OR LISTS OF SKY OBJECTS, IT IS ABOUT 1½° (90')

ANGULAR FIELD COLUMN IN TABLE SAYS A 35 IN. F.L. LENS WILL COVER 92 MIN. HENCE, IT WILL COVER THE PLEIADES (90')

Example 2 PEOPLE

ALLOW 8-FT. FIELD FOR FULL-LENGTH, 4 FT. FOR HALF-LENGTH AND 2 FT. FOR HEAD-AND-SHOULDERS

SUPPOSE YOU WANT HALF-LENGTH PORTRAIT AT 300 FT. *RUN YOUR EYE DOWN THE 300 FT. COLUMN TO LOCATE NEEDED FIELD OF 4 FT. (48"). IN LEFT-HAND COLUMN READ 70" FOCAL LENGTH*

DIRECT CALCULATION FOR APPROX. IMAGE SIZE OF ANY SKY OBJECT WHEN F.L. AND ANGULAR SIZE ARE KNOWN

Example: M13 CLUSTER IN HERCULES, 10 MIN. DIA. F.L. – 300 IN. (10 × 60 = 600 SEC.)

WRITE DOWN ANGULAR SIZE IN SECONDS 600
DIVIDE BY 2 300
MULTIPLY BY F.L. 300
POINT OFF 5 PLACES .... .90000

IMAGE SIZE = .9 IN.

F.L. OR E.F.L. OF LENS	ANGULAR FIELD	LINEAR FIELD (SHORT SIDE OF FILM ONLY) [1] AT 100 FT.	AT 200 FT.	AT 300 FT.	AT 500 FT.	AT 1/4 M.	AT 1/2 M.	AT 1 M.	AT 5 M.
2"	26.5° x 39.7°	47 FT.	94 FT.	141 FT.	235 FT.	621 FT.	414 YD.	828 YD.	4140 YD.
3"	17.8° x 26.7°	31 FT.	63 FT.	94 FT.	157 FT.	413 FT.	827 FT.	551 YD.	2756 YD.
4"	13.4° x 20.1°	24 FT.	47 FT.	70 FT.	117 FT.	310 FT.	620 FT.	413 YD.	2066 YD.
5"	11.0° x 16.5°	19 FT.	39 FT.	58 FT.	97 FT.	255 FT.	509 FT.	339 YD.	1696 YD.
6"	9.0° x 13.5°	16 FT.	31 FT.	47 FT.	79 FT.	208 FT.	415 FT.	830 FT.	1383 YD.
7"	7.7° x 11.5°	13 FT.	27 FT.	40 FT.	67 FT.	177 FT.	354 FT.	708 FT.	1180 YD.
8"	6.7° x 10.1°	12 FT.	24 FT.	35 FT.	59 FT.	156 FT.	313 FT.	626 FT.	1043 YD.
9"	6.0° x 9°	10 FT.	21 FT.	31 FT.	52 FT.	138 FT.	275 FT.	550 FT.	917 YD.
10"	5.4° x 8.1°	9 FT.	19 FT.	28 FT.	47 FT.	125 FT.	249 FT.	498 FT.	830 YD.
12"	4.5° x 6.8°	8 FT.	16 FT.	23 FT.	39 FT.	103 FT.	206 FT.	412 FT.	687 YD.
14"	231' x 347'	7 FT.	13 FT.	20 FT.	33 FT.	88 FT.	177 FT.	353 FT.	589 YD.
16"	202' x 303'	6 FT.	12 FT.	18 FT.	29 FT.	77 FT.	155 FT.	309 FT.	516 YD.
18"	180' x 270'	5 FT.	10 FT.	16 FT.	26 FT.	69 FT.	137 FT.	275 FT.	458 YD.
20"	162' x 243'	4½ FT.	9 FT.	14 FT.	23 FT.	62 FT.	124 FT.	247 FT.	412 YD.
25"	129' x 194'	45 IN.	90 IN.	11 FT.	19 FT.	50 FT.	100 FT.	198 FT.	993 FT.
30"	108' x 162'	38 IN.	75 IN.	9 FT.	15 FT.	41 FT.	83 FT.	165 FT.	827 FT.
35"	92' x 138'	32 IN.	64 IN.	97 IN.	13 FT.	35 FT.	70 FT.	142 FT.	708 FT.
40"	81' x 122'	28 IN.	56 IN.	84 IN.	12 FT.	31 FT.	62 FT.	124 FT.	620 FT.
45"	72' x 108'	25 IN.	50 IN.	75 IN.	10 FT.	28 FT.	55 FT.	110 FT.	550 FT.
50"	65' x 98'	23 IN.	45 IN.	68 IN.	9 FT.	25 FT.	50 FT.	99 FT.	496 FT.
60"	53' x 80'	19 IN.	38 IN.	56 IN.	94 IN.	21 FT.	41 FT.	83 FT.	414 FT.
70"	46' x 69'	16 IN.	32 IN.	48 IN.	81 IN.	18 FT.	35 FT.	71 FT.	354 FT.
80"	40' x 60'	14 IN.	28 IN.	42 IN.	71 IN.	16 FT.	31 FT.	62 FT.	310 FT.
90"	36' x 54'	12.5 IN.	25 IN.	38 IN.	63 IN.	14 FT.	28 FT.	55 FT.	276 FT.
100"	32' x 48'	11 IN.	23 IN.	34 IN.	56 IN.	12 FT.	25 FT.	50 FT.	248 FT.
125"	26' x 39'	9 IN.	18 IN.	27 IN.	45 IN.	10 FT.	20 FT.	40 FT.	199 FT.
150"	22' x 33'	7.5 IN.	15 IN.	23 IN.	38 IN.	99 IN.	17 FT.	33 FT.	165 FT.
175"	18' x 27'	6.4 IN.	13 IN.	19 IN.	32 IN.	85 IN.	14 FT.	28 FT.	142 FT.
200"	16' x 24'	5.6 IN.	11 IN.	17 IN.	28 IN.	74 IN.	12 FT.	25 FT.	124 FT.
250"	13' x 20'	4.5 IN.	9 IN.	14 IN.	23 IN.	60 IN.	10 FT.	20 FT.	99 FT.
300"	11' x 16'	3.8 IN.	8 IN.	11 IN.	19 IN.	50 IN.	99 IN.	17 FT.	83 FT.
400"	8' x 12'	2.8 IN.	5.6 IN.	8 IN.	14 IN.	37 IN.	74 IN.	12 FT.	62 FT.
500"	6.4' x 9.6'	2.3 IN.	4.5 IN.	6.8 IN.	11 IN.	30 IN.	60 IN.	10 FT.	50 FT.
600"	5.4' x 8'	1.9 IN.	3.8 IN.	5.6 IN.	9 IN.	25 IN.	50 IN.	99 IN.	41 FT
700"	4.6' x 7'	1.6 IN.	3.2 IN.	4.8 IN.	8 IN.	21 IN.	43 IN.	85 IN.	35 FT
800"	4.0' x 6'	1.4 IN.	2.8 IN.	4.2 IN.	7 IN.	19 IN.	37 IN.	74 IN.	31 FT
900"	3.6' x 5.4'	1.25 IN.	2.5 IN.	3.8 IN.	6 IN.	17 IN.	33 IN.	66 IN.	28 FT.
1000"	3.2' x 4.8'	1.13 IN.	2.3 IN.	3.4 IN.	5.6 IN.	15 IN.	30 IN.	60 IN.	25 FT.

[1] *THE LINEAR FIELD OF LONG SIDE OF 35mm FILM CAN BE CALCULATED BY MULTIPLYING FIELD OF SHORT SIDE (FROM TABLE) BY 1.5*

APPROX. FIELD OF OTHER FILM SIZES: *USE ANGULAR OR LINEAR FIELD GIVEN FOR SHORT SIDE OF 35mm FILM AND MULTIPLY BY THE SIZE OF YOUR FILM IN INCHES*

Fig. 14, is the common mounting method. It can be seen that the camera is bolted in place, and the camera in turn supports the objective lens. With focal lengths over 10 inches, it is more practical to mount the lens tube, and the tube then supports the camera. Fig. 15 is a typical mount of this kind; the adjustable-position feature is not for alignment since the camera can point in any direction, but simply to give you some choice of guide stars. All of this is explained in the chapter, "Shooting the Stars."

In the way of auxiliary lenses, the usual choice is standard anastigmats up to about 6 or 7 inches focal length. Over 8 inches f.l., achromats begin to function very well because of the comparatively narrow field angle involved. A 14 in. f.l. objective on a 35mm camera looks at a 7-degree field, and for this narrow angle an achromat is nearly as good as the most expensive anastigmat. Of course you don't get speed, the usual rating being about f/8. However, with the fast films now available, f/8 is more than enough aperture for most objects. Fig. 16 details the construction of a 14-inch f.l. telecamera. This is a very practical size for many land and sky objects; it can be hand-held for land objects if you are shooting at 1/500 second or faster. Its 7-degree field is just about the right size for many popular star groups. Fitted with adapter and 1-inch eyepiece, its 14x power is about all you can hold for rich field star-gazing. Coupled with a Barlow working at 3x, it zooms up to 42 inches equivalent focal length.

FIELD AND SCALE. With land objects, the photographer can get wide field or narrow field, big scale or small scale--all by the simple process of changing his position in relation to the target object. Not so in the sky. Every object in outer space has a fixed angular field, and your only control is by changing objectives. If you want to shoot wide-angle star fields, you need objectives of short focal length; if you want big images of small objects, you need a long f.l. glass or its equivalent in a compound optical system. You can get the angular size of any popular sky object from almost any list or catalog of stars. The rest is a matter of selecting the proper focal length. Sometimes you will be able to shoot direct with the telescope lens or mirror; other objects may require an auxiliary lens of shorter focal length. All of the needed data can be obtained from Tables 1 and 2. Image size is proportional to focal length; if you want data for, say, 14 inch f.l., simply read the values for 140 inch f.l. and point off one decimal place to the left.

TABLE 2

IMAGE SIZE

F.L. or E.F.L.	ANGULAR SIZE (SECONDS OR MINUTES OF ARC)								
	20"	40"	1'	5'	10'	30'	31'	40'	60'
10"	.001	.002	.003	.015	.029	.087	.090	.12	.18
20"	.002	.004	.006	.029	.058	.17	.18	.23	.35
30"	.003	.006	.009	.044	.087	.26	.27	.35	.52
40"	.004	.008	.012	.058	.12	.35	.36	.47	.70
50"	.005	.010	.015	.073	.15	.44	.45	.58	.87
60"	.006	.012	.017	.088	.18	.52	.54	.70	1.05
70"	.007	.014	.020	.10	.20	.61	.63	.81	1.22
80"	.008	.016	.023	.12	.23	.70	.72	.93	1.40
90"	.009	.018	.026	.13	.26	.79	.81	1.05	1.57
100"	.010	.020	.029	.15	.29	.87	.90	1.16	1.75
120"	.012	.024	.034	.17	.35	1.05	1.08	1.40	2.10
140"	.014	.028	.041	.20	.41	1.22	1.26	1.63	2.44
160"	.016	.032	.047	.23	.47	1.39	1.44	1.86	2.79
180"	.018	.036	.052	.26	.52	1.57	1.62	2.10	3.14
200"	.019	.039	.058	.29	.58	1.74	1.80	2.33	3.49
225"	.022	.044	.066	.33	.66	1.94	2.03	2.62	3.93
250"	.024	.048	.073	.37	.73	2.18	2.26	2.91	4.36
275"	.027	.053	.080	.40	.80	2.40	2.48	3.20	4.80
300"	.029	.058	.087	.44	.87	2.62	2.71	3.49	5.24
400"	.039	.078	.12	.58	1.16	3.49	3.61	4.66	6.98
500"	.049	.097	.15	.73	1.45	4.36	4.51	5.82	8.73
600"	.058	.12	.18	.88	1.75	5.24	5.41	6.98	10.48
700"	.068	.14	.20	1.02	2.04	6.11	6.31	8.15	12.22
800"	.078	.16	.23	1.16	2.33	6.98	7.22	9.31	13.97
900"	.087	.18	.26	1.31	2.62	7.85	8.12	10.48	15.71
1000"	.097	.19	.29	1.46	2.91	8.73	9.02	11.64	17.46

ALL DIMENSIONS IN INCHES

E.F.L. - EQUIVALENT FOCAL LENGTH

▲ *USE THIS COL. FOR SUN OR MOON*

ANGULAR SIZE OF SOME

ASTRO OBJECTS

SUN	32' *	M8 - LAGOON	60'
MOON	31' *	M13 - HERC. CLUSTER	10'
MOON OBJECTS	APPROX 1 SEC. PER MILE	M27 - DUMBBELL	8'
JUPITER	40" *	M31 - ANDR. NEB.	160'
SATURN BALL / RING	18" / 42" *	M42 - ORION NEB.	60'
VENUS	20" *	M44 - BEEHIVE	90'
MARS	8" *	M45 - PLEIADES	120'
M1 - CRAB NEB.	6'	M57 - RING NEB.	1½'

*AVERAGE

...for any camera

The AFOCAL SYSTEM

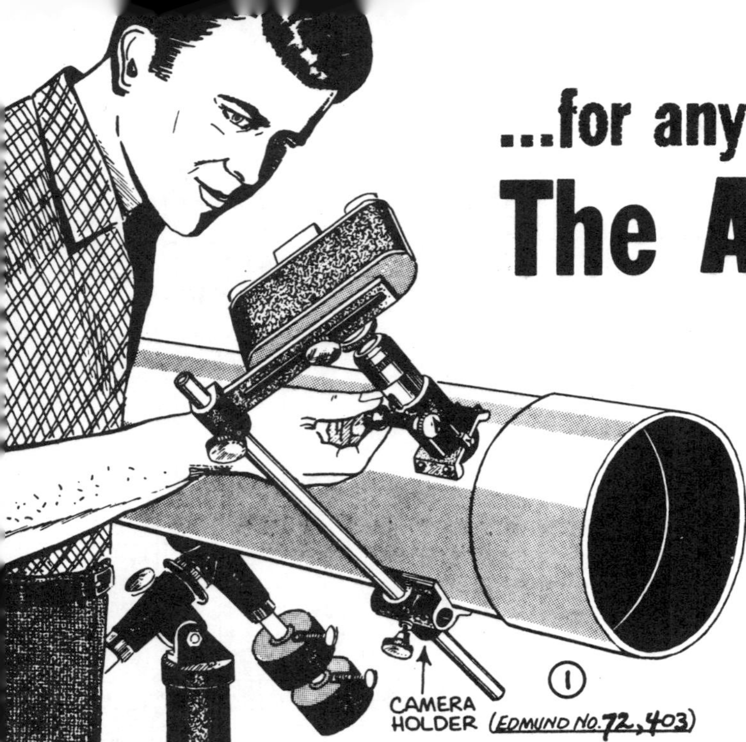

THE AFOCAL system consists of a telescope in front of a camera. It gets its name from the fact that the telescope eyepiece is located exactly its own focal length from the first image, while the camera lens is exactly its own focal length from the second image. Technically, the system is one of projection, the only distinction being that it is an afocal projection system as differing from other projection systems where object and image distances do not match the focal lengths of the lenses used. A telescope in focus for eye use is in afocal adjustment. This fact makes it possible to shoot pictures with a small telescope or binoculars in combination with a "blind" camera by simply focusing the telescope by eye and then putting it in front of the blind camera which is set at infinity. However, the precise focusing needed for long focal length telescopes is much too critical to permit this kind of guesswork.

Figs. 1 and 3 show a typical system. The telescope eyepiece and the camera lens must be in fair alignment, but they are not coupled. The small intervening space need not even be enclosed unless you are shooting outdoors in sunlight. The camera must be supported externally and gadgets for this purpose can be purchased or made as desired. The equivalent focal length of the whole system is the power of the telescope multiplied by the focal length of the camera lens. While the example shown in Fig. 3 is quite modest, it will be apparent that terrific magnification

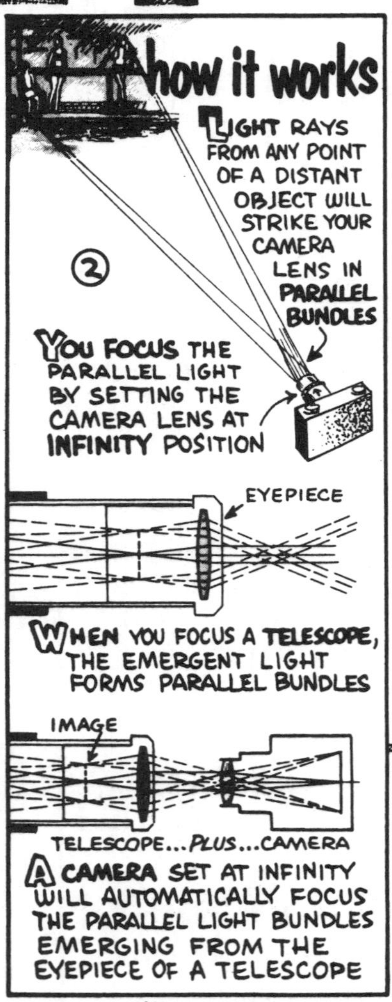

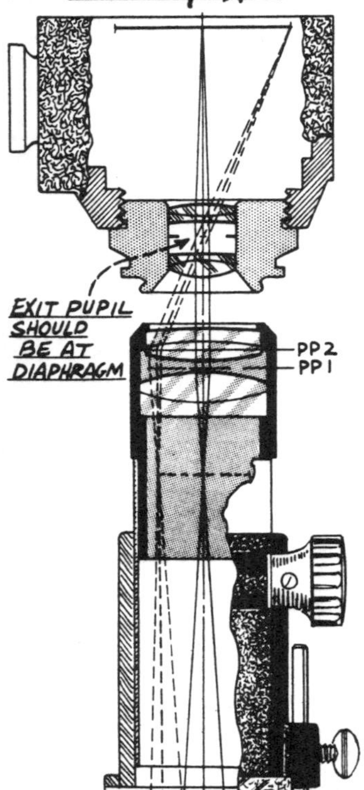

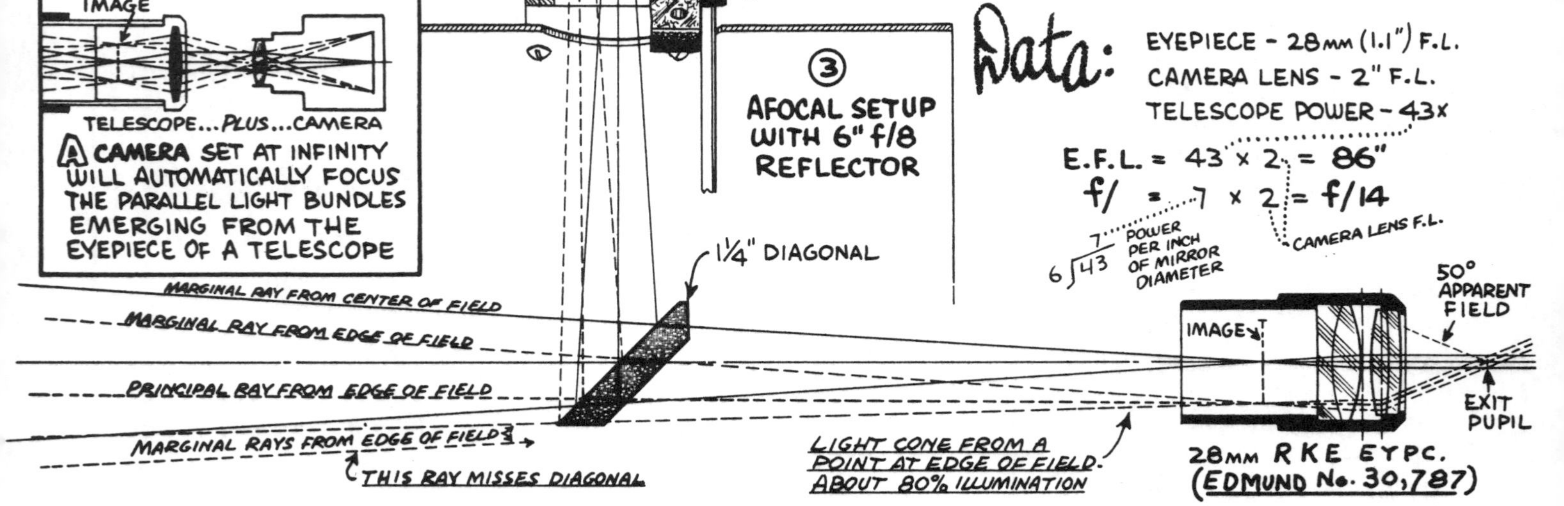

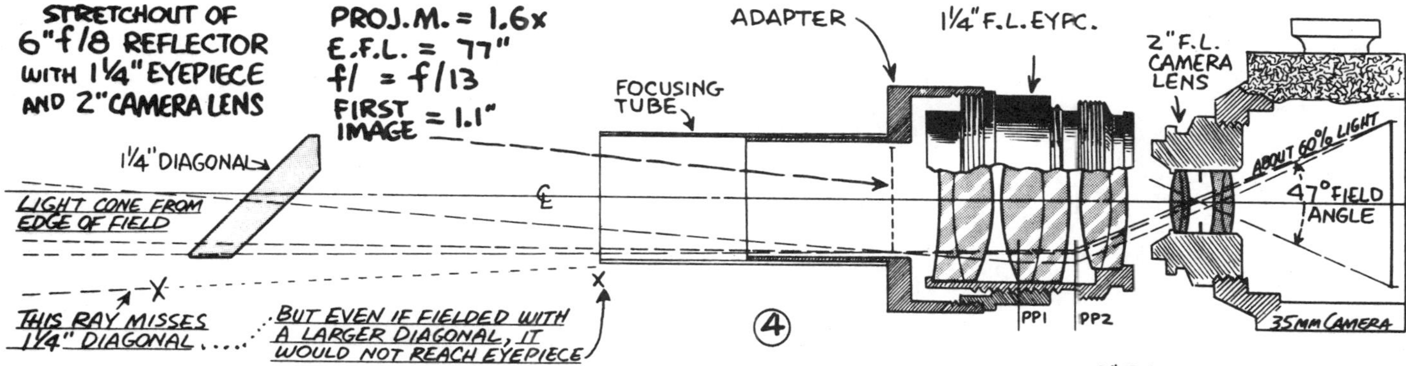

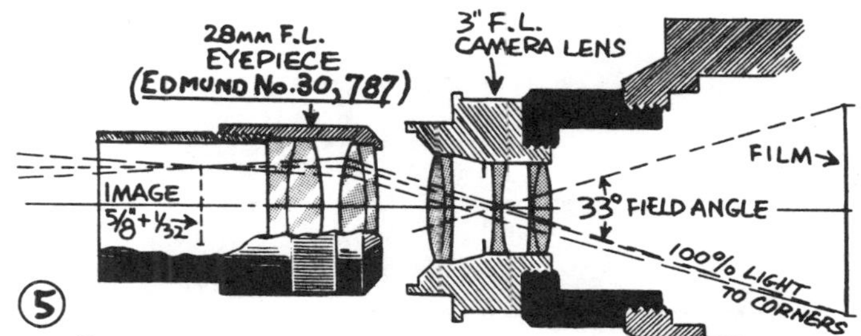

A 3" CAMERA LENS REDUCES FIELD ANGLE

E.F.L. = 129"
f/ = f/22 (WITH 6" f/8 REFLECTOR)

can be obtained by using an eyepiece of short focal length in combination with a long f.l. lens on the camera.

AFOCAL CALCULATIONS. Afocal systems are easy to calculate by using the formulas given in Table 3. Usually the power of the telescope with a certain eyepiece will be known, and in such case the calculations are based on this specification. However, it is often easier to calculate the system from the amount of Projection Magnification, and equations for doing this are given in the lower half of the table. In an afocal system, magnification is obtained by having the rear lens (the camera lens) of longer focal length than the front lens (the eyepiece). If your camera lens is 2 in. focal length and you use a 1 in. eyepiece, the Projection M. is 2x. When the eyepiece is the same focal length as the camera lens, the magnification is unity or 1x, that is, same size, no magnification.

FIELD ANGLE. In all afocal systems, the apparent field angle of the eyepiece should equal or exceed the field angle of the camera. If you are using a standard 35mm camera with 2-inch focal length objective, the field angle will be a maximum of 47 degrees, Fig. 4. Now, since what goes into the camera must come out of the eyepiece, you can see that light rays emerging from the eyepiece must embrace an angle of at least 47 degrees. If the eyepiece has a smaller apparent field than 47 degrees, it will not cover the 47 degree angle required by the film. This situation occurs frequently since most Huygens and Ramsden eyepieces, many Symmetricals and some Kellners will have no more than 40 degrees apparent field. The result will be slight vignetting at the corners of the film. Even when the eyepiece has the required wide field, vignetting will occur at 1.6x or less projection magnification because the small-diameter focusing tube will not field the big primary image which

TABLE 3

AFOCAL Calculations

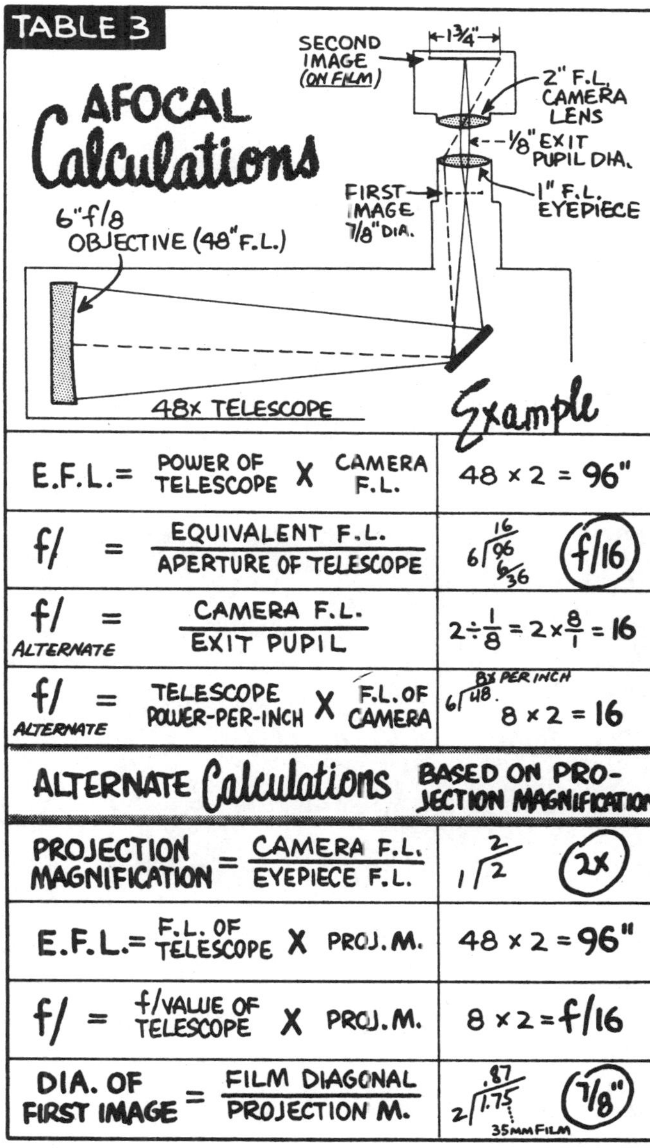

	Example
E.F.L. = POWER OF TELESCOPE X CAMERA F.L.	48 × 2 = 96"
f/ = EQUIVALENT F.L. / APERTURE OF TELESCOPE	96 ÷ 6 = 16 (f/16)
f/ (ALTERNATE) = CAMERA F.L. / EXIT PUPIL	$2 \div \frac{1}{8} = 2 \times \frac{8}{1} = 16$
f/ (ALTERNATE) = TELESCOPE POWER-PER-INCH X F.L. OF CAMERA	48 ÷ 6 = 8 PER INCH; 8 × 2 = 16
ALTERNATE Calculations BASED ON PROJECTION MAGNIFICATION	
PROJECTION MAGNIFICATION = CAMERA F.L. / EYEPIECE F.L.	2 ÷ 1 = 2 (2x)
E.F.L. = F.L. OF TELESCOPE X PROJ.M.	48 × 2 = 96"
f/ = f/VALUE OF TELESCOPE X PROJ.M.	8 × 2 = f/16
DIA. OF FIRST IMAGE = FILM DIAGONAL / PROJECTION M.	1.75 (35MM FILM) ÷ 2 = .87 (7/8")

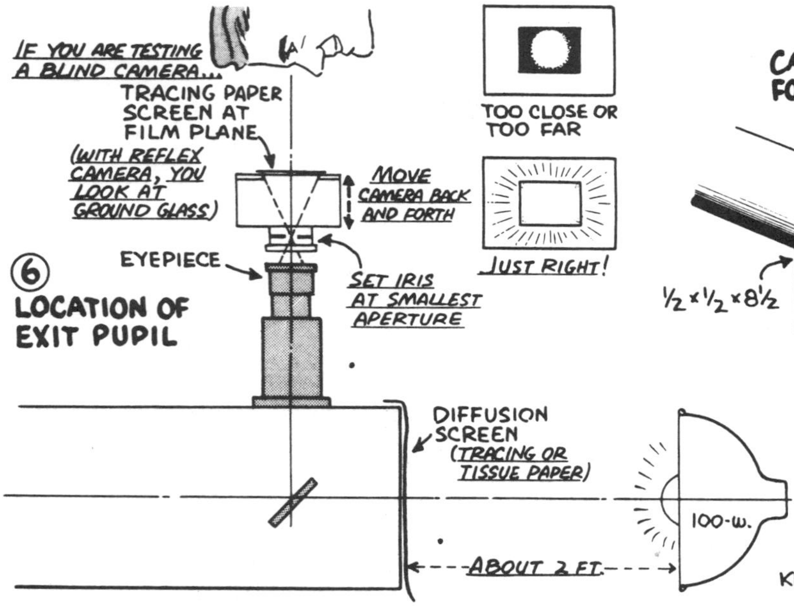

⑥ **LOCATION OF EXIT PUPIL**

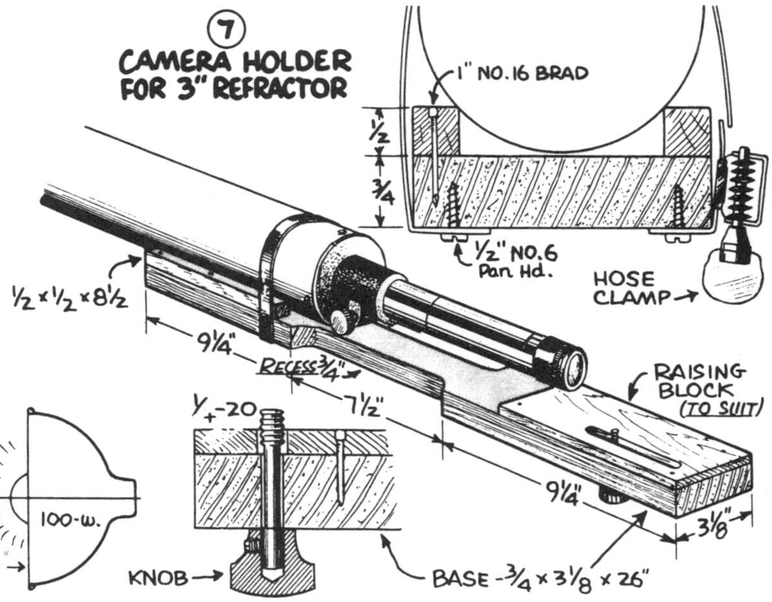

⑦ **CAMERA HOLDER FOR 3" REFRACTOR**

goes with low power. Fig. 4 is an example. On the other hand, high-power means that the first image will be small and the focusing tube is no longer a bottleneck.

If your camera permits changing lenses, it is sometimes advantageous to use a slightly longer focal length, Fig. 5. As can be seen, this reduces the field angle of the camera, which in turn means that the eyepiece has an easy job. Such a system works with no vignetting whatever, and also is non-critical as regards the location of the exit pupil of the telescope.

LOCATION OF EXIT PUPIL. The space between eyepiece and camera is optically free space, meaning that a little more or less space will have no effect on the power of the system or the location of the final image. But the camera lens must pass all of the required cone of light, and the best insurance to meet this requirement is to locate the exit pupil of the telescope at or near the iris of the camera lens. The proper position is obtained automatically with a simple visual test, Fig. 6. If you push the camera back and forth, you will note that the corners of the ground glass (or tracing paper screen) will show black when the camera is too close or too far from the eyepiece. Determine by this test some position where the lighting is uniform over the whole film area; make a note of the spacing for future use in mounting the camera.

Most shooters will use the afocal system with

BIG PLATE SCALE... IS GRAPHICALLY PORTRAYED WHEN REGULAR 2-INCH F.L. OF 35MM CAMERA LENS IS COMPARED WITH SAME VIEW SEEN WITH A 6-INCH REFLECTOR TELESCOPE OF 48" F.L. ALL PICS TAKEN FROM INDOORS THRU OPEN WINDOW. THE SCENE, ABOVE, 1/200 SECOND AT f/11 ON PLUS-X FILM

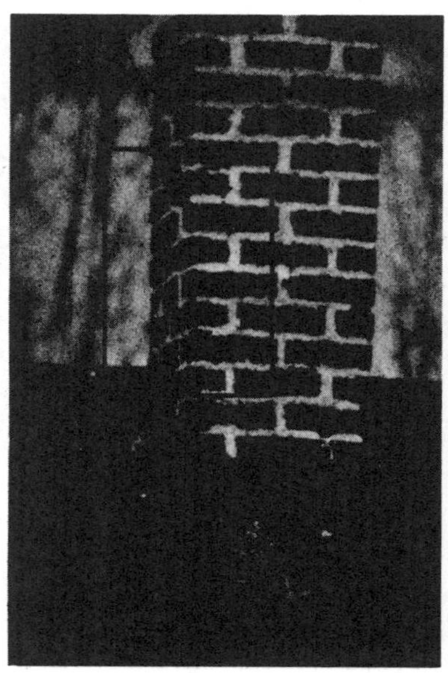

6" REFLECTOR AT FOCUS. PANATOMIC-X, 1/100 SEC. AT f/8

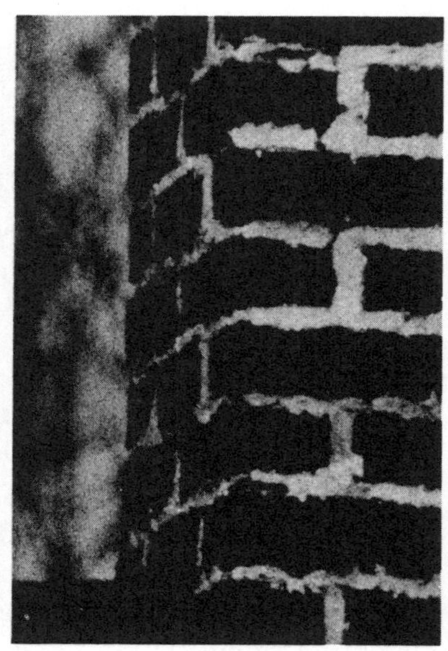

2.7x BARLOW (GOODWIN) ON 6-INCH REFLECTOR. 1/10 SEC. AT f/22

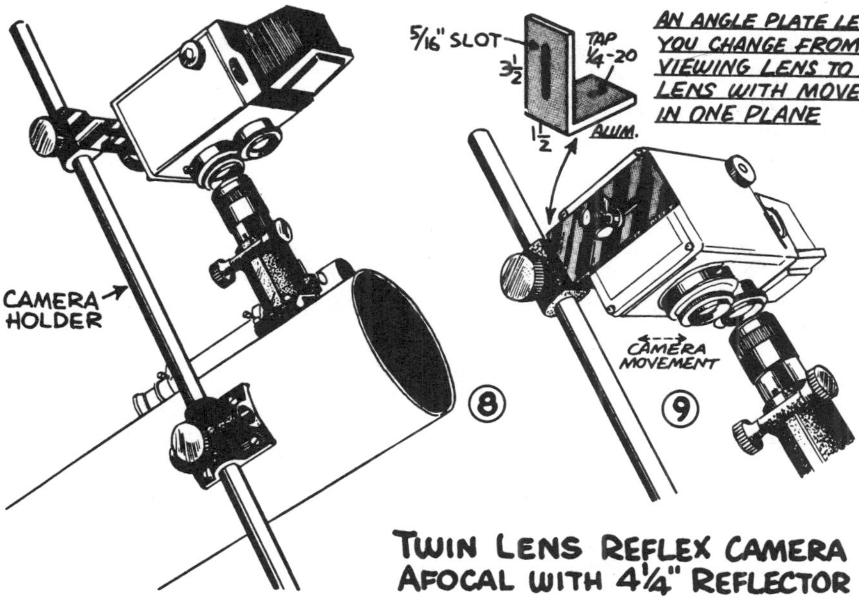

TWIN LENS REFLEX CAMERA AFOCAL WITH 4¼" REFLECTOR

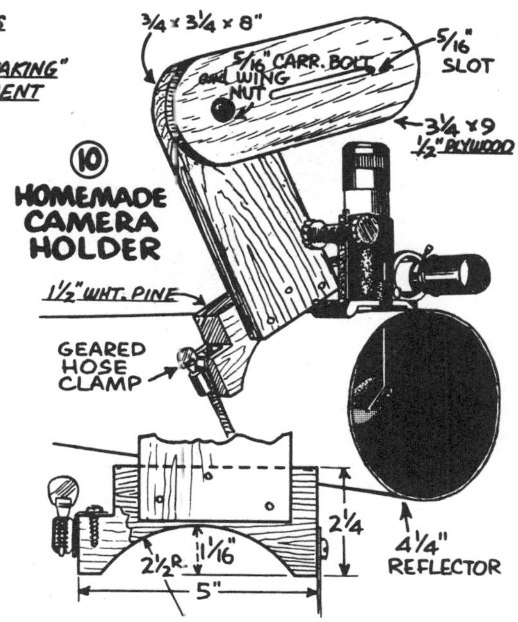

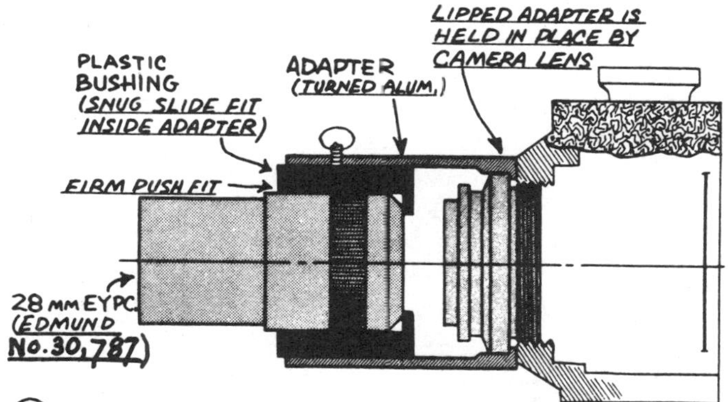

⑪ **1.8× COUPLED AFOCAL ATTACHMENT**

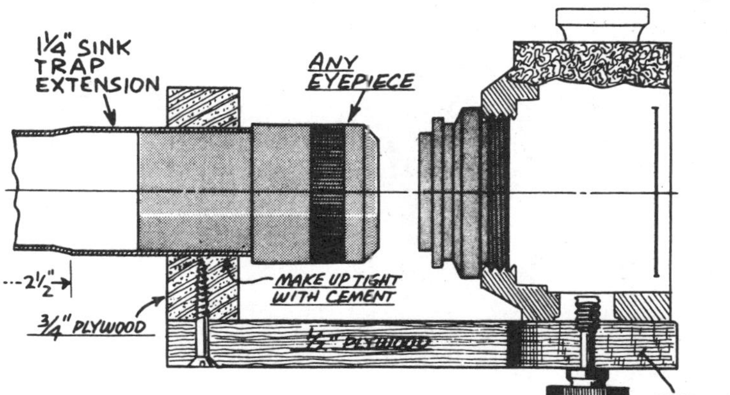

⑫ **COUPLED AFOCAL ON OPEN BASE**

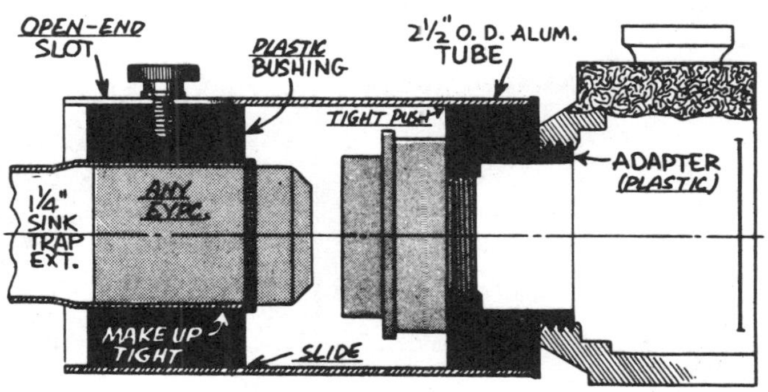

⑬ **COUPLED AFOCAL WITH 3" F.L. CAMERA LENS**

the iris wide open, just as insurance that all the light gets through. However, if you have the exit pupil exactly at the iris, you can stop down as desired. If you can stop down below the rated f/value of the telecamera, the new f/value can be read directly from the f/scale. As noted, however, you must have the exit pupil exactly at the iris and concentric with it to make use of this feature. Probably the best standard practice is to stop down to about f/8. The idea, of course, is simply that the stopped-down iris makes a good glare stop.

MOUNTING THE CAMERA. If you are using a refractor, a simple camera mount is a board clamped to the main tube, Fig. 7. A twin lens job is more of a problem, since with ordinary mounting bracket, Fig. 8, it is clumsy and time-consuming to set the camera for the viewing and taking positions. One partial solution is to use an angle plate, Fig. 9, which confines the needed movement to a single plane. The single lens reflex is easy to mount and almost any kind of wood or metal bracket can be used; Fig. 10 design is pivoted at the center in order to get around the finder which is usually in the way of a straight support made of wood.

If you like the afocal system, coupled mounting is worth consideration. The idea here is that the combined eyepiece and camera is free-standing, supported only by the focusing tube in much the same manner as a long and bulky eyepiece. A nice feature of this setup is that you can rotate the camera as needed to square the picture with the side of the film; with a fixed bracket, the only way you can do this is by rotating the telescope.

positive and negative PROJECTION Systems

ALMOST any telescope can be used as a projection system by merely extending the eyepiece a little from the normal afocal position. Even the Galilean telescope may be used in this manner, although the negative lens projection is more generally recognized as a Barlow or telephoto system. The usual positive projecting eyepiece is simply a telescope eyepiece--it gets tagged as a "projection" eyepiece only from the manner in which it is used. Regular slide projector lenses are sometimes used as well as short focal length camera lenses. The most practical focal lengths are the same as for visual use--in the neighborhood of 1 inch--and eyepieces can be used interchangeably for looking and shooting. The comfortable magnifying range runs from 2x to 6x projection magnification. This builds up the equivalent focal length rather quickly since the telescope itself is a long focus lens. With a 6-inch f/8 reflector of 48 in. focal length, a 3x projection system makes the e.f.l. 144 inches.

PROJECTION WITH A POSITIVE LENS

In all projection systems, magnification is obtained by making the image distance greater than the object distance. In the telescope setup, the "object" to be projected is the real image formed by the telescope objective, while the "image" is the real second image formed on the film. Fig. 1 shows a basic type of lens spacing which results in an image the same size as the object. When the lens or eyepiece is one focal length from the object, no real image is formed, Fig. 2. This is the afocal position, the way an eyepiece is adjusted for visual use. Between

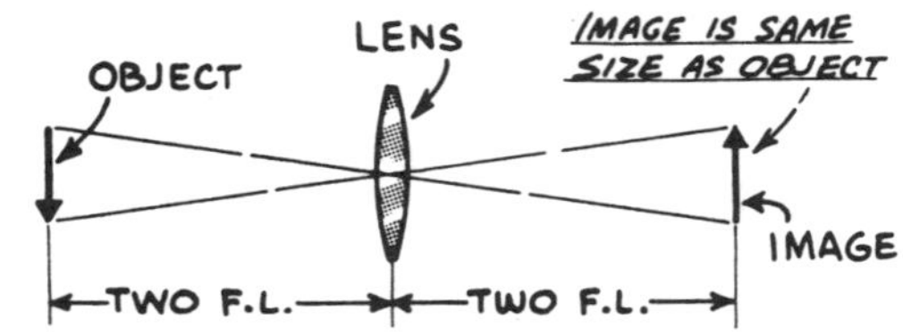

① UNITY (1x) MAGNIFICATION

② AFOCAL (NO IMAGE)

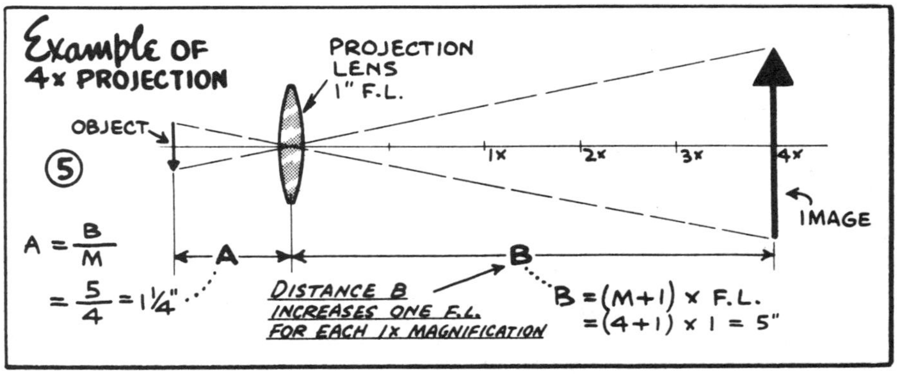

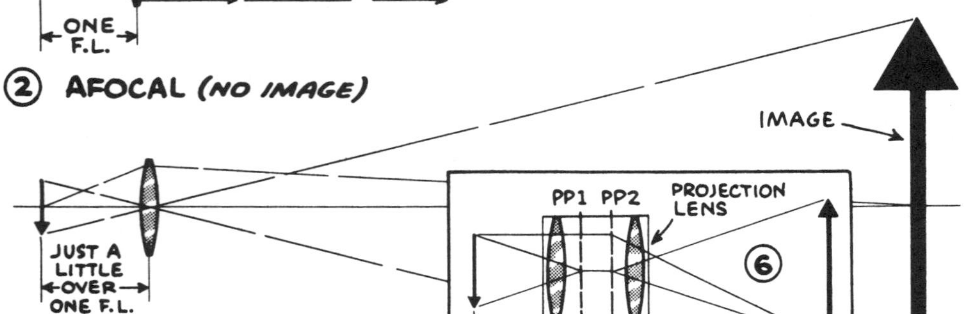

③ GREATEST MAGNIFICATION

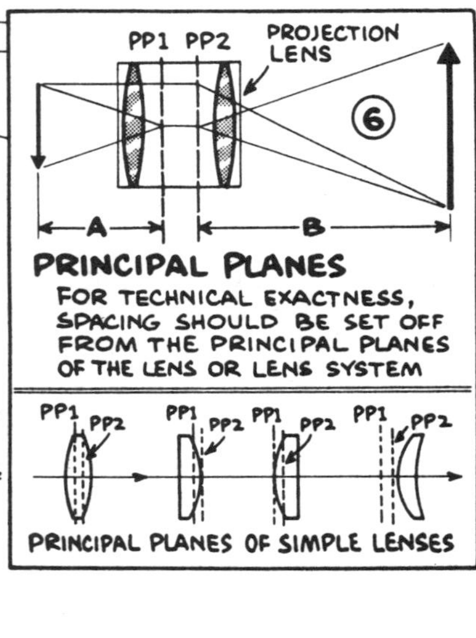

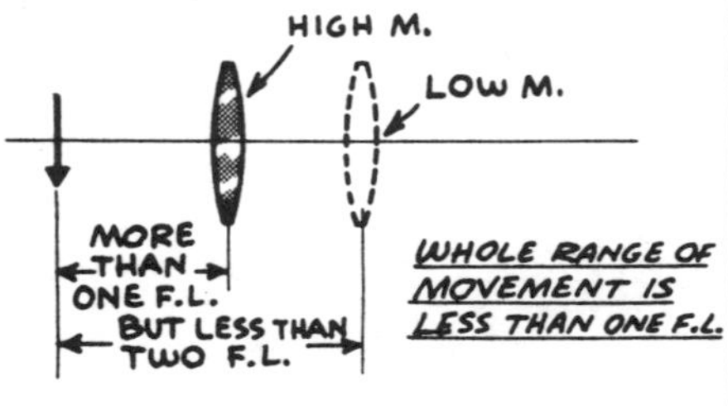

④ WHOLE RANGE

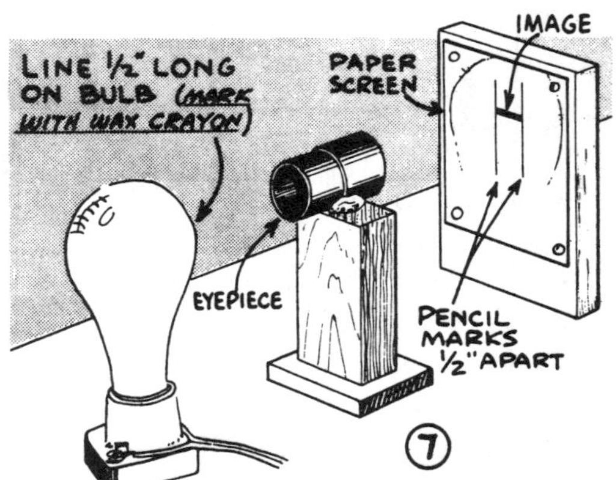

⑦ LOCATING THE PRINCIPAL PLANES

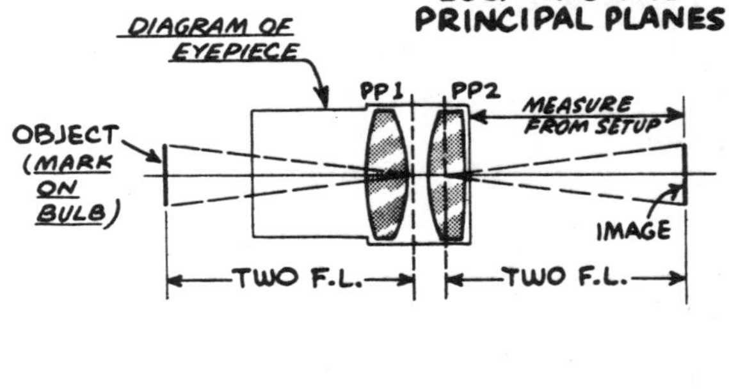

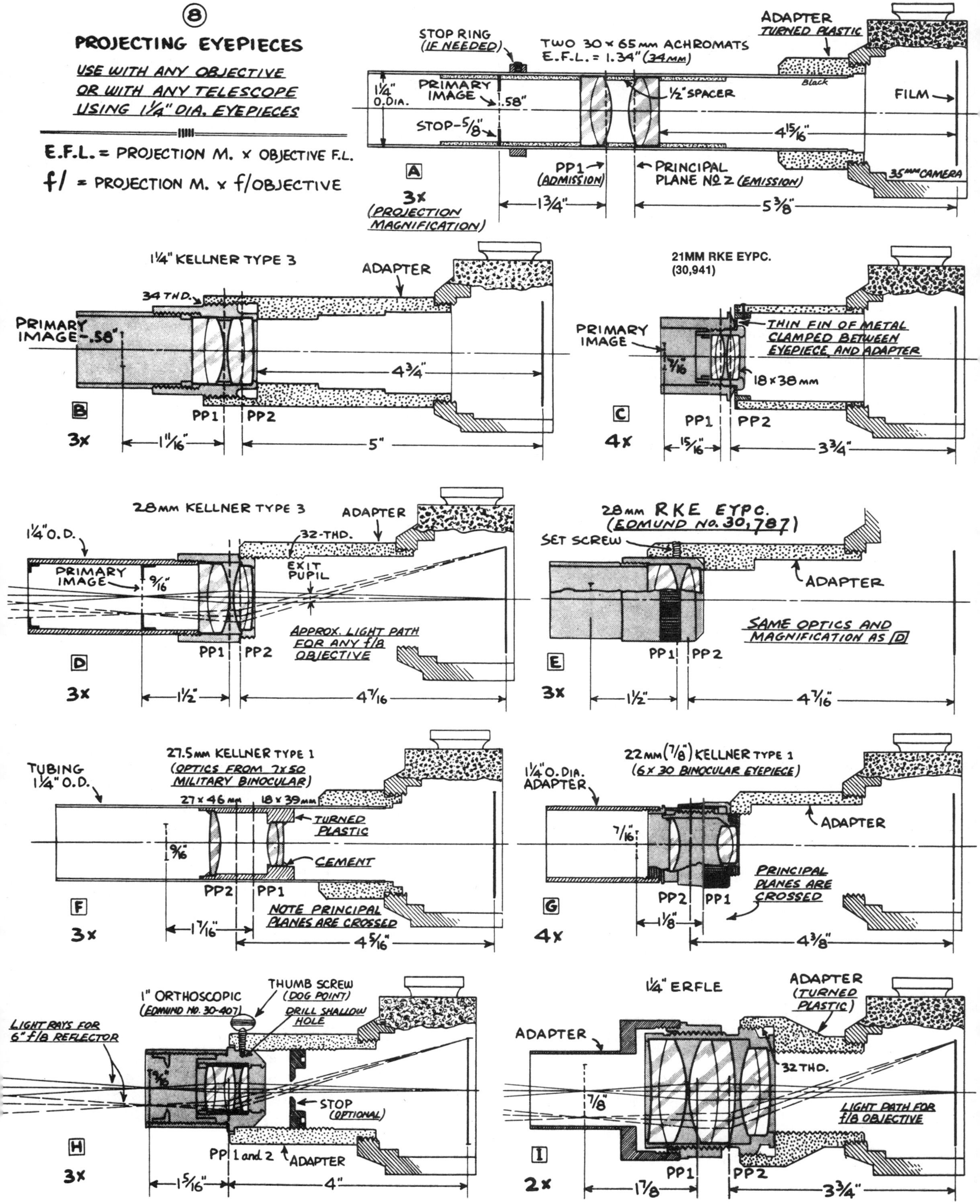

8
PROJECTING EYEPIECES
USE WITH ANY OBJECTIVE OR WITH ANY TELESCOPE USING 1¼" DIA. EYEPIECES
E.F.L. = PROJECTION M. x OBJECTIVE F.L.
f/ = PROJECTION M. x f/OBJECTIVE
STOP RING (IF NEEDED)
TWO 30 x 65 mm ACHROMATS E.F.L. = 1.34" (34 mm)
ADAPTER TURNED PLASTIC
Black
1¼" O.DIA.
PRIMARY IMAGE
.58"
½" SPACER
FILM
STOP-5/8"
4 15/16"
PP1 (ADMISSION)
PRINCIPAL PLANE No 2 (EMISSION)
35mm CAMERA
A
3x (PROJECTION MAGNIFICATION)
1¾"
5⅜"
1¼" KELLNER TYPE 3
ADAPTER
34 THD.
PRIMARY IMAGE-.58"
4¾"
B
PP1
PP2
3x
1 11/16"
5"
21MM RKE EYPC. (30,941)
PRIMARY IMAGE
THIN FIN OF METAL CLAMPED BETWEEN EYEPIECE AND ADAPTER
7/16"
18 x 38 mm
C
PP1
PP2
4x
15/16"
3¾"
28mm KELLNER TYPE 3
ADAPTER
1¼" O.D.
32-THD.
PRIMARY IMAGE
9/16"
EXIT PUPIL
APPROX. LIGHT PATH FOR ANY f/8 OBJECTIVE
D
PP1
PP2
3x
1½"
4 7/16
28mm RKE EYPC. (EDMUND No. 30,787)
SET SCREW
ADAPTER
SAME OPTICS AND MAGNIFICATION AS D
E
PP1
PP2
3x
1½"
4 7/16"
TUBING 1¼" O.D.
27.5mm KELLNER TYPE 1 (OPTICS FROM 7x50 MILITARY BINOCULAR)
27 x 46 mm
18 x 39 mm
TURNED PLASTIC
9/16"
CEMENT
PP2
PP1
NOTE PRINCIPAL PLANES ARE CROSSED
F
3x
1 7/16"
4 5/16"
1¼" O. DIA. ADAPTER
22mm (7/8") KELLNER TYPE 1 (6x30 BINOCULAR EYEPIECE)
ADAPTER
7/16"
PRINCIPAL PLANES ARE CROSSED
PP2
PP1
G
4x
1⅛"
4⅜"
LIGHT RAYS FOR 6" f/8 REFLECTOR
1" ORTHOSCOPIC (EDMUND No. 30-407)
THUMB SCREW (DOG POINT)
DRILL SHALLOW HOLE
3/16"
STOP (OPTIONAL)
PP 1 and 2
ADAPTER
H
3x
1 5/16"
4"
1¼" ERFLE
ADAPTER
ADAPTER (TURNED PLASTIC)
32 THD.
7/8"
LIGHT PATH FOR f/8 OBJECTIVE
I
PP1
PP2
2x
1⅞
3¾"

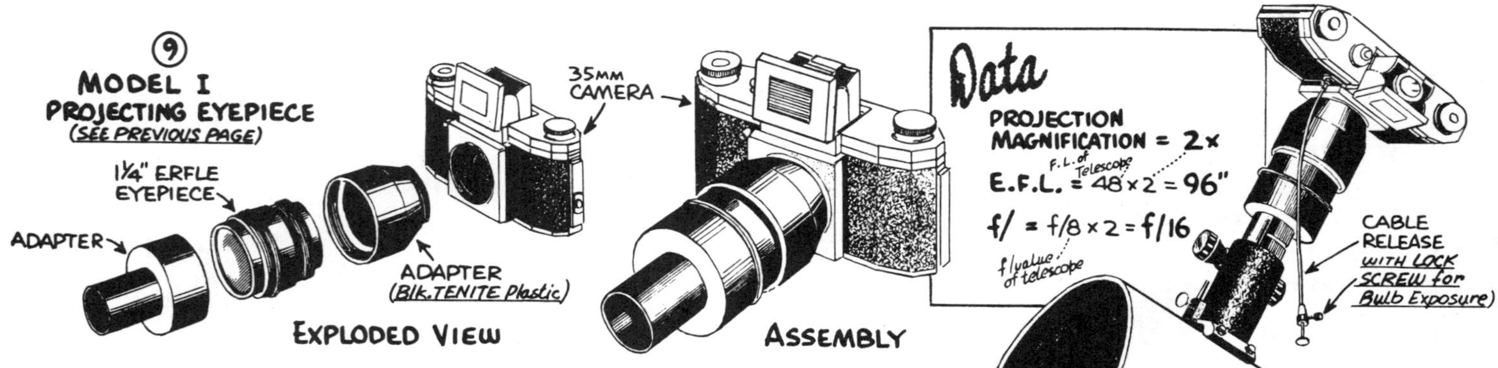

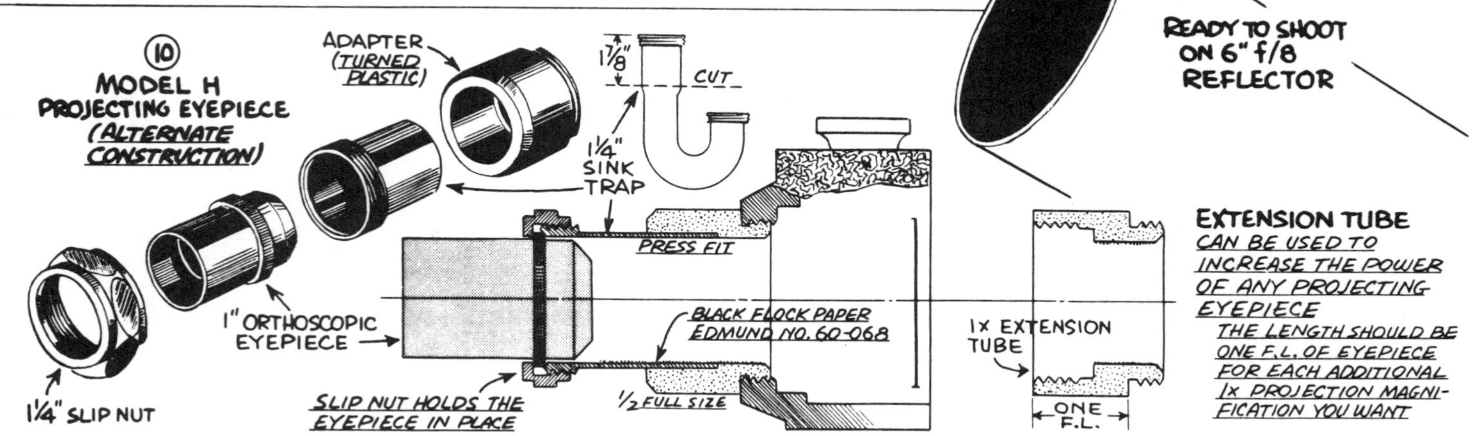

the 1x position and the afocal position it is possible to obtain a complete range of magnifications, from no magnification to infinite magnification--all with a lens adjustment of less than one focal length, Fig. 4.

The required spacing for any degree of projection magnification can be calculated from the simple formulas given in Fig. 5. Dimension B is the controlling dimension--it is the same as the "throw" distance of a slide projector. Being a comparatively long dimension, it is not unduly affected by slight errors. In the usual practice, spacing distances are set off from the center of the lens or lens system, but if you want to be technically correct, the spacing should be measured from the principal planes, Fig. 6. If you know the focal length of any lens or eyepiece, you can locate the principal planes with the simple setup shown in Fig. 7. The general idea is to juggle the eyepiece and screen back and forth until the image formed on the screen is exactly in focus and exactly 1/2 inch long, the same as the object. Then, if you measure two focal lengths from the object, you will locate PP1, the plane of admission. Since the setup is 1x, the plane of emission, PP2, is also two focal lengths distance from the image. Once known, the principal planes of any eyepiece can be used as measuring points for any application of the eyepiece. If you make light ray diagrams, draw the rays to PP1, then parallel with the axis to PP2, and then to the image. Such rays do not show the actual path through the glass, but are accurate as regards entering and exit surfaces, which is all you need to know.

PROJECTING EYEPIECES. Typical projecting eyepieces are shown in Fig. 8. These are standard telescope eyepieces, the adaptation being mainly a matter of an extension tube which sets the eyepiece a fixed distance from the film plane. You can change the magnification of any design by simply changing the spacing. Low power is the most difficult to obtain since this demands a big field lens. Of the designs shown, only Model I can work at less than 2x.

Fig. 9 shows the construction and mounting of Model I projecting eyepiece. Like most projecting systems, this is a compact unit requiring no additional support other than the normal mounting in the focusing tube of the telescope. Fig. 10 shows an alternate construction for Model H projecting eyepiece which you may prefer to drilling a hole in the eyepiece.

PROJECTION CALCULATIONS. Much of the arithmetic connected with projection systems can be eliminated by using Table 4 which gives spacing distances for most of the lenses you are likely to use. If you use some focal length not listed, find, from table, the spacing for a 1-inch f.l. lens at the desired magnification. Then, multiply these values by the f.l. of the lens you plan to use. Alternately, use Table 5, which

TABLE 4

OBJECT-IMAGE SPACING for Projection Systems

PROJECTION WITH A POSITIVE LENS

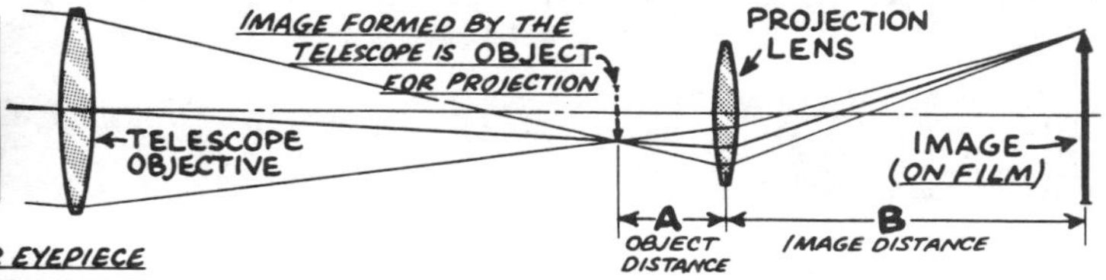

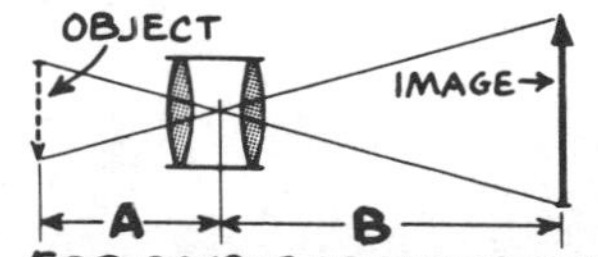

FOR DOUBLE LENSES OR ANY KIND OF EYEPIECE, MEASURE FROM CENTER (SEE TEXT)

FOCAL LENGTH OF PROJECTION LENS OR EYEPIECE

F.L.		1¼x	1½x	2x	3x	4x	5x	6x	7x	8x	9x	10x	12x	15x	20x
3/8" 9.5mm	A→	.67"	.62"	.56"	.50"	.47"	.45"	.44"	.43"	.42"	.42"	.41"	.40"	.40"	.39"
	B→	.84	.93	1.12	1.50	1.87	2.25	2.62	3.00	3.37	3.75	4.12	4.87	6.00	7.87
1/2" 12.7mm	A→	.90	.83	.75	.67	.62	.60	.58	.57	.56	.55	.55	.54	.53	.52
	B→	1.12	1.25	1.50	2.00	2.50	3.00	3.50	4.00	4.50	5.00	5.50	6.50	8.00	10.50
5/8" 15.9mm	A→	1.12	1.04	.93	.83	.78	.75	.73	.71	.70	.69	.68	.67	.66	.65
	B→	1.40	1.56	1.87	2.50	3.12	3.75	4.37	5.00	5.62	6.25	6.87	8.12	10.00	13.12
3/4" 19.1mm	A→	1.34	1.25	1.12	1.00	.94	.90	.88	.86	.84	.83	.82	.81	.80	.79
	B→	1.68	1.87	2.25	3.00	3.75	4.50	5.25	6.00	6.75	7.50	8.25	9.75	12.00	15.75
7/8" 22.2mm	A→	1.58	1.46	1.31	1.16	1.09	1.05	1.02	1.00	.98	.97	.96	.95	.93	.92
	B→	1.97	2.19	2.63	3.50	4.38	5.25	6.13	7.00	7.87	8.75	9.62	11.37	14.00	18.37
1" 25.4mm	A→	1.80	1.66	1.50	1.33	1.25	1.20	1.17	1.14	1.13	1.11	1.10	1.08	1.07	1.05
	B→	2.25	2.50	3.00	4.00	5.00	6.00	7.00	8.00	9.00	10.00	11.00	13.00	16.00	21.00
1⅛" 28.6mm	A→	2.02	1.87	1.68	1.50	1.40	1.35	1.31	1.28	1.27	1.25	1.24	1.22	1.20	1.18
	B→	2.53	2.81	3.37	4.50	5.62	6.75	7.87	9.00	10.13	11.25	12.37	14.62	18.00	23.62
1¼" 31.8mm	A→	2.25	2.08	1.87	1.67	1.56	1.50	1.46	1.43	1.41	1.39	1.37	1.35	1.33	1.31
	B→	2.81	3.12	3.75	5.00	6.25	7.50	8.75	10.00	11.25	12.50	13.75	16.25	20.00	26.25
1⅜" 34.9mm	A→	2.47	2.29	2.06	1.83	1.72	1.65	1.60	1.57	1.55	1.53	1.51	1.49	1.47	1.44
	B→	3.09	3.43	4.13	5.50	6.87	8.25	9.62	11.00	12.37	13.75	15.12	17.87	22.00	28.87
1½" 38.1mm	A→	2.69	2.50	2.24	2.00	1.88	1.80	1.76	1.71	1.68	1.66	1.64	1.62	1.60	1.58
	B→	3.37	3.74	4.50	6.00	7.50	9.00	10.50	12.00	13.50	15.00	16.50	19.50	24.00	31.50

PROJECTION WITH A NEGATIVE (BARLOW) LENS

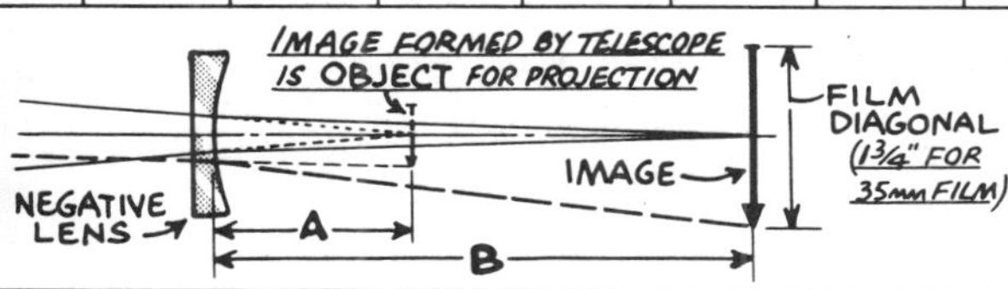

WHOLE TeleCamera: POSITIVE OR NEGATIVE SYSTEM

E.F.L. = F.L. OBJECTIVE X PROJECTION M.

f/ = f/OBJECTIVE X PROJECTION M.

DIAMETER OF FIRST IMAGE = FILM DIAGONAL / PROJECTION M.

F.L.		1¼x	1½x	2x	2½x	3x	3½x	4x	4½x	5x	6x	7x	8x	9x	10x
-1" 254mm	A→	.20"	.33"	.50	.60"	.67"	.71"	.75"	.78"	.80"	.83"	.86"	87"	89"	90"
	B→	.25	.50	1.00	1.50	2.00	2.50	3.00	3.50	4.00	5.00	6.00	7.00	8.00	9.00
-2" 50.8mm	A→	.40	.67	1.00	1.20	1.33	1.43	1.50	1.55	1.60	1.67	1.71	1.75	1.78	1.80
	B→	.50	1.00	2.00	3.00	4.00	5.00	6.00	7.00	8.00	10.00	12.00	14.00	16.00	18.00
-3" 76.2mm	A→	.60	1.00	1.50	1.80	2.00	2.14	2.25	2.33	2.40	2.50	2.57	2.62	2.67	2.70
	B→	.75	1.50	3.00	4.50	6.00	7.50	9.00	10.50	12.00	15.00	18.00	21.00	24.00	27.00
-4" 102mm	A→	.80	1.33	2.00	2.40	2.67	2.86	3.00	3.11	3.20	3.33	3.43	3.50	3.55	3.60
	B→	1.00	2.00	4.00	6.00	8.00	10.00	12.00	14.00	16.00	20.00	24.00	28.00	32.00	36.00
-5" 127mm	A→	1.00	1.66	2.50	3.00	3.33	3.57	3.75	3.88	4.00	4.17	4.28	4.37	4.44	4.50
	B→	1.25	2.50	5.00	7.50	10.00	12.50	15.00	17.50	20.00	25.00	30.00	35.00	40.00	45.00
-6" 152mm	A→	1.20	2.00	3.00	3.60	4.00	4.29	4.50	4.67	4.80	5.00	5.14	5.25	5.33	5.40
	B→	1.50	3.00	6.00	9.00	12.00	15.00	18.00	21.00	24.00	30.00	36.00	42.00	48.00	54.00
-7" 178mm	A→	1.40	2.33	3.50	4.20	4.67	5.00	5.25	5.44	5.60	5.83	6.00	6.12	6.22	6.30
	B→	1.75	3.50	7.00	10.50	14.00	17.50	21.00	24.50	28.00	35.00	42.00	49.00	56.00	63.00
-8" 203mm	A→	1.60	2.66	4.00	4.80	5.33	5.71	6.00	6.22	6.40	6.67	6.85	7.00	7.11	7.20
	B→	2.00	4.00	8.00	12.00	16.00	20.00	24.00	28.00	32.00	40.00	48.00	56.00	64.00	72.00
-9" 229mm	A→	1.80	3.00	4.50	5.40	6.00	6.43	6.75	7.00	7.20	7.50	7.71	7.87	8.00	8.10
	B→	2.25	4.50	9.00	13.50	18.00	22.50	27.00	31.50	36.00	45.00	54.00	63.00	72.00	81.00
-10" 254mm	A→	2.00	3.33	5.00	6.00	6.67	7.14	7.50	7.78	8.00	8.33	8.57	8.75	8.88	9.00
	B→	2.50	5.00	10.00	15.00	20.00	25.00	30.00	35.00	40.00	50.00	60.00	70.00	80.00	90.00

TABLE 5

The Arithmetic of PROJECTION SYSTEMS

POSITIVE LENS PROJECTION

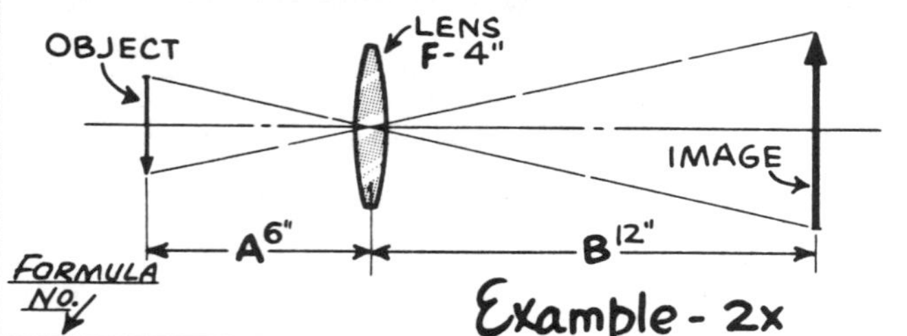

Example - 2x

1	$B = (M+1) \times F$	$B = (2+1) \times 4 = 3 \times 4 = 12''$
2	$B = \frac{F \times A}{A - F}$	$B = \frac{4 \times 6}{6 - 4} = \frac{24}{2} = 12''$
3	$B = A \times M$	$B = 6 \times 2 = 12''$
4	$A = \frac{B}{M}$	$A = \frac{12}{2} = 6''$
5	$A = \frac{F}{M} + F$	$A = \frac{4}{2} + 4 = 2 + 4 = 6''$
6	$A = \frac{F \times B}{B - F}$	$A = \frac{4 \times 12}{12 - 4} = \frac{48}{8} = 6''$
7	$M = \frac{B}{A}$	$M = \frac{12}{6} = 2x$
8	$M = \frac{F}{A - F}$	$M = \frac{4}{6 - 4} = \frac{4}{2} = 2x$
9	$M = \frac{B - F}{F}$	$M = \frac{12 - 4}{4} = \frac{8}{4} = 2x$
10	$F = \frac{A \times M}{M + 1}$	$F = \frac{6 \times 2}{2 + 1} = \frac{12}{3} = 4''$
11	$F = \frac{B}{M + 1}$	$F = \frac{12}{2 + 1} = \frac{12}{3} = 4''$
12	$F = \frac{A \times B}{A + B}$	$F = \frac{6 \times 12}{6 + 12} = \frac{72}{18} = 4''$

NEGATIVE LENS PROJECTION

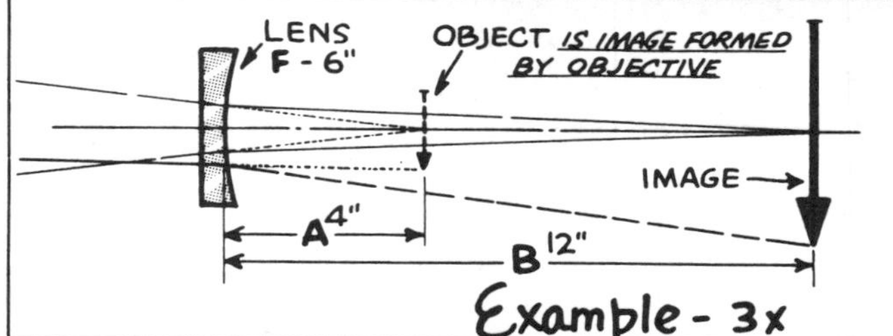

Example - 3x

1	$B = (M-1) \times F$	$B = (3-1) \times 6 = 2 \times 6 = 12''$
2	$B = \frac{F \times A}{F - A}$	$B = \frac{6 \times 4}{6 - 4} = \frac{24}{2} = 12''$
3	$B = A \times M$	$B = 4 \times 3 = 12''$
4	$A = \frac{B}{M}$	$A = \frac{12}{3} = 4''$
5	$A = F - \frac{F}{M}$	$A = 6 - \frac{6}{3} = 6 - 2 = 4''$
6	$A = \frac{F \times B}{F + B}$	$A = \frac{6 \times 12}{6 + 12} = \frac{72}{18} = 4''$
7	$M = \frac{B}{A}$	$M = \frac{12}{4} = 3x$
8	$M = \frac{F}{F - A}$	$M = \frac{6}{6 - 4} = \frac{6}{2} = 3x$
9	$M = \frac{B + F}{F}$	$M = \frac{12 + 6}{6} = \frac{18}{6} = 3x$
10	$F = \frac{A \times M}{M - 1}$	$F = \frac{4 \times 3}{3 - 1} = \frac{12}{2} = 6''$
11	$F = \frac{B}{M - 1}$	$F = \frac{12}{3 - 1} = \frac{12}{2} = 6''$
12	$F = \frac{A \times B}{B - A}$	$F = \frac{4 \times 12}{12 - 4} = \frac{48}{8} = 6''$

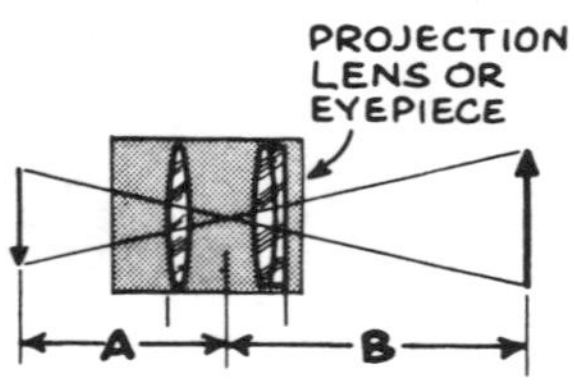

FOR DOUBLE LENSES OR ANY TYPE OF EYEPIECE, MEASURE FROM CENTER.
THIS IS NOT EXACT BUT CAUSES NO GREAT ERROR, ESPECIALLY AS APPLIED TO DIMENSION "B"

THE PROPER MEASURING POINTS FOR SPACING ARE THE PRINCIPAL PLANES OF THE LENS OR EYEPIECE

SYMBOLS:

- M - *MAGNIFICATION OBTAINED BY PROJECTION*
- F - *FOCAL LENGTH OF THE PROJECTION LENS OR EYEPIECE*
- A - *DISTANCE FROM PROJECTION LENS TO OBJECT, THE "OBJECT" BEING THE PRIME IMAGE FORMED BY TELESCOPE OBJECTIVE*
- B - *DISTANCE FROM THE PROJECTION LENS TO IMAGE, THE "IMAGE" BEING THE FINAL IMAGE, COINCIDENT WITH FILM PLANE*

Formula INDEX

IF YOU KNOW	YOU CAN FIND
F AND A	M.....8 B..... 2
F AND B	M.....9 A.....6
F AND M	A.....5 B..... 1
M AND A	F.....10 B.....3
M AND B	F.....11 A.... 4
A AND B	M.... 7 F.... 12

WHOLE TELE-Camera SYSTEM

AFTER FINDING PROJECTION M, THE WHOLE SYSTEM CAN BE CALCULATED:

E.F.L. = F.L. OF OBJECTIVE × PROJECTION M
f/ = OBJECTIVE f/VALUE × PROJECTION M
f/ ALTERNATE = E.F.L. ÷ CLEAR DIA. OF OBJECTIVE
DIA. OF FIRST IMAGE = FILM DIAGONAL ÷ PROJECTION M

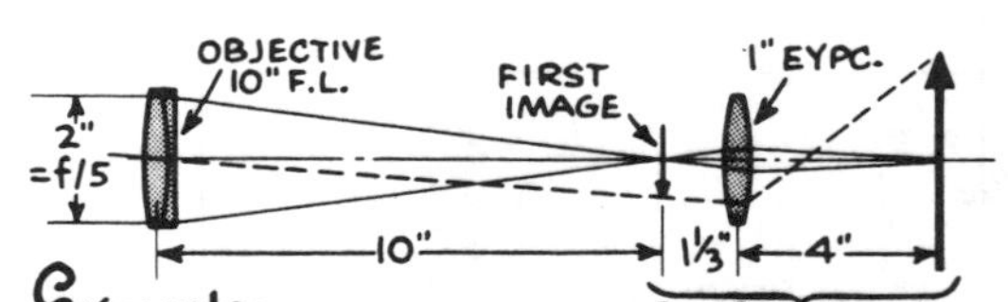

Example:

E.F.L. = 10 × 3 = **30"**
f/ = f/5 × 3 = **f/15**
f/ ALTERNATE = 30 ÷ 2 = **f/15**
FIRST IMAGE = $\frac{1.75}{3}$ = .58"
FOR 35MM FILM

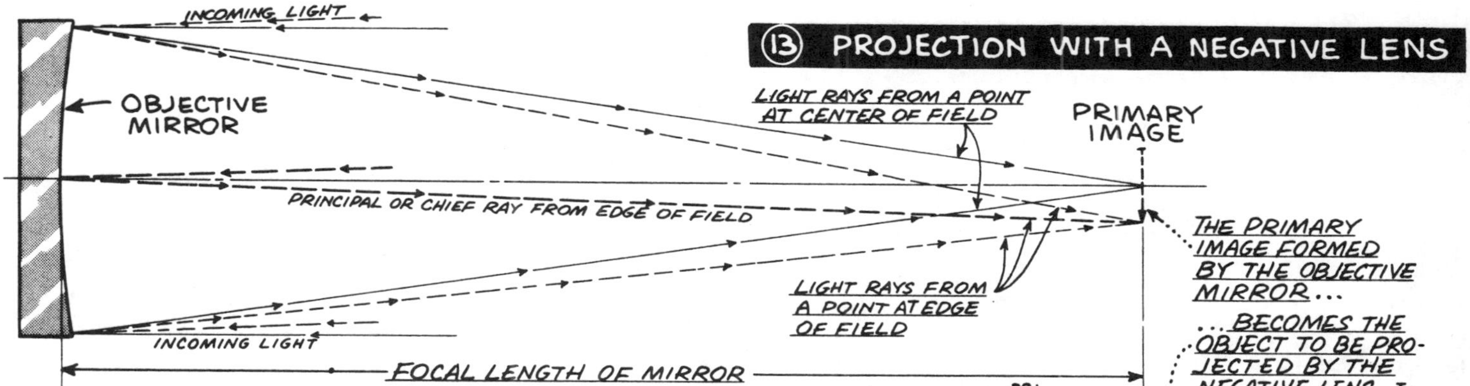

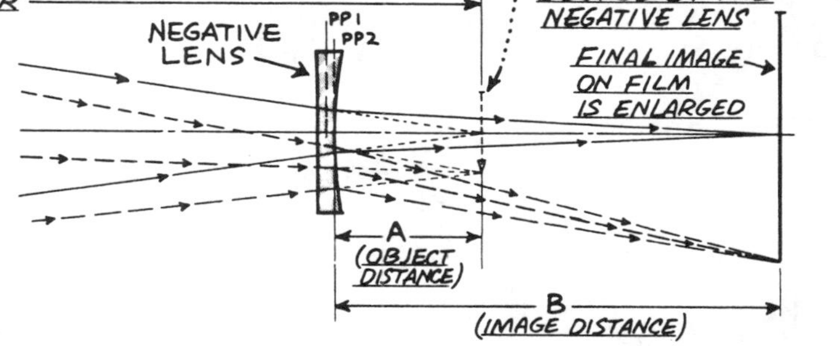

⑬ PROJECTION WITH A NEGATIVE LENS

covers all of the various equations and transpositions needed to calculate any system.

NEGATIVE LENS PROJECTION

Whether you prefer a negative lens or a positive lens for a photographic system is largely dependent on the quality of the optics. Assuming good optics, the negative lens is preferable, most practical. The outstanding feature of projection by using a negative lens is compactness, the lens being located inside the primary focus instead of outside as is the case for projection with a positive lens. Practically all of the big satellite tracking cameras and super telephoto lenses use some form of this optical system. As can be seen in Fig. 13, the primary image formed by the telescope objective becomes the object to be projected. It is, therefore, a virtual object, that is, it does not exist at all in reality but it is quite valid for diagrams and calculations. The whole range of lens movement from 1x magnification to infinite magnification is contained in one focal length, Fig. 14.

Good negative achromats--especially big ones--are hard to find because they have limited uses in optical instruments, and also because they can be fully corrected only for one specific application. Available lenses are usually designed for visual use in a telescope, working in an f/8 light cone at 2x magnification. All of which is

TABLE 4-A OBJECT-IMAGE SPACING for EDMUND BARLOW LENSES

F.L.		PROJECTION MAGNIFICATION							
		1½x	2x	2½x	3x	3½x	4x	5x	6x
1.31" [A]	A→	.43"	.66"	.79"	.88"	.93"	.98"	1.05"	1.09"
	B→	.66	1.31	1.96	2.62	3.28	3.93	5.24	6.55
1.73" [B]	A→	.58	.86	1.04	1.15	1.23	1.30	1.38	1.44
	B→	.87	1.73	2.60	3.46	4.32	5.19	6.92	8.65
1.83" [C]	A→	.61	.92	1.10	1.23	1.30	1.37	1.46	1.52
	B→	.92	1.83	2.74	3.66	4.58	5.49	7.32	9.15
PRIMARY* IMAGE DIA.→		1.17"	.87"	.70"	.58"	.50"	.44"	.35"	.29"

[A] SINGLE CROWN [B] GOODWIN ACHROMAT [C] EDMUND ACHROMAT

*for 35mm FILM

BASIC Calculations	Example: ASSUME YOU WANT 4x PROJECTION M. ASSUME F.L. OF NEG. LENS IS 2"
$B = (M-1) \times F$	$B = (4-1) \times 2 = 6''$
$A = \frac{B}{M}$	$A = \frac{6}{4} = 1.5''$

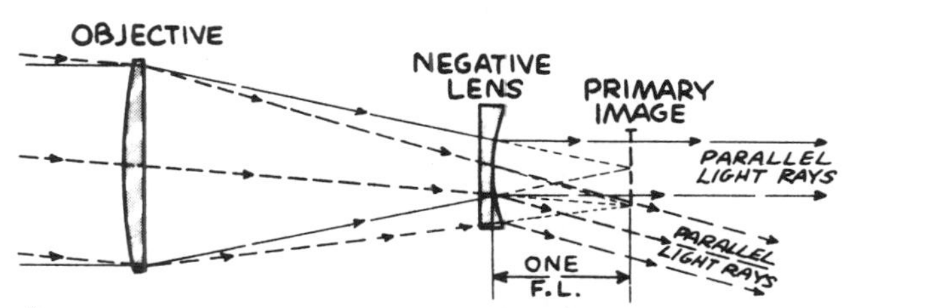

AFOCAL - *EMERGENT RAYS DO NOT COME TO A FOCUS - NO IMAGE IS FORMED. THIS SPACING IS A GALILEAN TELESCOPE*

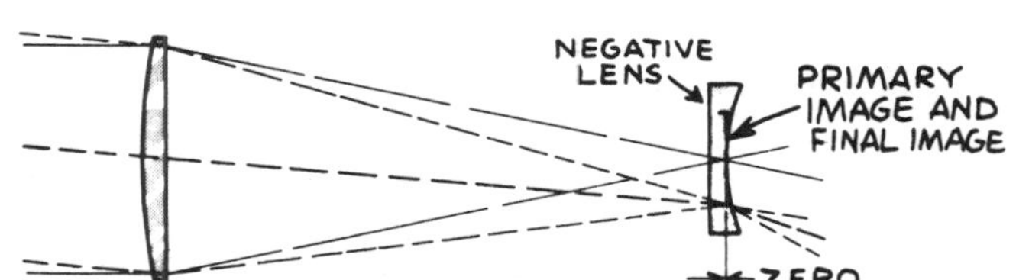

UNITY MAGNIFICATION (1x) - *THE FINAL IMAGE COINCIDES WITH THE PRIMARY IMAGE - MAGNIFICATION IS UNITY OR SAME SIZE (1x)*

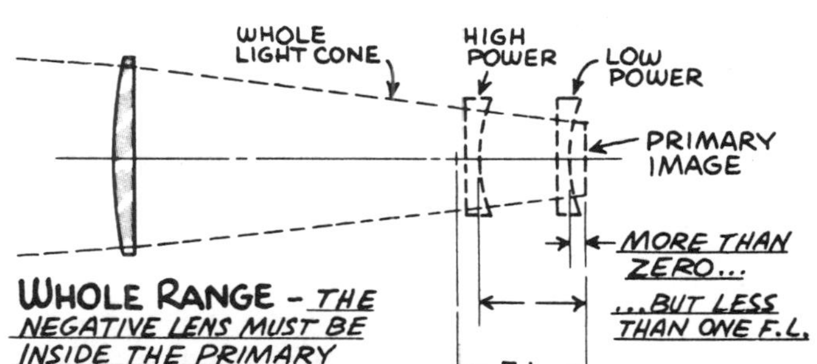

WHOLE RANGE - *THE NEGATIVE LENS MUST BE INSIDE THE PRIMARY FOCUS, BUT THIS DISTANCE MUST BE LESS THAN ONE FOCAL LENGTH*

⑭ SPACING LIMITS

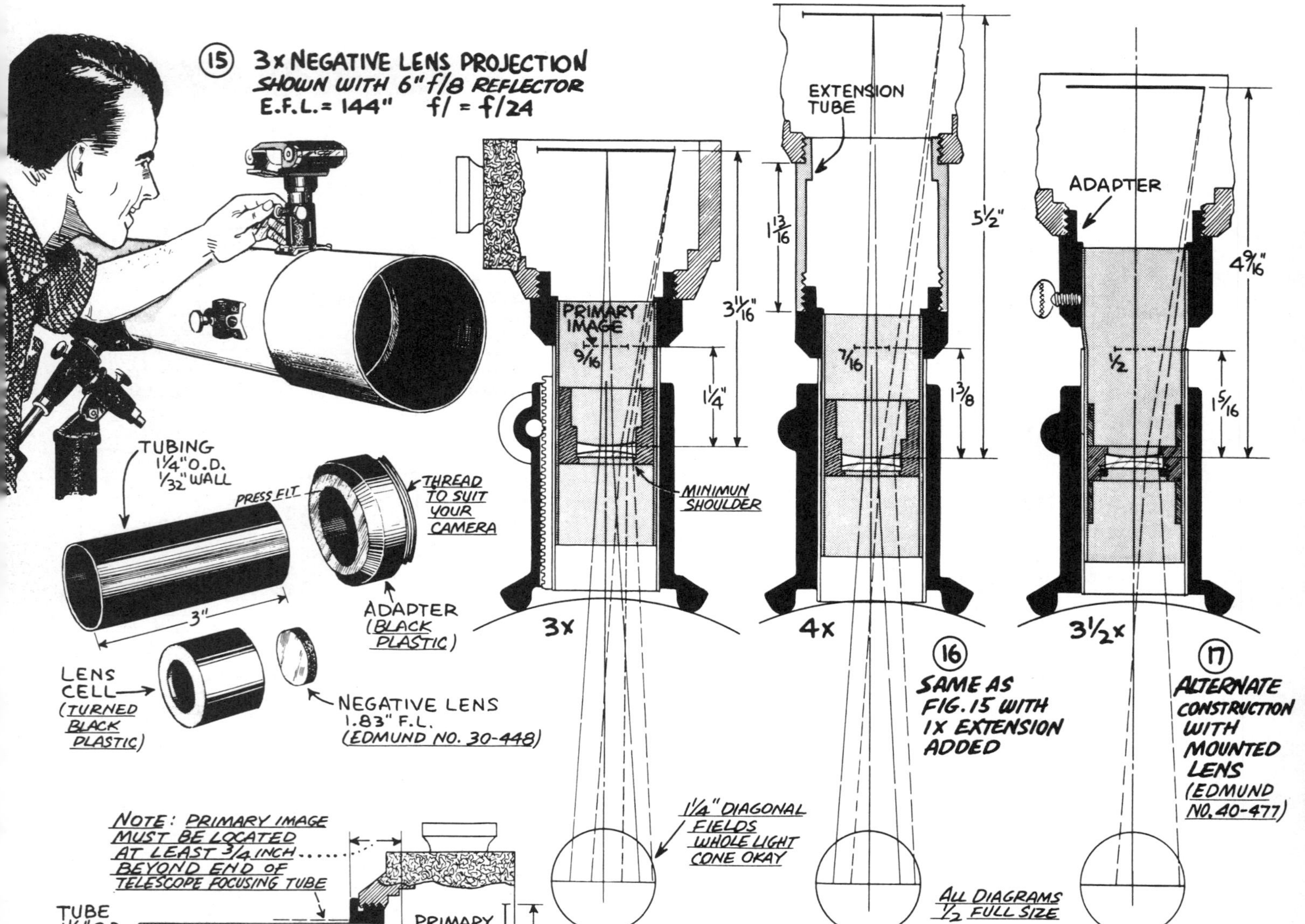

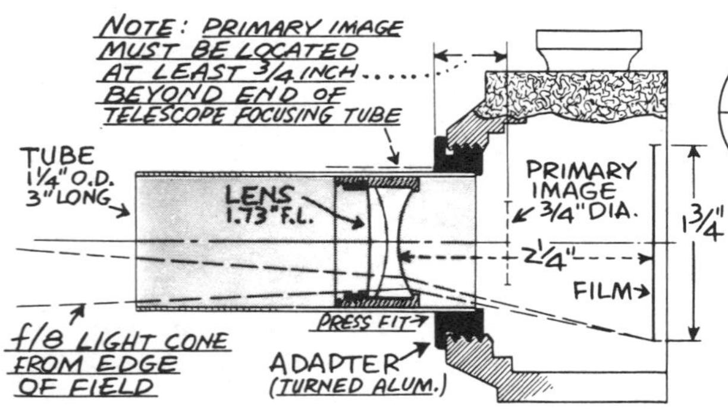

(18) 2.3x WITH GOODWIN LENS

to say that the linear field is limited to about 1/2 inch diameter; further, the lens will work best at 2x, although it will perform fairly well over a wide range of higher magnifications. The lenses are usually sold as "Barlow" lenses, named after Peter Barlow, English physicist and mathematician, first to use such a lens to increase the power of a telescope. Color-wise, the visual Barlow is perfectly satisfactory for photographic use. In fact, today's films have so nearly the same color response as the human eye, it can be said in general that any optical system that works well visually will perform equally well photographically over the same size of field.

Fig. 15 shows a 3x Barlow system mounted on a 6-inch reflector--note how compact it is compared to positive. However, like the positive lens system, 1x additional magnification calls for an extension tube of one focal length, Fig. 16. The Edmund Barlow lens can also be purchased mounted and the whole unit used photographically, Fig. 17, requiring only an adapter. Unmounted, this lens is modestly priced, and it works just as good as other more expensive Barlow lenses. It is too small for low power, but if you use 3x or more projection magnification, the lighting will be uniform to the corners of 35mm film. A 3x Barlow is often added to a finderscope for use as a tracking telescope.

One of the larger of the visual Barlows is the Goodwin design, and this can be used down to 2.3x as shown in Fig. 18. This requires a shallow adapter, and even so you may have to shift the main mirror forward a little to put the primary image at least 3/4 inch beyond the end of the eyepiece holder. All optical systems should have dead black interiors, and this goes double for the Barlow because of the way the lens pitches the light rays outward after transmission.

STAR TRAILS

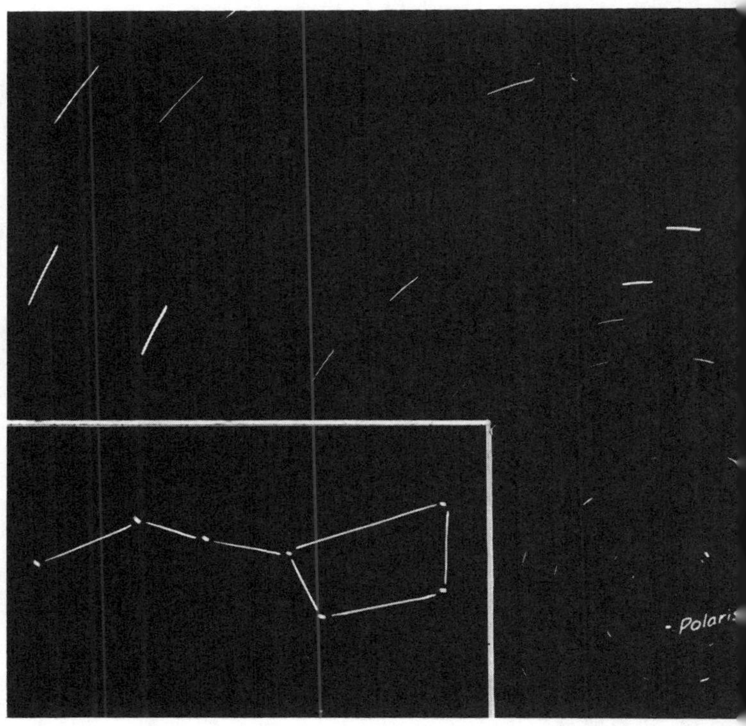

A STAR TRAIL is made by pointing any fixed camera at any part of the sky and exposing for 10 minutes or more. The stars, of course, keep moving right along, making a pattern of light streaks.

One common difficulty is field coverage. The average camera has a field of about 50 degrees. This means that if you center on Polaris, the half-field angle will reach down to about 65°N. declination, not quite reaching the Dipper. One way to capture the Dipper is by putting Polaris at one corner of the film, as shown in photo. This is a 25 min. exposure on 5x7 Tri-x film, using a 6-inch Metrogon lens at f/5.6. The bright streaks at top left are the Dipper stars.

Sky fog is another problem. Every minute you expose, the sky background becomes lighter. In 3 hrs. or so, sky fog may wash out the star trails. There is no simple solution to this except the obvious one that if you make a short exposure, the sky will stay black. The photo insert shows the Dipper stars as seen with a twin lens reflex working at f/4.5 with Tri-x film, the exposure being 2 minutes. Big star images were obtained by putting them slightly out of focus. One way to tackle sky fog is to make the trails brighter. This takes big aperture, such as the sniperscope objective shown in drawing.

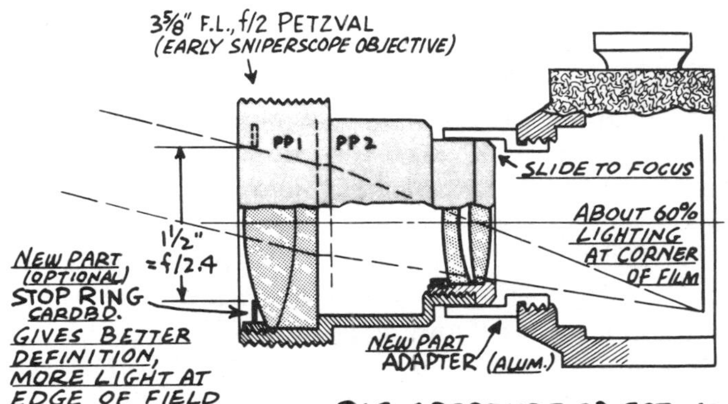

BIG APERTURE OBJECTIVE

STAR TRAIL Formulas	DECLINATION... EITHER NORTH OR SOUTH								
	0°	10°	20°	30°	40°	50°	60°	70°	80°
1 LENGTH OF STAR TRAIL (INCHES) FOR 1 MINUTE EXPOSURE TIME	F times .0044	F times .0043	F times .0041	F times .0038	F times .0033	F times .0028	F times .0022	F times .0015	F times .0008
for 1 SECOND EXPOSURE TIME	× .00007	× .00007	× .00007	× .00006	× .00006	× .00005	× .00004	× .00002	× .00001
2 LENGTH OF STAR TRAIL (INCHES) FOR ANY EXPOSURE (MINUTES)	$\frac{F \times Time}{229}$	$\frac{F \times T}{232}$	$\frac{F \times T}{244}$	$\frac{F \times T}{264}$	$\frac{F \times T}{300}$	$\frac{F \times T}{358}$	$\frac{F \times T}{458}$	$\frac{F \times T}{674}$	$\frac{F \times T}{1320}$
3 EXPOSURE TIME (MINUTES) NEEDED FOR A SPECIFIED STAR TRAIL LENGTH	$\frac{L \times 229}{F}$	$\frac{L \times 232}{F}$	$\frac{L \times 244}{F}$	$\frac{L \times 264}{F}$	$\frac{L \times 300}{F}$	$\frac{L \times 358}{F}$	$\frac{L \times 458}{F}$	$\frac{L \times 674}{F}$	$\frac{L \times 1320}{F}$
4 MAXIMUM EXPOSURE (SECONDS) WHICH WILL NOT SHOW A TRAIL BASED ON PERMISSIBLE MOVEMENT OF .004"	$\frac{55}{F}$	$\frac{56}{F}$	$\frac{59}{F}$	$\frac{63}{F}$	$\frac{72}{F}$	$\frac{86}{F}$	$\frac{110}{F}$	$\frac{162}{F}$	$\frac{316}{F}$

F... IS FOCAL LENGTH OR EQUIVALENT F.L. IF YOU PLAN TO ENLARGE, MULTIPLY F.L. BY ENLARGEMENT TO GET WHOLE PRINT E.F.L.

Example: YOUR CAMERA LENS IS 3" F.L. and YOU ARE USING 2X PROJECTION SYSTEM. YOU ALSO PLAN 3X ENLARGEMENT. SO, WHOLE PRINT E.F.L. = 3 × 2 × 3 = 18"

PROBLEM: YOU WANT 1½" (ON PRINT) STAR TRAILS FOR STARS ON EQUATOR. WHAT EXPOSURE IS NEEDED?

SOLUTION: FORMULA 3 EXPOSURE = $\frac{1.5 \times 229}{18} = \frac{343.5}{18} = 19.1$ MIN.

Example: YOU ARE SHOOTING STARS AT 40° N. DECLINATION WITH 6" F.L. LENS. PRINTS WILL BE ENLARGED 3X. (E.F.L. = 3×6 = 18")

PROBLEM: FIND MAX. EXPOSURE THAT WILL RETAIN ROUND STAR IMAGE

SOLUTION: FORMULA 4 $\frac{72}{18} = 4$ SECONDS

NOTE: TWO OR THREE TIMES THE SPECIFIED EXPOSURE WILL SHOW ONLY SLIGHT ELONGATION

FORMULA NO. 4 CAN BE USED FOR MOON PICS.... IF ON CLOCK DRIVE, USE FACTOR OF 1200 AND DIVIDE BY PRINT E.F.L.

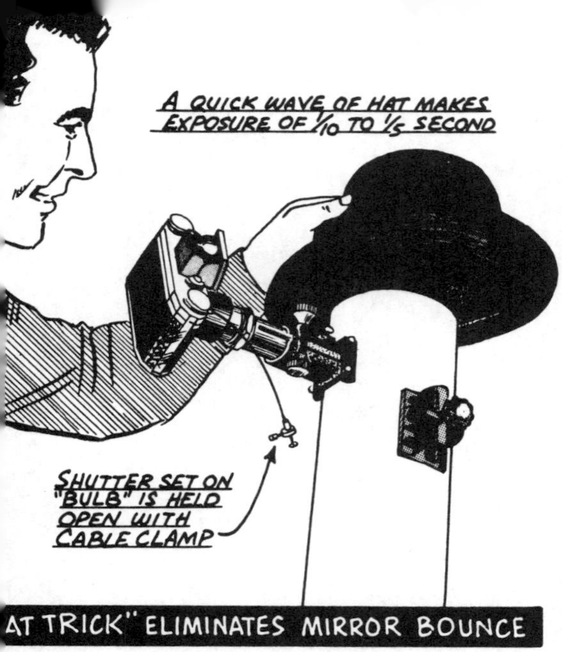

shooting the MOON

6-DAY OLD MOON AS SEEN WITH 6-inch REFLECTOR. PANATOMIC-X FILM, FAST "HAT TRICK" EXPOSURE. PICTURE AT RIGHT IS 3x BARLOW (GOODWIN) PROJECTION. PIC AT LEFT IS PROJECTION WITH 1-inch ORTHOSCOPIC EYEPIECE, ABOUT 2½X. DRUGSTORE PROCESSING OF 35MM FILM WITH 3X ENLARGEMENT. REPRODUCTION ABOUT FULL SIZE BUT CROPPED SLIGHTLY

BIGGEST, fastest, brightest object in the night sky, the moon is also the most photographed. Point your refractor or reflector her way on any clear night, and you are almost certain to get at least some kind of picture. Experts also give a look at the calendar because the moon "goes south" every month and when low in the sky may be a poor target. Follow the recommendations in table below. The sample pictures are not blue ribbon winners but are exactly the kind of pictures you will get the very first time out if you follow a few simple rules. The pics shown are all from 35mm film with drugstore processing, standard 3x enlargment; shown here full-scale but prints are cropped.

A few general rules on exposure will cover most situations. In general the full moon is simply a distant object, front-lighted by the sun. As such it takes about the same exposure as any distant land object in sunlight. The various phase pictures are no less bright at the limb (edge), but since you usually want shadow detail at the terminator, the average phase picture is exposed for the shadows, about 4 times the exposure of a full moon shot.

The full moon is most popular with beginners although it is a difficult object to photograph due to the flat lighting. The only really good cure for this is to use a high contrast film, such as High Contrast Copy or any of the various process films. High Contrast Copy was formerly known as Micro-file, and you will run across many full moon pics of bygone years taken with this film. High Contrast Copy film is a copy film intended for copying clippings, checks and other printed material; it is not intended for continu-

WHEN to shoot [1] **HOW to shoot**

PHASE	GOOD	FAIR		SUITABLE FILM	EXPOSURE [3]	REMARKS
FIRST QUARTER	JAN. thru JUNE MARCH is best	JULY thru DEC. SEPT. poorest	PHASE	PANATOMIC-X	1/10 SEC. AT f/16	USE FAST "HAT TRICK" EXPOSURE
			PHASE	HI CONTRAST [2]	1 SEC. AT f/16	TELESCOPE PREFERABLY ON CLOCK DRIVE
LAST QUARTER	JULY thru DEC. OCT. is best	JAN. thru JUNE MARCH poorest	PHASE	TRI-X	1/100 SEC. AT f/16	EXPOSURE BY SHUTTER OK FOR F.L. TO 100"
			PHASE	PLUS-X	1/25 SEC. AT f/16	WIDE LATITUDE IN EXPOSURE TIME
FULL MOON	OCT. thru MAR. DECEMBER is best month	APRIL thru SEPT. JUNE poorest	FULL	HIGH CONTRAST COPY [2]	1/5 SEC. AT f/16	"HAT TRICK" EXPOSURE. USE CLOCK DRIVE IF YOU HAVE
			FULL	TRI-X	1/400 SEC. AT f/16	FAST EXPOSURE PERMITS USE OF SHUTTER. YEL. FILTER IMPROVES CONTRAST

[1] BASED ON ALTITUDE OF MOON ... HIGH MOON IS BEST
[2] FORMERLY MICRO-FILE. OTHER HIGH CONTRAST FILMS CAN BE USED
[3] SEE TABLE OF EQUIVALENT EXPOSURES FOR OTHER f/VALUES

MOON PICTURES AT FIRST FOCUS ALL 3X ENLARGEMENTS FROM 35MM FILM, SHOWN FULL SCALE BUT CROPPED

3" REFRACTOR f/15 TRI-X K-2 FILTER | **4¼" REFLECTOR** TRI-X | **6" REFLECTOR** PLUS-X

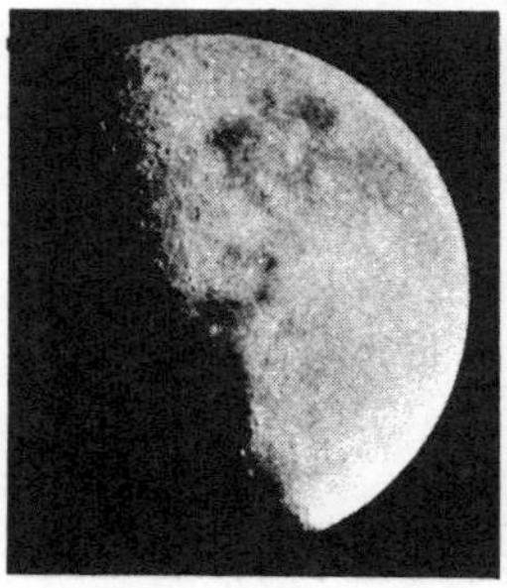

OVER-EXPOSED AT 1/25 SEC.

TOO THIN AT 1/500 SEC.

JUST RIGHT AT 1/50 SEC.

1/200 SEC. AT f/11

1/500 SEC. AT f/8

ous tone subjects at all, and hence has no ASA number applying to such situations. On the basis of comparative exposures, it can be rated at about ASA 5.

Many moon pics are not effective unless you capture the full diameter. This sets a limit to the focal length. Since the moon image is roughly 1/100 the focal length of telescope, and35mm film is 1 inch wide, it can be seen that about 100 inches of focal length is the most you can squeeze into the 1 inch width of 35mm film. Actually since a little leeway is needed, 80 or 90 inches f.l. is about tops for a full moon picture. For a starter, the average stock refractor or reflector of about 50 inches focal length does nicely. The resulting pictures are interesting (see above) but hardly big enough to show detail. However, if you use a fine-grain film, enlargements to 20x are practical if the negative itself is sharp and clear.

For detail pictures, 2x projection is a popular starter. For such work it is often permissible to rotate the camera to put the terminator the long way of the film. With 2x projection and 3x enlargement, the print e.f.l. of a 6-inch reflector is 288 inches. You are getting up there where vibration of any kind - - even the snap of a focal plane shutter--will result in loss of definition at the image plane. A common cure for the shutter problem is the familiar "hat trick" exposure. This in turn poses the problem of the moon drifting out of the field during the interval between camera exposure and the actual "hat trick." However, with a little practice, this can be timed nicely. On clock drive, of course, this is no problem at all since the moon once centered will stay put as long as you like.

In general, without clock drive, the best technique is to use fast film in order to reduce exposure time to a minimum; if you use clock drive, longer "hat trick" exposures are practical, and you can make good use of the slower films which are usually somewhat better in graininess, resolution and contrast.

2x PROJECTION WITH 1¼" ERFLE EYEPIECE ON 3" EDMUND REFRACTOR. K-2 FILTER. TRI-X. 1/25 SEC. AT f/32

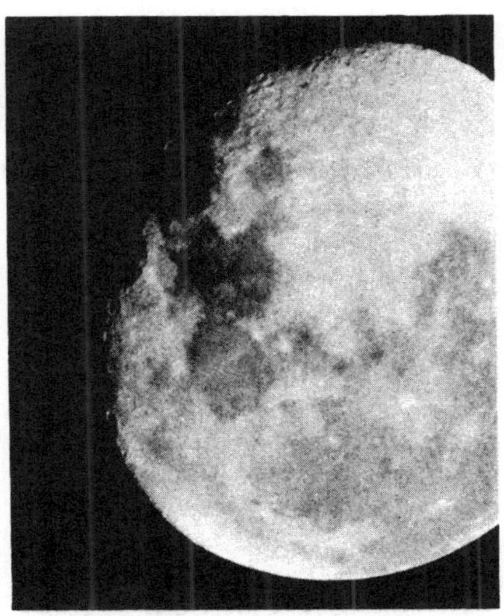

2X PROJECTION WITH 1¼" ERFLE EYEPIECE ON 4¼" REFLECTOR. PAN-ATOMIC-X. CAPPED EXPOSURE 1/5 SEC.

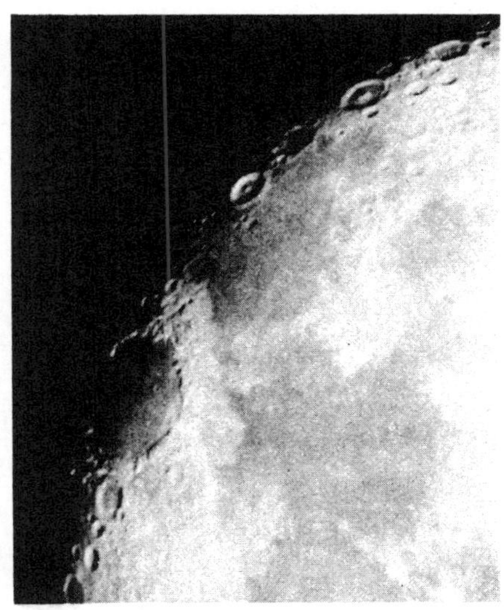

3x PROJECTION WITH EDMUND ACHROMATIC BARLOW ON 3" REFRACTOR. K-2, TRI-X. CAPPED EXPOSURE ABOUT 1/5 SEC.

SHOOTING THE STARS

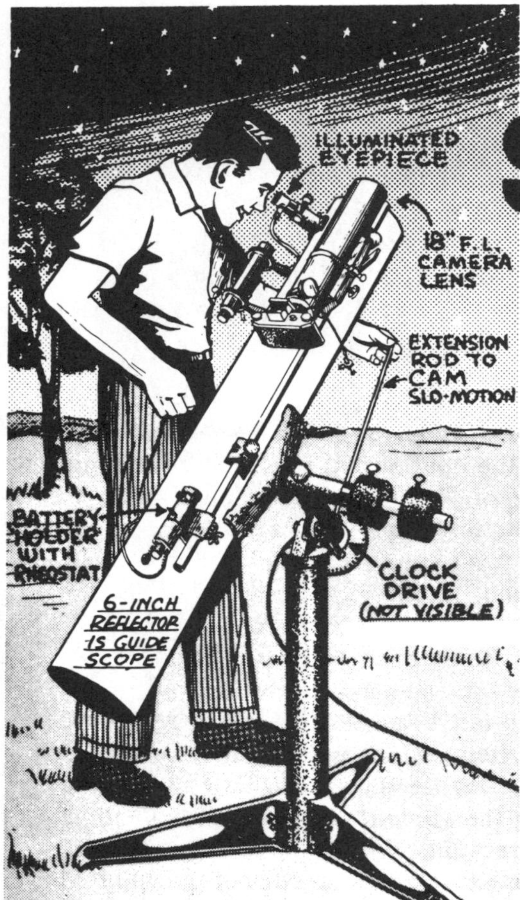

① A TYPICAL SETUP FOR ASTRO-PHOTOGRAPHY

SHOOTING pictures of the stars is probably the most fascinating phase of astro photography. Objects which you see dimly or not at all are revealed clearly in time exposures. Picture-taking equipment need not be expensive; you can get nice image quality with ordinary achromats as objectives. The one "must" in equipment is a clock drive, plus a compensating slow-motion on the clock drive itself. You will hear tales of star shooters getting good pics using ordinary manual guiding of the telescope, but such feats are never enjoyable and are rarely repeated.

GUIDING. Star pictures require time exposures from 1 minute to 30 minutes or more. During the exposure period, the telescope must be guided. This is a continuous operation, somewhat like steering a car down a road where you keep on the right track with dozens of almost imperceptible movements of the steering wheel. Many beginners get the idea that the clock motor drives the telescope automatically. To a degree, it does, but not with the exact precision needed for taking pictures.

In the usual "starter" outfit, you will be shooting with a camera of modest focal length, using the telescope itself as a guide telescope, Fig. 1. An illuminated, grid-type reticle is an aid since it is

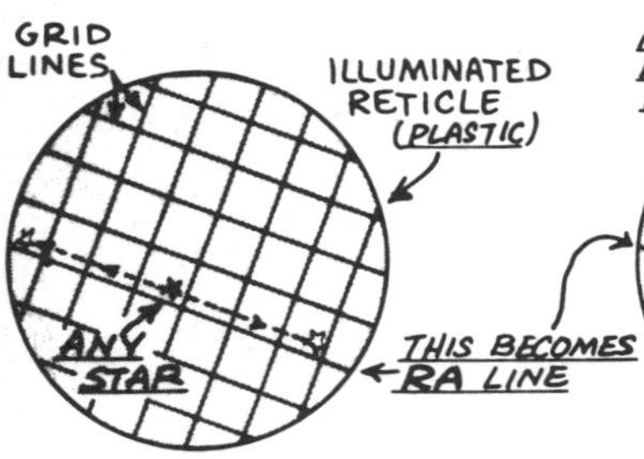

② ORIENTATION OF RETICLE

MOVE TELESCOPE IN R.A. AND ROTATE EYE-PIECE TO MAKE GRID LINES FOLLOW THE STAR

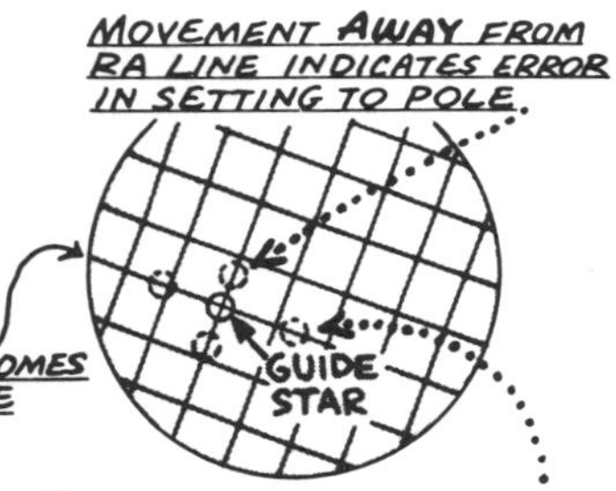

③ MOVEMENT OF GUIDE STAR

ANY MOVEMENT ALONG THE RA LINE IS A DRIVE ERROR WHICH YOU CAN CORRECT WITH SLOW-MOTION CONTROL

ORION. 5 MIN. EXPOSURE ON TRI-X WITH 6" F.L. LENS

18" F.L. CAMERA LENS GIVES A CLOSER VIEW OF NEB M42. 30 MIN. EXP. SHOWS FAINT STARS

④

HOW TO FIND
GUIDING TOLERANCE

YOU START WITH THE GENERAL IDEA THAT A STAR MOVEMENT OF .003" (1/333") IN ANY DIRECTION FROM CENTER OF STAR IMAGE WILL BE PRACTICALLY INVISIBLE IN A PHOTO PRINT

FOLLOW THESE RULES:

PRINT E.F.L. = F.L. CAMERA × PT. ENLARGEMENT

GUIDE SCOPE M. = $\frac{\text{GUIDE SCOPE E.F.L.}}{\text{PRINT E.F.L.}}$

GUIDING TOLERANCE = .003 × GUIDE SCOPE M.

Calculations for GUIDING TOLERANCE

6" TIMES 3x
PRINT E.F.L. = 18"
GUIDE SCOPE M. = $\frac{48}{18}$ = 2.7x (6" REFLECTOR)
TOLERANCE = .003 × 2.7 = .008"

18" TIMES 3x
PRINT E.F.L. = 54"
GUIDE SCOPE M. = $\frac{48}{54}$ = .89x
TOLERANCE = .003 × .9 = .0027"

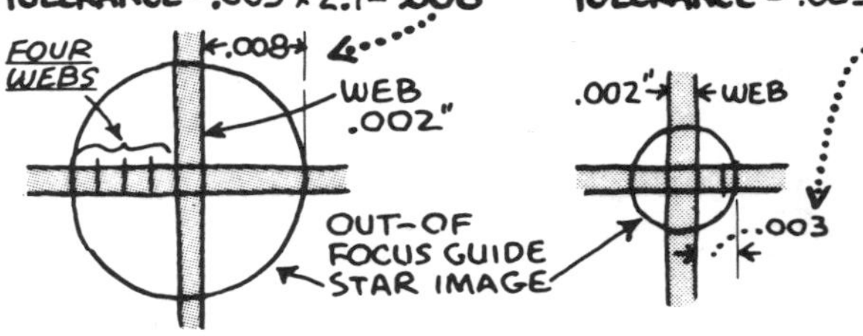

YOU KNOW THICKNESS OF RETICLE CROSS-HAIRS (WEB) AND WITH THIS AS GUIDE YOU CAN MAKE OUT-OF-FOCUS GUIDE STAR IMAGE THE PROPER SIZE

easily visible and allows you to position the guide star at any of the multiple intersections of cross lines. Do some practice guiding before attempting actual shooting. Pick up any bright star and locate it near the center of field. Now, move the telescope back and forth by hand in RA only and note the movement of the star. Rotate the eyepiece to make any reticle grid line parallel with this movement, as shown in Fig. 2. Fix in your mind that this is the RA line. Any movement of the guide star along the RA line, Fig. 3, means the drive is too fast or too slow, and you can correct this with the compensating slow-motion. Any movement away from the RA line means an error in setting the telescope to the pole. The ideal way to correct such an error is to have a slow-motion on the declination shaft. Lacking this you can sometimes get the needed correction by moving the telescope in declination by hand, preferably stopping the exposure while doing this and resuming exposure after the guide star is seen to be riding right on the wire. The drift away from the RA line is always a slow, steady movement, always one way, requiring correction at long intervals. The preferable way to solve declination drift is by accurate setting to the pole position; if the polar axis of your telescope is no more than 1/2 degree off the pole, you can make exposures up to 30 minutes without appreciable elongation of the star images.

Practice until you can guide smoothly, never once letting the bead off the wire. Actually, the way the guiding tolerance is applied, you do not let the out-of-focus guide star encroach on the reticle line; a crossover is a definite guiding error in excess of the permissible tolerance. The guiding tolerance is about .003 inch as applied to a photo print viewed at a distance of 10 inches, this being about the limit of visual acuity, corresponding to about 1 minute of arc. Fig 4 explains how this tolerance

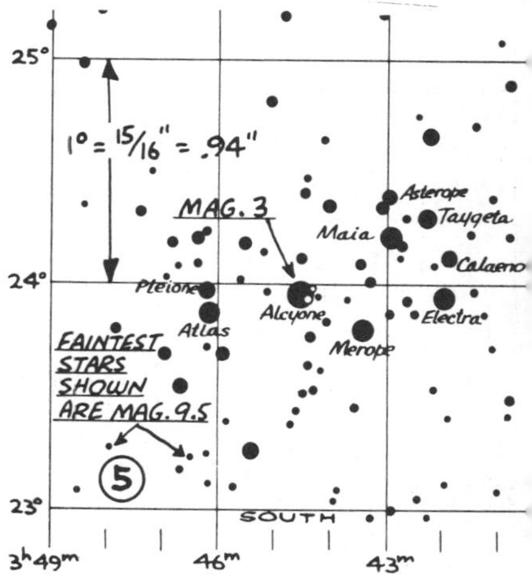

THE PLEIADES. PICS BELOW TAK
WITH 18"F.L., 2¼" DIA. ACHROMAT
35MM TRI-X. PRINTS ENLARGED 3X
AND SHOWN HERE FULL-SIZE
BUT CROPPED ABOUT 50%

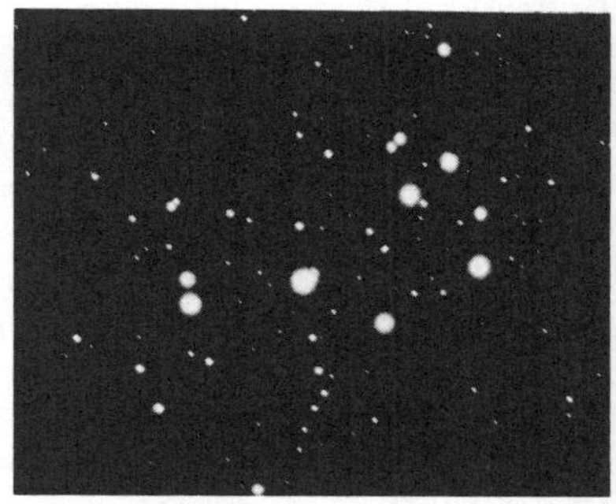

▲ 30 MIN. EXPOSURE ▼ 12 MIN.

▲ 3 MIN. ▼ 1 MIN.

TABLE 6 — MAGNITUDE of Faintest Star

APERTURE (CAMERA OR TELESCOPE)	VISUAL WITH TELESCOPE	NORMAL PHOTOGRAPHIC RANGE — EXPOSURE TIME: 1 MIN.	3 MIN.	9 MIN.	27 MIN.	81 MIN.	FOG LIMIT: f/3	f/5	f/7
½"	7.3	4.5	5.5	6.6	7.6	8.6	10.7	11.4	12.0
¾"	8.2	5.4	6.4	7.4	8.5	9.5	11.4	12.3	13.0
1"	8.8	6.0	7.0	8.0	9.1	10.1	12.0	12.9	13.5
1¼"	9.3	6.5	7.5	8.5	9.6	10.6	12.4	13.5	14.0
1½"	9.7	6.9	7.9	8.9	10.0	11.0	12.8	13.9	14.4
1¾"	10.0	7.2	8.2	9.3	10.3	11.3	13.2	14.2	14.8
2"	10.3	7.5	8.5	9.6	10.6	11.6	13.5	14.5	15.1
3"	11.2	8.4	9.4	10.4	11.5	12.5	14.4	15.4	16.0
4"	11.8	9.0	10.1	11.1	12.1	13.2	15.1	16.0	16.6
5"	12.3	9.5	10.5	11.5	12.6	13.6	15.7	16.6	17.1
6"	12.7	9.9	10.9	11.9	13.0	14.0	16.1	17.0	17.4
8"	13.3	10.5	11.5	12.6	13.6	14.6	16.6	17.4	17.9
10"	13.8	11.0	12.0	13.1	14.1	15.1	16.8	17.7	18.1
12"	14.2	11.4	12.4	13.4	14.5	15.5	17.0	17.9	18.3
	BASED ON ABILITY OF NAKED EYE TO SEE MAG. 6.2	FOR AVERAGE FILMS AND CLEAR ATMOSPHERE. WITH VERY FAST FILM AND "GOOD SEEING," YOU MAY GAIN ONE OR TWO MAGNITUDES FAINTER					EXPOSURE TIME LIMIT IS REACHED WHEN SKY BACKGROUND BECOMES NEARLY AS BRIGHT AS STAR IMAGES		

⑥ CONVERSION OF PLATE SCALE TO E.F.L.*		
SCALE GIVEN	Formula for E.F.L.	Examples
SECONDS OF ARC PER MILLIMETER	E.F.L. (Inches) = 8127 / SEC. OF ARC PER MM	EX: 1100 SEC. PER MM E.F.L. = 8127/1100 = 7.4" (inches)
INCHES PER DEGREE	E.F.L. (inches) = LINEAR SIZE OF 1° IN INCHES × 57.3	EX: 1° = .13" (inches) E.F.L. = .13 × 57.3 = 7.45" (inches)
...to FIND PLATE SCALE WHEN E.F.L. IS KNOWN		
You WANT TO FIND	FORMULA	Example (AS ABOVE)
PLATE SCALE IN SEC. OF ARC PER MM.	SEC. OF ARC PER MM = 8127 / E.F.L. (Inch)	8127 / 7.4 = 1100" (Seconds)
PLATE SCALE IN INCHES PER DEGREE	INCHES PER DEGREE = E.F.L. / 57.3	7.45 / 57.3 = .13" (Inches)
PLATE SCALE OF SOME POPULAR MAPS		
MAP OR ATLAS	PLATE SCALE	Corresponding E.F.L.*
NORTON'S STAR ATLAS	1° = .13 inch	.13 × 57.3 = 7.45"
BECVAR SKALNATE PLESO ATLAS OF THE HEAVENS	1° = .3 inch	.3 × 57.3 = 17.19"
BECVAR ATLAS ECLIPTICALIS	1° = .8 inch	.8 × 57.3 = 45.85"
VEHRENBERG PHOTOGRAPHIC STAR ATLAS	1° = .6 inch (ALL APPROX.)	.6 × 57.3 = 34.38"

* E.F.L. CAN BE ANY COMBINATION OF CAMERA F.L. times PRINT ENLARGEMENT

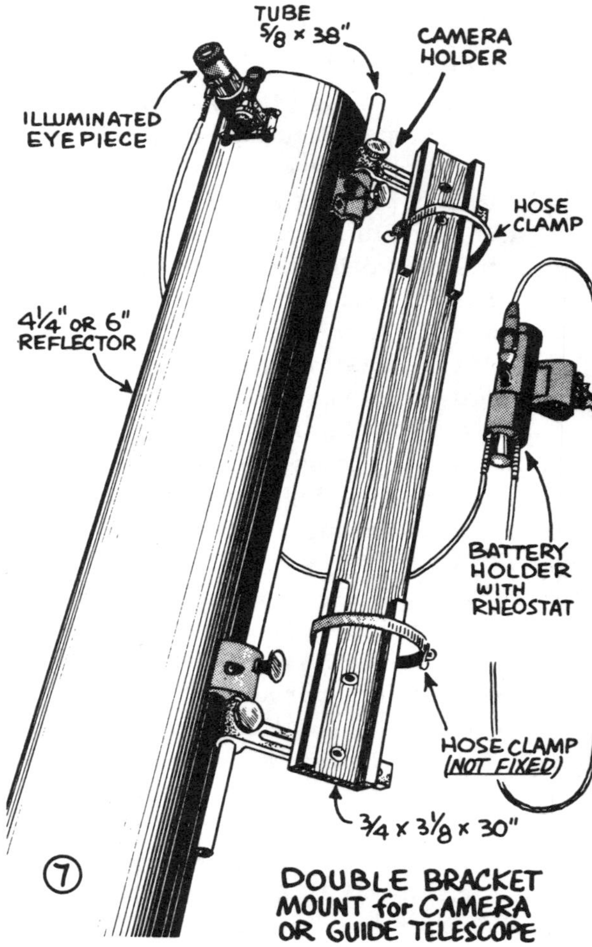

⑦ DOUBLE BRACKET MOUNT for CAMERA OR GUIDE TELESCOPE

is applied to the movement of the guide star itself. Previously, you should measure the lines or wires of the eyepiece reticle, using a measuring magnifier. This known dimension can then be used to measure the approximate diameter of an out-of-focus star image as well as its movement. Putting the guide star slightly out of focus has the added advantage of making it easier to see--the minimal point image of a star can very easily get lost behind a thick crosswire or grid line.

Preferably the guide scope should be of longer focal length than the print e.f.l. In any case, the guide scope magnification (see Fig. 4 and examples) should not be allowed to drop much below unity, i.e., you should not try to guide with a guide scope of shorter f.l. than the print focal length of the actual photo prints. In the setup shown in Fig. 1 where the camera is 18 in. f.l., the print e.f.l. will be 54 inches when prints are enlarged 3x. This is greater than the focal length of the guide telescope, which in this case is the reflector itself of 48 in. f.l., a little shy of unity M but still workable. Preferably for this setup, a positive or negative projection system would be used to increase the guide scope magnification, a 2x Barlow being most common. Image quality in a guide star is of no importance, making it practical to use extreme projection or even inferior optics as needed, the sole aim being to increase the equivalent focal length of the guide telescope.

FAINTEST STAR. Table 6 shows about what you can expect in star pictures with various apertures and exposures. The general idea is that increased exposure will bring faint stars into the picture while making bright stars bigger. This is illustrated by the top and bottom pics of the Pleiades on previous page. It will be noted from the table that an exposure increase of about three times is needed to gain one additional star magnitude. That is, if with any outfit and any film you are able to capture 8th magnitude stars with an exposure of 1 minute, you will need 3 minutes exposure to capture stars of 9th magnitude, and 9 minutes to capture 10th magnitude.

You can reduce exposure time with faster film or larger aperture. When you are shooting pictures of extended objects, the light pick-up of a lens depends solely on its f/value, i.e., the ratio of lens aperture to focal length. Any lens rated, say, f/4, will pick up exactly as much light as any other lens rated f/4 even if one is much larger than the other in aperture. On the

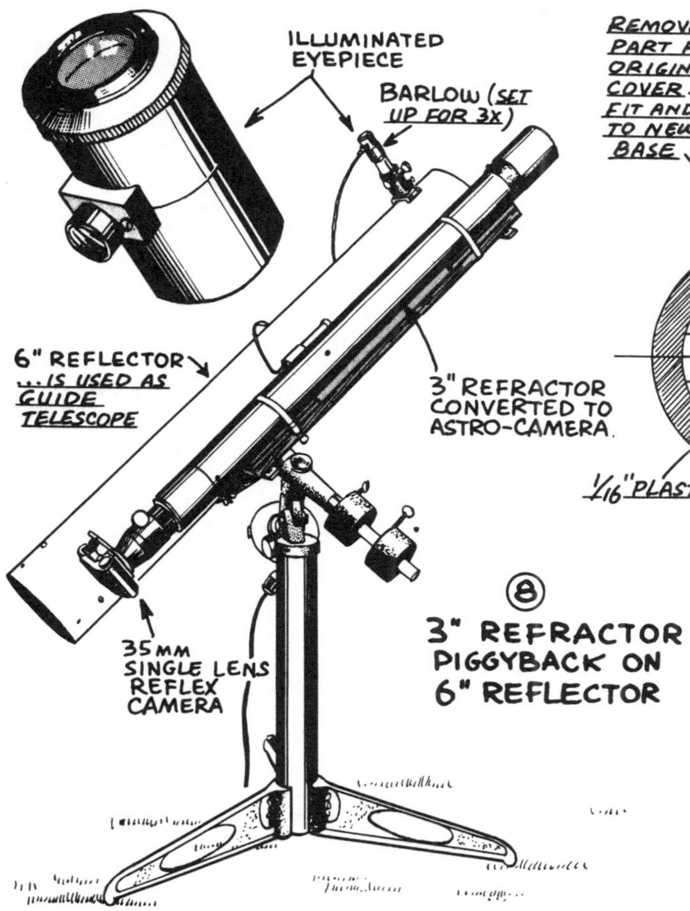

⑧ 3" REFRACTOR PIGGYBACK ON 6" REFLECTOR

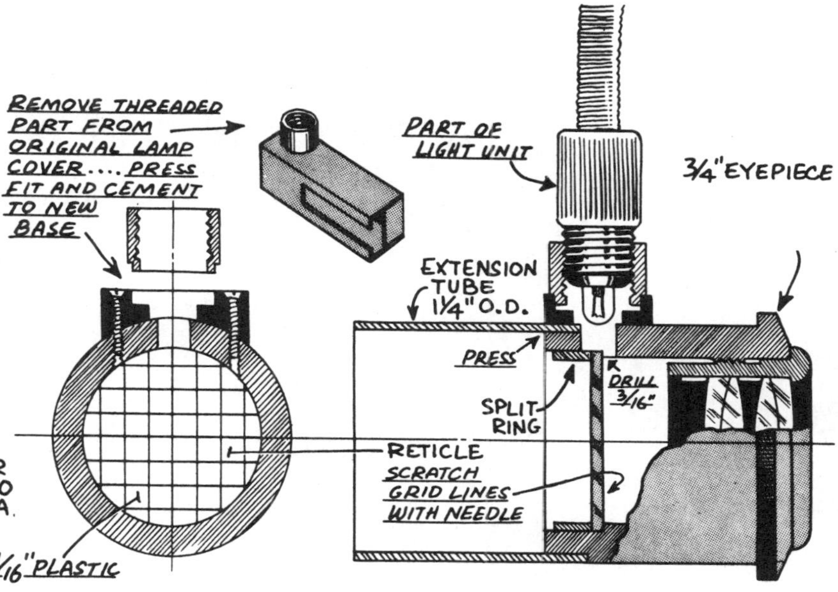

⑨ ILLUMINATED EYEPIECE

other hand, the light pick-up from luminous point objects (stars) depends entirely on the aperture of the lens. A fairly fast f/4 camera lens of 2 inches focal length is a comparatively "slow" lens for star photography for the simple reason it is only 1/2 inch diameter. Most star pics are taken with lenses of 1 inch or more aperture; 2 in. or a little more is a comfortable size for average astro cameras like the one shown in Fig. 1. In any case, the lens will pick up just as much starlight whether it is f/32 or f/4.

PLATE SCALE. Most star pictures must be planned. You must know what you are going to shoot, how to find it, what size it will be on film, how long to expose, etc. In particular, you must be sure that the target object will fit the film area. Tables 1 and 2 will be found useful. The plate scale of a map or photo often enters into the problem; this can be determined by applying the formulas given in Fig. 6. The scale of most star maps is not large. If you plan 3x enlargement, the scale of Norton's atlas can be obtained with a lens of only 2-1/2 inches focal length. One of the larger atlases in physical size is Becvar's Atlas Eclipticalis covering 30 degrees north and south from the celestial equator. Even its comparatively huge scale of about 4/5 of an inch per degree is exceeded by the modest outfit shown in Fig. 1 when negatives are enlarged 3x as is standard procedure for 35mm film.

EQUIPMENT. A good starter is an astro camera of 12 to 20 in. focal length and 2 in. or more clear aperture. This is the kind of outfit shown in Fig 1. If eventually you plan to use bigger equipment, the double bracket support shown in Fig. 7 may be installed at the start since it will handle both large and small astro cameras. One common combo has a 3-inch refractor piggyback on a 6-inch reflector, Fig. 8. In this drawing, the reflector is the guide telescope while the "camera" is the converted 3-inch refractor used as a direct objective. With the standard 3x print enlargement, the print e.f.l. becomes 135 inches. This is matched by the 3x Barlow on the reflector, making the two scopes practically a 1:1 match--the guiding tolerance reverts to the basic .003 inch. It is also practical to use the refractor with elbow erector as the guide scope, shooting with the reflector. For this setup, the reflector main mirror must be set forward to make the image plane accessible.

A full-size section of a typical illuminated eyepiece is shown in Fig. 9. This uses a standard purchased eyepiece which must be machined to accomodate the plastic reticle and lamp. The lamp is a grain-of-wheat lamp operating from a single 1-1/2 volt (size D) battery. This lamp is made for 3 volts and at 1-1/2 volts will burn dim yet bright enough for reticle illumination. The battery case specified has a rheostat which is needed to fade out the light a little in order to see the star images. The reticle must be in the focal plane of the eyepiece; check this with or

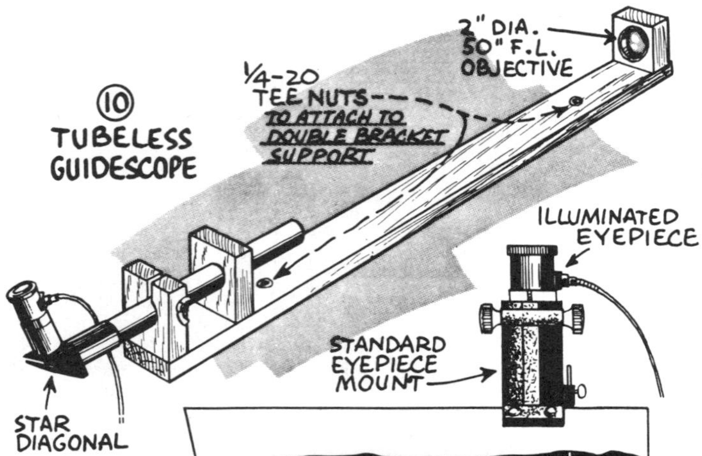

(10) TUBELESS GUIDESCOPE

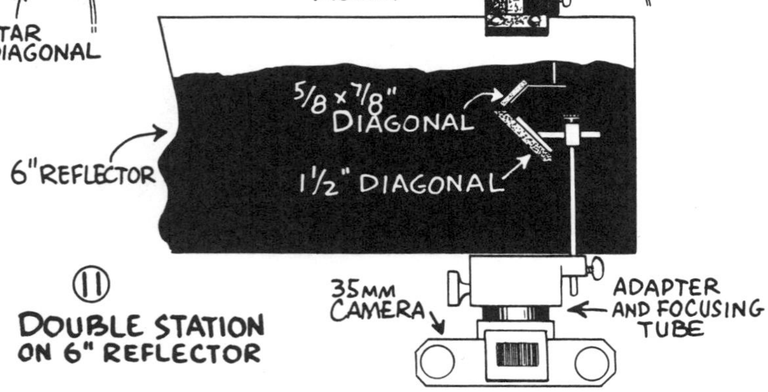

(11) DOUBLE STATION ON 6" REFLECTOR

without your glasses, duplicating your regular method of using the eyepiece in actual observing. Recheck this in actual use--the star image should stay put on the reticle line when you move your head from side to side.

A long focal length guidescope for use with a reflector can be tubeless, as shown in Fig. 10. This is conveniently mounted on the double bracket support shown on a previous page. Another common setup with a reflector is the double station, Fig. 11, which has separate optical systems for guiding and shooting.

SHOOTING HINTS. In most cases you will have no trouble finding a suitable guide star. It is not necessary to use the object being photographed as the guide star. In fact, you can point the guidescope at random to any star in the vicinity and it will guide just as perfectly as on the object itself. However, it is always more convenient to guide on the same object you shoot for the simple reason that the guide scope then serves as a finder--you know that whatever you see in the eyepiece is duplicated at the camera. This is sometimes a must when objects are too dim to be seen on the camera ground glass. Most 35mm cameras have a built-in magnifier and this is always used in focusing. Sharp focus is just a matter of making the star image as small as possible. If the object is too dim to focus easily, turn the camera toward a bright star and focus on that.

In guiding, always let the drive run several minutes to take out any lost motion. Try to avoid jockeying back and forth in guiding. With short f.l. cameras, the guiding tolerance is comfortable and you can even let the drive run unattended for several minutes. Anything over 100 inches print e.f.l. needs constant and careful guiding. You can get some nice pictures with as little as 5 minutes exposure and almost all of the open star clusters can be photographed in a half hour or less, using ordinary ASA 400 film.

equivalent EXPOSURES

TABLE 7 is three tables in one. The main body of the table gives equivalent exposures for a wide range of f/values. The two right hand columns give the linear resolution of a perfect optical system. The third table is a general exposure guide based on ASA speed index.

EQUIVALENT EXPOSURES are mathematically exact. The bug in any such table is "reciprocity failure," which is a technical term applied to film emulsions. It means simply that for extreme cases the film does not respond or reciprocate in exact mathematical ratio to the amount of light received. You can rely on equivalent exposures over a moderate range, say, about from f/4 to f/64. Beyond these limits, additional exposure time may be necessary.

RESOLUTION can be related to f/value, and the resolution columns in the table give the lines per millimeter which a perfect lens of a specified f/value is able to resolve. Most films also have a resolution value in lines per millimeter. The general idea is that a film should be selected which is capable of registering the degree of detail which the camera lens is able to resolve. Obviously there is no point in using a film of high resolving power, say, 200 lines per mm., if the camera lens itself is able to resolve only, say, 24 lines per mm. On the other hand, if the astro camera has high resolving power, it becomes necessary to use film of high resolution in order to realize the full potential of the astro camera.

It should be noted that the tabulated resolution values are for perfect optical systems where the resolution is limited only by the physical nature of light. Most films--even the fast ones--will resolve 60 lines per mm., and this kind of resolution is satisfactory for most astro cameras and subjects. The human eye can resolve (barely) 6 lines per millimeter. Hence, a print of this resolution will appear sharp because the eye can't quite resolve its graininess of 6 lines per millimeter. At 3 lines per millimeter, you can see the "pattern" and the picture looks soft or "wooly." Since pictures are usually enlarged from the negative, the negative must show greater resolution. For example, a negative resolving 18 lines per mm. will be 6 lines per mm. when enlarged three times. As already mentioned, this is the borderline case for true sharpness. From the table you can see this limits astro camera systems to about f/80 if you plan 3x enlargements.

ASA NUMBERS are keyed into the table at about the level currently recommended by film manufacturers. This leads to a minimum exposure value--more exposure time may be required. An easy way to relate ASA numbers to exposure is that the exposure at f/16 is equal to the ASA number expressed as a fraction. This is for a daytime object in bright sun. Example: ASA 400 means 1/400 second at f/16 for a distant object in bright sun.

TABLE 7

f/	EQUIVALENT EXPOSURES IN SECONDS FOR VARIOUS f/VALUES — USE ANY VERTICAL COLUMN. Example*: 1/400 SEC. AT f/8 = 1/20 SEC. AT f/36													MAXIMUM RESOLUTION: LINES PER MM	LINES PER INCH
3.2	1/24000	1/16000	1/12000	1/9600	1/7200	1/4800	1/3800	1/2400	1/2000	1/1500	1/1200	1/1000	1/800	458	11600
4	1/16000	1/10000	1/8000	1/6400	1/4800	1/3200	1/2600	1/1600	1/1300	1/1000	1/800	1/600	1/500	366	9300
4.5	1/13000	1/8300	1/6400	1/5000	1/3800	1/2600	1/2000	1/1300	1/1000	1/800	1/600	1/500	1/400	325	8300
5.6	1/8000	1/5200	1/4000	1/3200	1/2400	1/1600	1/1300	1/800	1/600	1/500	1/400	1/300	1/250	261	6650
6.3	1/6400	1/4200	1/3200	1/2500	1/1900	1/1300	1/1000	1/600	1/500	1/400	1/300	1/250	1/200	232	5900
8	1/4000	1/2600	1/2000	1/1600	1/1200	1/800	1/600	1/400*	1/300	1/250	1/200	1/160	1/125	183	4650
9	1/3200	1/2100	1/1600	1/1300	1/1000	1/600	1/500	1/300	1/250	1/200	1/160	1/125	1/100	163	4125
11.3	1/2000	1/1300	1/1000	1/800	1/600	1/400	1/300	1/200	1/160	1/125	1/100	1/80	1/60	130	3300
12.5	1/1600	1/1000	1/800	1/600	1/500	1/300	1/250	1/160	1/125	1/100	1/80	1/60	1/50	117	2950
16	1/1000	1/650	1/500	1/400	1/300	1/200	1/160	1/100	1/80	1/60	1/50	1/40	1/30	91	2325
18	1/800	1/500	1/400	1/320	1/240	1/160	1/125	1/80	1/60	1/50	1/40	1/30	1/25	81	2050
22.6	1/500	1/325	1/250	1/200	1/150	1/100	1/80	1/50	1/40	1/30	1/25	1/20	1/16	67	1650
25	1/400	1/260	1/200	1/160	1/120	1/80	1/60	1/40	1/30	1/25	1/20	1/16	1/13	59	1500
28	1/320	1/200	1/160	1/125	1/100	1/60	1/50	1/30	1/25	1/20	1/16	1/12	1/10	52	1325
32	1/250	1/150	1/125	1/100	1/75	1/50	1/40	1/25	1/20	1/16	1/12	1/10	1/8	46	1175
36	1/200	1/125	1/100	1/80	1/60	1/40	1/30	1/20*	1/16	1/13	1/10	1/8	1/6	41	1025
40	1/160	1/100	1/80	1/60	1/50	1/30	1/25	1/16	1/13	1/10	1/8	1/6	1/5	37	925
45	1/125	1/80	1/60	1/50	1/35	1/25	1/20	1/12	1/10	1/8	1/6	1/5	1/4	32	825
50	1/100	1/65	1/50	1/40	1/30	1/20	1/16	1/10	1/8	1/6	1/5	1/4	1/3	29	750
55	1/80	1/50	1/40	1/32	1/24	1/16	1/13	1/8	1/6	1/5	1/4	1/3	2/5	27	675
60	1/70	1/45	1/35	1/28	1/20	1/14	1/11	1/7	1/5	1/4	1/3	1/3	1/2	24	625
64	1/60	1/40	1/30	1/24	1/18	1/12	1/10	1/6	1/5	1/4	1/3	2/5	1/2	23	575
71	1/50	1/30	1/25	1/20	1/15	1/10	1/8	1/5	1/4	1/3	1/2	1/2	2/3	21	500
80	1/40	1/25	1/20	1/16	1/12	1/8	1/6	1/4	1/3	2/5	1/2	2/3	4/5	18	475
90	1/30	1/20	1/15	1/12	1/9	1/6	1/5	1/3	2/5	1/2	2/3	4/5	1	16	400
100	1/25	1/16	1/12	1/10	1/8	1/5	1/4	1/2	1/2	2/3	1	1	1 1/3	14	375
128	1/15	1/10	1/7	1/6	1/5	1/3	1/2	2/3	5/6	1	1 1/3	1 2/3	2	11	300
	ASA 1000	ASA 650	ASA 500	ASA 400	ASA 300	ASA 200	ASA 160	ASA 100	ASA 80	ASA 64	ASA 50	ASA 40	ASA 32		

THE ASA NUMBERS ARE KEYED INTO THE TABLE FOR A BRIGHT SUN, AVERAGE SUBJECT, FRONT LIGHTED, AT MEDIUM LONG DISTANCE (20 TO 100 YARDS)

BLACK and WHITE FILM DATA 1966

		ASA SPEED INDEX	RES. LINES PER MM	AVAILABILITY: 35 MM	ROLL	SHEET	MAIN FEATURE
AGFA	ISOPAN IFF	25	130	✓			FINE GRAIN
	ISOPAN IF	100	100	✓			GOOD CONTRAST
	ISOPAN ISS	200	80	✓			ALL-PURPOSE
	ISOPAN ULTRA	400	70	✓			ALL-PURPOSE
	ISOPAN RECORD	1250	60	✓			VERY HIGH SPEED
ANSCO	VERSAPAN	125	100	✓	✓	✓	FINE GRAIN
	SUPER HYPAN	500	80	✓	✓	✓	HIGH SPEED
KODAK	HI CONTRAST COPY [1]	5 [2] ESTIMATED	220	✓			GOOD FOR FULL MOON
	PANATOMIC-X	32	120	✓			FINE GRAIN
	PLUS-X	125	100	✓	✓	✓	FINE GRAIN
	VERICHROME	125	100		✓		ALL-PURPOSE
	INFRARED	50 [3]	NO DATA	✓			INFRA RED
	TRI-X	400	80	✓	✓	✓	HIGH SPEED
	ROYAL-X	1250	60		#120 ✓ ONLY		VERY FAST

[1] FORMERLY "MICRO-FILE." [2] NOT FOR GENERAL WORK AND HAS NO ASA NO. FOR SUCH [3] WITHOUT FILTER

Action of FILTERS

WRATTEN NUMBER	COLOR OF FILTER	EXP.* FACTOR	Action of FILTERS
K1	LIGHT YELLOW	1.5	ABSORBS SOME ULTRAVIOLET AND BLUE-VIOLET. NOT EFFECTIVE WITH ACHROMATS
K2	YELLOW	2	BETTER THAN K1 IN ABSORBING BLUE-VIOLET. PENETRATES HAZE. GOOD WITH ACHROMATS
AERO 1	LIGHT YELLOW	1.5	ABSORBS U-V, VIOLET AND SOME BLUE. PENETRATES SLIGHT HAZE
AERO 2	YELLOW	2	LIKE A1 BUT STRONGER. GOOD HAZE PENETRATION. GOOD WITH ACHROMATS
K3	DEEP YELLOW	2	REMOVES MORE BLUE THAN K2. CUTS THROUGH HAZE. GOOD FOR FAR OBJECTS
G	DEEP YELLOW	3	ALMOST COMPLETE ELIMINATION OF BLUE. EXCELLENT HAZE PENETRATION
X1	LIGHT GREEN	3	ABSORBS VIOLET, SOME BLUE AND SOME RED. GOOD FOR NATURAL COLOR TONES
X2	GREEN	5	SIMILAR TO X1 BUT ABSORBS MORE RED, HENCE MAKES RED OBJECTS DARKER
B	GREEN	8	ABSORBS VIOLET, MOST BLUE AND MOST RED LIGHT. TRANSMITS GREEN AND YELLOW
A	RED	8	TRANSMITS RED AND ABSORBS BLUE AND GREEN. PENETRATES HAZE
C5	BLUE	5	ABSORBS RED, YELLOW AND GREEN. TRANSMITS BLUE... INCREASES HAZE

*FOR PAN FILM AND DAYLIGHT, MULTIPLY NORMAL EXPOSURE BY FACTOR

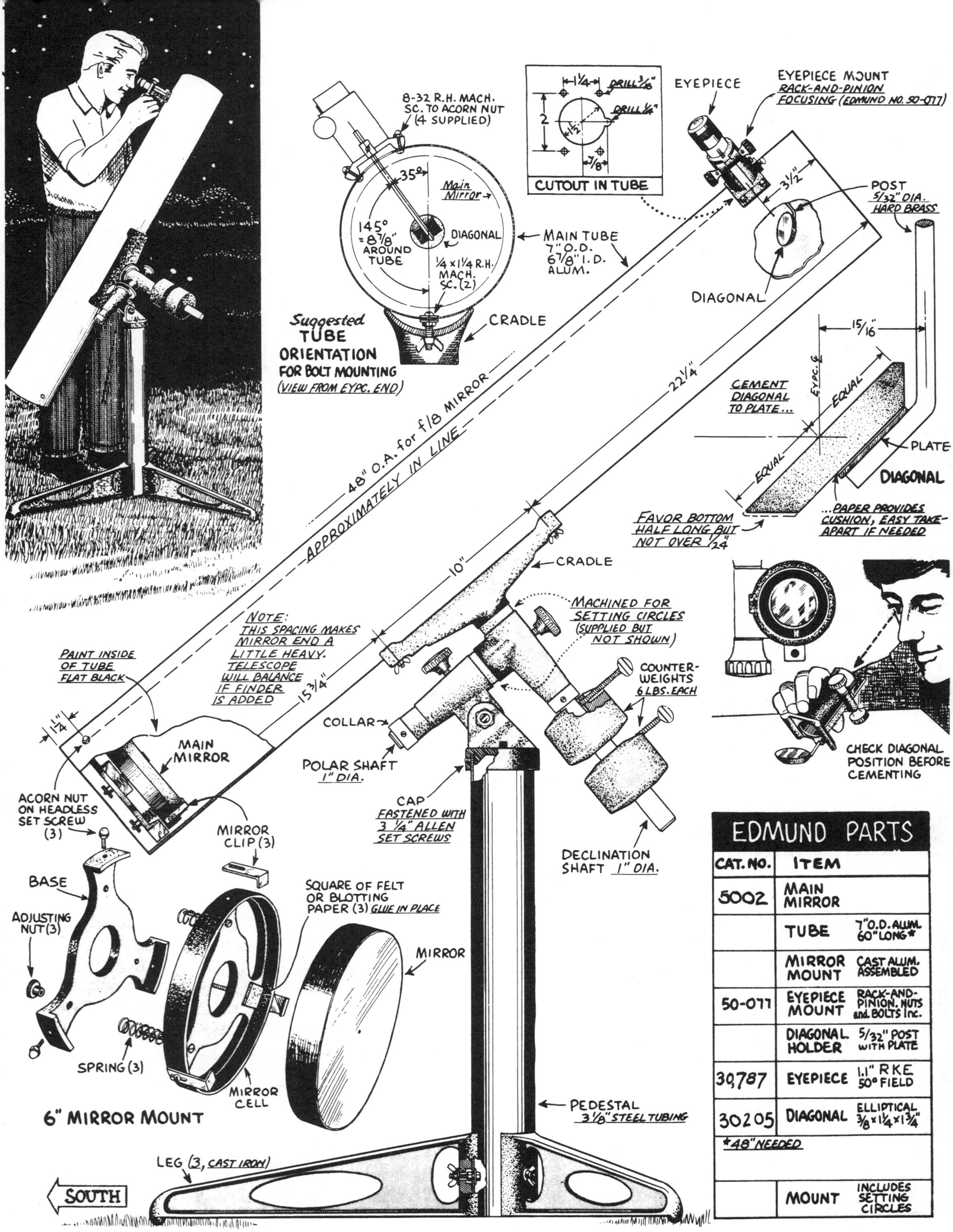

EDMUND PARTS

CAT. NO.	ITEM	
5002	MAIN MIRROR	
	TUBE	7" O.D. ALUM. 60" LONG*
	MIRROR MOUNT	CAST ALUM. ASSEMBLED
50-077	EYEPIECE MOUNT	RACK-AND-PINION, NUTS and BOLTS Inc.
	DIAGONAL HOLDER	5/32" POST WITH PLATE
30,787	EYEPIECE	1.1" RKE 50° FIELD
30205	DIAGONAL	ELLIPTICAL 3/8×1 1/4×1 3/4"
*48" NEEDED		
	MOUNT	INCLUDES SETTING CIRCLES

section 2

Telescopes You Can Build

standard 6 reflector

THE POPULAR 6-inch f/8 reflector with all purchased parts is shown on the opposite page. Very few builders buy all of the separate parts for the simple reason such purchases will total about 15% more than the price of the same identical telescope in a complete factory assembly. But the stock model provides a useful guide.

For the telescope itself, most builders buy all of the parts except possibly the main tube, which can often be purchased locally at a lower price. You still have the work of cutting the hole for the eyepiece and painting the tube inside and out. The eyepiece hole is not difficult if you have the tools, either a hole saw or a saber saw, but it is something of a chore if you have to do it the hard way by drilling many small holes around the opening and then filing smooth. The inside of the tube should be painted a flat black. You can buy locally spray cans of flat black paint (for ironwork) which goes on smoothly and dries dead flat. White enamel is the usual thing for the outside.

Apart from the tube, the only other job of assemblying a telescope from purchased parts is the cementing of the small flat mirror to the diagonal plate. This is not a hard job but it is a bit on the fussy side. Before you cement, make a dry assembly, as shown in drawing. Note that the diagonal post is pushed into the eyepiece mount to locate the mirror close to the mount for comfortable viewing. If you are using an elliptical diagonal, it should appear exactly concentric with the eyepiece tube, as shown in the detail. The actual cementing should be done with a piece of glazed paper (magazine cover) at the joint. This provides a slight cushion and also makes it easy to remove the mirror from the plate should this become necessary. For a cement, you can use practically any kind of glue. Needless to say, you handle the mirror only at the edges to avoid fingerprints on the aluminized surface. However, if the mirror gets dirty, it can be cleaned by any method used to clean eyeglasses.

The equatorial mount shown is a completely finished product and requires only minor assembly. The tilt of the polar axis should be the same as your latitude (the drawing is 40 degrees). Adjustment is made by loosening the single nut at cap lug, and then tightening very securely after the proper setting has been made. Only approximate accuracy is needed unless you are making use of setting circles or doing star photography.

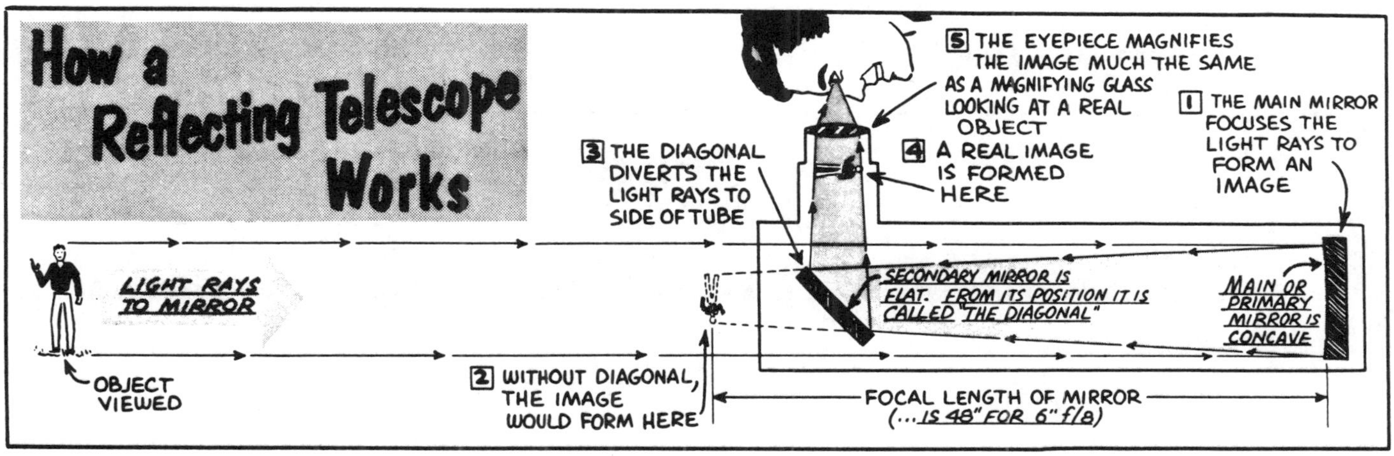

UTILITY SIX REFLECTOR

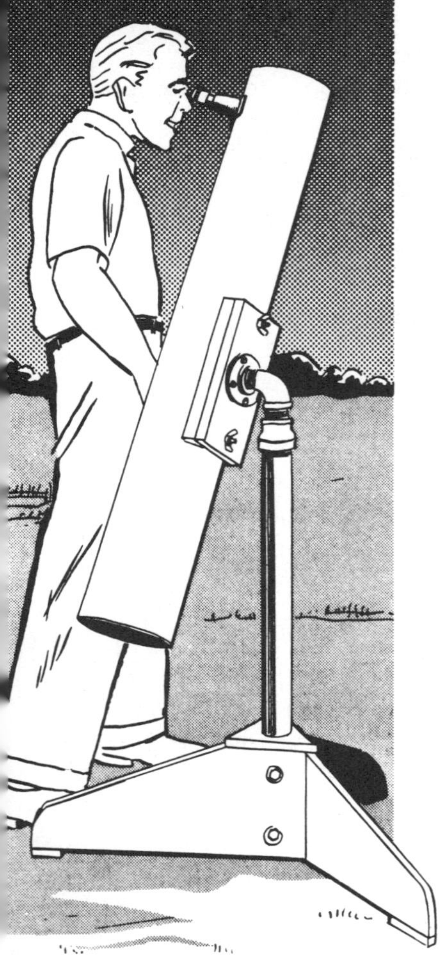

IF YOU are building a reflector on a tight budget, the best plan is to use top quality optics and save what you can on the mount and homemade parts. The result will be something like the design shown with simple turn-on-threads altazimuth mount and slide focusing. You will find this scope solid and satisfying, especially if it is a first instrument.

The mount uses a 2-inch pipe pedestal about 30 in. long. A coupling reduces to 1-1/2 in. pipe for the mount itself. This is a substantial mount; definitely you do not need larger and heavier pipe as suggested by some writers. The wood legs can be pre-drilled and are fitted to the pedestal one at a time; you put together, take apart as often as needed to mark hole positions and drill the tap holes in the pedestal. This procedure assumes the drilling is done on a drill press. With normal care in fitting, the final leg will lock in place and secure the whole assembly.

It is assumed the purchased mirror will run a little overlength in focus, i.e., it will probably be closer to 48-1/2 in. focal length than the basic 48 inches. Note in drawing below that the distance between the main mirror and the diagonal is the focal length minus about 6-1/2 inches. If you are a myope, it is best to increase this to 6-3/4 in. or even 7 inches, the reason being that a near-sighted person will focus a telescope "in" a bit more than normal; hence, you have to put the image plane "out" this extra distance to get the "in" focusing movement needed. Figure it out yourself! In any case, the worst that can happen is that the focal plane will be inaccessible and you will have to remount the main mirror.

The mirror mount is a simple job for the band saw and drill press. The mirror clips are made by partly flattening 1-inch angle irons. The plastic slide focus at the eye end is not as convenient as rack-and-pinion, but it works and you can always switch to R&P later. The worst job is cutting the hole in the main tube under the eyepiece holder. Actually this is a fairly easy job if you have a hole saw of the required size. Or a saber saw. More often, you will have to resort to the tedious process of drilling many small holes and then smoothing by filing. A square hole with rounded corners is entirely practical.

A first eyepiece should be 1 in. or a little more in focal length. Edmund No. 30,787 of 1.1 inch f.l. is a good choice.

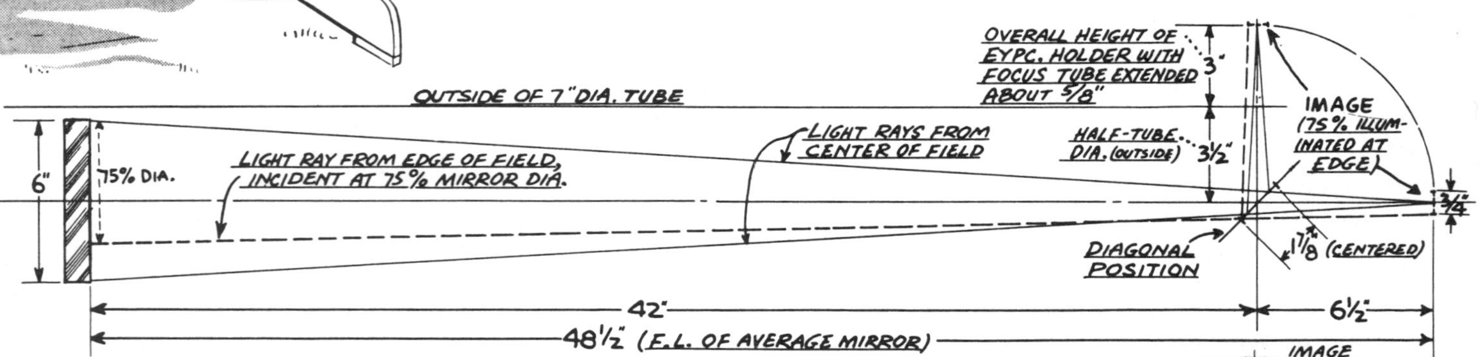

OPTICAL LAYOUT AND MAIN DIMENSIONS

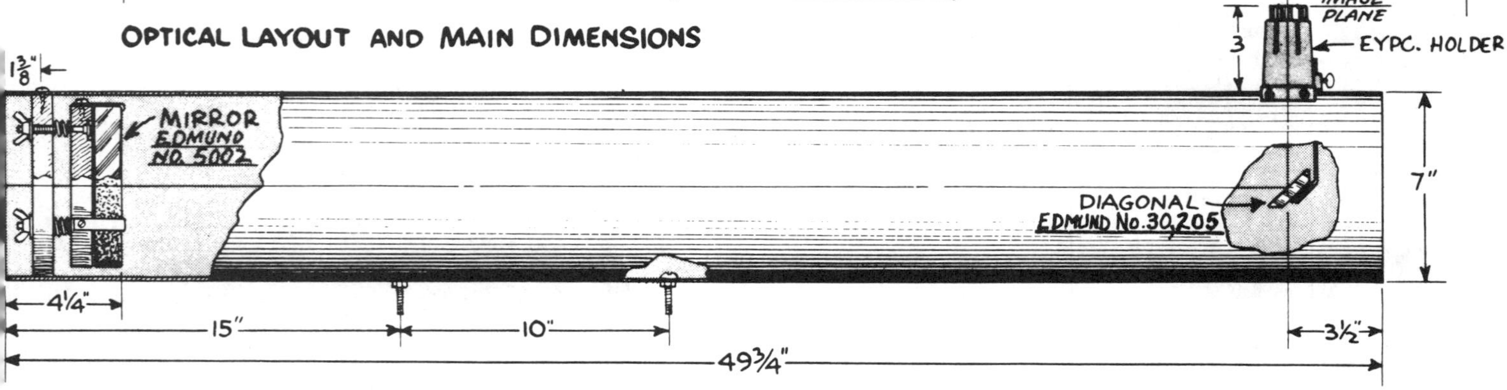

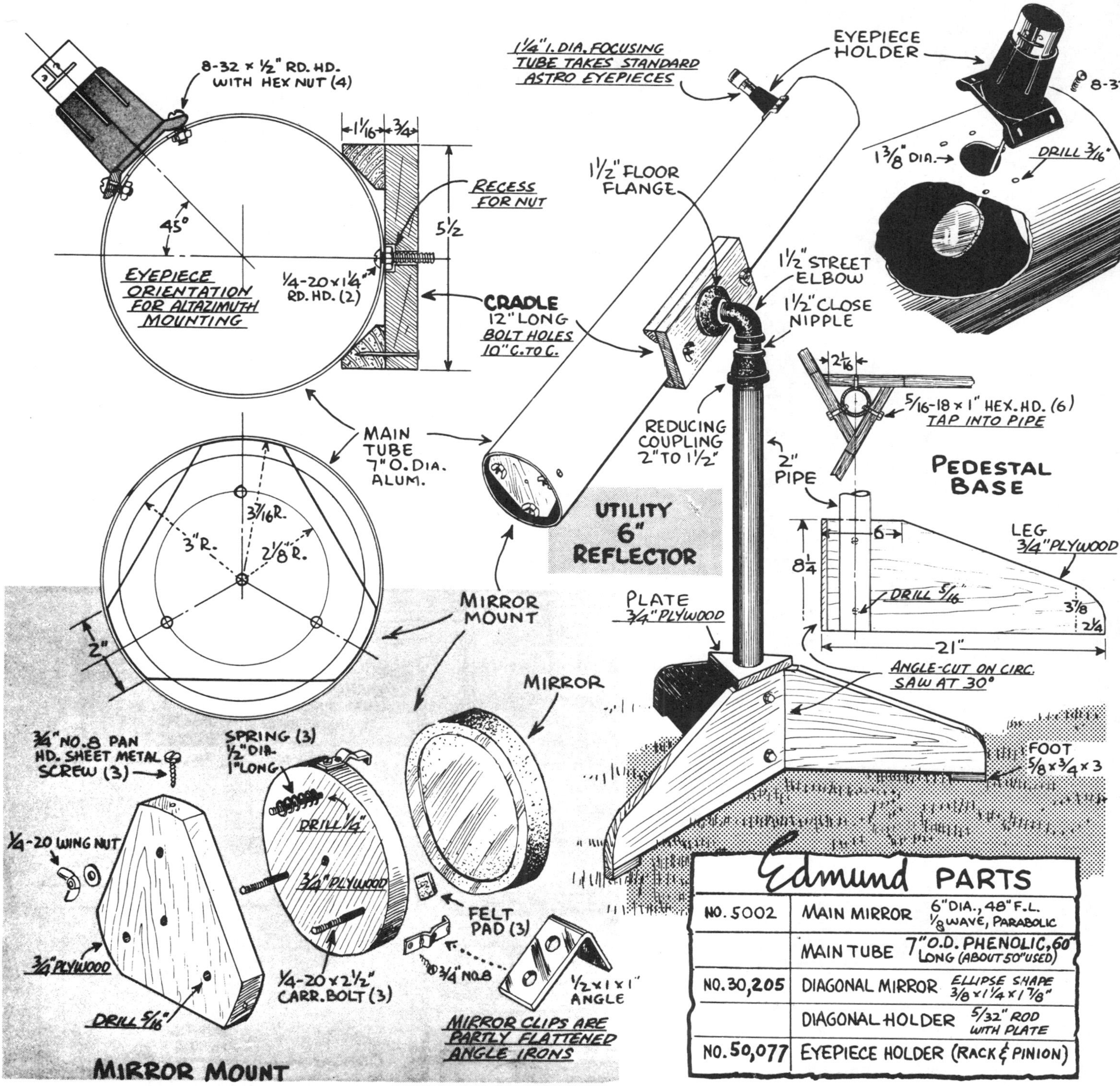

Edmund PARTS

NO. 5002	MAIN MIRROR	6" DIA., 48" F.L. ⅛ WAVE, PARABOLIC
	MAIN TUBE	7" O.D. PHENOLIC, 60" LONG (ABOUT 50" USED)
NO. 30,205	DIAGONAL MIRROR	ELLIPSE SHAPE ⅜ × 1¼ × 1⅞"
	DIAGONAL HOLDER	5/32" ROD WITH PLATE
NO. 50,077	EYEPIECE HOLDER (RACK & PINION)	

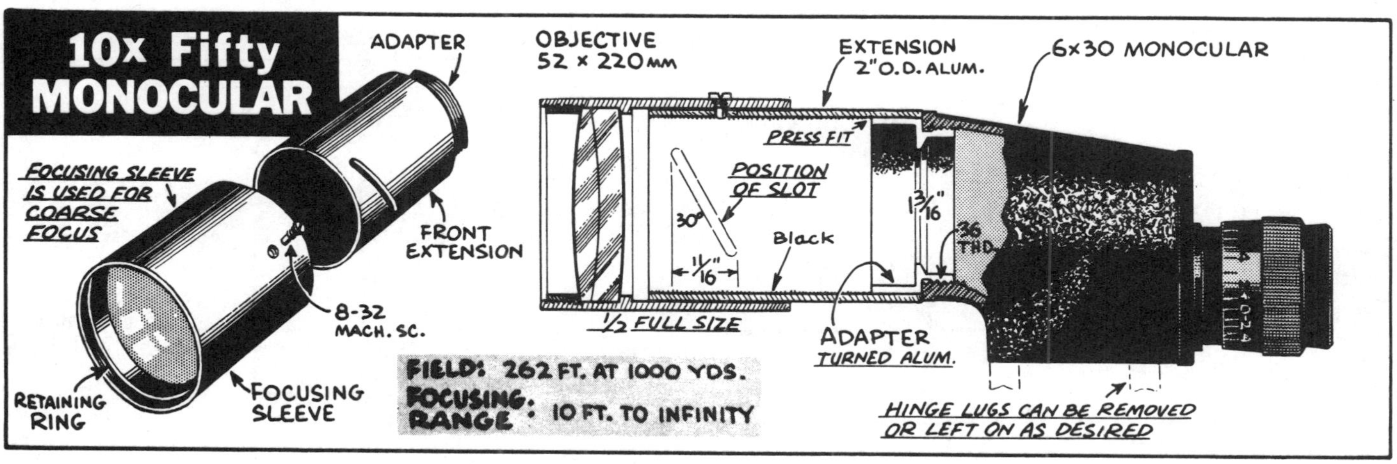

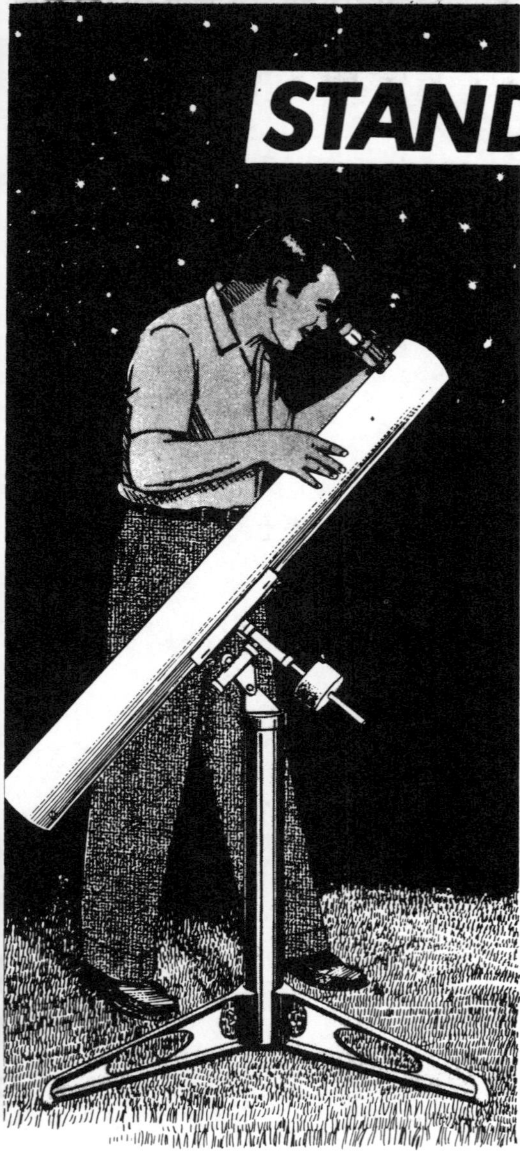

STANDARD 4¼" REFLECTOR

MOST scope builders buy some parts and make the rest. When it comes to a case of buying all readymade parts, you are money ahead to buy a finished telescope. Despite its financial weakness, the standard scope is useful as a how-to design in that it shows general construction.

Except for painting the main tube, you can build this standard design in one evening. The work is simply a matter of drilling the needed holes in the main tube and then making the assembly. The tube layout shown will handle mirrors from 45 to 45-3/4 inch focal length. If your f.l. is longer than this, it is best to set the mirror back a little more in order to take up the excess in focal length. The average spherical-ground mirror will show about 45-1/2 inches f.l. The main idea of spacing, of course, is to keep the image plane as close as possible to the diagonal for maximum light pick-up. This is not a great item in this case because the elliptical diagonal is ample size.

Simple bolt mounting is shown. The suggested orientation angles the eyepiece 35 degrees to the left as viewed from the eyepiece end of the tube. You will find this gives comfortable viewing in almost all positions. It favors an operating position east of pedestal for all south sky objects. If you view with your right eye (the usual case), the east-of-pedestal position is the most convenient because you can look through the telescope and then look at the sky directly with minimum head movement. To better the all-East viewing position for south sky objects, a double collar may be used at the head of the declination shaft to gain a little more swing through the meridian.

4¼" rich-field

MOST telescope nuts go for high power, but once this is out of your system you may get an even bigger kick from the brilliance and rich field of a low-power telescope. The one shown works at f/5 with a magnification of 19x showing 300 to 400 stars in a single view. Of course you have to grind your own mirror, but if you have done this just once before you will find no new problems except the work is more exacting.

MIRROR CORRECTION - 4¼" DIA. MIRRORS to ¼ WAVE

f/VAL.	F.L.	Radius	EDGE ZONE ¼ WAVE UNDER	EDGE ZONE PERF. PARABOLA	EDGE ZONE ¼ WAVE OVER	70% ZONE ¼ WAVE UNDER	70% ZONE PERF. PARAB.	70% ZONE ¼ WAVE OVER	ERROR OF SPHERICAL MIRROR
f/3	12¾"	25½"	.128"	**.132"**	.136"	.083"	**.086"**	.089"	7½ WAVES
f/4	17	34	.091	**.099**	.107	.059	**.064**	.069	3 WAVES
f/4.5	19⅓	38⅔	.077	**.088**	.099	.050	**.057**	.064	2 WAVES
f/5	21¼	42½	.066	**.080**	.094	.042	**.051**	.060	1½ WAVE
f/6	25½	51	.048	**.066**	.084	.031	**.043**	.055	⅞ WAVE
				r = 1.84"			r = 1.48"		

4¼" f/5 rich-field

M. - 19x	WITH EDMUND NO. 30787 EYPC.
FIELD - 2½ DEG.	
DIAG. OBST. -	9% with 1¼" / 6.4% with 1⅟16 (shown)
overall LENGTH - 21¾"	

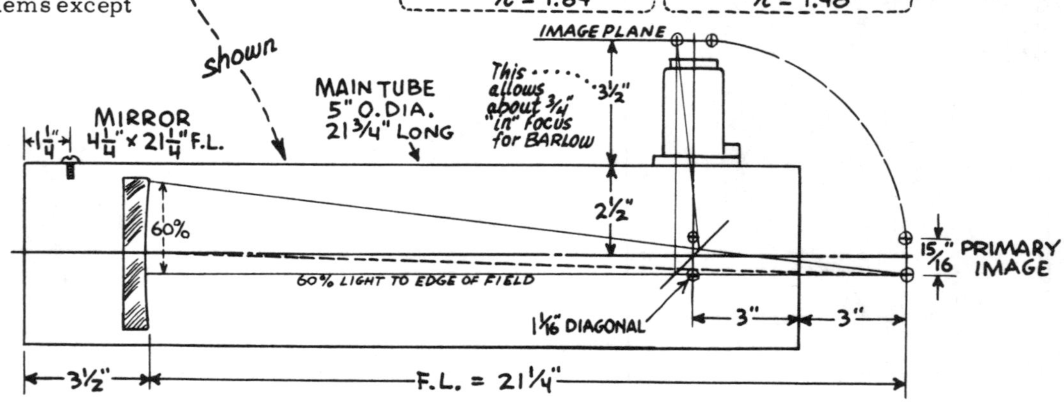

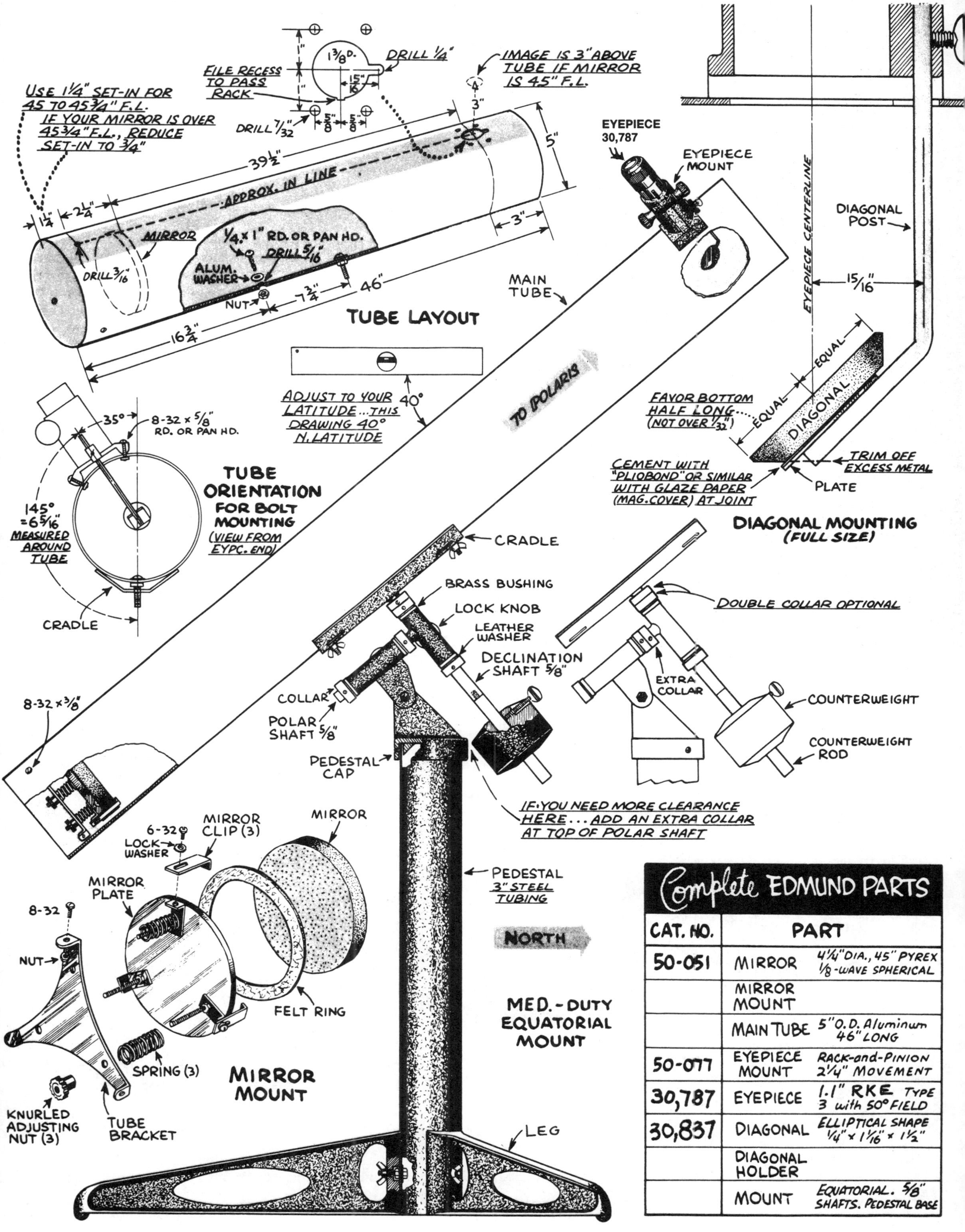

Complete EDMUND PARTS

CAT. NO.	PART	
50-051	MIRROR	4¼" DIA., 45" PYREX ⅛-WAVE SPHERICAL
	MIRROR MOUNT	
	MAIN TUBE	5" O.D. Aluminum 46" LONG
50-077	EYEPIECE MOUNT	Rack-and-Pinion 2¼" Movement
30,787	EYEPIECE	1.1" RKE Type 3 with 50° FIELD
30,837	DIAGONAL	ELLIPTICAL SHAPE ¼" x 1 1/16" x 1½"
	DIAGONAL HOLDER	
	MOUNT	EQUATORIAL. 5/8" SHAFTS. PEDESTAL BASE

$4\frac{1}{4}$" REFLECTOR in Square Plywood Tube

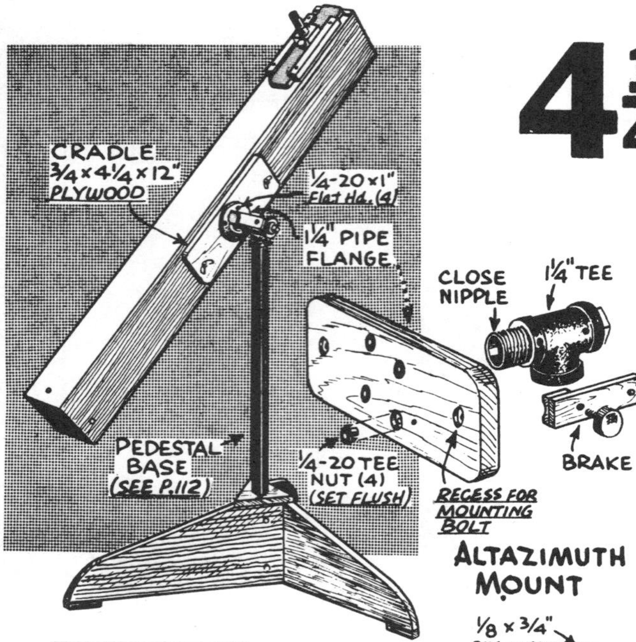

PLYWOOD TUBES are practical for telescopes, the main fault being that the common square shape is a bit more bulky than a round tube of cardboard or metal. A good feature is low heat conduction which means less air disturbance. The flat sides of the square tube are made-to-order for focusing devices working lengthwise, such as the one shown. This permits 6 in. of travel--plenty for all eyepieces and most attachments.

Assemble the tube with small nails and let it remain in this form until you are sure everything fits. Then, make the permanent assembly with glue, plus screws or nails as needed. The glue is the important part since this is the best way to get rigid joints. The turn-on-threads altazimuth mount is a good starter, readily converted to equatorial by adding a 45-deg. street elbow. If you use an equatorial mount, it is somewhat more convenient to have the eyepiece at the side of the tube.

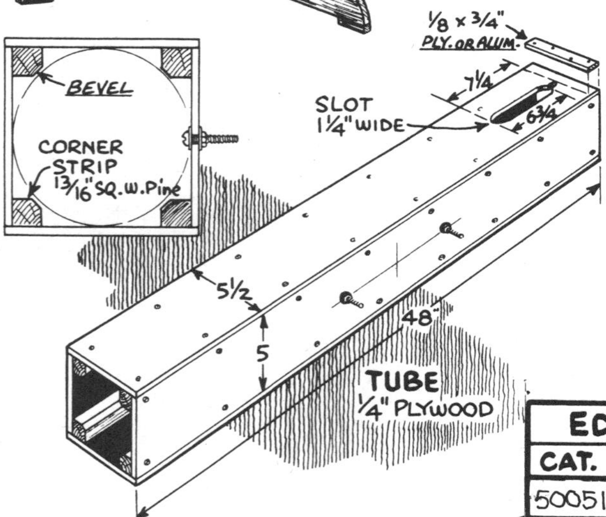

EDMUND PARTS	
CAT.	PART
50051	4 1/4" f/10 MIRROR
32454	DIAGONAL 21 x 30
40195	RACK
40197	PINION SHAFT

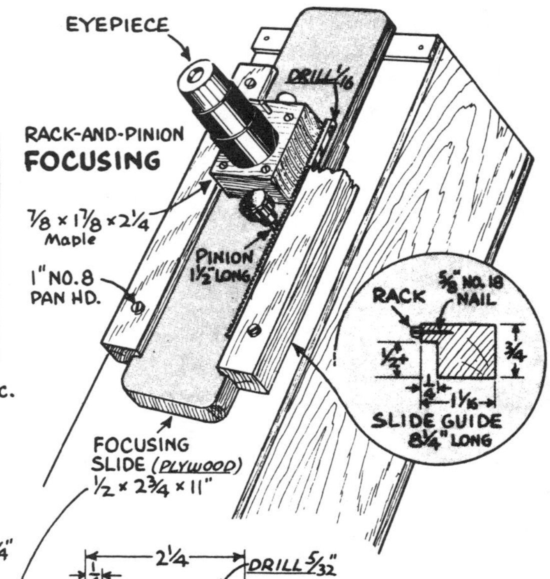

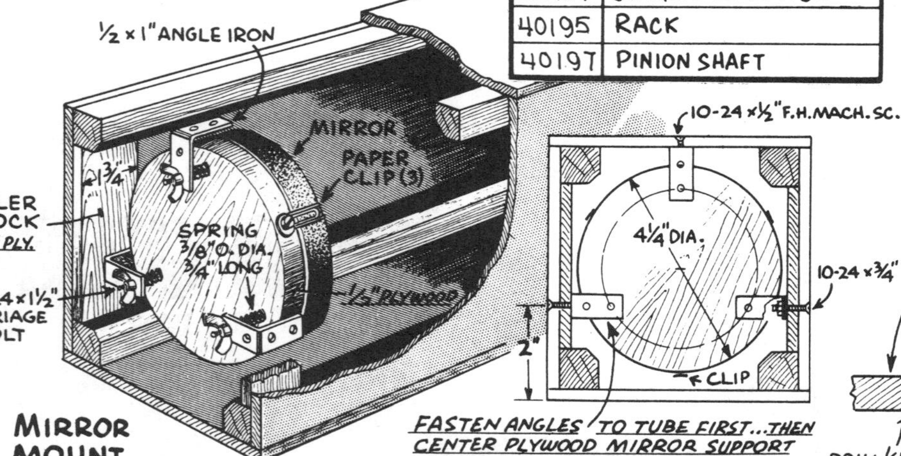

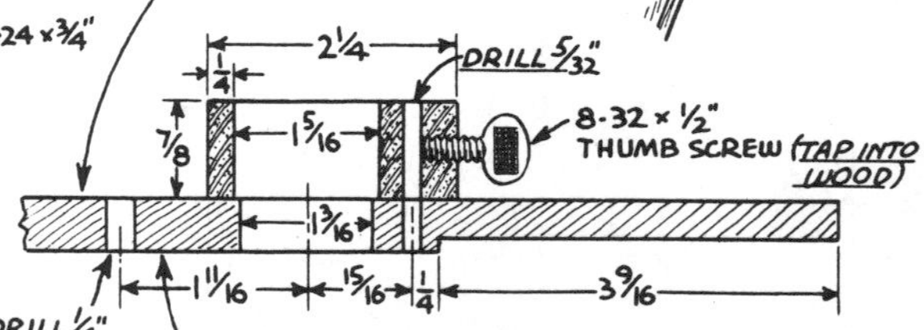

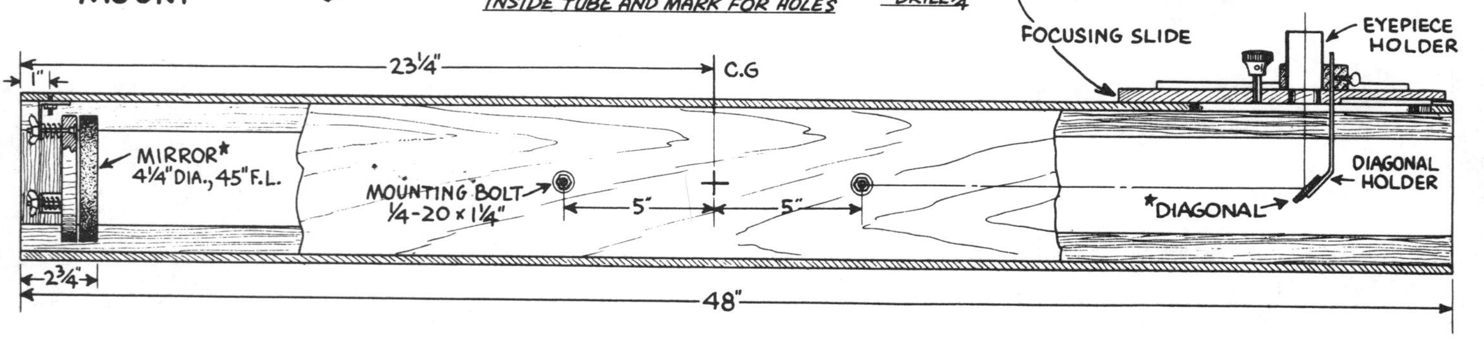

smallest practical size
2½" REFLECTOR

ALTHOUGH it has a diagonal obstruction of twice the recommended maximum, this mini 2-1/2-in. reflector has a sharp eye for bright objects in the night sky. The light loss of 12% is hardly noticed, while the more-than-normal obscuration of the blind spot is annoying only when you use a low-power eyepiece for daytime viewing. The telescope weighs in at less than 2 lbs., is easy to cradle in your arms, but is at its best on a light-duty mount at 30x or more.

Light-duty lock washers are used under the plastic eyepiece holder to provide a diagonal adjustment. Do this before you mount the main mirror. Center your eye on the focusing tube minus the eyepiece. The condition you want to see is that the inside and end of the main tube appears to be a straight continuation of the focusing tube.

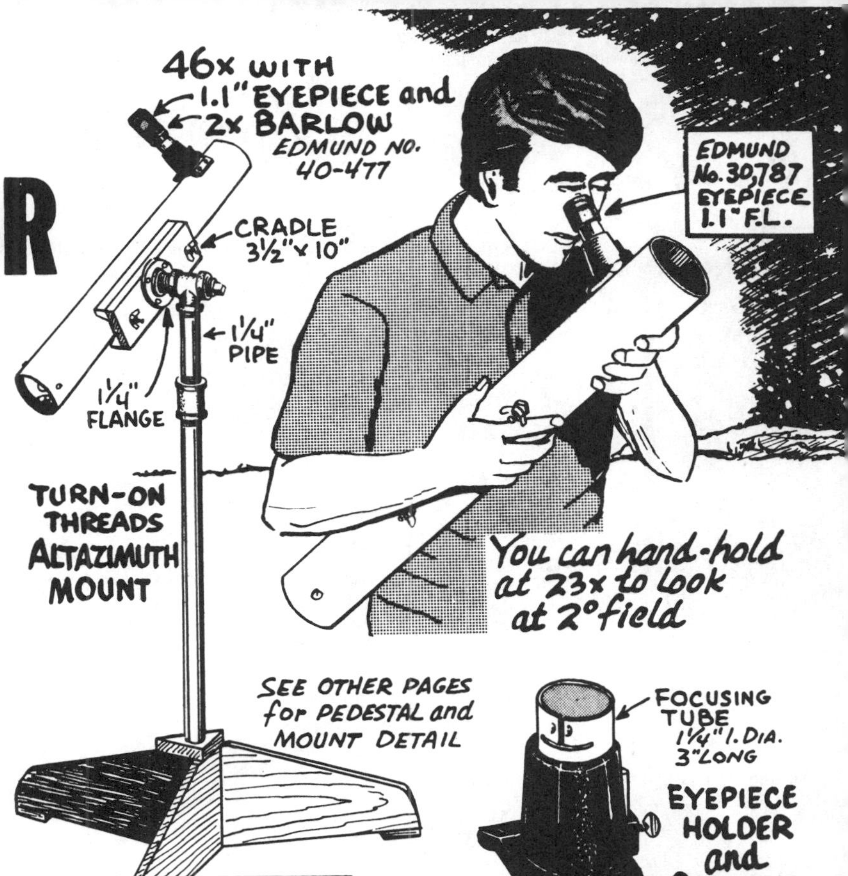

The BLIND SPOT in a REFLECTOR

MIRROR DIA.	MIRROR AREA	DIAGONAL WIDTH	DIAGONAL AREA	% BLIND SPOT
8"	50.3	1½"	1.77	3.5%
6"	28.3	1¼	1.23	4.3%
4¼"	14.2	1¹⁄₁₆	.80	6.3%
3"	7.1	⅞	.60	8.5%
2½"	4.9	⅞	.60	12.2%
	SQ.IN.		SQ.IN.	

MORE THAN RECOMMENDED TOLERANCE *

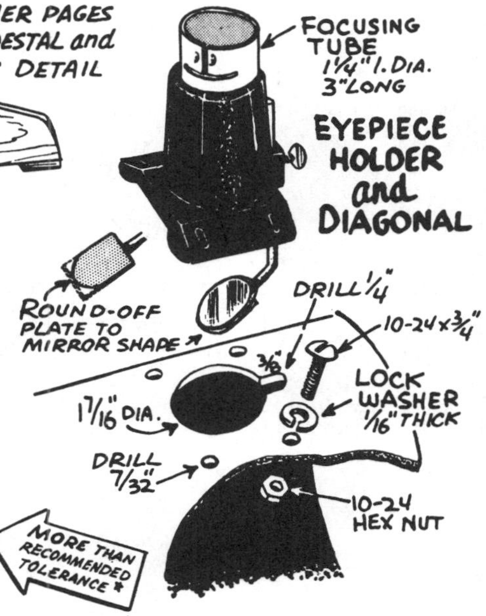

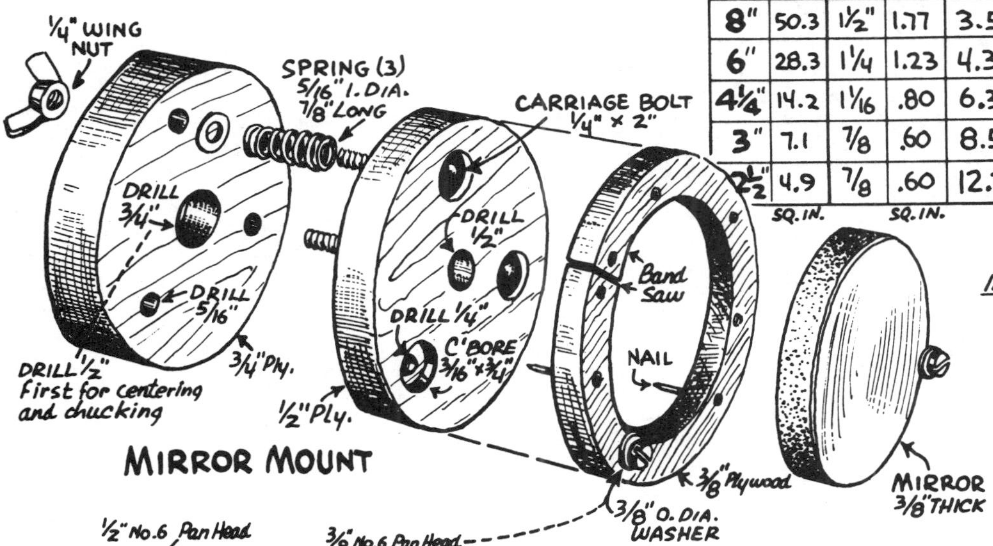

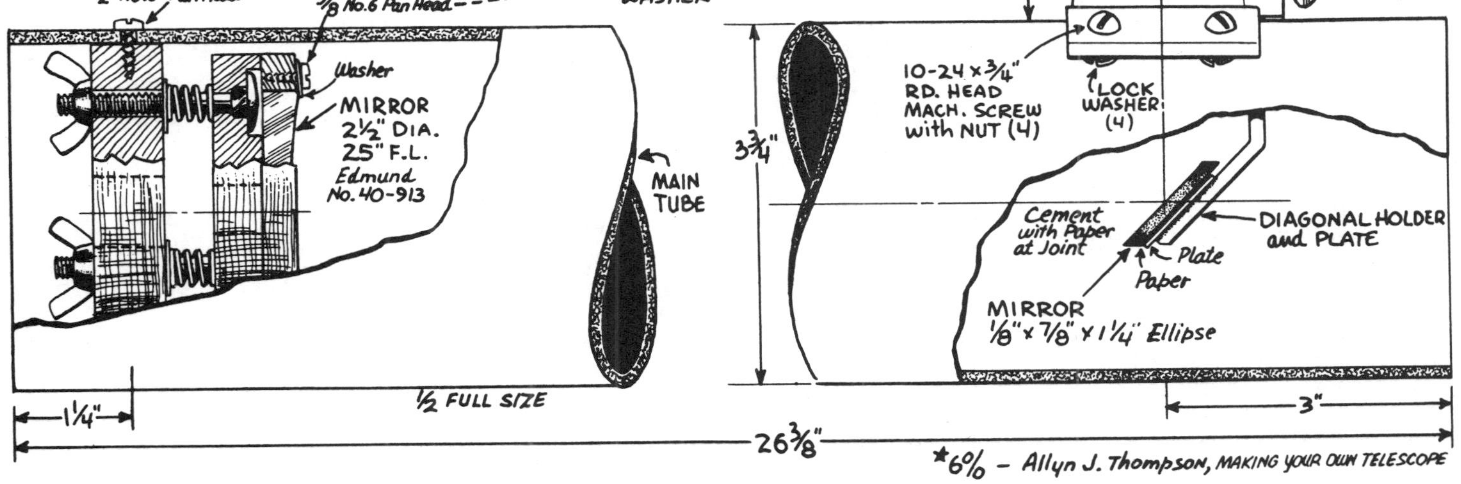

*6% - Allyn J. Thompson, MAKING YOUR OWN TELESCOPE

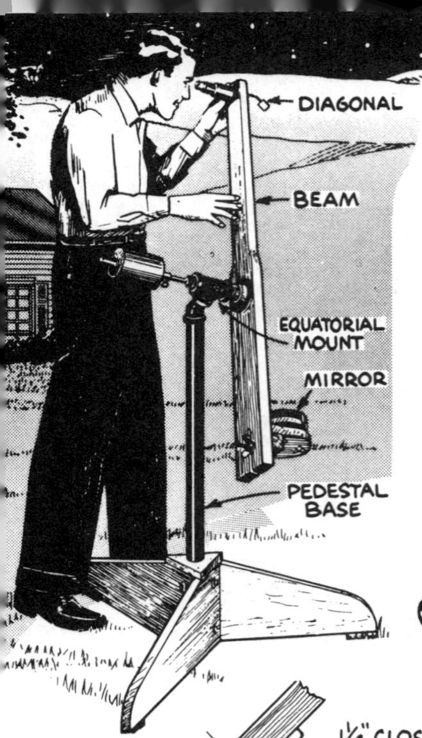

4¼-inch "SKY BEAM"

USED in the dark and shielded from outside light, the astro telescope does not actually require a tube. Hundreds of scope builders seeking a way to make a cheap reflector have taken advantage of this fact, mounting the mirror and eyepiece holder at opposite ends of a single beam of wood. The design shown is typical "homemade" construction with equatorial mount made of pipe fittings. The fixed polar axis of 45 degrees is satisfactory for latitudes between 40 and 50 degrees, and as a matter of fact can be used with some success at any latitude. The movement on both axes is obtained simply by letting the pipe fittings turn on their own threads. The 1-1/16 inch Ramsden gives 43X. The whole outfit loaded for stars packs about 20 lbs., three-fourths of this being in the mount.

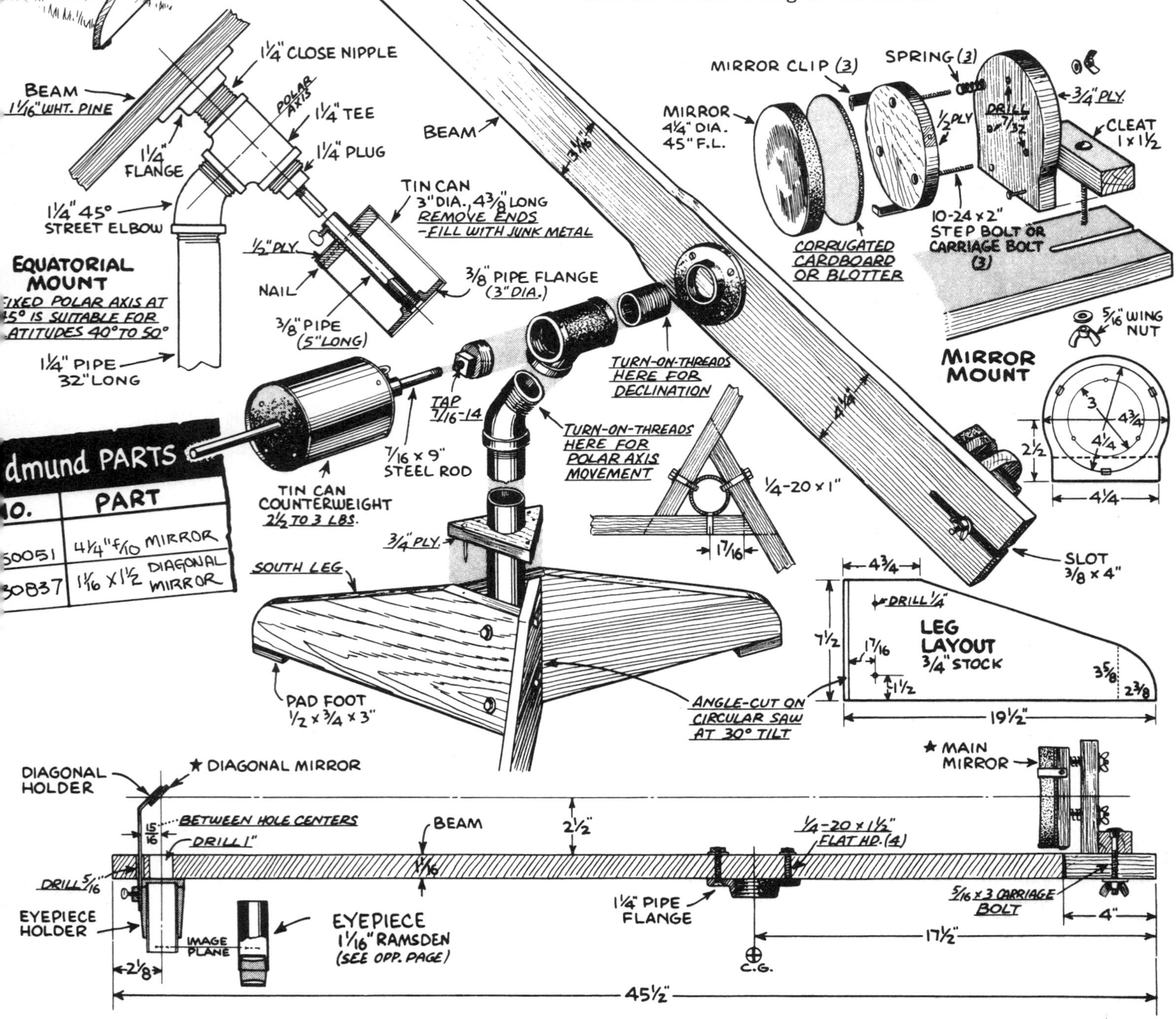

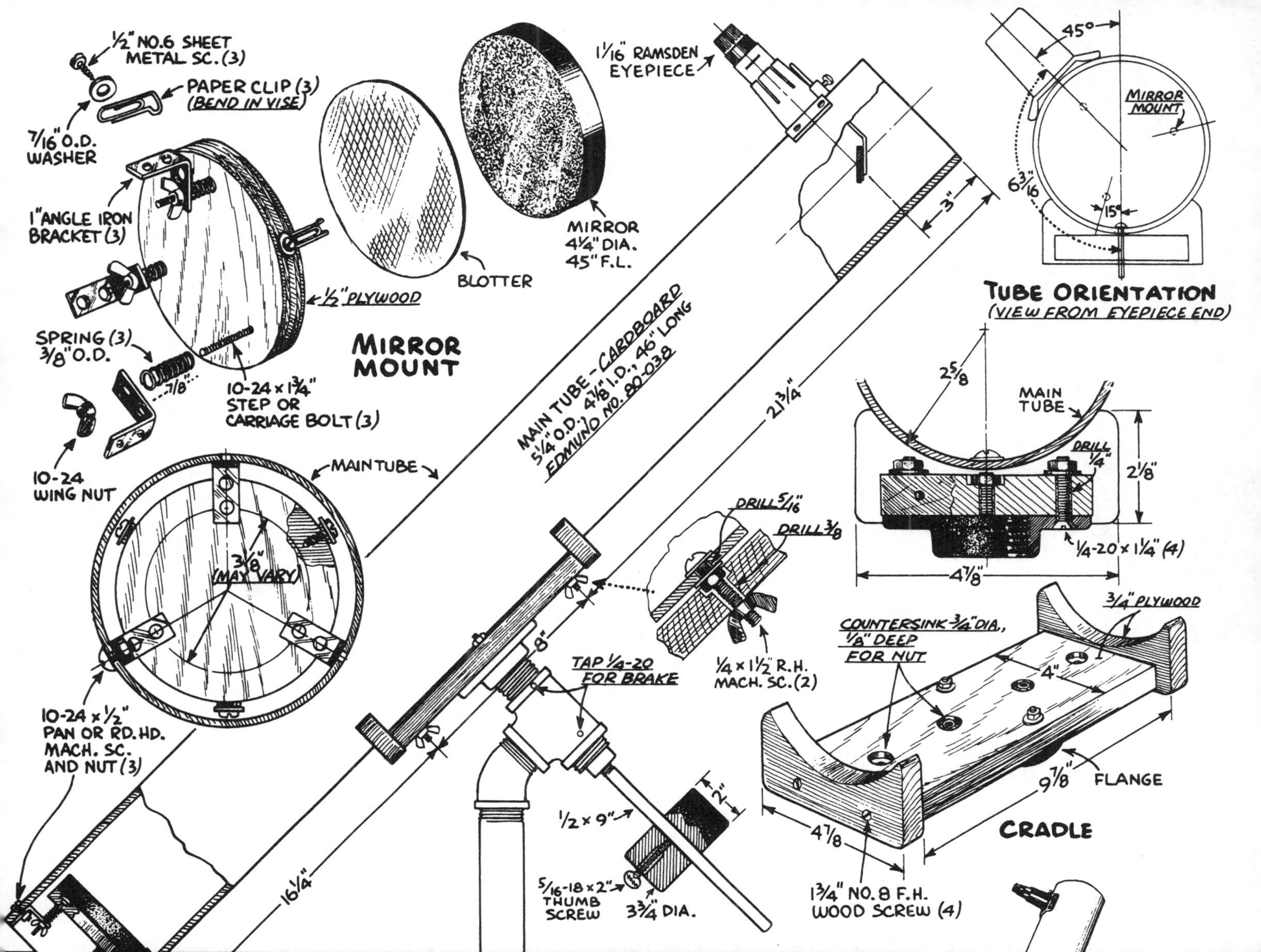

Easi-bilt 4¼" REFLECTOR

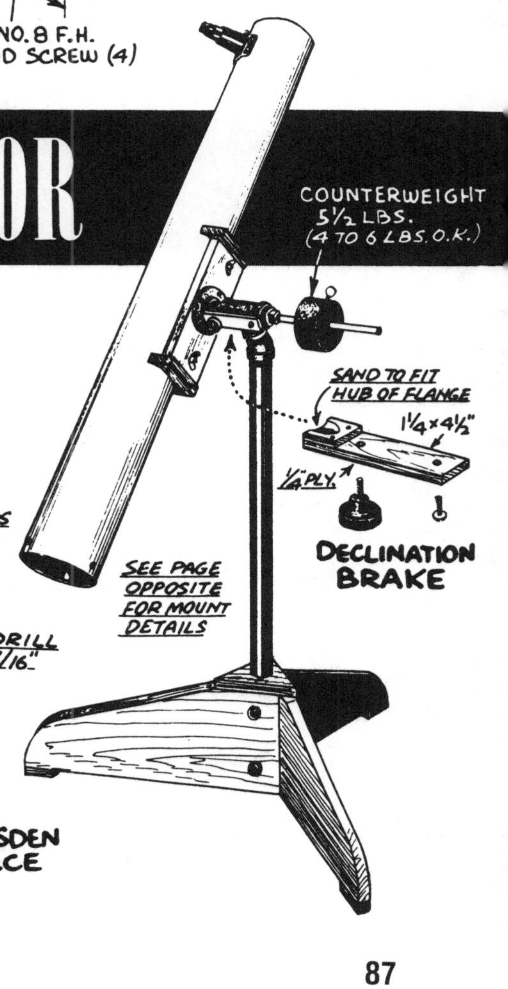

IF YOU want a good telescope on a lean budget and minimum work, **this is it. You use the same Edmund telescope kit specified on opposite page, plus a cardboard tube which adds a few dollars to the cost, but worth every penny for better appearance and protected optics. The mirror mount is an inexpensive design costing about a dollar for materials. The equatorial mount is the same as opposite page with a handy declination brake added.**

The weight of this outfit is about 22 lbs., mount and all. You will need lenses to make a 1-1/16in. Ramsden eyepiece giving 43x.

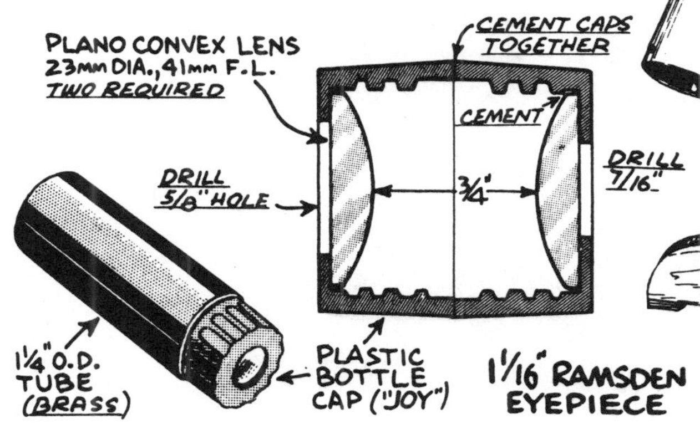

16x SKY SWEEPER

YOU CAN add to the versatility of any small reflector by using a sliding mirror fitted in a separate tube. Such an arrangement allows the telescope to be used as a regular astroscope as well as a photographic telescope, a prism scope, etc., the sliding mirror providing for air-path adjustment and coarse focusing. A fine focus can be obtained at the eyepiece, either by sliding it in the eyepiece plate, or, by using a focusing eyepiece, such as the eyepiece from a 7 x 50 binocular or monocular. The 16x power rating is obtained with the 27.5mm eyepiece from the 7x monocular, or a 28mm RKE, Edmund No. 30,787.

Mainly, this design is intended for low-power sweeping and can be hand-held satisfactorily if the power is not over 20x. You may need an adapter to use the focusing eyepiece from a 7x binocular or monocular, although some designs are 1-1/4 inch diameter at the thread and so fit the 1-1/4 inch eyepiece plate. The eyepiece plate has a brass tension sleeve which can be adjusted just right to grip standard 1-1/4 inch diameter eyepieces firmly but loose enough to permit slide focusing.

The 3-inch mirror of 18 in. focal length works at f/6 and is preferably parabolized although a spherical shape works fairly well at low power. The Edmund mirror specified is parabolized; if you make your own you can suit yourself. Much of the long focusing travel provided by the slide tube is needed only for a prism erector. If you do not use this kind of erector, the sliding tube could be omitted entirely. In such case, the eyepiece plate would be made an inch or more thick for added slide focus adjustment. If used, the sliding tube should be a neat slide fit to maintain proper collimation.

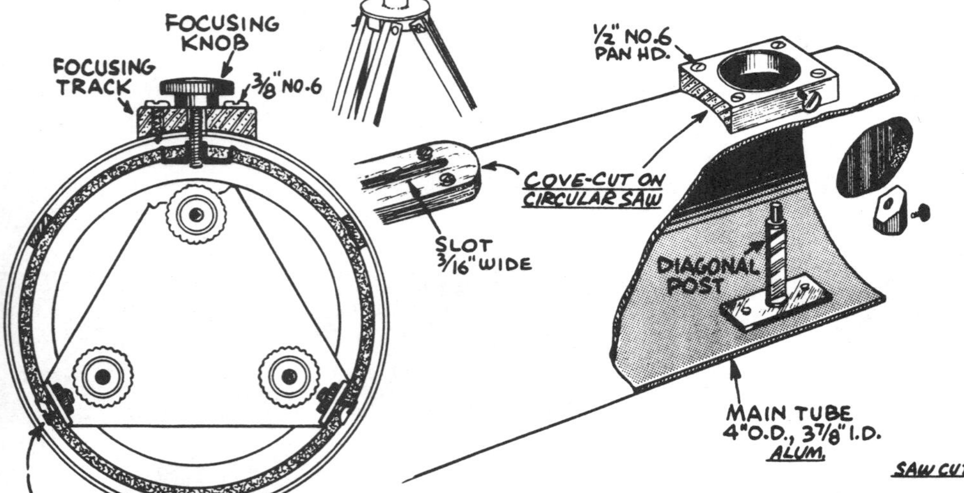

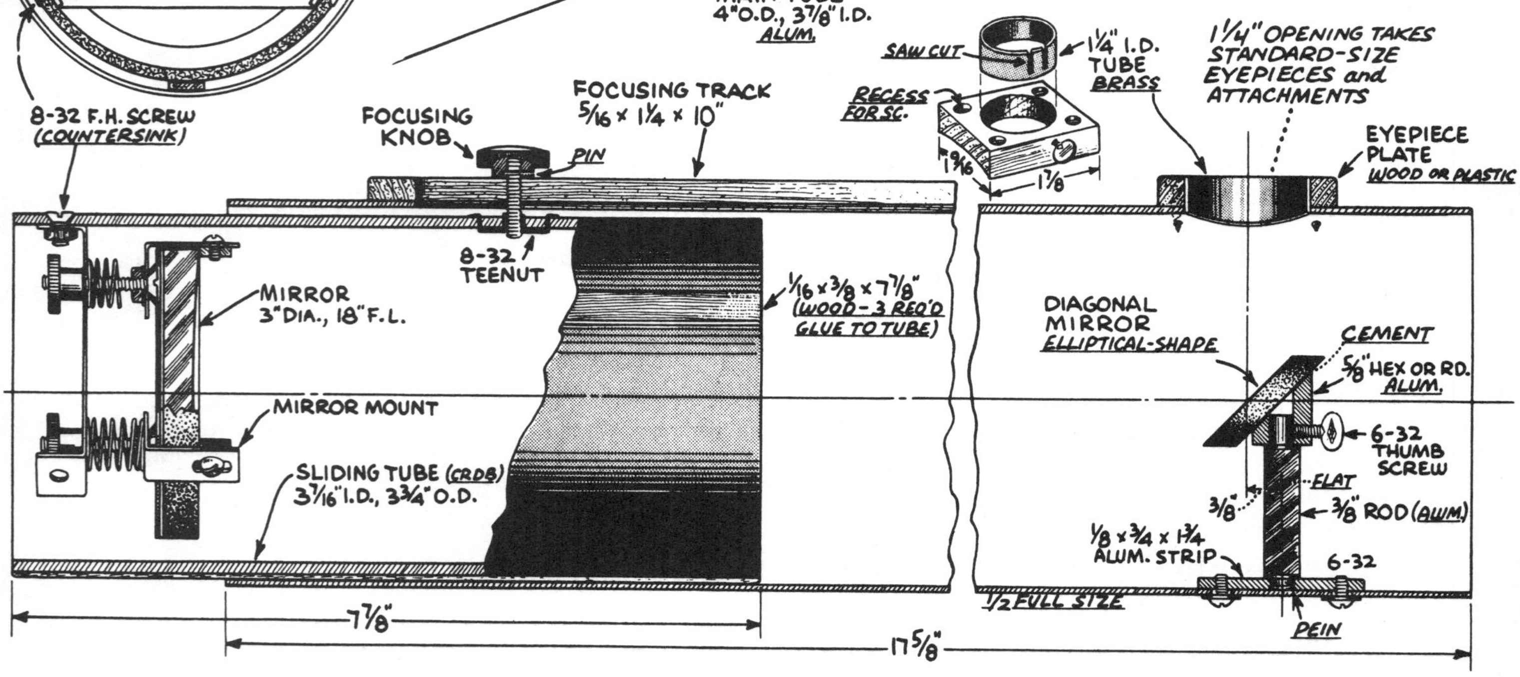

16x 'Bazooka' SPOTSCOPE

ONE OF the more popular methods of making an erect-image reflecting telescope is to use a roof prism in the manner shown for this 16x design. The on-shoulder mounting is comfortable, steady; the field is bright and covers 2.6 degrees, equal to 136 ft. at 1000 yards.

The construction is mainly a matter of making the prism support. This is satisfactory if made of wood. You can saw to shape on the band saw, and the holes needed can be drilled on the drill press except possibly the large one needed to accommodate the threads of the monocular eyepiece.

The telescoping main tube provides plenty of adjustment for focusing near objects; fine focus is obtained with the focusing adjustment of the eyepiece. The Edmund 3-in. mirror is now supplied parabolized. Of course it costs more, but the image is much sharper than with a spherical shape as supplied originally for this design.

Edmund PARTS

CAT. NO.	PART
31400	MAIN MIRROR 3" f/6 17½" F.L.
	MIRROR MOUNT 3-POINT SPRING ADJUSTMENT. Fits tube below
32454	DIAGONAL MIRROR 21 X 30 mm
	MAIN TUBE
3002	ROOF PRISM CORRECTED ROOF. 21MM (.83") FACE
41394	EYEPIECE ...from 7x Monocular. 27.5MM F.L., focusing mount

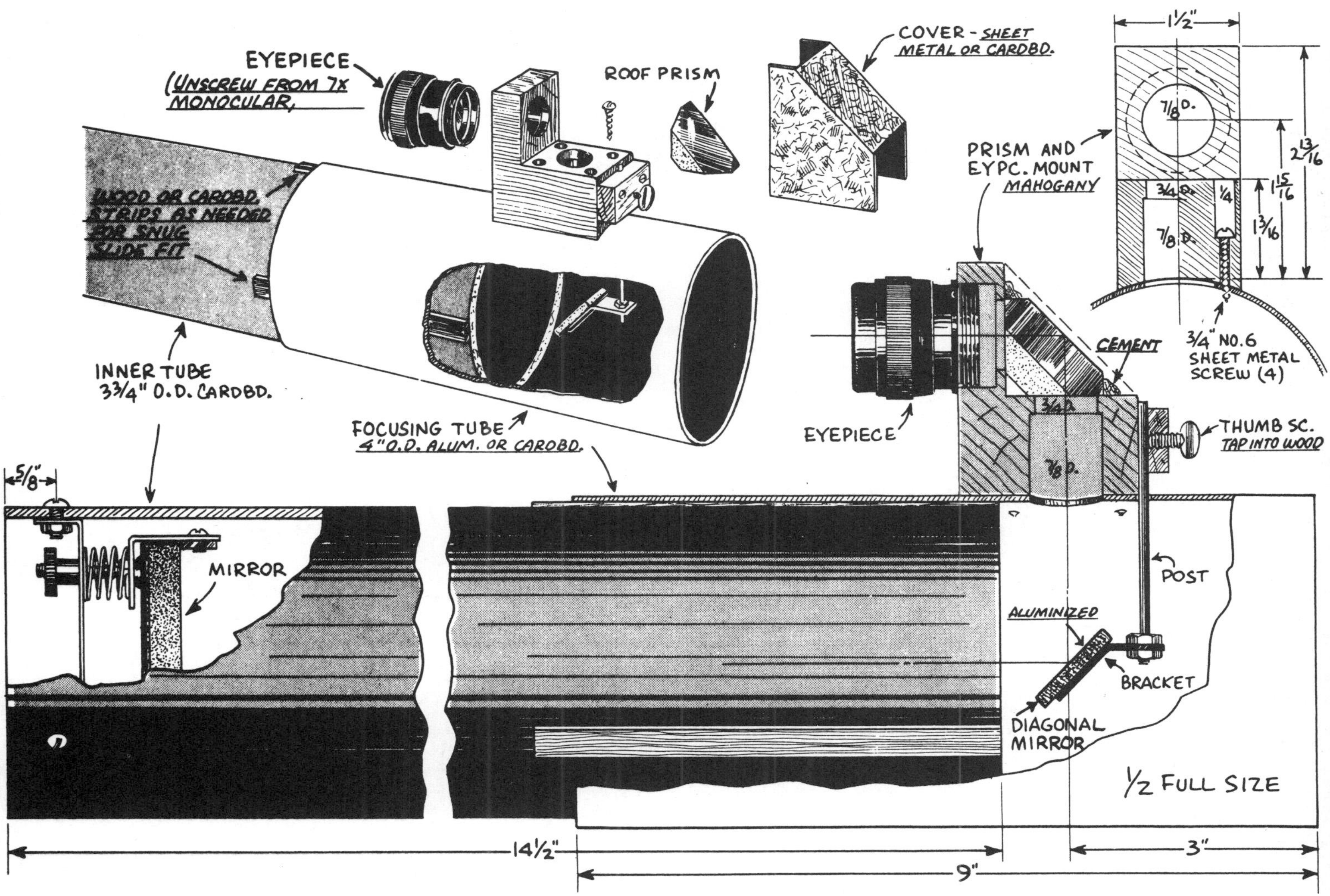

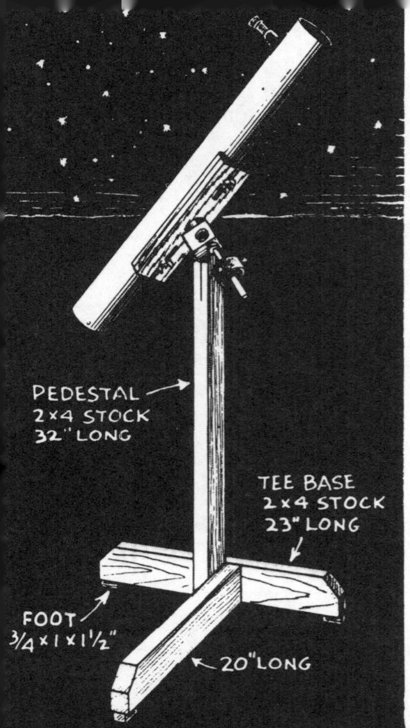

3-inch Reflector

This 3-inch reflector is easy to build, being only a job of putting the various parts together. A scope of this size is excellent for moon and planet viewing. With an erecting eyepiece as shown on opposite page you can see daytime land objects with amazing brightness and clarity.

A minimum-size diagonal mirror is used, and it must be located squarely under the

EDMUND PARTS

CAT. NO.	ITEM
50082	MIRROR KIT
	MIRROR MOUNT
	MAIN TUBE
	EYEPIECE HOLDER
30875	DIAGONAL MIRROR

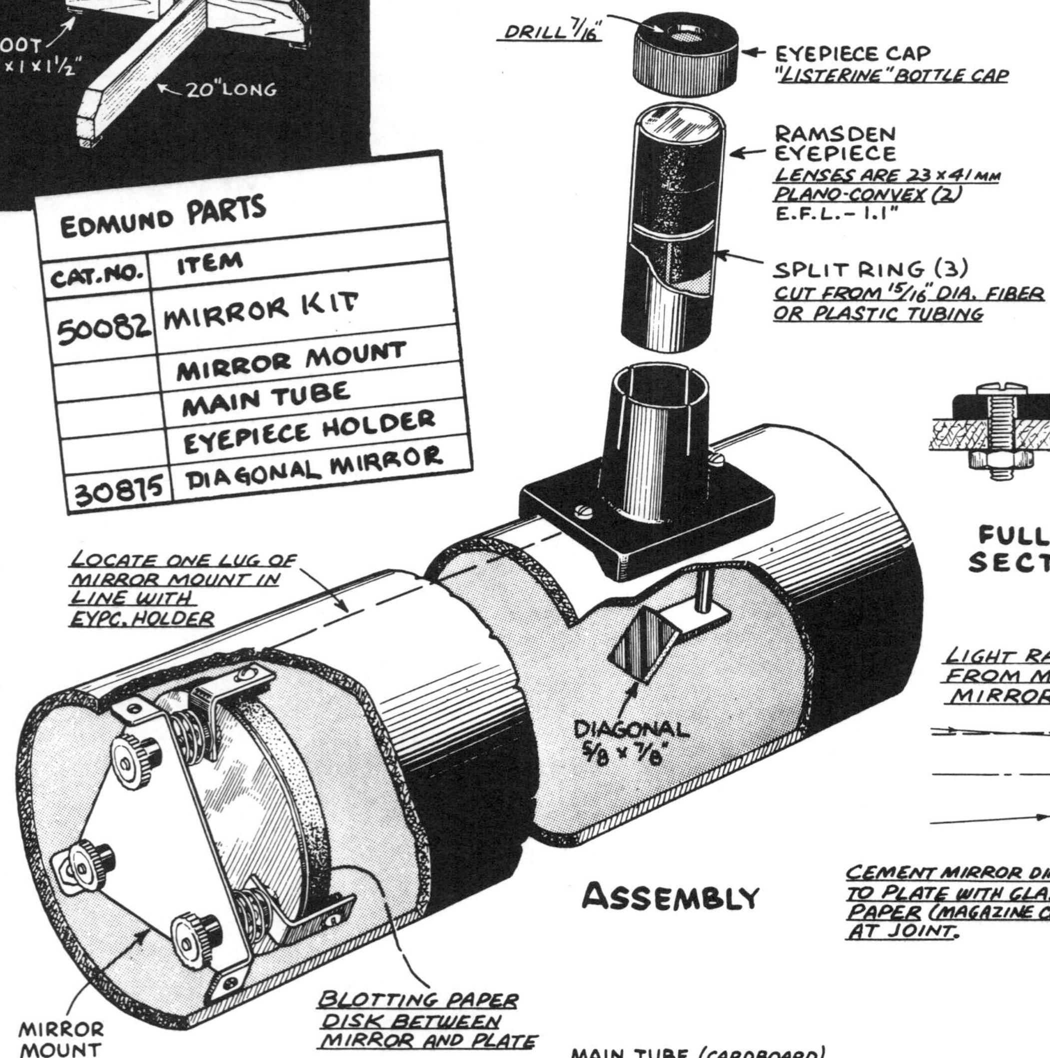

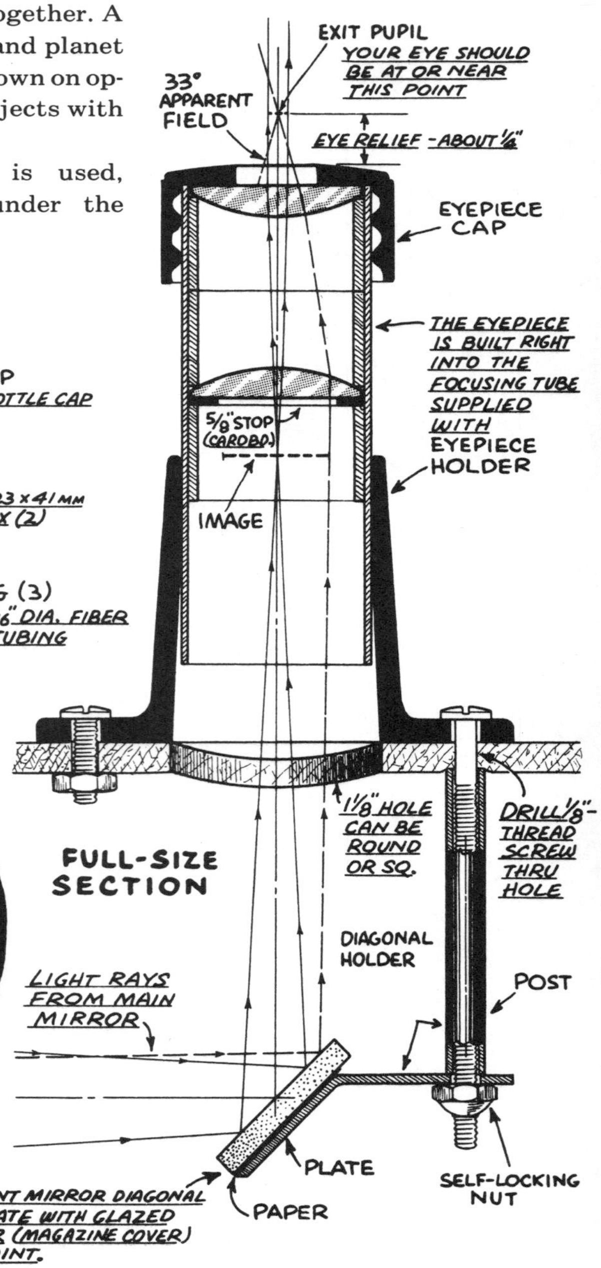

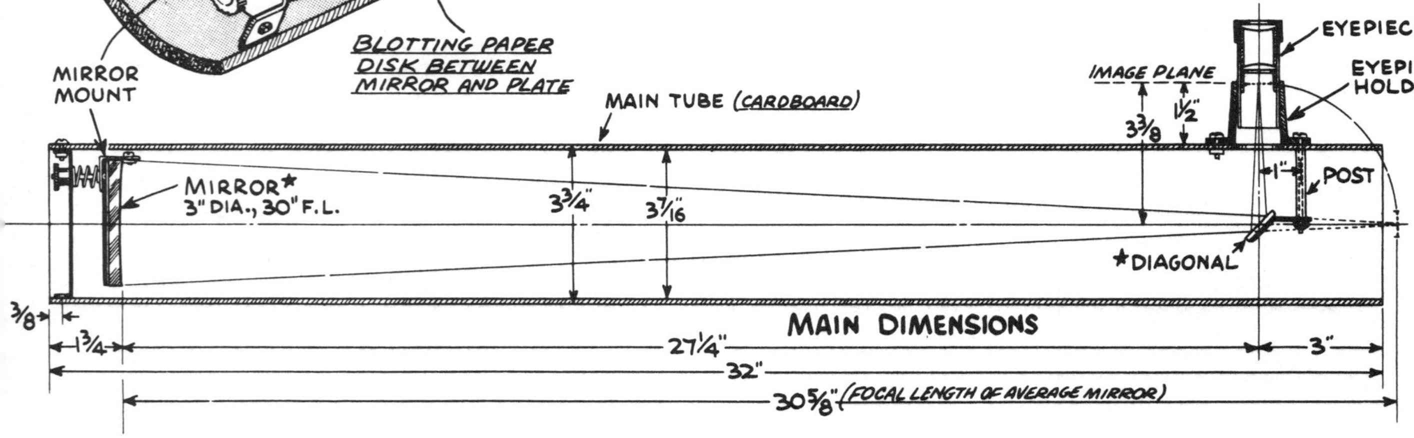

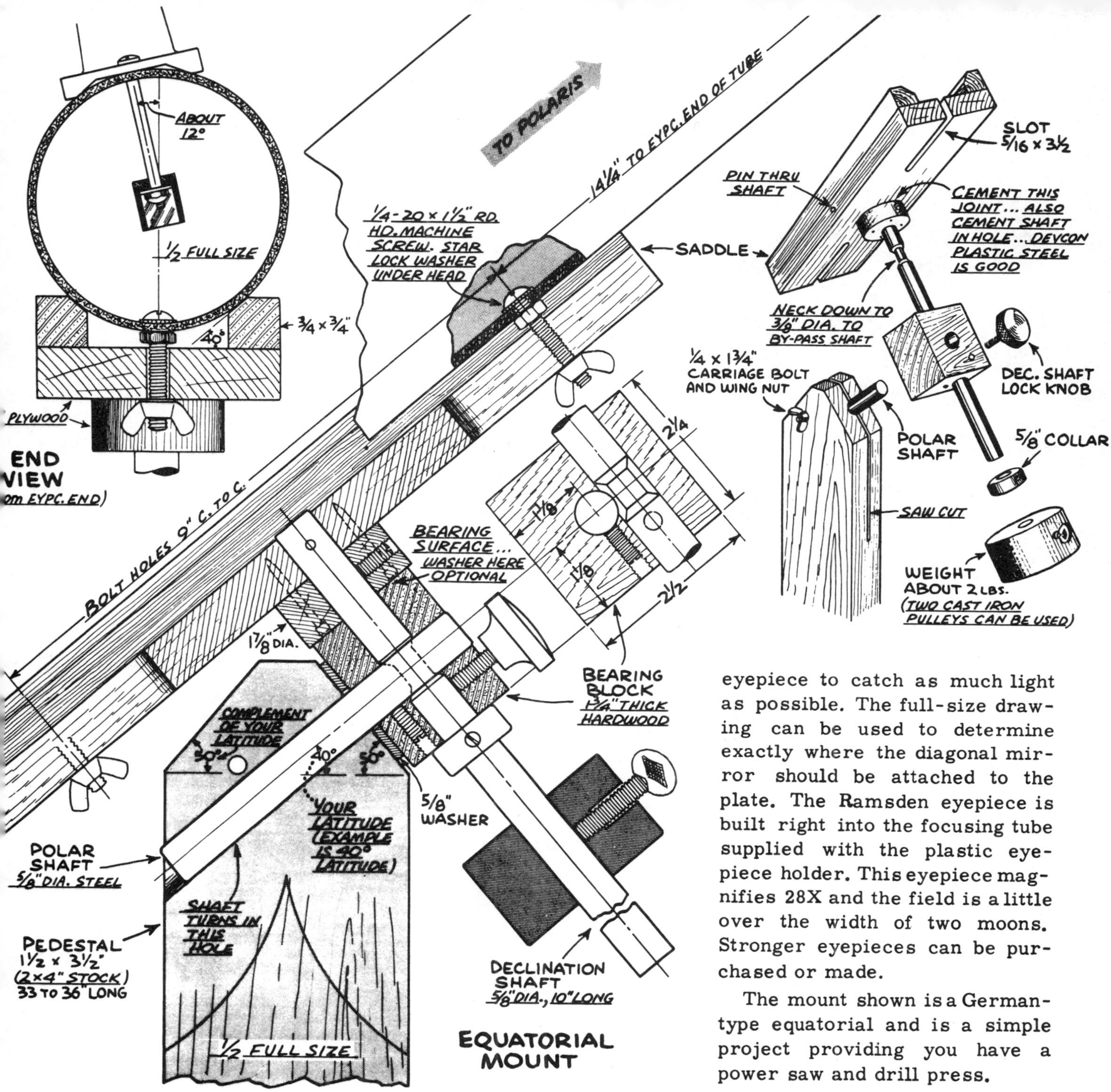

eyepiece to catch as much light as possible. The full-size drawing can be used to determine exactly where the diagonal mirror should be attached to the plate. The Ramsden eyepiece is built right into the focusing tube supplied with the plastic eyepiece holder. This eyepiece magnifies 28X and the field is a little over the width of two moons. Stronger eyepieces can be purchased or made.

The mount shown is a German-type equatorial and is a simple project providing you have a power saw and drill press.

Erecting Eyepiece for your 3" Reflector

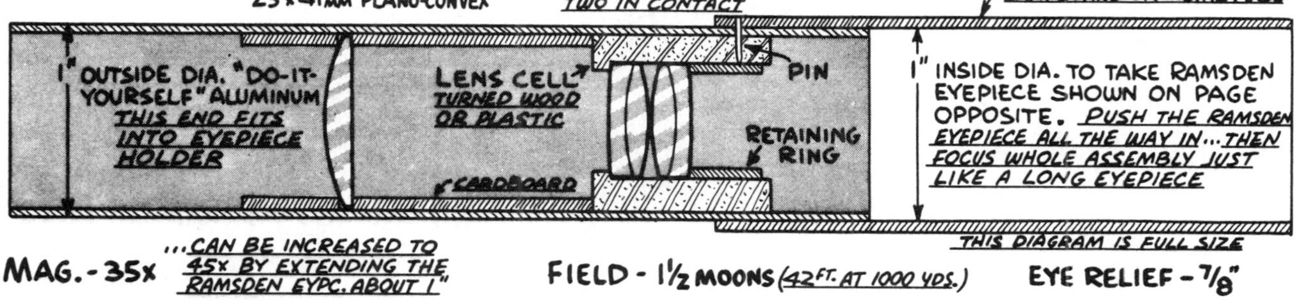

MAG. - 35x ...CAN BE INCREASED TO 45x BY EXTENDING THE RAMSDEN EYPC. ABOUT 1" FIELD - 1 1/2 MOONS (42 FT. AT 1000 YDS.) EYE RELIEF - 7/8"

erect-image 3" REFLECTOR

FOR moon-gazing or land-gazing you will like this 3-inch reflector with built-in erecting system. Two erecting systems are shown. The one using the 30mm diameter lenses is easiest to make because this size is an exact fit for the tubing used. The alternate design has a trifle flatter field, is longer.

The main feature of a built-in erector is that it puts the narrowest part of the light cone close to the diagonal. The small diagonal specified is more than big enough to field all of the light rays, whereas the same flat in a standard reflector misses about half of the edge-of-field rays.

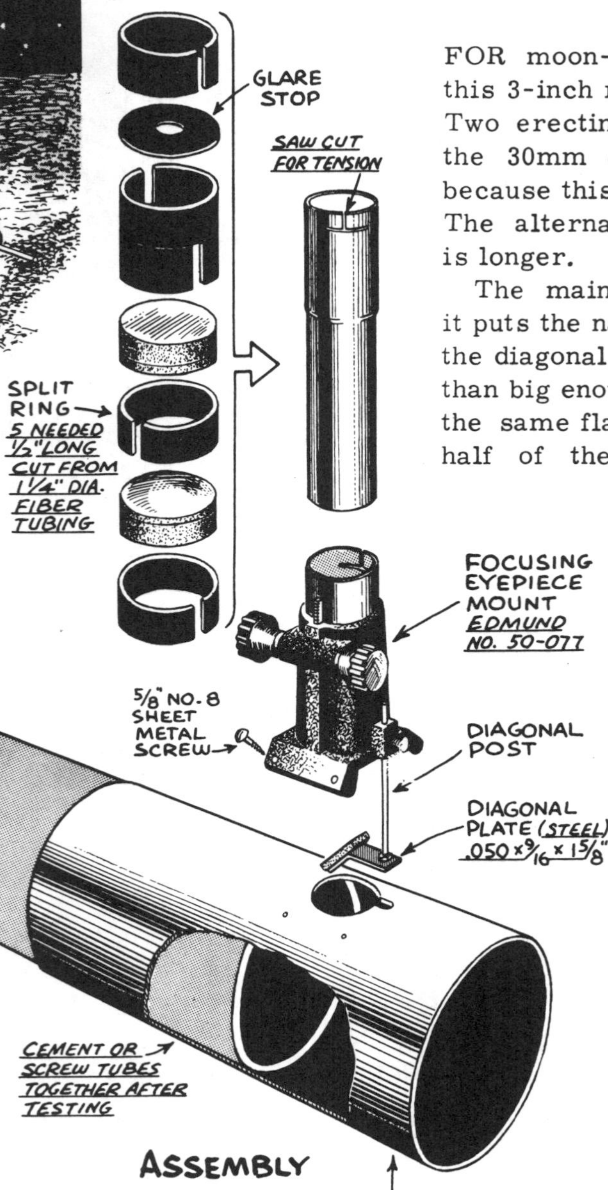

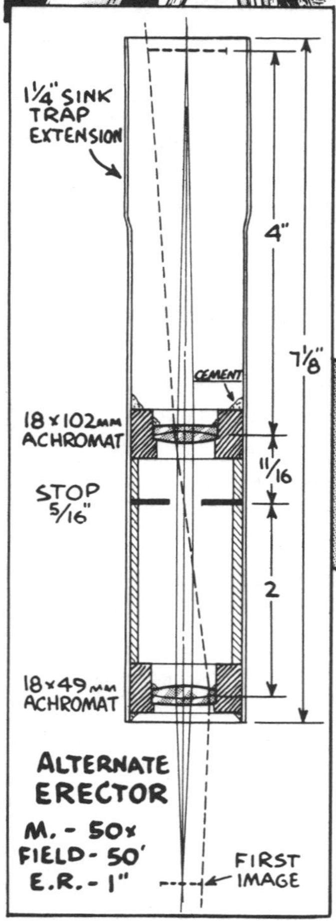

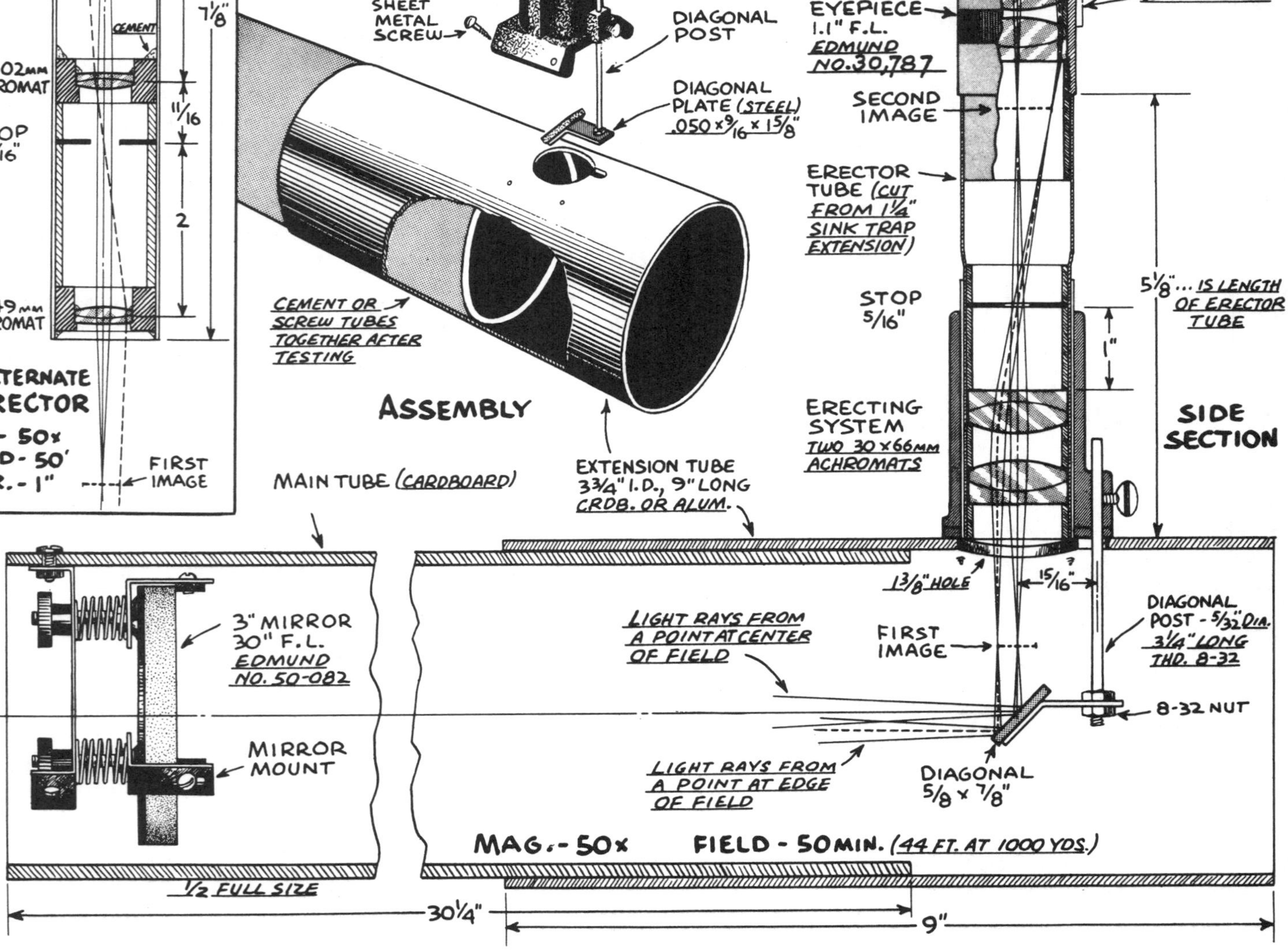

Altazimuth or Equatorial

Convertible 3-inch REFLECTOR

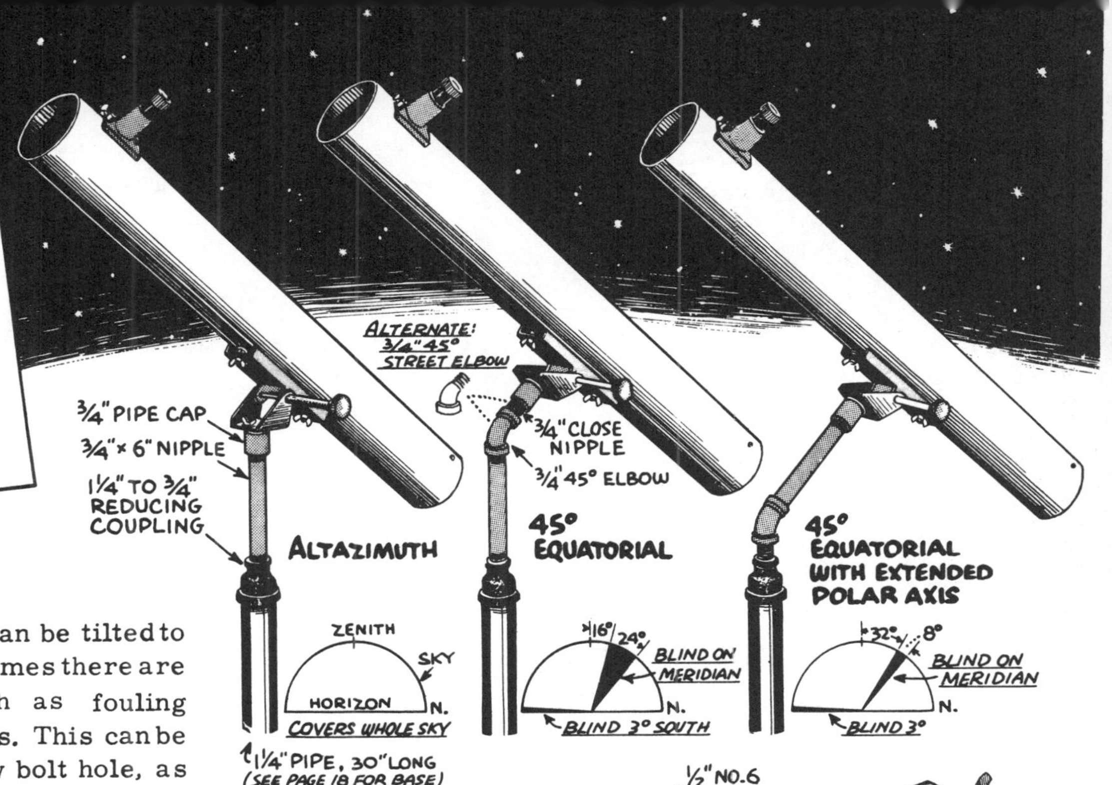

MOST altazimuth mounts can be tilted to make an equatorial. Sometimes there are minor complications, such as fouling at one of the mounting bolts. This can be corrected by drilling a new bolt hole, as shown at bottom center.

A minor fault of the short fork as an equatorial is that part of the north sky on the meridian is not accessible. This can be partly corrected by extending the polar axis, as in right-hand drawing above. The blind area is only on the meridian; you can see all objects in this area when they are one hour or more on either side of the meridian.

The mounting arrangement shown is easily converted from altazimuth to equatorial by changing the pipe fittings.

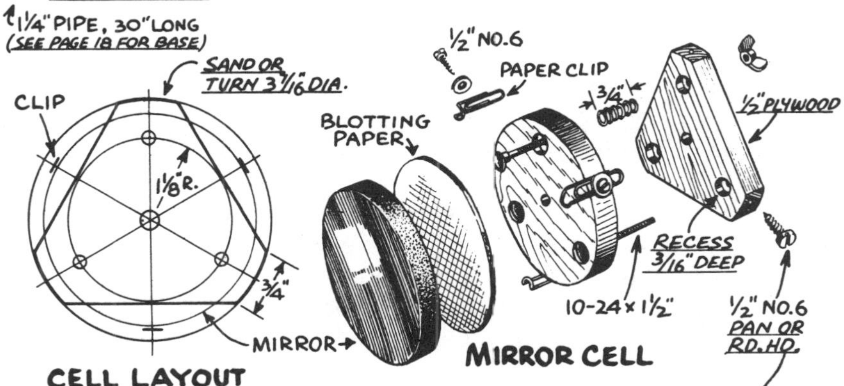

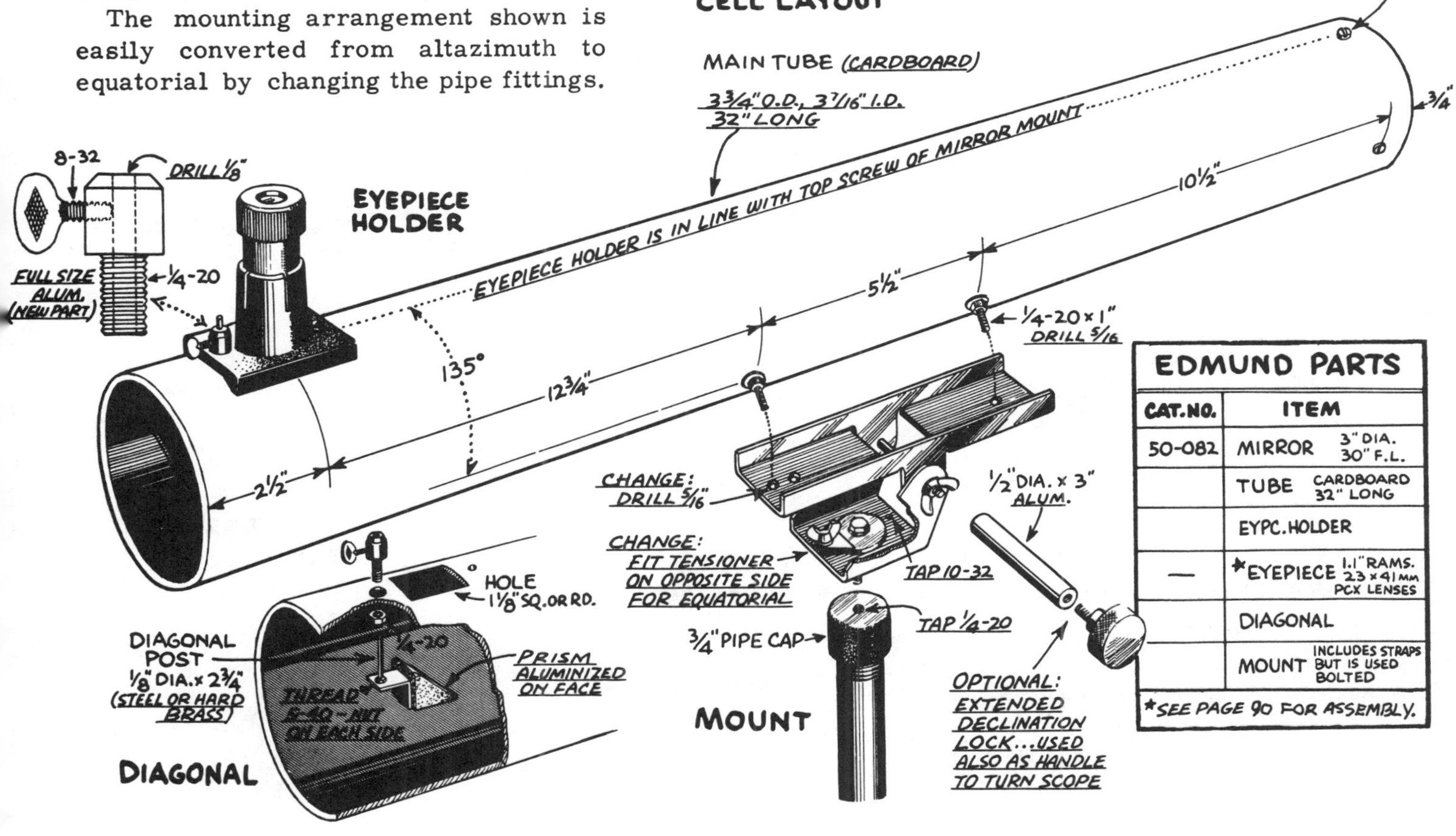

EDMUND PARTS

CAT. NO.	ITEM	
50-082	MIRROR	3" DIA. 30" F.L.
	TUBE	CARDBOARD 32" LONG
	EYPC. HOLDER	
—	*EYEPIECE	1.1" RAMS. 23 × 41 mm PCX LENSES
	DIAGONAL	
	MOUNT	INCLUDES STRAPS BUT IS USED BOLTED

*SEE PAGE 90 FOR ASSEMBLY.

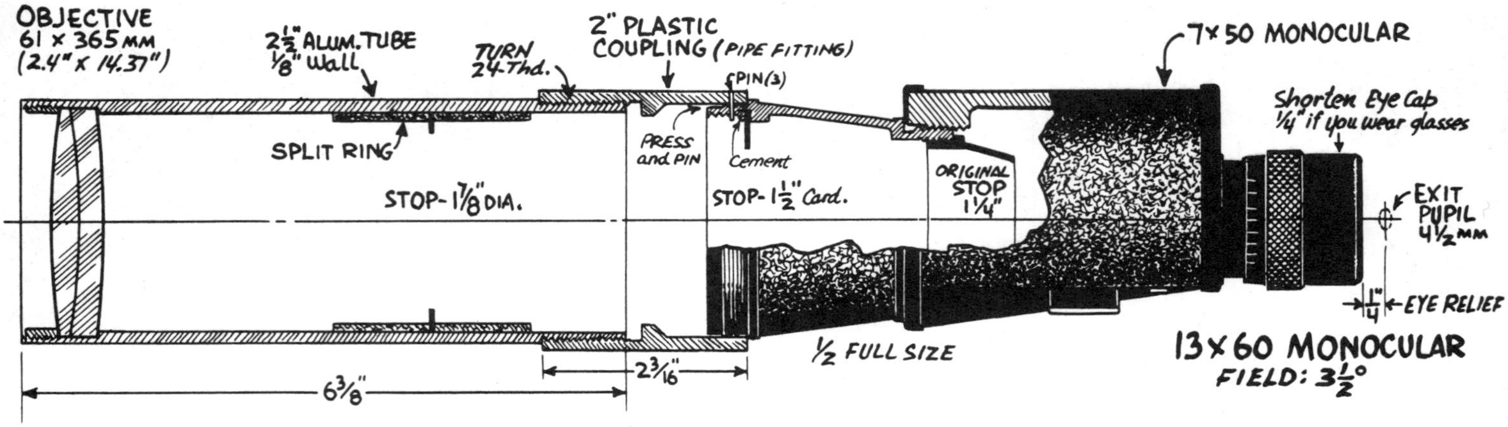

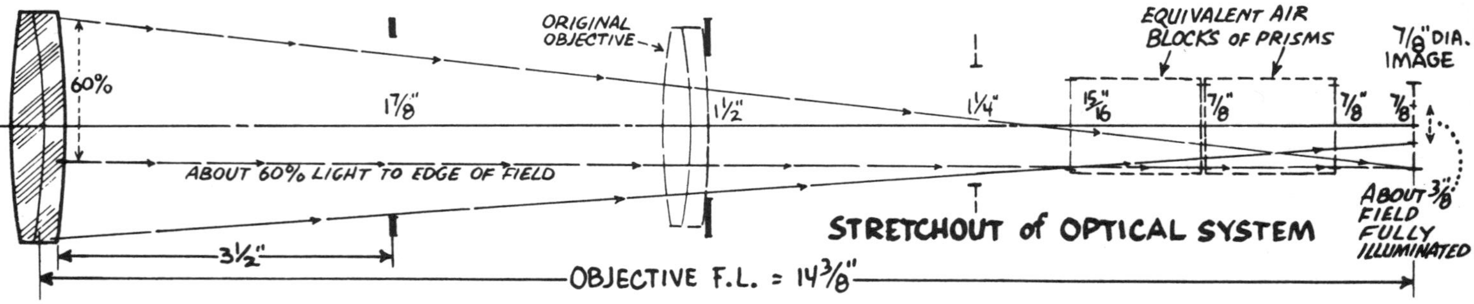

7x50 Conversion

13x60 MONOCULAR

REBUILDING the front end of a 7 x 50 monocular has long been a favorite project among telescope builders blessed with a metal lathe. The conversion shown here uses a 61 x 365mm objective, which itself is from a prismatic instrument and so admirably suited for the job. Of course, the original 52 x 193mm objective and its cell is removed--it makes a fine 4x Galilean with 25 x 48mm double concave lens as the eyepiece.

The new objective is mounted in a length of thick-wall aluminum tube. The link between the monocular and main tube is a plastic pipe coupling in the 2-inch size. Without machining, you can press this over the threaded end of the SELSI monocular, but it is best to reinforce the joint with cement and pins. A threaded connection is to be provided between the plastic coupling and the main tube; this can be used for extra "out" focusing travel if needed for near land objects. With careful fitting, the 3/8 in. travel provided by the original focusing eyepiece will put you on target for any sky object as well as land objects at moderate to long range. The instrument makes an excellent finderscope for a larger telescope--its 3-1/2 deg. field more than covers the pole star orbit.

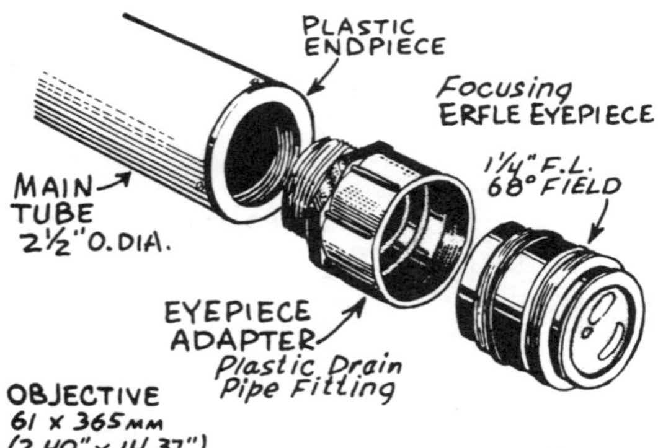

6° Wide-field Refractor

The popular 1-1/4 inch Erfle eyepiece is well-known for its wide field. A minor fault is a slight amount of lateral color. The focusing travel of nearly 3/8 inch is all you need for astro viewing; extra movement for near objects can be obtained by unscrewing the eyepiece adapter from the endpiece. The plastic pipe fitting used for the eyepiece adapter can be obtained at plumbing or hardware stores.

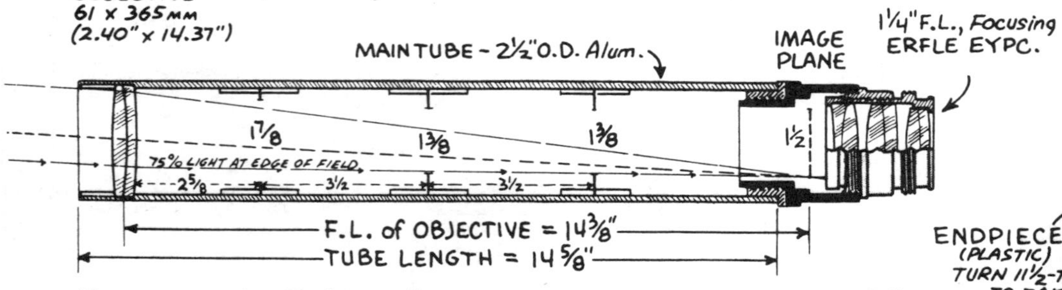

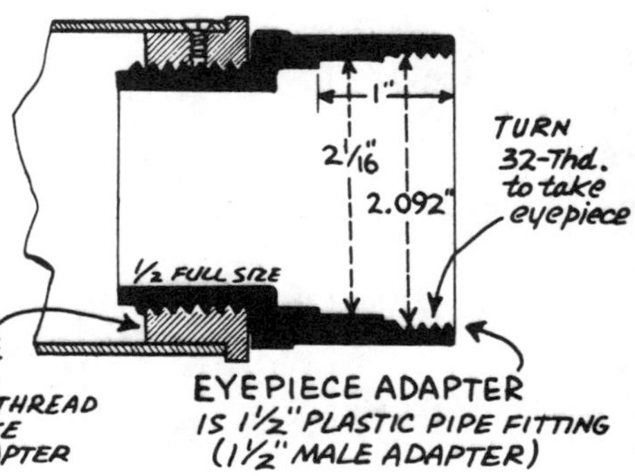

Mag. - 11.5x Field - 6° Exit Pupil - 5MM Eye Relief - 5/8"

8x *Terrestrial* TELESCOPE

A VARIETY of compact, good quality terrestrial telescopes can be made from available surpl... lenses at low cost. The one shown here is a typical design, and its construction is a plain job of lathe turning. If you use a split ring for the objective cell support, it should be faced-off in the lathe after riveting. The objective cell is a push fit inside the main tube--it can be wrapped with a turn of masking tape if too loose. The erecting eyepiece by itself may be used with any other objective, or it may be used with any astro telescope accommodating 1-inch diameter eyepieces.

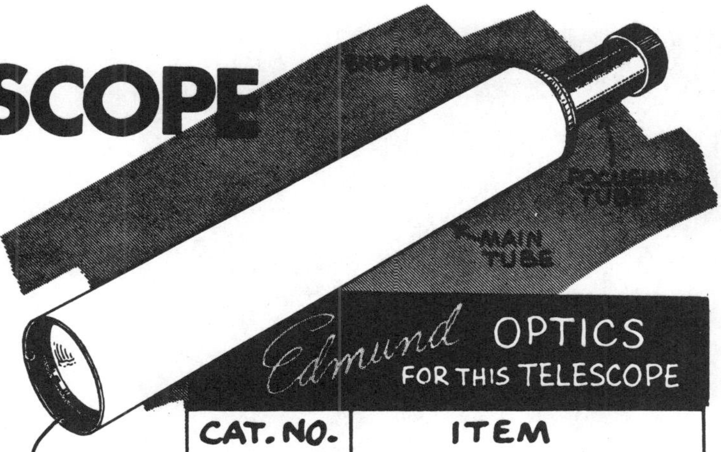

Edmund OPTICS FOR THIS TELESCOPE

CAT. NO.	ITEM
32,397	OBJECTIVE
	23 x 41 MM PLANO-CONVEX 3 REQ'D.
	15 x 41 MM ACHROMAT 2 REQ'D.

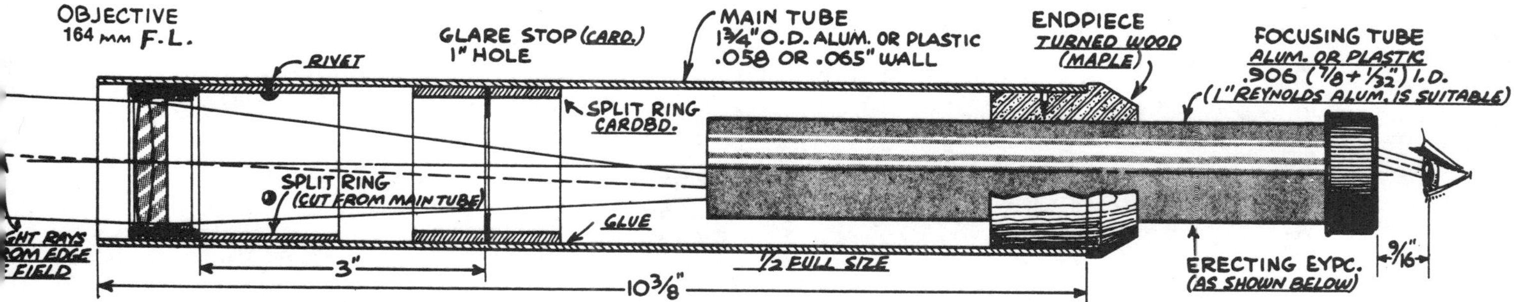

8x TERRESTRIAL TELESCOPE FIELD = 8 MOONS (4°) 102 FT. AT 1000 YDS. **FOCUS RANGE:** 25" TO INFINITY **EYE RELIEF:** 9/16"

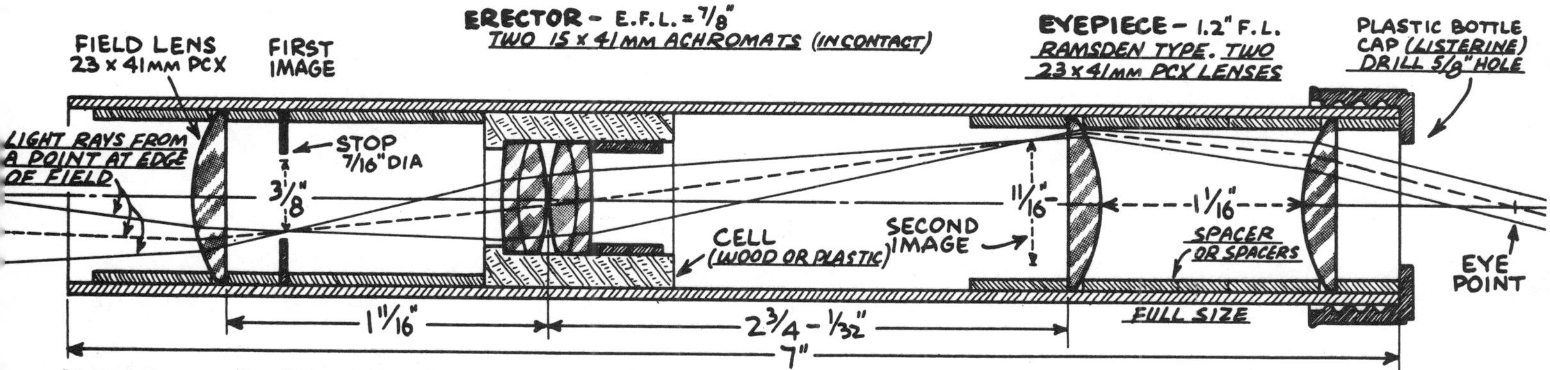

TERRESTRIAL (*ERECTING*) EYEPIECE (*E.F.L.* = *.9 INCH*)

1x ERECTOR TUBE

FITS ANY SCOPE USING 1¼" DIA. EYEPIECES... *FINE FOR "LAND-GAZING"*

USE THIS WITH ANY TELESCOPE TO GET ERECT IMAGE SAME SIZE AS EYEPIECE USED ALONE *BUY THE CHROME TUBE AT SEARS.*

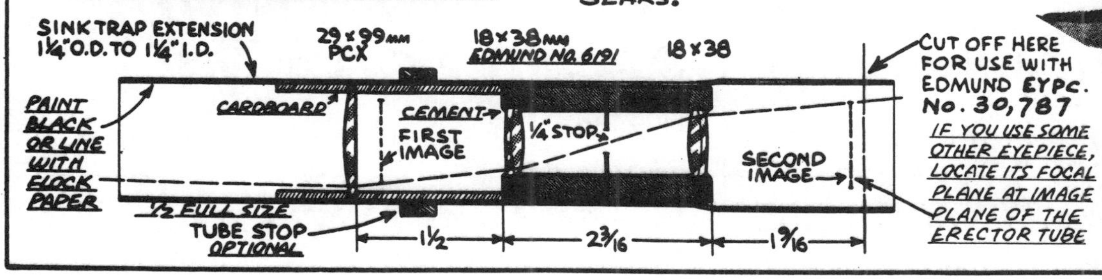

WIDE-FIELD 33x Prismatic

THIS is a neat prismatic telescope, only fairly bright because of the rather small objective, but clean and sharp over a wide field. The image is erect; you can use this scope day or night.

Machining is required in adapting the mounted roof prism. This can be easily disassembled into two parts, one of which is the prism mount proper, while the other is a round turret on which you do a little turning and threading. The best way to chuck it is to turn a stub mandrel on which to mount the flange. In case you are using some less expensive eyepiece than the big Erfle, the machining could be varied to suit, or, the turret can be discarded, mounting the prism mount directly to a piece of wood which is fastened to the rear end of the wood barrel.

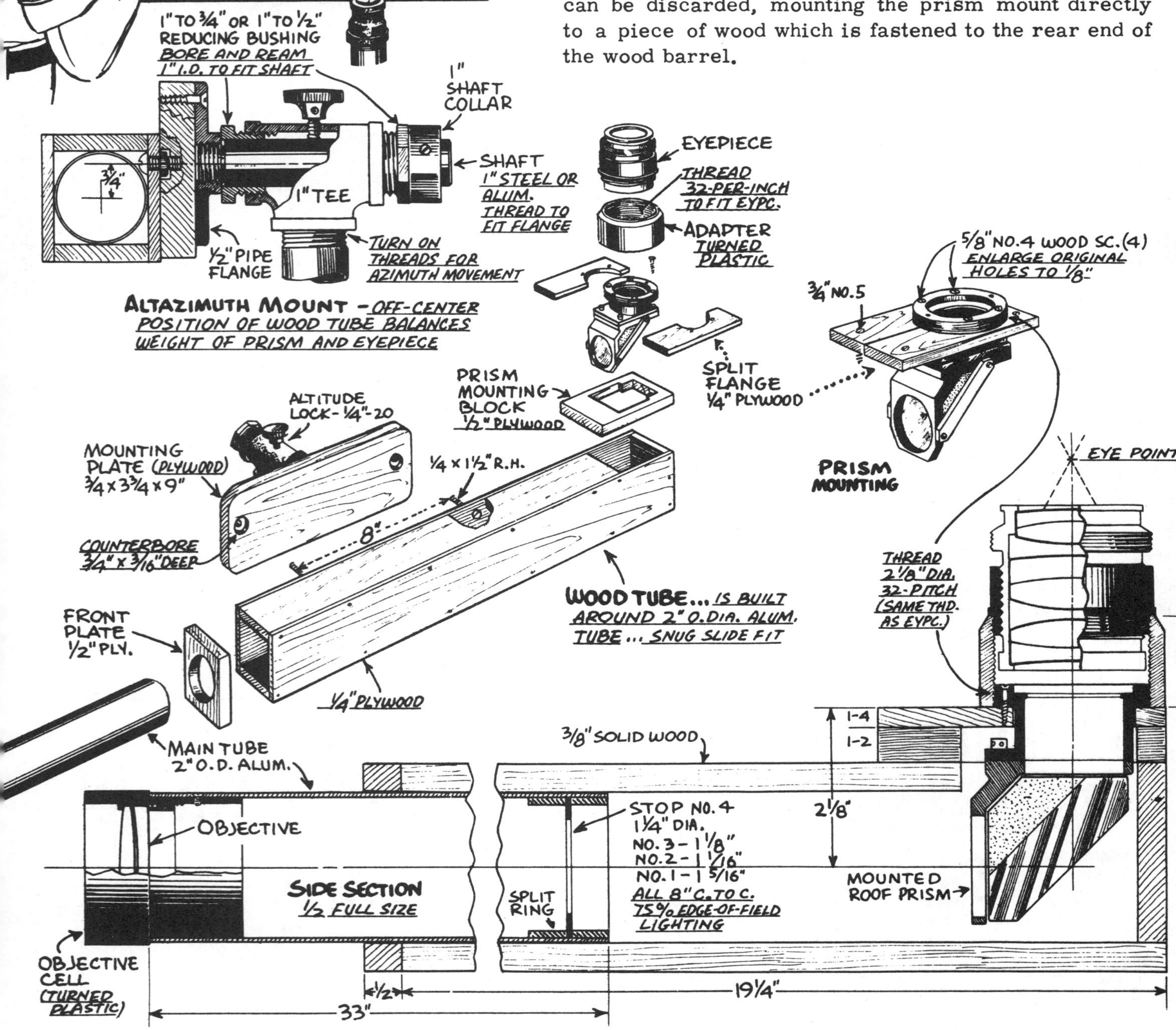

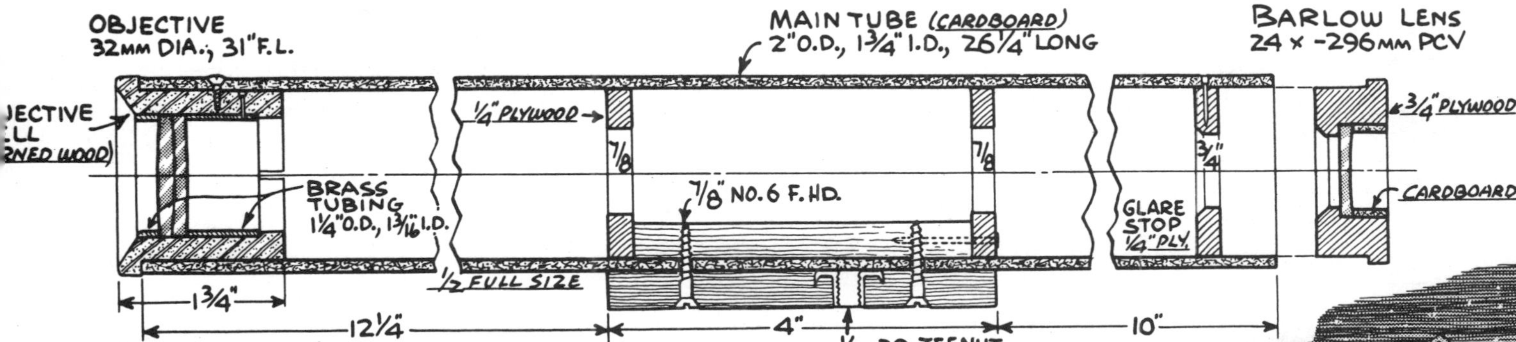

MAIN TUBE - 31"F.L. OBJECTIVE

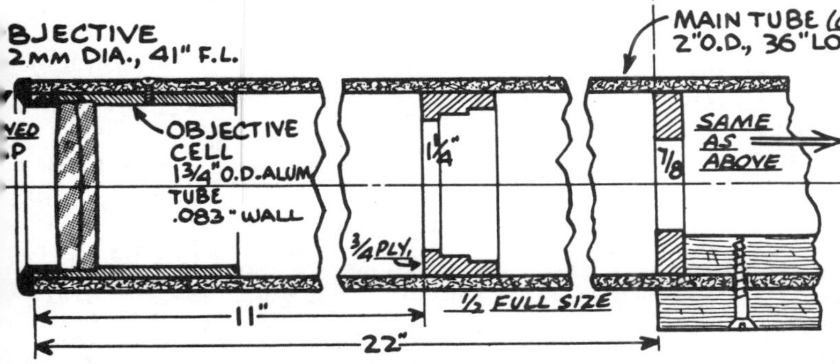

MAIN TUBE - 41"F.L. OBJECTIVE

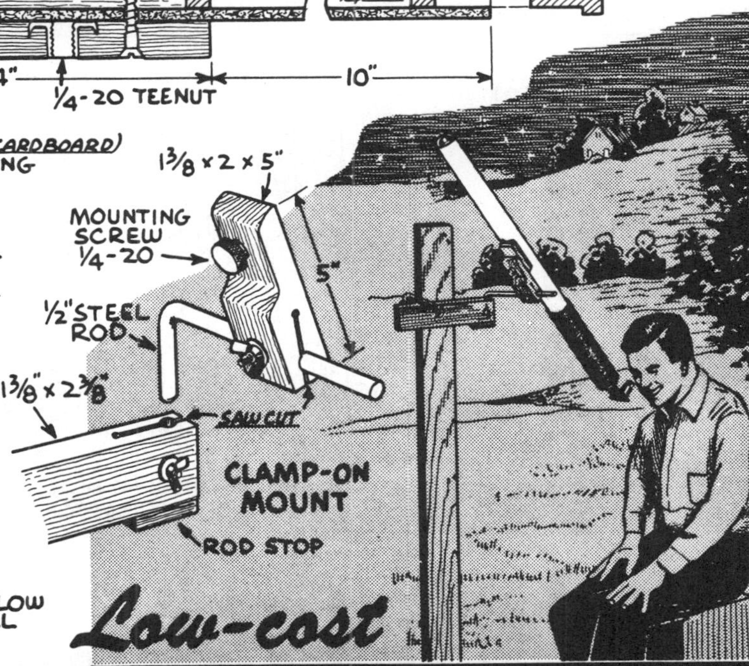

Low-cost REFRACTORS

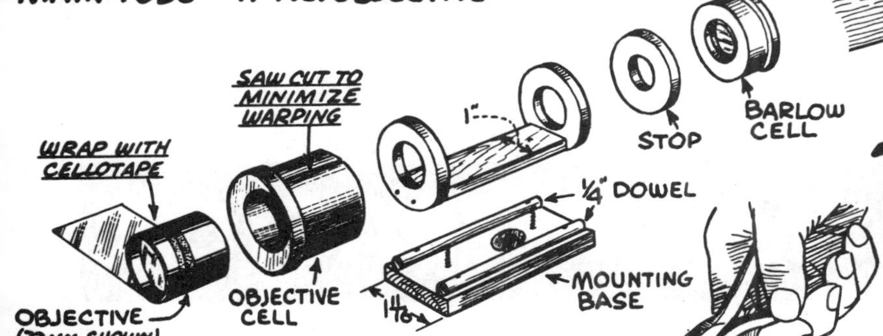

MAIN TUBE PARTS

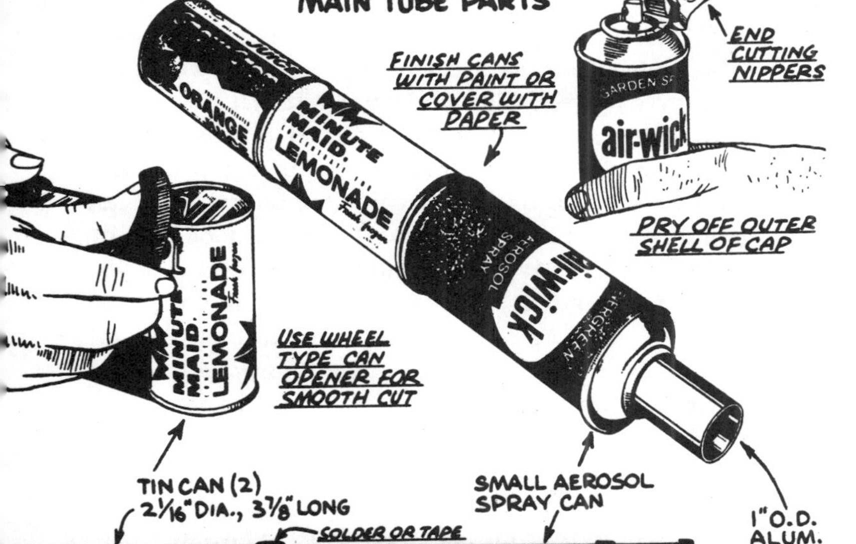

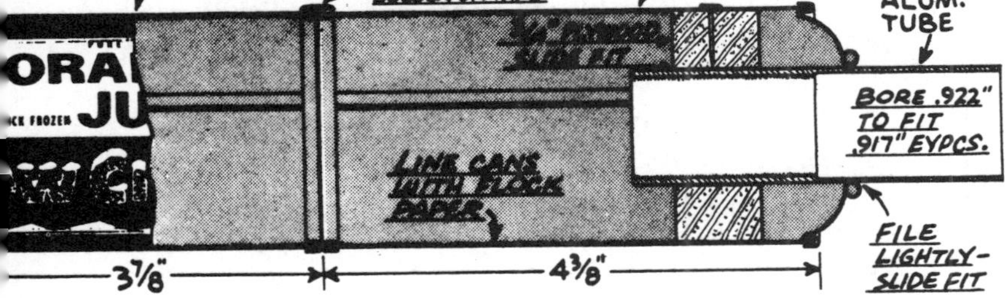

FOCUSING TUBE - *FITS OUTSIDE MAIN TUBE*

YOU WILL enjoy making and using either of the two low-cost refractors shown on this page. Both use the same construction and differ only in objective mounting and main tube length.

The mechanical design departs from the conventional by using a focusing tube which fits outside the main tube. This is made from three tin cans as shown and is lined with black flock paper for a smooth slide fit over the 2-inch diameter main tube. The outside focusing tube provides a rough adjustment to suit various accessories, eyepieces, etc., while a fine focus is provided in the usual manner with a slide-fit eyepiece tube. Much of the long focusing travel is needed for the simple lens Barlow; this is spaced for 2X amplification and needs about 6 inches out-focus travel.

A 1-inch eyepiece is recommended as basic equipment. This will give 31X and 41X for the two designs. The Barlow system will double the power. A simple clamp-on mount is shown which can be attached to any post, fence or garage door.

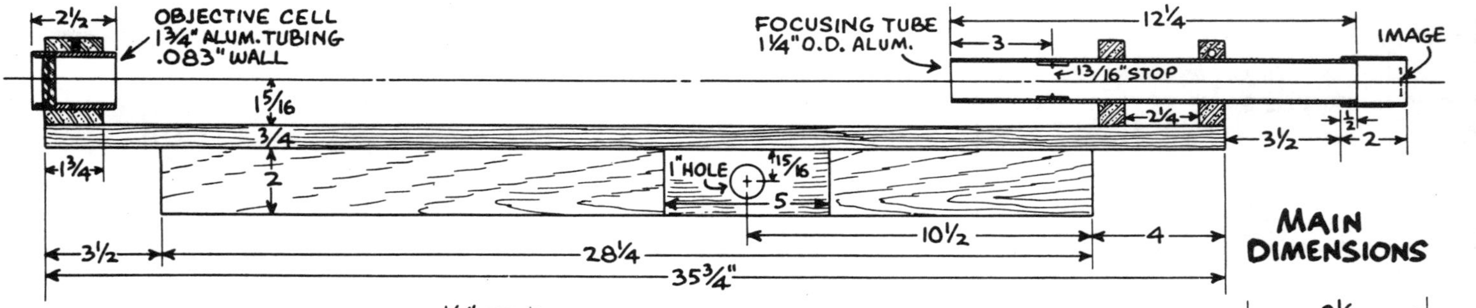

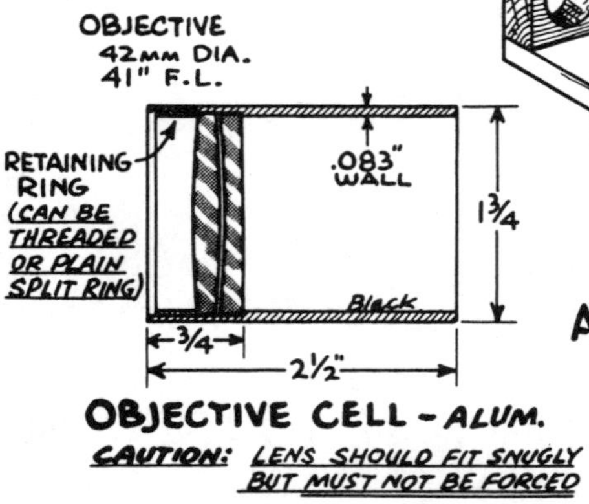

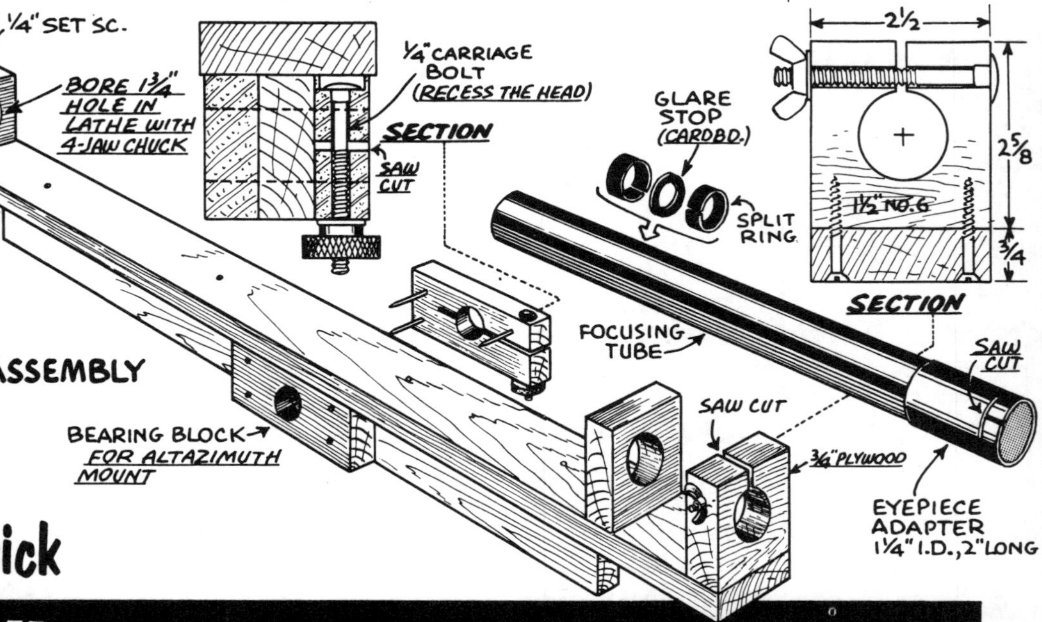

Telescope-on-a-Stick

1.6" Tubeless REFRACTOR

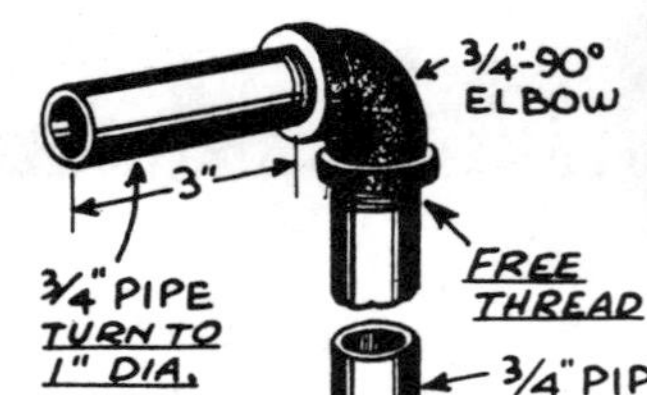

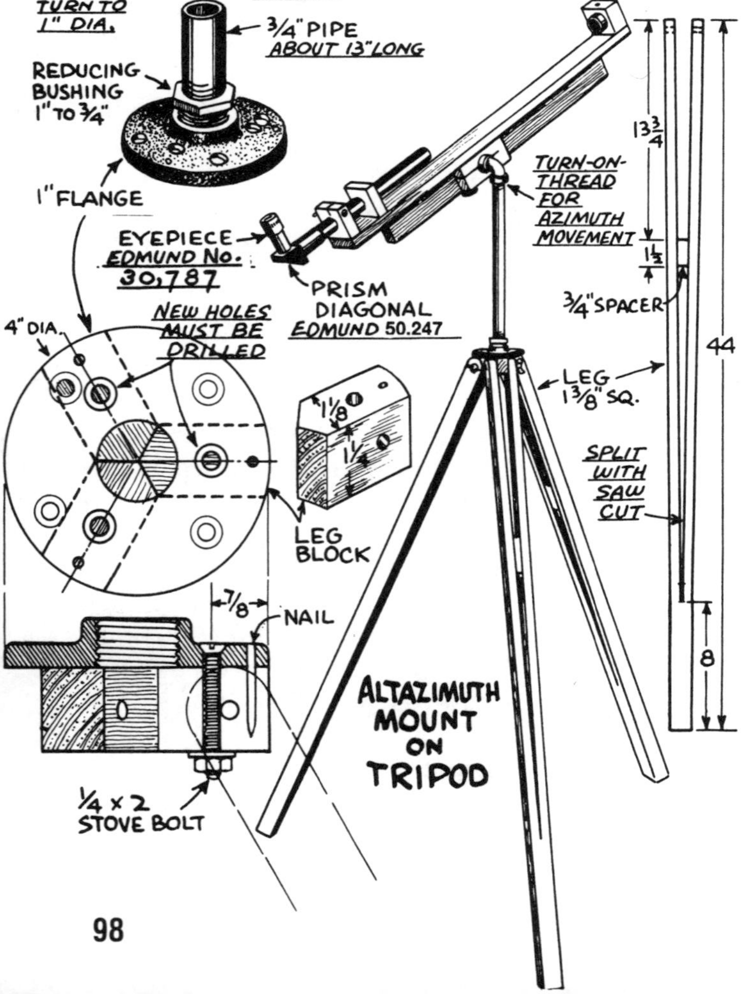

IF YOU want a cheap but good small refractor, you will like this tubeless job. Used with a 28mm eyepiece (Edmund No. 30,787) it magnifies 38X over a 2-1/2-moon field. A tubeless telescope of this kind is practical for astro work if shielded from outside light. Obviously, it is not much of a trick to span a light cardboard tube between objective and focusing tube block if you want to make use of the scope for day objects.

Make the focusing tube from a piece of Reynolds do-it-yourself aluminum, available at many hardware stores. The Edmund 1-1/4" inside diameter tubing is a press fit over this to provide the right size for standard eyepieces and accessories. Alternately, you can save money by using the smaller .917" eyepiece size, and in such case an adapter should be made to fit inside the focusing tube.

The leg blocks for the tripod head should be select hardwood. These take a lot of strain and it is a good idea to augment the mechanical fastening with Devcon Plastic Steel adhesive. Round and drill the tripod legs before making the splitting saw cuts.

2" refractor-on-a-stick

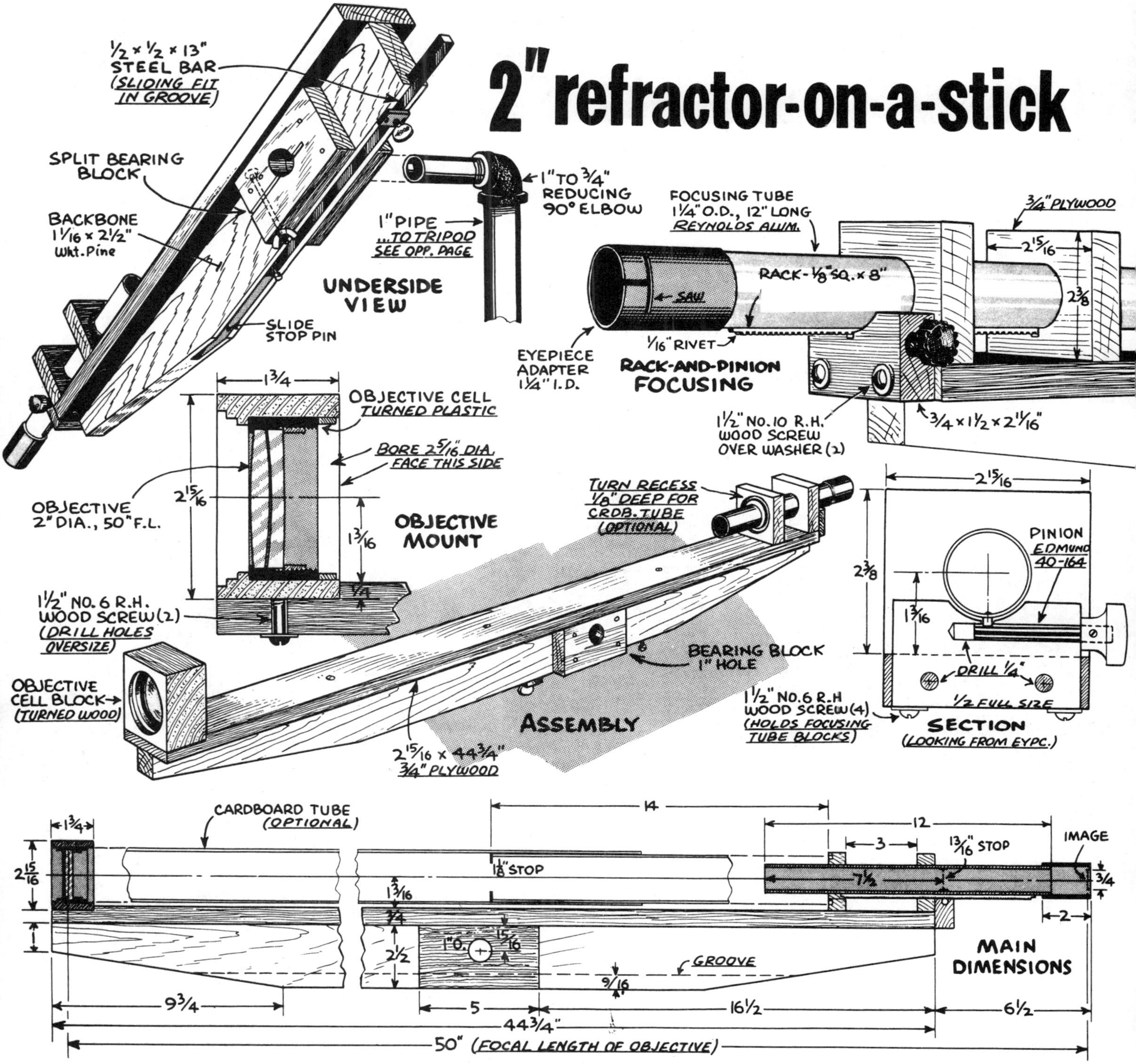

THIS IS a bigger version of the tubeless design shown on opposite page, with rack-and-pinion focusing added. The 2-in. objective is air-spaced, coated on all four surfaces, and is a good glass, assembled with three small bits of gummed foil equally spaced around the glass. A sliding counter-weight in the form of a bar of 1/2-in. square steel achieves a fair degree of balance around the altitude axis.

A job like this usually starts by mounting the objective. This is housed in a turned plastic cell, which is then fitted inside a block of wood turned and faced on a wood or metal lathe. Turning is also required for the two blocks which hold the focusing tube; these should be nailed together and bored in one operation. If you plan to use a cardboard tube, turn a recessed ring for this, as shown. An alternate cover would be light cardboard or heavy paper creased to a U-section and tacked over the wood members. It will be obvious that the wood backbone must be perfectly straight along its top edge.

The finished scope is mounted by slipping it over the altitude shaft of the mount. The azimuth rotation is obtained by the familiar method of letting the elbow turn on its own threads.

1.6" Altazimuth REFRACTOR

82x with ½" Eyepiece

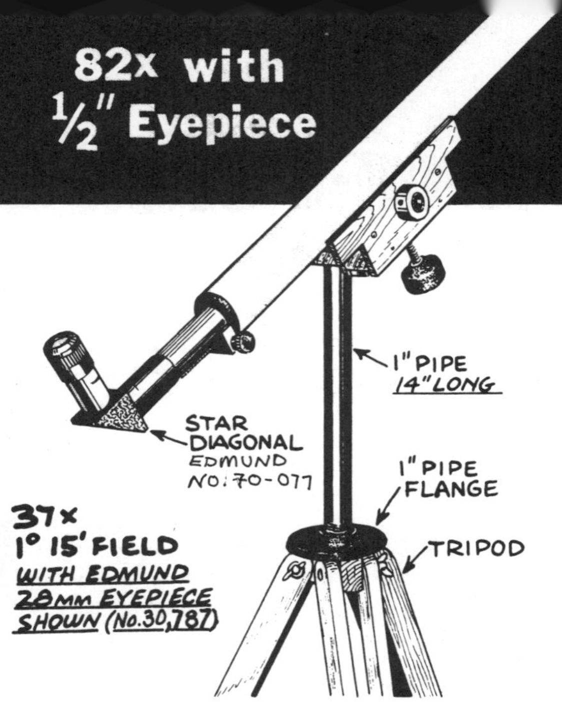

EDMUND 42mm air-spaced objective is a good glass, and when you house it in a metal or plastic tube with rack-and-pinion focusing, you have a top-quality telescope. Making your own R&P endpiece is not too difficult, as can be seen in the detail below. Plastic for the objective cell is readily obtained in various plastic pipe fittings. A metal cell is easily made by press-fitting together two short lengths of thick-wall aluminum tubing. Most objective lenses are rather brittle; they should fit snugly but you never try to squeeze them in.

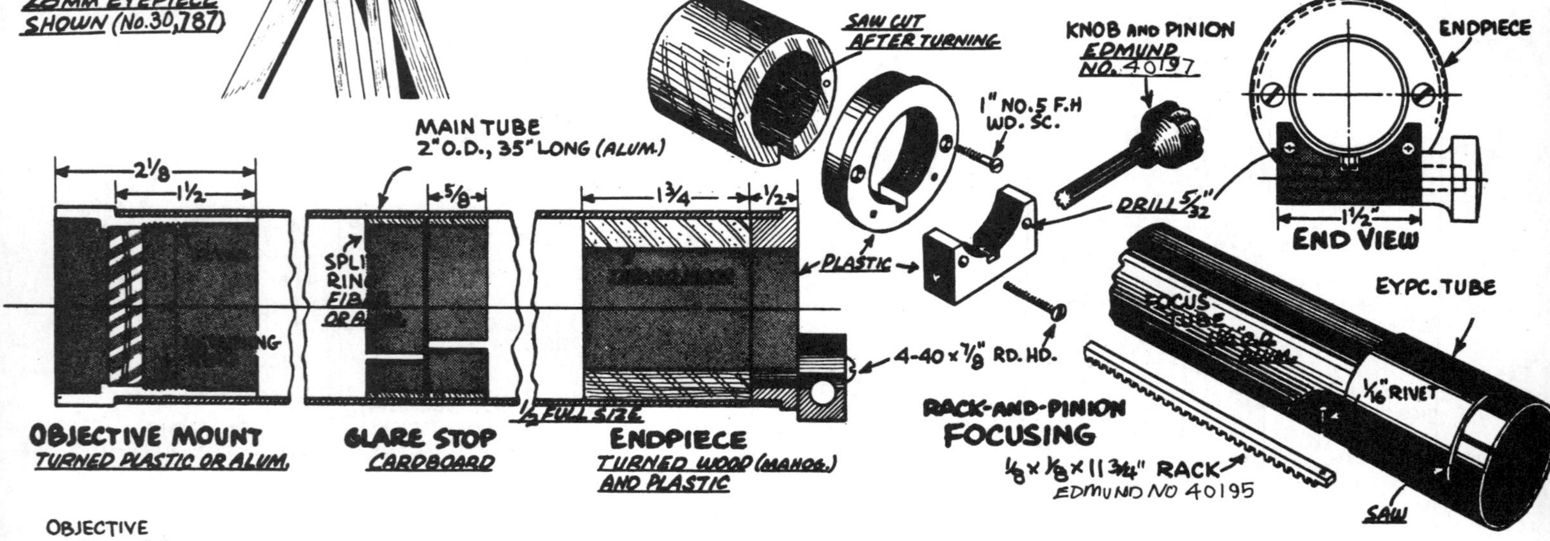

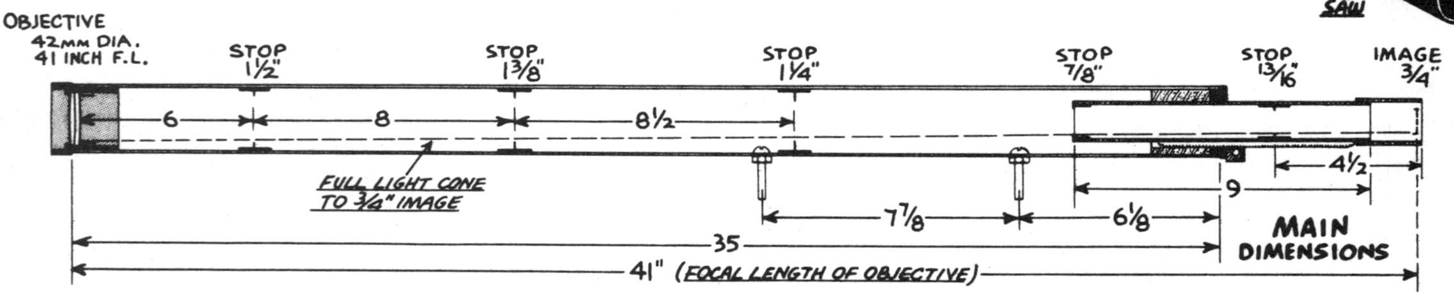

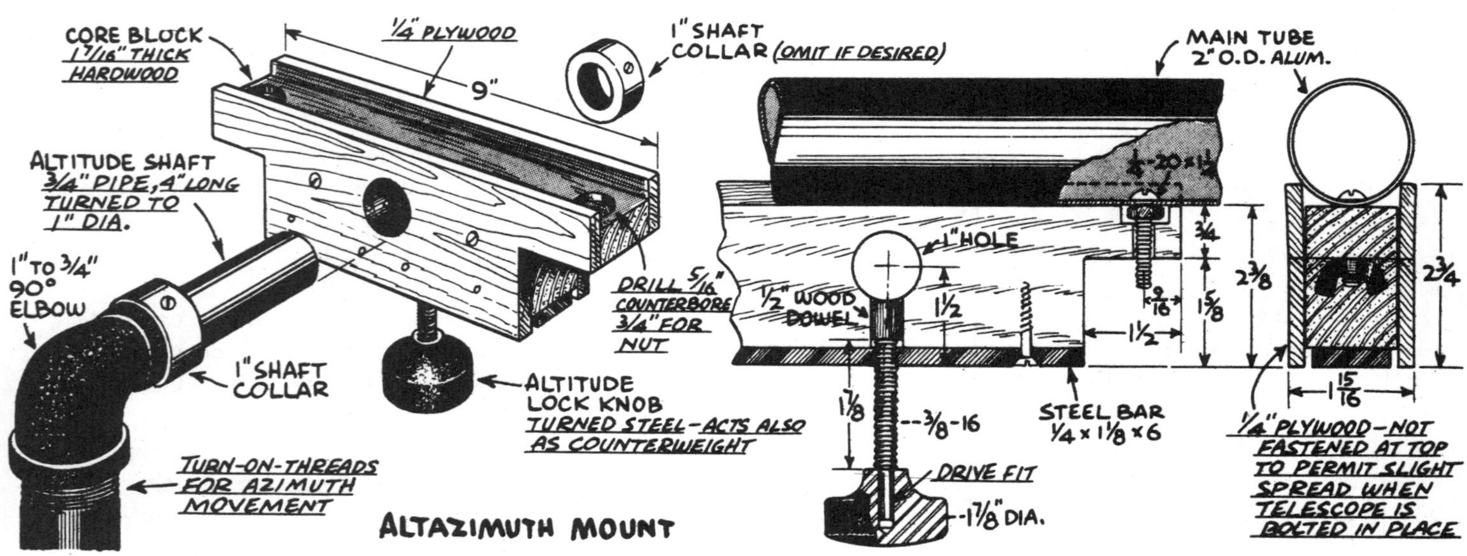

KIDS LIKE THIS SIMPLE LENS REFRACTOR

8x "Sky Starter"

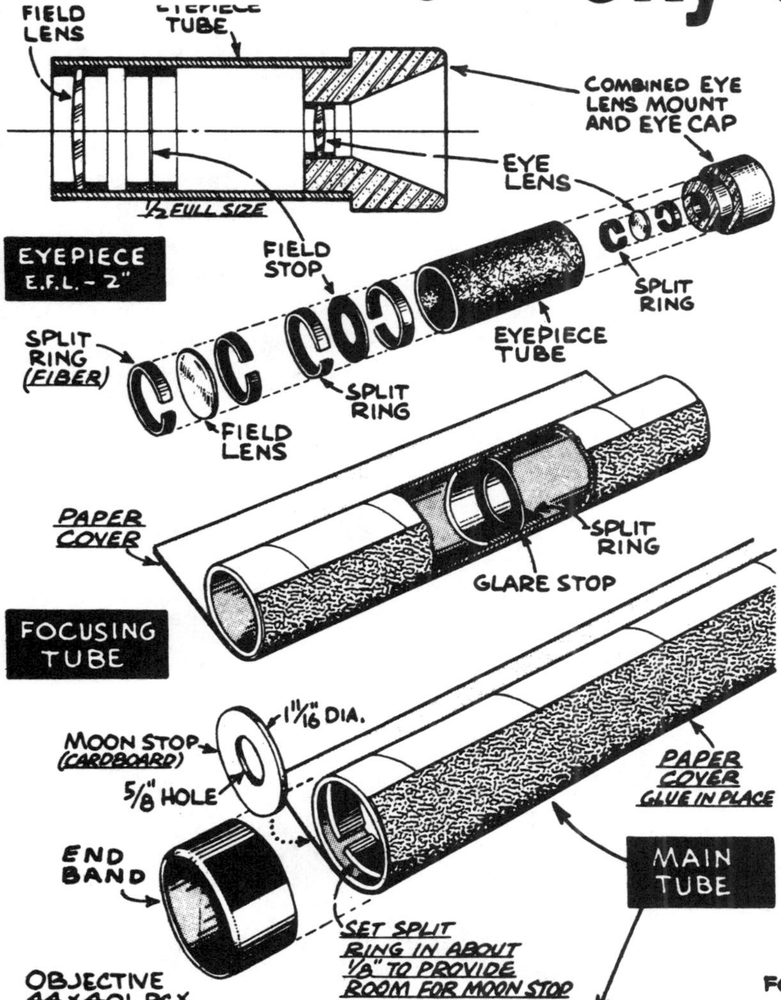

EDMUND simple lens telescope kit includes all the glass and parts to make an 8X astronomical refractor telescope. Of course the simple lens objective will show some false color and the image sharpness is not the best. However, the scope is good enough to show craters on the moon in sharp detail, and it will reveal hundreds of stars not visible to the naked eye.

All of the parts can be put together in a dry assembly in less than ten minutes. The scope so assembled will stay put for testing, after which you can make the permanent assembly by touching the various split rings with a dab of glue and applying the cover paper and end bands.

On a bright object like the moon, a smaller aperture at the objective lens will still admit plenty of light. It is suggested that you take advantage of this situation by using the cardboard "moon stop" shown. With this in place you will see moon craters in sharp detail, while the color fringe will vanish. Full aperture should be used for stars since you need all the light you can get.

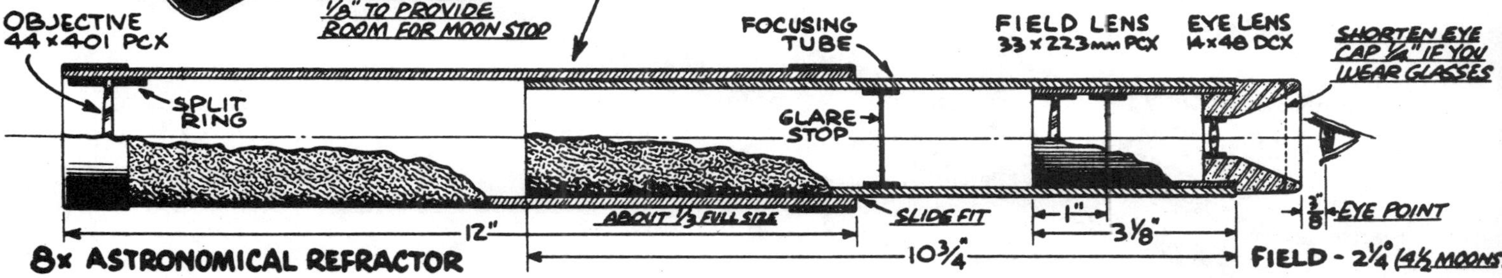

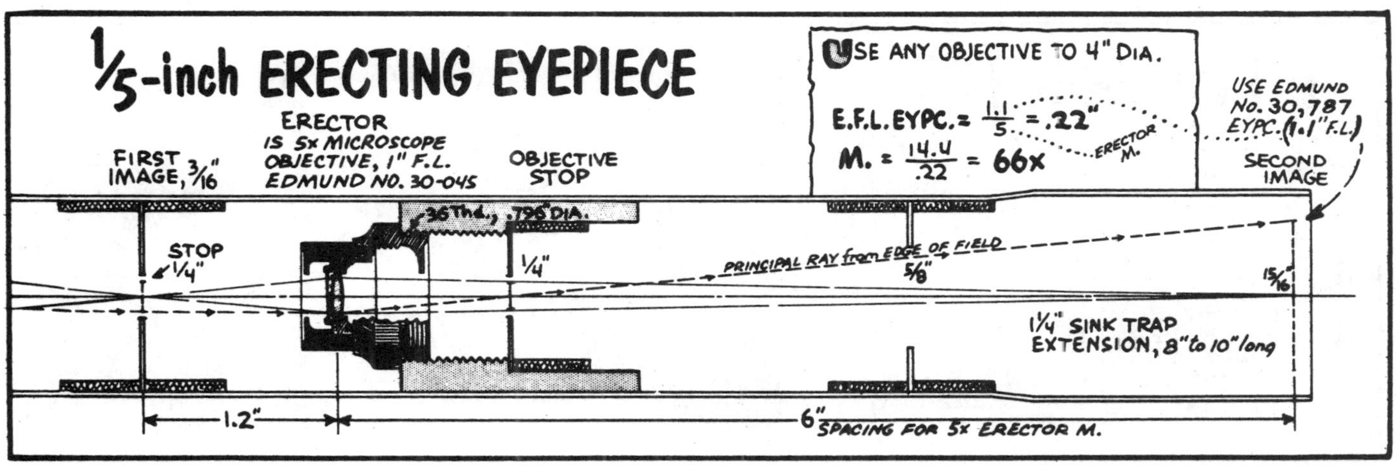

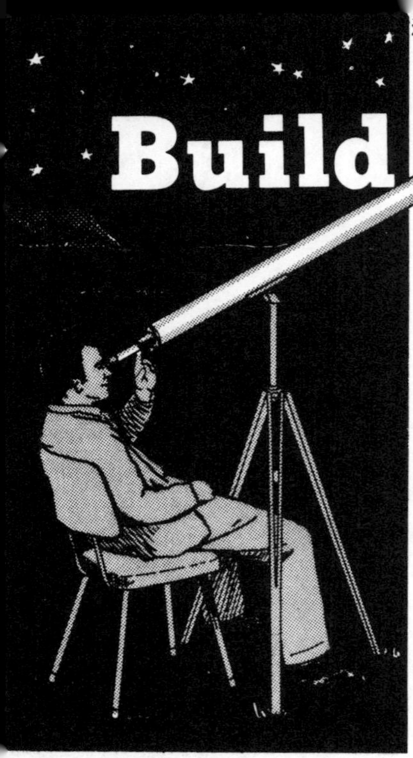

Build this 2" REFRACTOR

THE 2-inch diameter, 50 inch focal length coated air-spaced objective, makes a good hi-power refractor at low cost. The construction shown makes use of a purchased focusing eyepiece holder, but you can substitute your own homemade if desired. The main tube is cut 46-1/2 inches long to put the image plane fairly well out at the end of the focusing movement, as can be seen in lower right detail. You will then have enough forward focusing travel available to accommodate a star diagonal or prismatic erector, such as the one shown in box below. The glare stop cage is made of plywood rings with the stringers inletted, the whole an easy slide fit inside the main tube. An additional glare stop is fitted at the end of the focusing tube, as shown. You will get 45x with Edmund **No. 30,787 eyepiece**, which is just about the ideal power for most sky-gazing. Shorter f.l. eyepieces can be used for higher power; you can also get high M. with a Barlow lens.

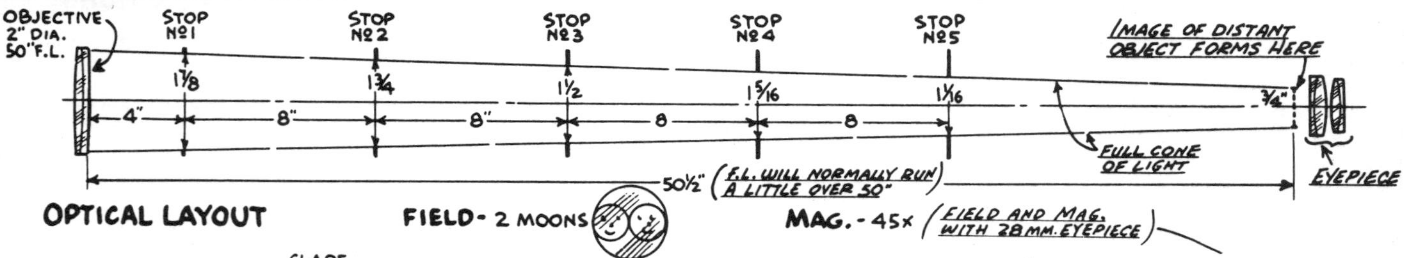

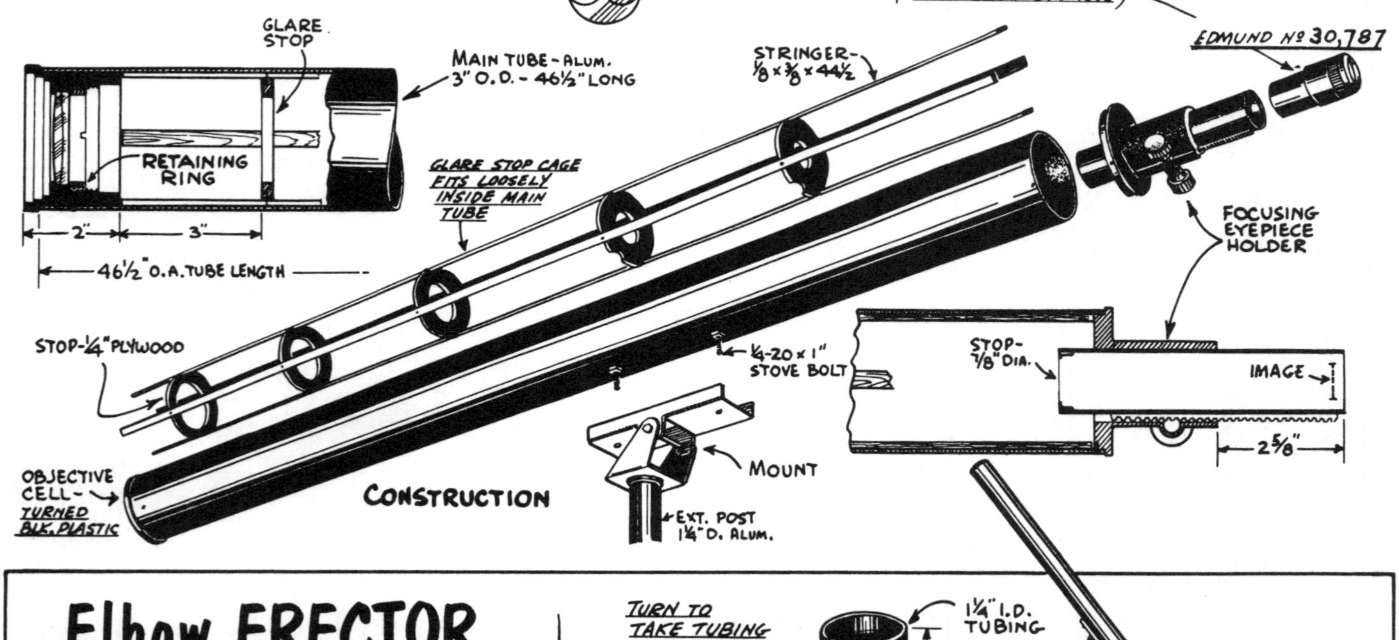

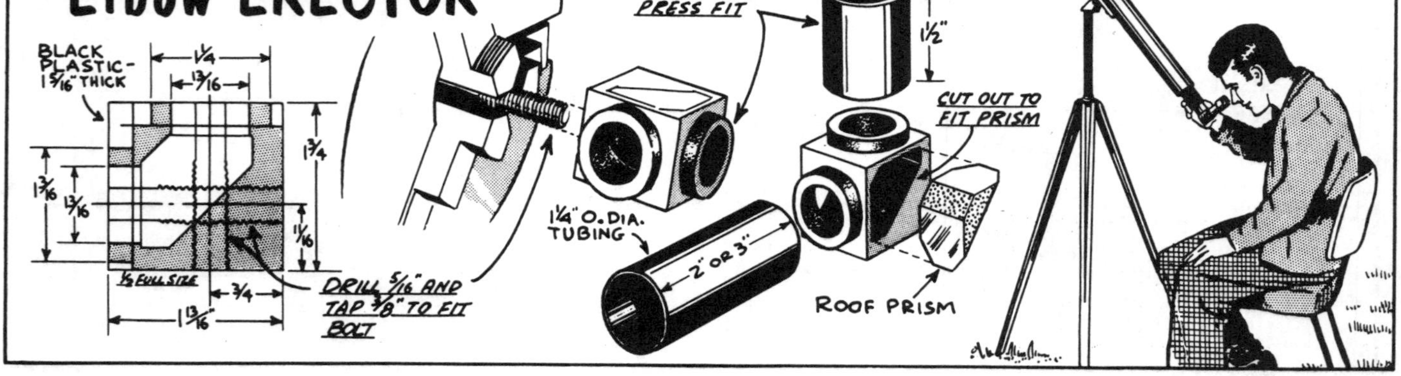

EDMUND PARTS		
CAT.	PART	
	OBJECTIVE	2" DIA., 50 in. F.L. COATED, AIR-SPACED
30-286	MIRROR	5 × 25 × 35 MM FIRST SURFACE
30-286	MIRROR	(OTHER SIZES CAN BE USED)
	EYPC. TUBE	1¼" I. DIA., Black BRASS, 2" LONG
40195	RACK	
40197	PINION	
30,787	EYEPIECE 28MM R.K.E.	

2" Folded Refractor

A FOLDED refractor makes a compact telescope which handles very much like a reflector. The image is the usual astro inverted.

The objective could be mounted directly in the wood end block, but it is preferable to use a separate cell of aluminum or plastic, as shown. The end block is then mounted in a 4-jaw chuck in the lathe and is bored and faced to take the objective cell. 25 x 35mm is about the minimum size for the two mirrors needed; larger sizes can be used. The required collimation is easily done by adjusting the mirror while looking into the eyepiece tube without the eyepiece.

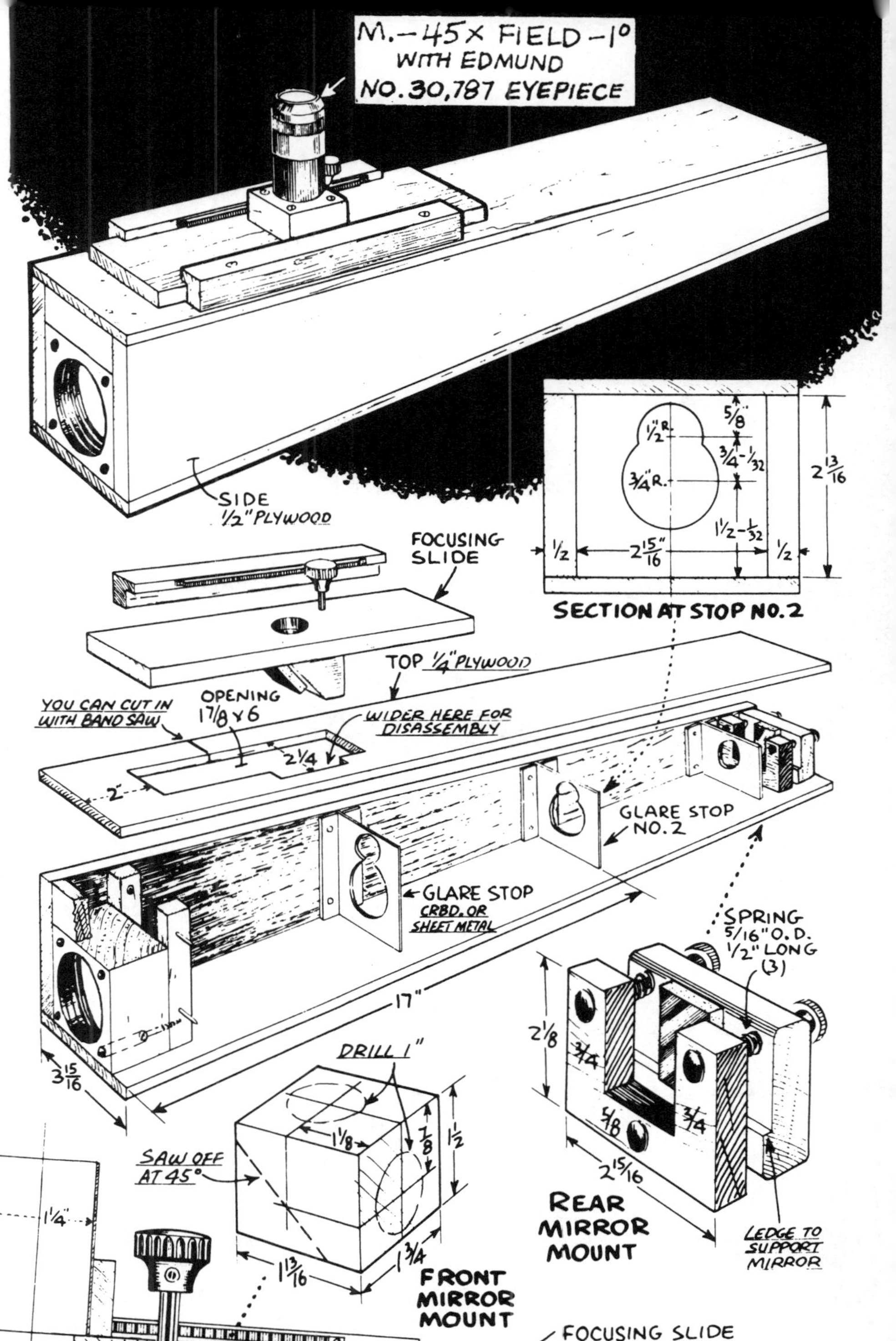

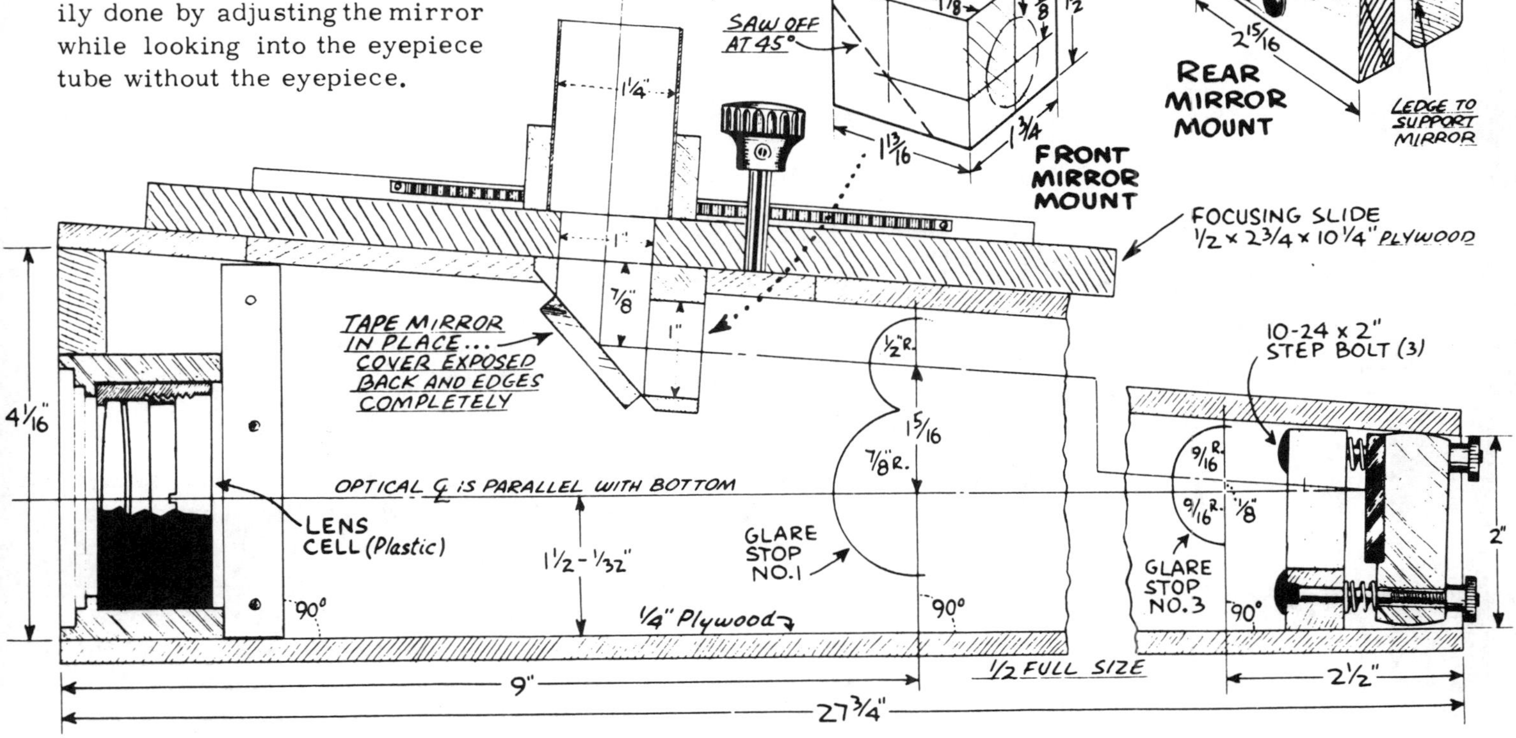

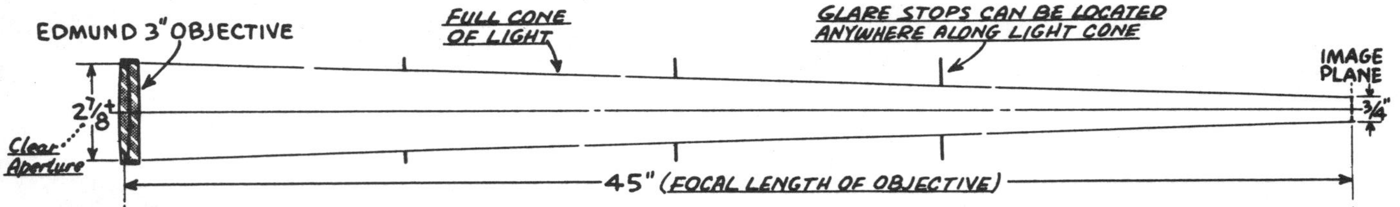

OPTICAL LAYOUT MAG.-41x FIELD- 1° 13' (WITH 28MM EYEPIECE)

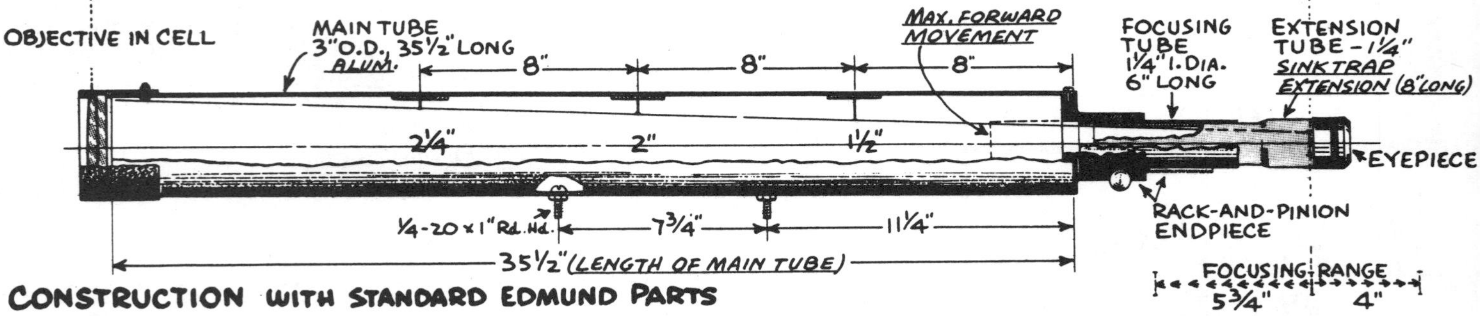

CONSTRUCTION WITH STANDARD EDMUND PARTS

3-inch REFRACTOR

easy to build with EDMUND Parts

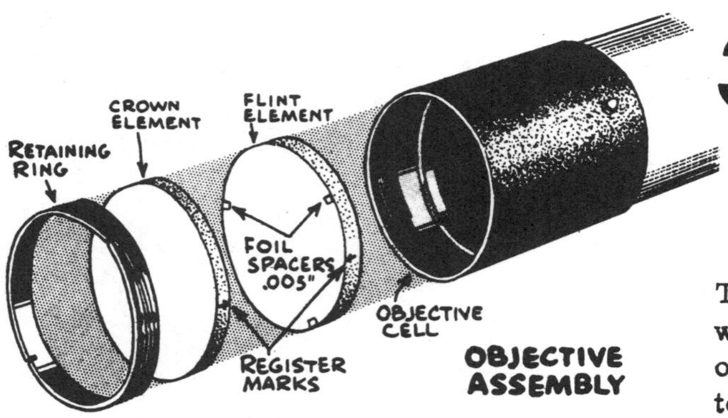

OBJECTIVE ASSEMBLY

THE REFRACTOR is an easy scope to build, especially when you buy the main parts readymade. Then, the only problem is how long to make the main tube in order to have enough focusing adjustment for a camera, star diagonal, Barlow lens, etc. This plan specifies 35-1/2 inches long for the main tube and this length combined with an extension focusing tube will give you a 10-inch focusing range--enough for almost any attachment except possibly a very long Porro prism erector.

The main tube is cut to the specified length and is squared-up by filing or grinding. The objective cell is fastened with pan head screws and chrome acorn nuts (supplied), while the endpiece is fastened with machine screws tapped into the flange of this part.

Do not neglect the glare stops--they are pesky things to fit but worth the effort. If you buy the 48-inch length of 3-inch aluminum tubing, you will have enough left over to make split rings for glare stop retainers. The glare stops can be cut from sheet metal or cardboard. A tin can or a turned piece of wood of the right size to fit loosely inside the tube should be used as a "pusher" to ram the split rings into the tube to the required position. Glare stops need only be approximately square with the tube.

Use care and caution when fitting the air-spaced objective in place. First, cut three tiny squares of adhesive foil (supplied) and space equally around the inside surface (hollow side) of the flint element. Put the crown element over the flint with register arrows pointing together, as shown in the drawing. Hold the assembled objective with a clean facial tissue and push it up into the objective cell. This is a top-quality object glass and you will be pleased with its performance.

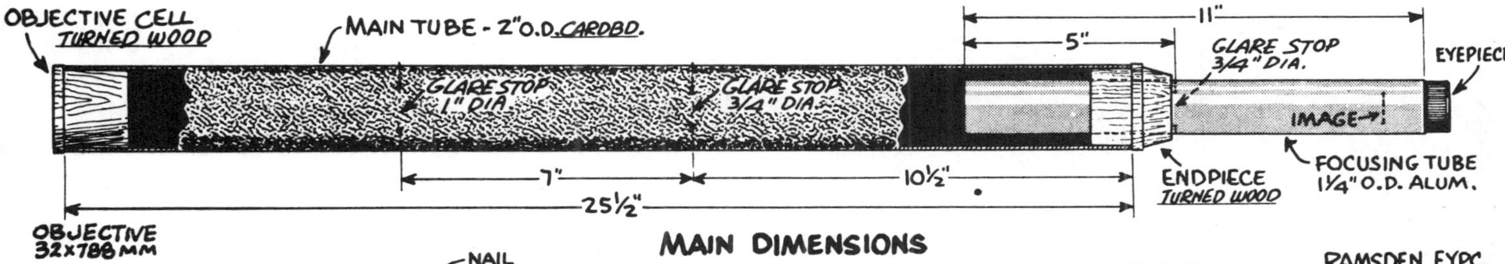

MAIN DIMENSIONS

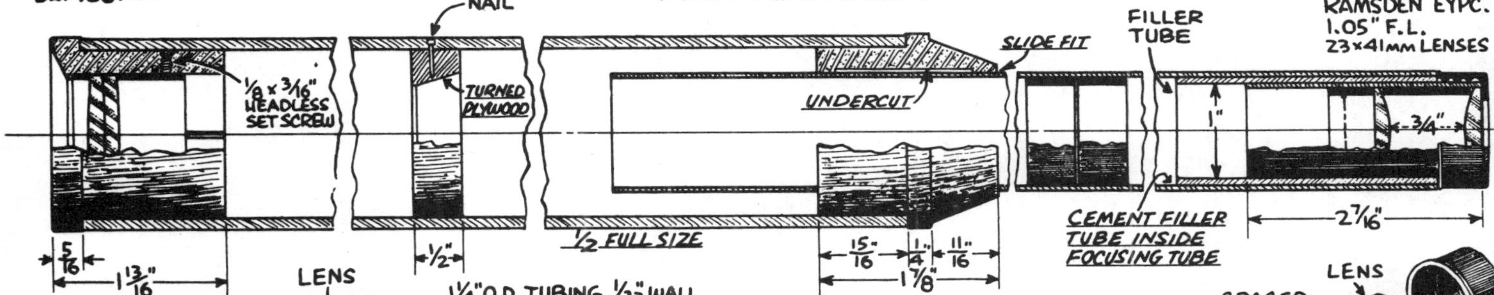

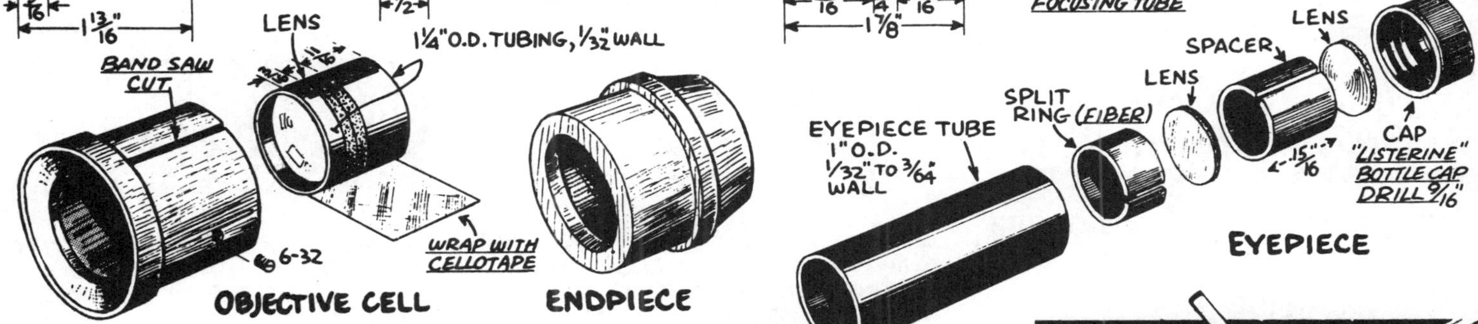

30x BABY Refractor

THE BABY of the astronomical telescope family is a refractor with a 1-1/4 inch objective. A small object glass like this will show stars to 9th magnitude; it is swell for the moon and bright planets, but poor for dim sky objects. With 1.05 inch focal length eyepiece, the power is an even 30X, enough to show a wealth of moon detail and the ring around Saturn. The image sharpness is excellent. With Ramsden eyepiece shown, the field is a comfortable 70 minutes of arc or a little over the width of two full moons. With a field this size in a small refractor, you can get along fine without a finder.

This is a cardboard-and-wood job and is easy sailing providing you have a wood lathe. The focusing tube is a length of Reynolds craft aluminum. For a mount, the simple clamp-on-altazimuth shown can be clamped to any garage door, fence post, etc. The hose clamp is a geared type used in the assembly of plastic drain pipes --you can buy at most hardware stores and plumbing concerns.

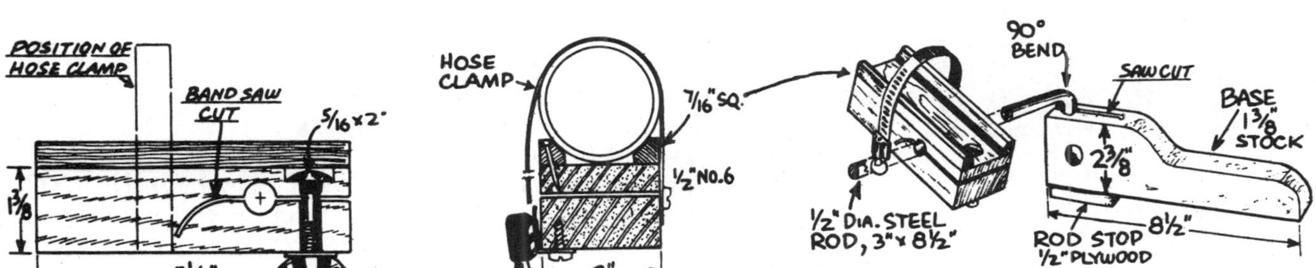

CLAMP-ON MOUNT

ALTAZIMUTH was First for Many Years

IN THE 200 YEARS FOLLOWING THE DISCOVERY OF THE TELESCOPE IN 1608, PRACTICALLY ALL TELESCOPES WERE CONFINED TO THE SIMPLE ACTION OF THE ALTAZIMUTH MOUNT

CIRCA 1650

A TYPICAL ALTAZIMUTH MOUNT for a SMALL REFRACTOR

ALTITUDE + AZIMUTH = ALTAZIMUTH

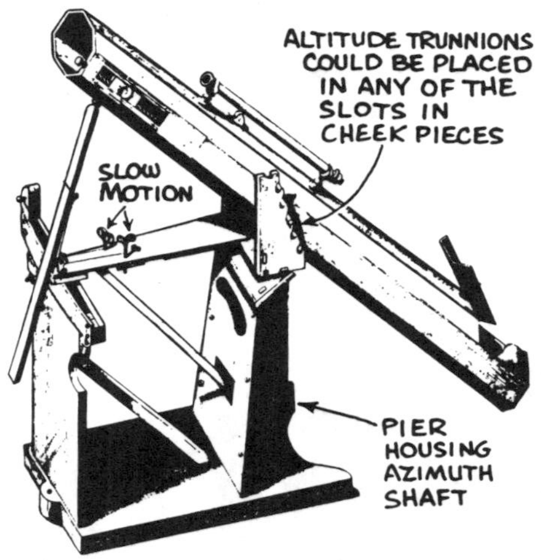

1722 HADLEY'S 6" REFLECTOR with PARABOLIZED MIRROR WAS MOUNTED ON A NEAT ALTAZIMUTH MOUNT. SLOW-MOTION WAS OBTAINED BY STRINGS WOUND ON PEGS

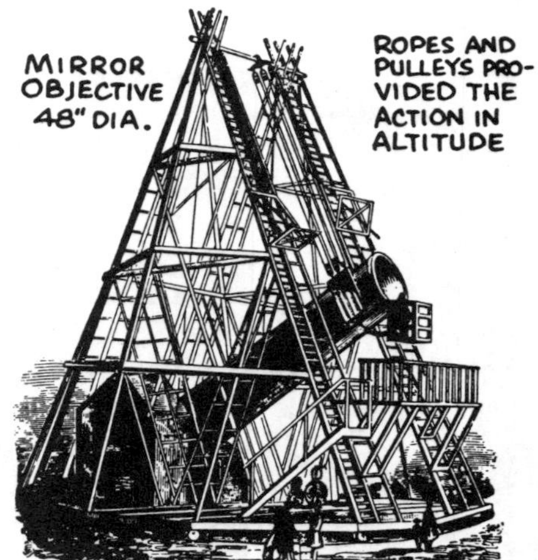

1789 HERSCHEL'S 40-FOOT REFLECTOR HAD A TILTED MIRROR TO PERMIT OBSERVING FROM FRONT OF TUBE. THE MOUNT ITSELF WAS AN ALTAZIMUTH

The ORIGINAL GERMAN EQUATORIAL

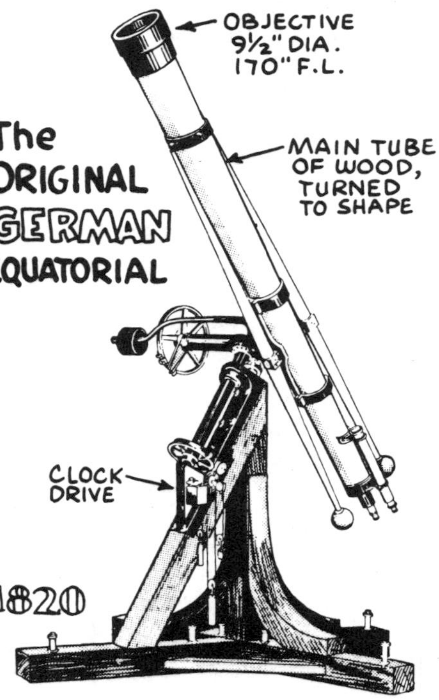

1820

9½" DORPAT REFRACTOR

WAS LOCATED AT UNIV. OF DORPAT (*Tartu, U.S.S.R.*). OBJECTIVE AND MOUNT BY JOSEPH VON FRAUNHOFER, GERMAN OPTICIAN AND PHYSICIST. TODAY, ANY MOUNTING OF THIS POPULAR STYLE IS CALLED A "GERMAN TYPE" EQUATORIAL

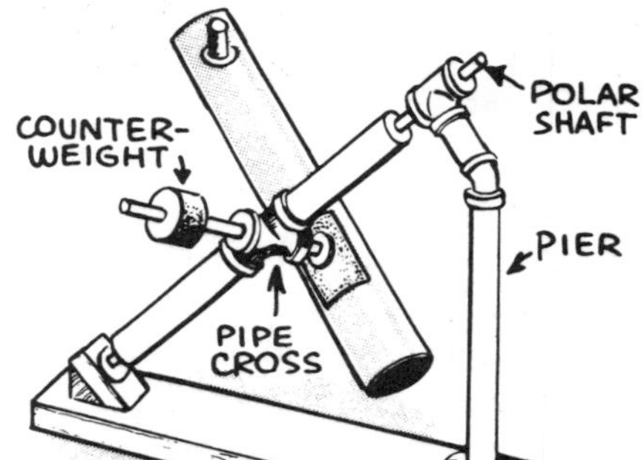

ABOUT 1824

English TYPE - **CROSS AXIS**

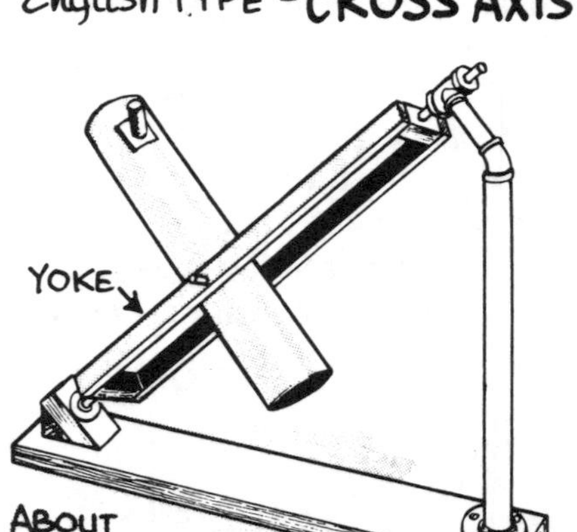

ABOUT 1825

English TYPE - **The English YOKE**

NO COUNTERWEIGHT NEEDED. POOR FEATURE: THE NORTH SKY AROUND POLARIS IS NOT ACCESSIBLE

Telescope Mounts

PRACTICALLY all of the early mounts were altazimuths. Such a mount has a primary axis in a vertical position, while the branch or secondary axis is horizontal. Movement around the primary axis moves the telescope to the right or left, technically known as a movement in azimuth; movement around the horizontal secondary axis moves the telescope up or down in altitude.

English inventor, John Hadley (1682-1744), was first to parabolize a telescope mirror, which he fitted in a wood tube supported on a neat and clever altazimuth mount, complete with slow motions worked by cords wrapped around pegs.

The equatorial mount made the scene in 1820 when German optician, Joseph von Fraunhofer, supplied both the objective and mount for the Dorpat refractor (Tartu, U.S.S.R.). This 9-1/2-inch refractor was the first telescope of any importance to be mounted equatorially with a clock drive. Fraunhofer was a German--the mounting has since been called the German type. It is used today for most portable telescopes, and is also used for practically all refractors from the smallest to Yerkes.

The Dorpat refractor was a little shaky, causing English astronomers to concentrate on a mounting with the polar shaft solidly supported at both ends. Early users of this kind of mount included John Herschel (son of William) and Adm. Wm. A. Smyth, who used it to mount his 5-inch refractor, laying the groundwork for his popular book, "A Cycle of Celestial Objects," published in 1844.

The basic English equatorial has a single solid beam for the polar axis, with the declination shaft cross-axis with counterweight in the manner of the German mount. The English Yoke came a little later. Both kinds of English equatorials are very much in use today--they support big reflectors all around the world.

It is not hard to see the evolution from yoke to fork mounting. Since long fork arms tend to wobble, the fork is usually compacted, with tube balance maintained by means of extra weight at the end of main tube. Another fairly obvious transition occurred at the polar shaft itself, which gradually changed from straight shaft to taper shaft, and finally to flat plate, as shown below, of which the shaft portion is only a minor appendage.

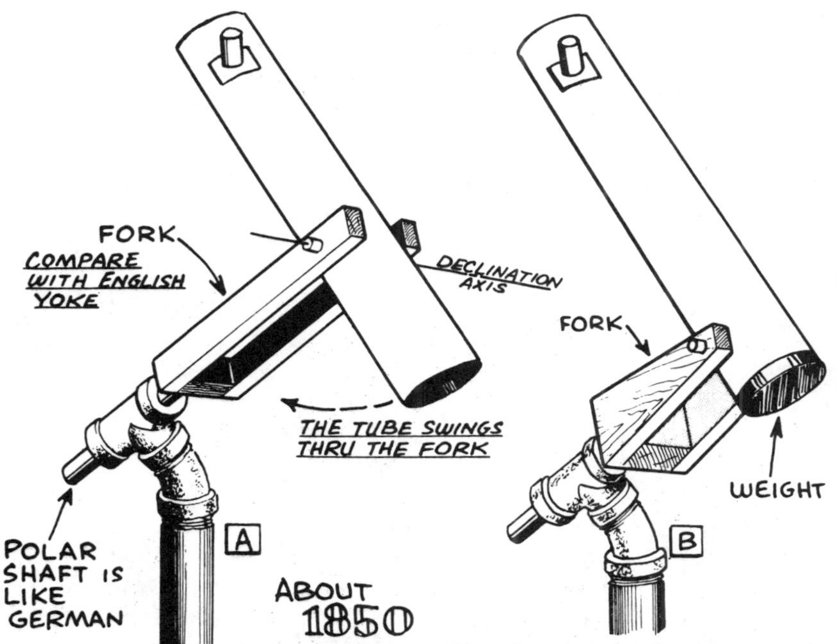

ABOUT 1850
FORK EQUATORIAL MOUNT

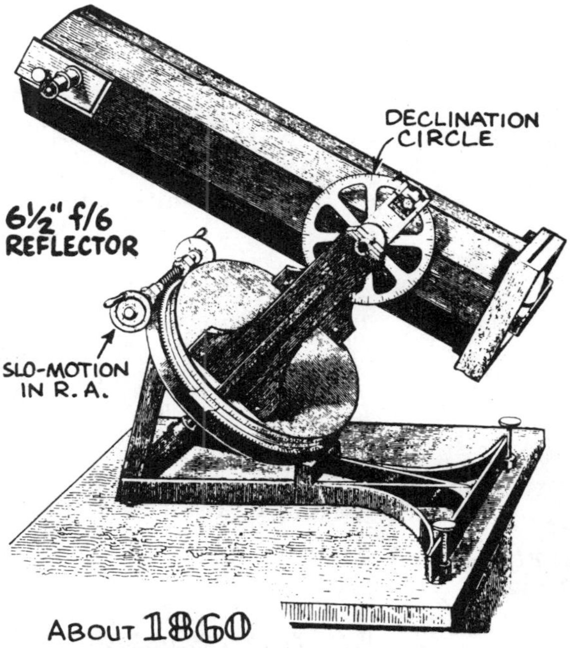

ABOUT 1860
FORK MOUNT ON PLATE

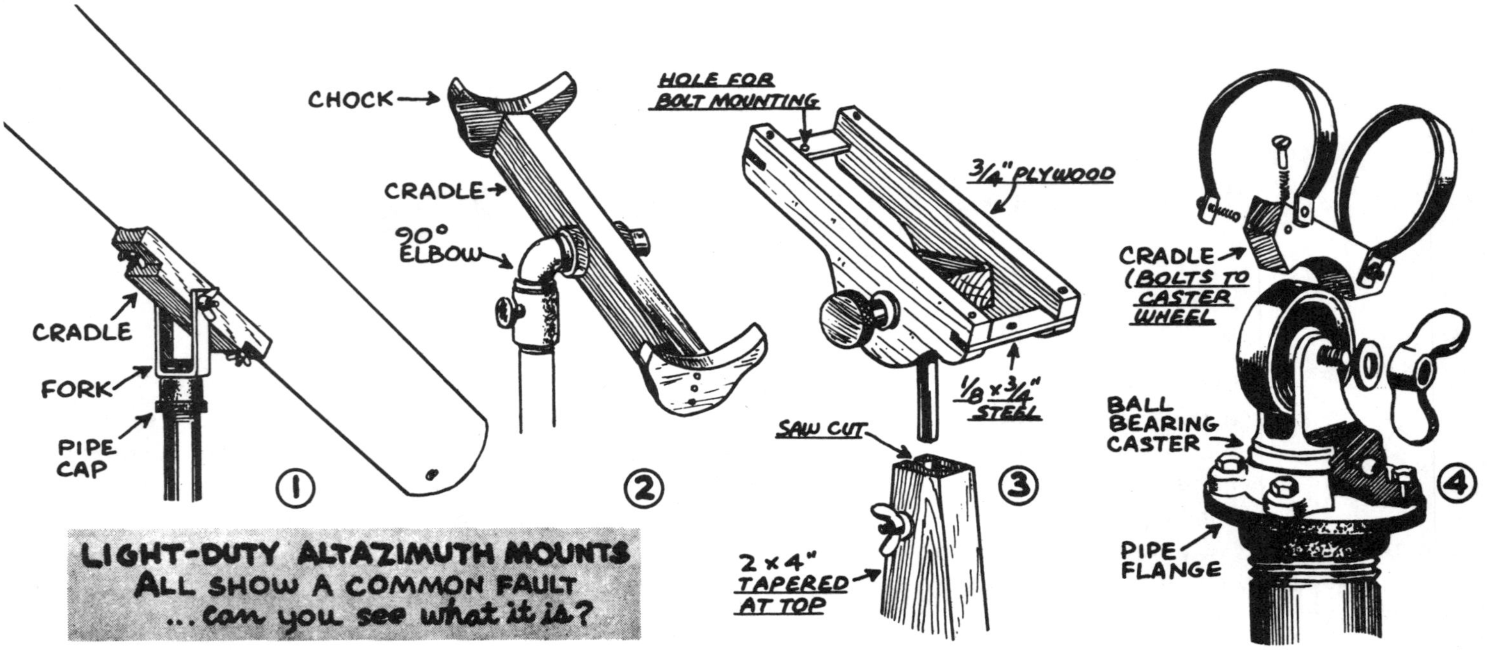

LIGHT-DUTY ALTAZIMUTH MOUNTS ALL SHOW A COMMON FAULT ... can you see what it is?

What Mount is Best?

A MOUNT must be steady. The test of this is when the telescope is in motion since, after all, a mount would have to be very flimsy to shake and tremble without even being touched. If your mount trembles and vibrates when you move it around gently at about 40x magnification, it is not a good mount. In addition to the "steady as a rock" feature, you will want the movements of the mount to be smooth and convenient, covering the whole sky. If, also, the mount is to be portable, the specified features must be obtained without excessive weight or bulk. You can learn something about mounts by reading, but what it really takes is build-and-try to get an intimate knowledge of the problem.

IMBALANCE IN AN ALTAZIMUTH. The cheapest, easiest to build mount is the altazimuth. Most of these are quite satisfactory for small telescopes, especially when the telescope is in a horizontal position, as it is for most daytime viewing of land objects. But the altazimuth sometimes runs into trouble when it is tilted, Fig.5. This is the common fault of the four designs shown at top page.

Altazimuth mounts can be classified according to the relation of the telescope to mount. Fig. 5 shows top mounting, while Fig. 6 shows side mounting. The meaning of the terms will be clear from the drawings. The true fork mount might be described as "center mounting," Fig. 7. It will be apparent that side and center mountings are balanced in all positions of the telescope tube, while any style of mount with the tube in top position, as in Figs. 1 to 5, will be imbalanced when the telescope is tilted. Top mounting is satisfactory only for very lightweight telescopes. A trifling amount of imbalance is easily correct-

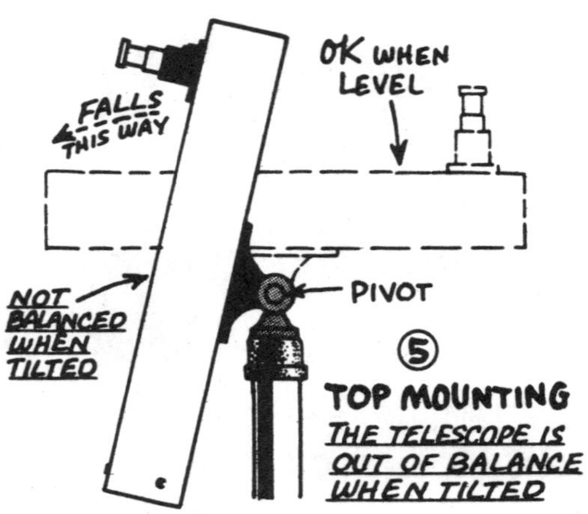

TOP MOUNTING
THE TELESCOPE IS OUT OF BALANCE WHEN TILTED

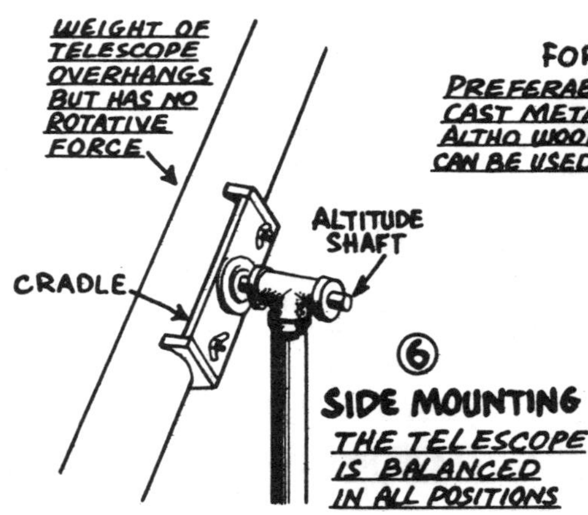

SIDE MOUNTING
THE TELESCOPE IS BALANCED IN ALL POSITIONS

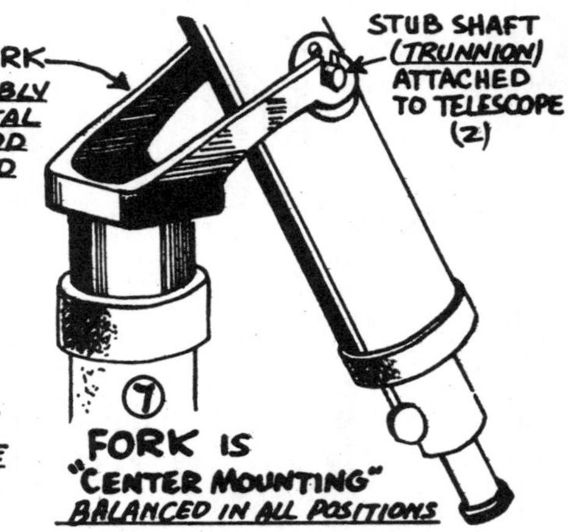

FORK IS "CENTER MOUNTING" BALANCED IN ALL POSITIONS

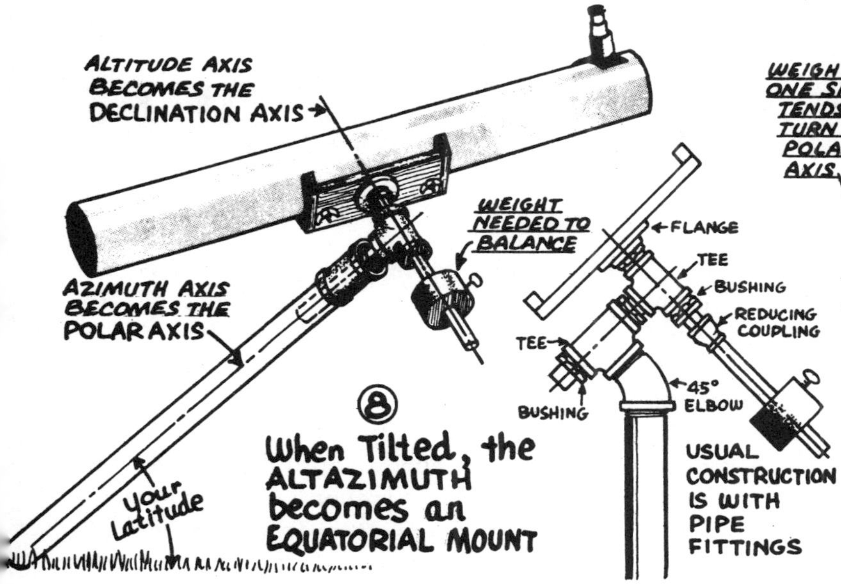

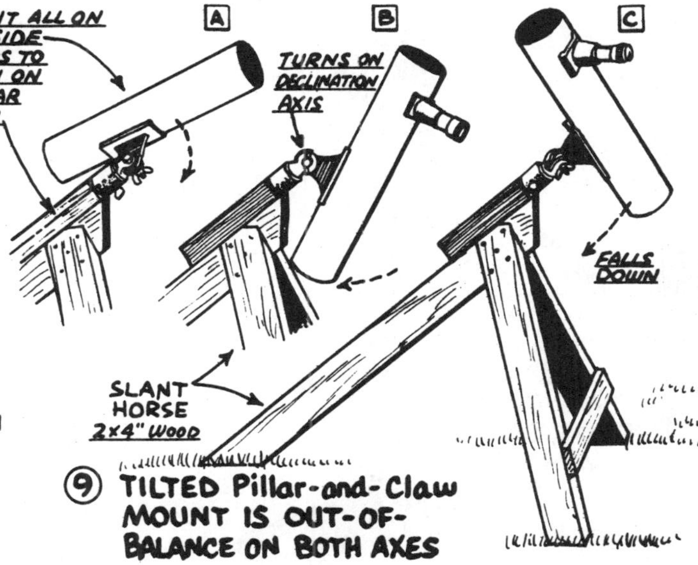

ed with a small amount of tension applied to the moving part. Excessive clamping pressure makes the telescope movement erratic.

THE GERMAN EQUATORIAL. If a conventional altazimuth is tilted, it becomes an equatorial mount of the German type, Fig. 8. A counterweight is needed to correct the imbalance. The pillar-and-claw type of altazimuth converts to equatorial fairly well, Fig. 9, and this mono-axis construction is used successfully with many small telescopes. Properly, it needs a counterweight, as can be seen in the drawings, but this is not readily fitted since it would interfere with the movement of the mount.

FORK AND YOKE MOUNTS. A main talking point of all fork and yoke mounts is that you can swing through the meridian, that is, you can go from east horizon to west horizon in one continuous sweep. As you may know, you can't do that with a German type, but instead have to make the movement with two different settings, the telescope being east of pedestal for west sky, and west of pedestal for east sky. Quite often both the fork and yoke mounts are compacted, as shown in Figs. 10 and 11. While this simplifies the construction it causes a loss of a large area of the north sky around Polaris.

THE BEST MOUNT. Most builders agree that the German equatorial is the best mount for a portable reflector. Second choice is the fork mount. In both cases, the construction is preferably all metal, although wood lends itself well to fork and yoke mounts for light duty. A pedestal base minimizes the operational fault of the German equatorial, permitting a fair degree of swing through the meridian.

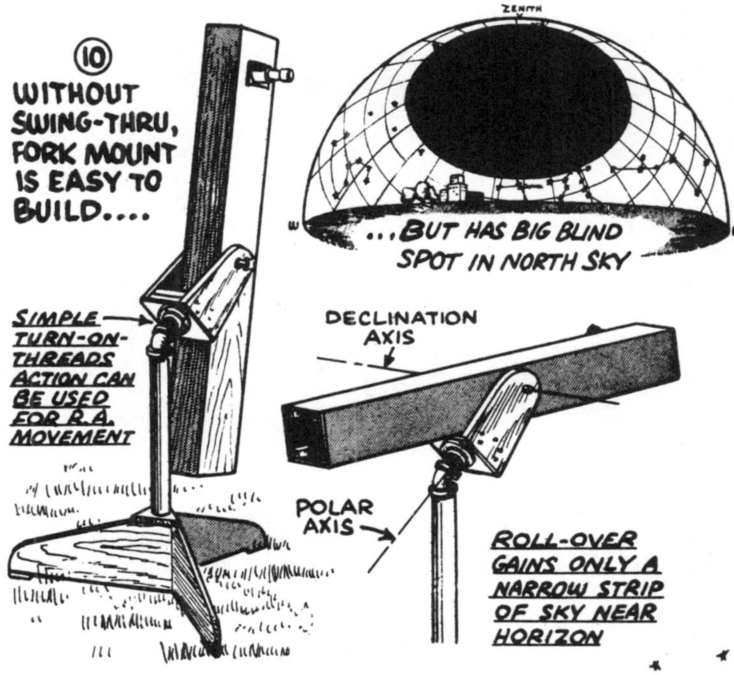

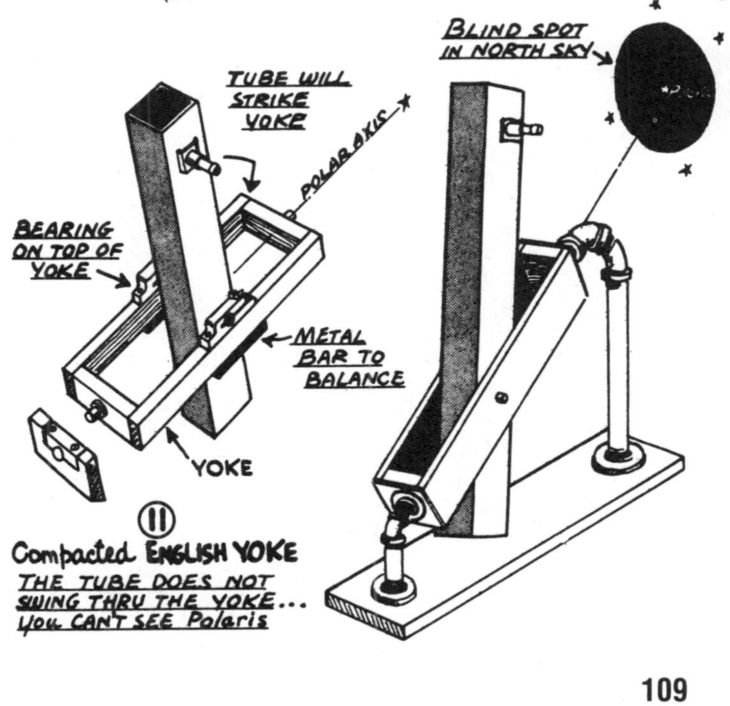

the German Equatorial

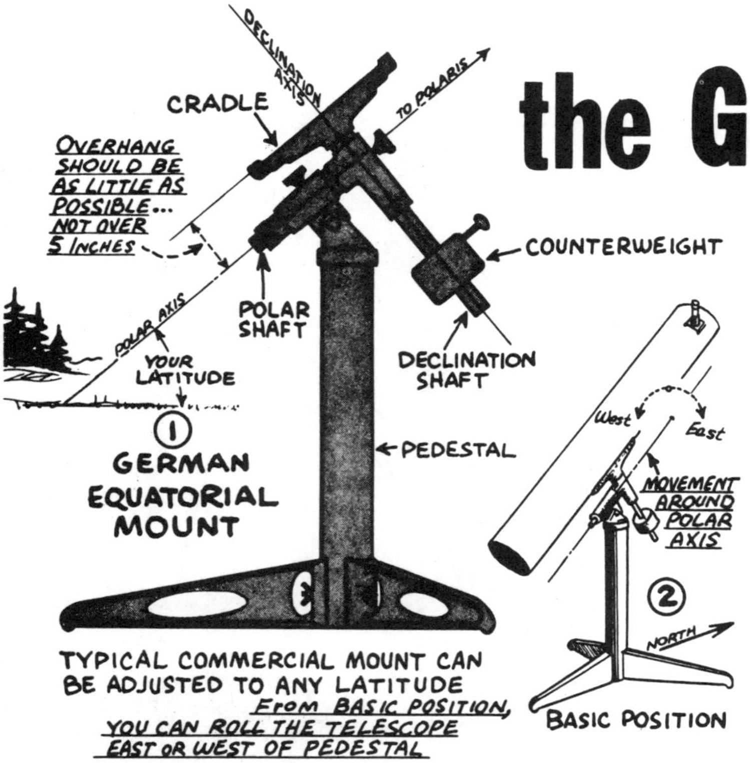

THE GERMAN equatorial is usually drawn as shown in Fig. 1, with the cradle directly over the polar shaft. Further, the view is usually from the east, looking west, as shown, although this orientation is by no means universal.

From the basic position shown in Fig. 2, you can swing the telescope to either east side or west side of pedestal, the movement being around the polar shaft. Fig. 3 shows the telescope moved to west side of pedestal and then swung toward the south to point the telescope near the zenith. In this position, you will be aware of the lone operational fault of the German equatorial: it does not permit movement through the meridian. Study Fig. 3. The telescope is west of pedestal. If you try to move into the western sky, the lower end of telescope tube will strike the pedestal. However, you have freedom unlimited in the eastern sky, right down to the horizon. Fig. 4 shows the same situation with telescope east of pedestal. Thus, the general rule for using a German equatorial is that the telescope should be east of pedestal for west sky objects, and west of pedestal for east sky objects. However, when you move southward from the zenith, you pick up a little more freedom, so that in the equatorial region of the sky, the tube is able to sneak around the pedestal for a good range of movement through the meridian. Fig. 5.

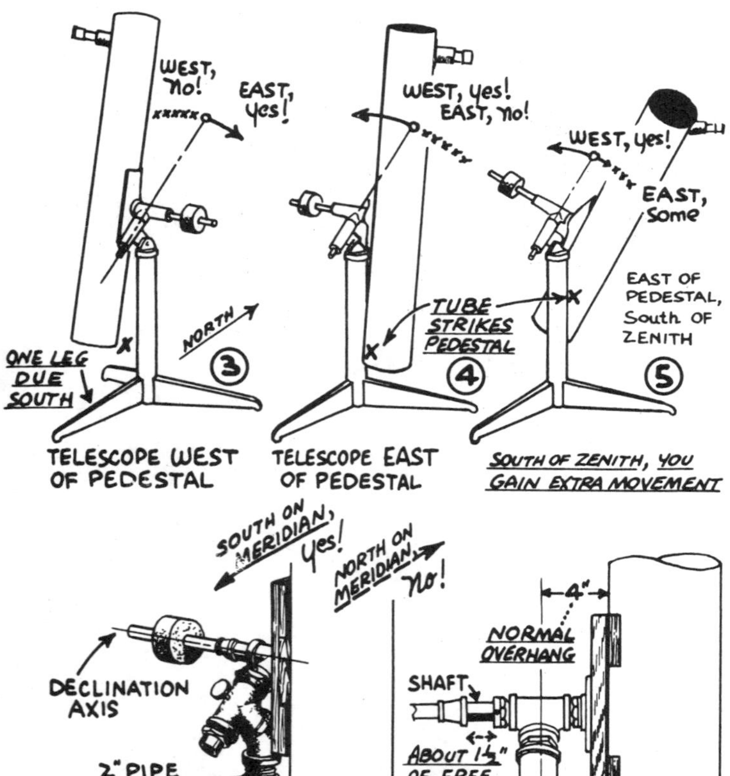

The movement of the German mount is much more restricted when you use a tripod. The example shown in Figs. 6 and 7 is typical, the telescope being unable to move at all in the zenith position. Of course, there is an easy cure for this--you give the telescope more overhang, and it will then have all the clearance you want. But this cost is a heavier counterweight, and usually the mount is less stable. The end result is that most experienced builders will take a little loss of sky rather than increase the overhang.

With a pedestal mount, one leg is usually pointed due south. With this setup, you stand in the pocket formed by south leg and east leg for most of your observations. With a tripod, however, the "pocket" is used to get the most freedom of movement for the telescope itself, which means one leg must face north for best possible use of the available clearance.

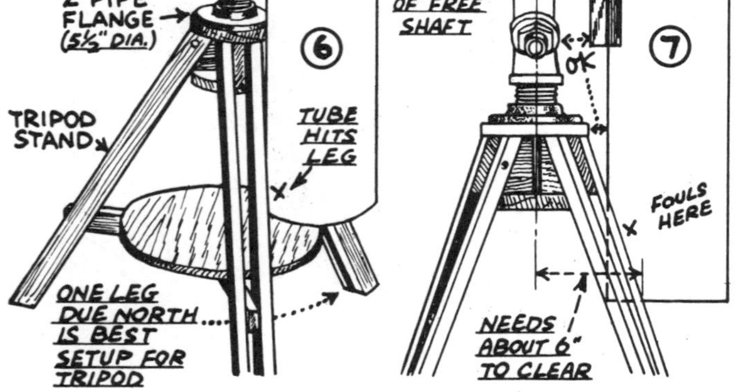

MANY WAYS TO BUILD. There are a thousand-and-one ways to build a German equatorial. Every book on the subject stresses substantial

many ways to build

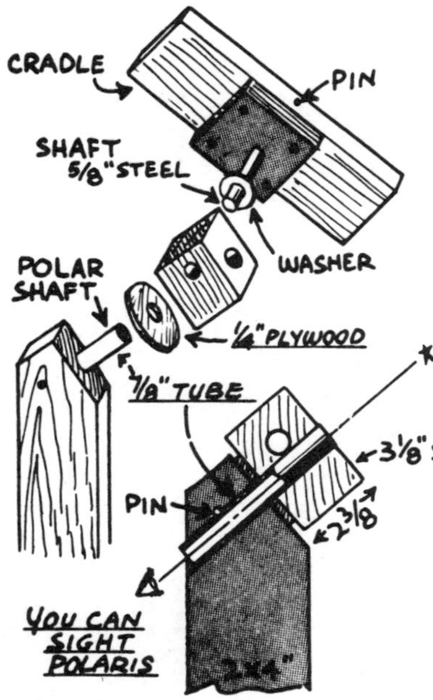

⑧ A LIGHT-DUTY MOUNT with WOOD BEARING BLOCK

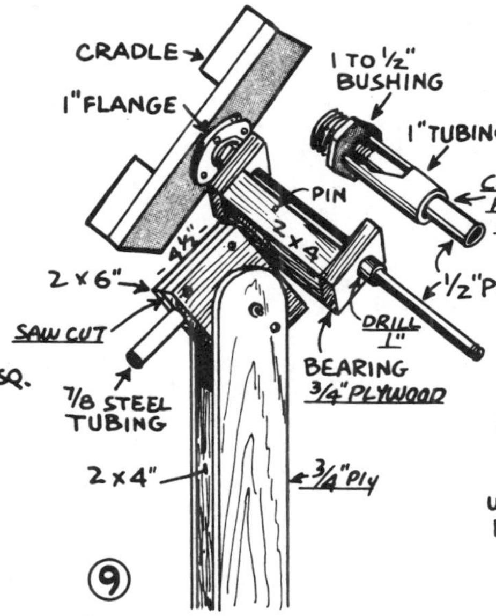

⑨ ANOTHER SIMPLE DESIGN WITH WOOD BEARINGS

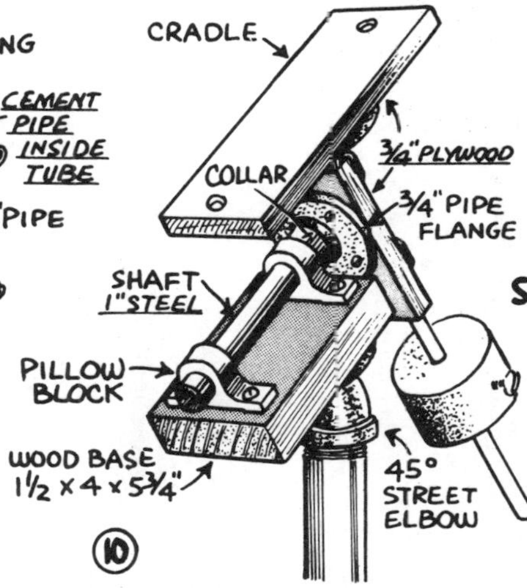

⑩ MANY KINDS OF METAL BEARINGS CAN BE PURCHASED

construction and close-fitting bearings. But in the end you build your own mount your own way and then take it outdoors to see if it is really so. In general, it can be said that anything made of wood will wobble, yet some quite workable mounts are made of wood, the classic in this medium being the plain wood block with a couple holes bored through it, as shown in Fig. 8.

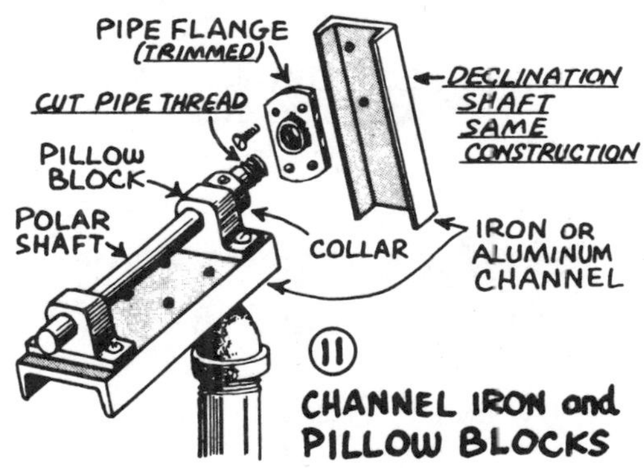

⑪ CHANNEL IRON and PILLOW BLOCKS

Fig. 9 construction shows a way to retain the useful thread of the pipe shaft while gaining the roundness and smoothness of tubing for bearing surfaces. Fig. 10 is a popular way to build, using either homemade or purchased pillow blocks; you go "whole hog" on metal by using base supports of channel iron, as shown in Fig. 11. Many builders like plate bearings instead of the more conventional shaft, Fig. 12. A nice feature of this construction is that with a little extra work the polar stud can be made hollow and so serve as a sighting tube or even house a polar telescope.

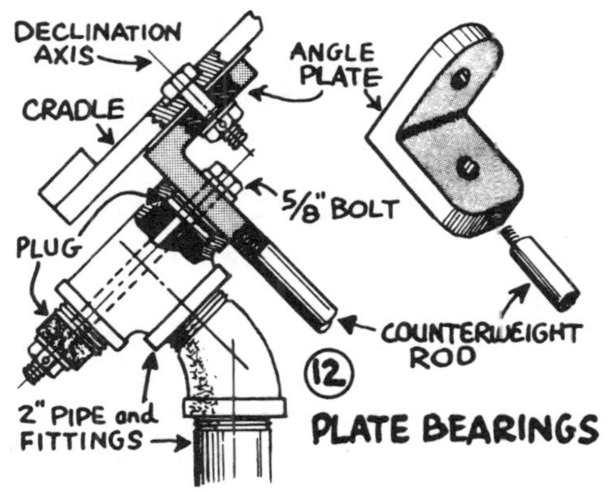

⑫ PLATE BEARINGS

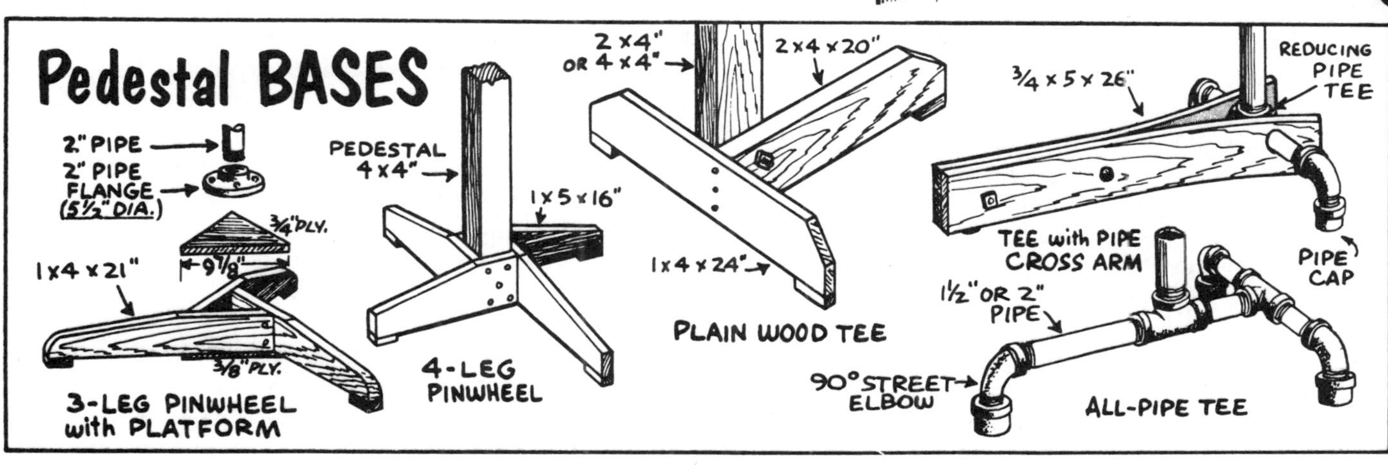

① ALTAZIMUTH
(SEE [A] ON OPP. PAGE)

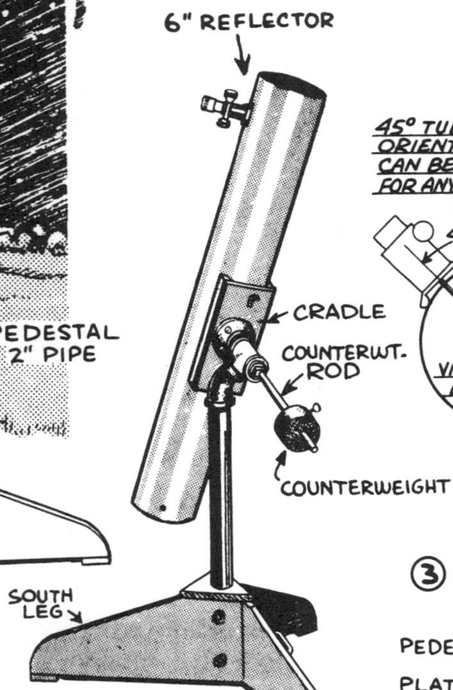

② EQUATORIAL
(SEE [D] ON OPP. PAGE)

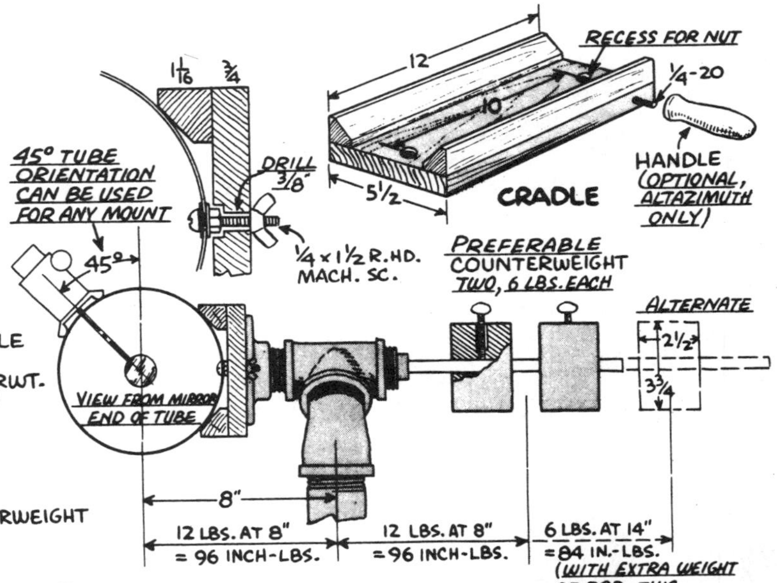

③ EQUATORIAL - VIEW FROM SOUTH

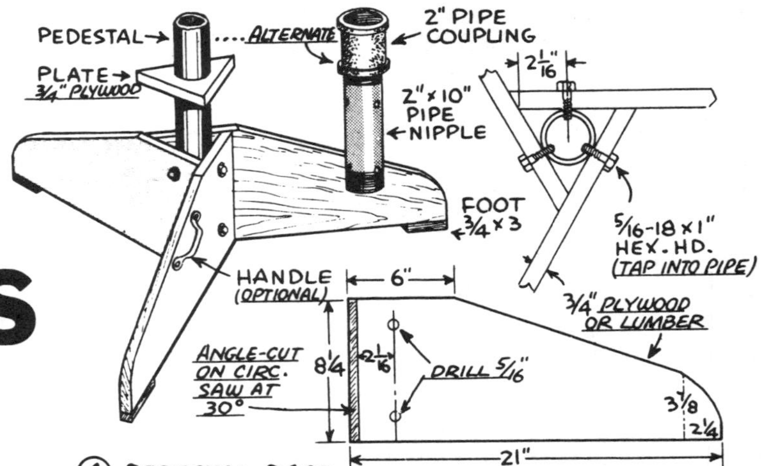

④ PEDESTAL BASE

PIPE MOUNTS

YOU MAY eventually make something much fancier, but for a starter there is no better telescope mount than the turn-on-threads variety made of pipe fittings. With this kind of mount you get the needed mechanical movements by simply leaving the pipe fitting a little loose, free to turn. Much the same construction is used for both altazimuth and equatorial mounts. Both have two axes at right angles, but the primary axis of the altazimuth stands erect, Fig. 1, while the equatorial is tilted, Fig. 2.

Pipe and pipe fittings for a turn-on-threads mount suitable for a 6-inch reflector should be the 2-inch size, although in a pinch you can use 1-1/2-inch. Fig. 5-A shows the basic altazimuth made with the fewest number of parts. If you can't locate a street elbow, an ordinary elbow with a close nipple can be used, Fig. 5-B. It is always a good idea to start with the simpler altazimuth mount, but if you have your eye on an equatorial, then construction like 5-C lets you move right along to the equatorial, Fig. 5-D. All of these mounts have the center of gravity considerably off-center, but this is not an item of great importance. Sometimes you will see a turn-on-threads equatorial fitted with a gooseneck, Fig. 5-E, to put the weight directly over the pedestal, but on the whole this is over-building.

The free threads should be lapped-in for a smooth turning fit, this operation being done by applying a mixture of oil and fine abrasive (400-grit) and then applying your muscle to screwing and unscrewing the joint many times until you gain at least two full threads. The sludge is then removed with paint thinner, but you don't have to be too thorough since a little abrasive left in the joint is helpful for its mild braking action.

Ten years ago, the favored base for a portable telescope was the familiar wood tripod, but today 90% of all small reflectors are supported by a pedestal base. The homemade version, Fig. 4, has three wood legs fastened to the pipe pedestal with machine screws. If you work this one leg at a time, spotting holes, drilling and tapping, you will get a tight assembly which needs no other fastenings. How high to make the pedestal is always a problem. A pedestal 32 in. high for an altazimuth or 28 in. for an equatorial is about average, but since people come in assorted lengths it is not possible to suit everybody. In any case, you favor staying fairly close to the ground for

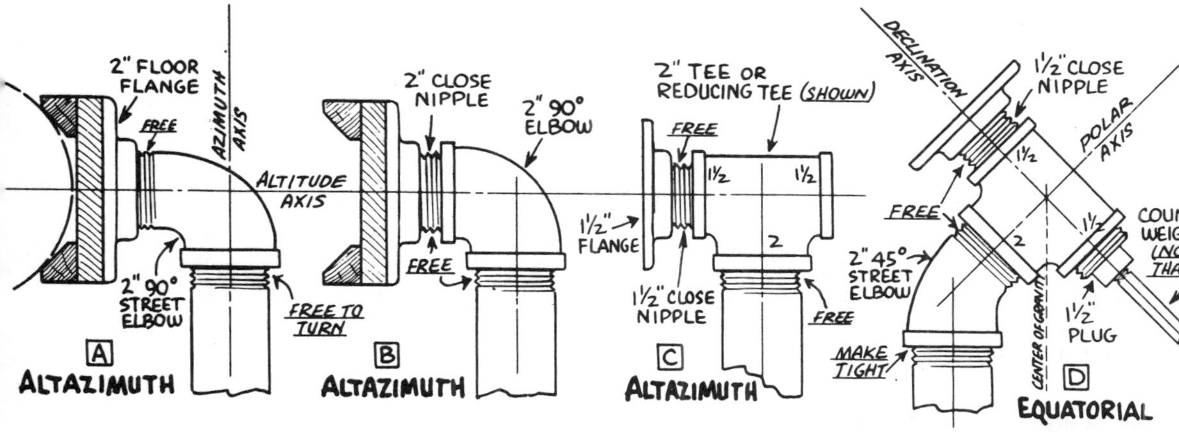

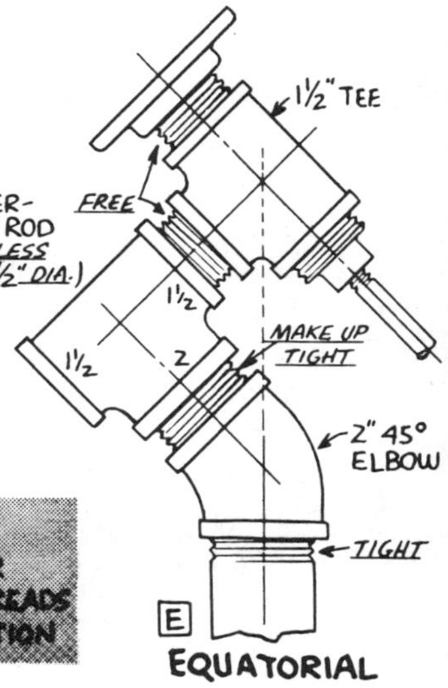

5 "TURN-ON-THREADS" PIPE MOUNTS
USE MALLEABLE IRON (STEEL) FITTINGS IN 1½" OR 2" SIZE FOR 6" REFLECTOR. THE FREE PIPE THREADS SHOULD BE LAPPED FOR SMOOTH TURNING ACTION

best stability. One good solution to the height problem is the all-purpose base shown in Fig. 4 detail--you can switch pedestals without tearing the whole assembly apart.

The equatorial shown in Fig. 2 is the popular German-type with primary polar axis providing a single outboard support for the secondary declination axis, which in turn provides a single outboard support for the telescope cradle. The other popular type of mounting is the fork type which provides a double support for the secondary axis. The fork principle can be applied to a turn-on-threads mount in the manner shown in Fig. 5-F. In use, the free thread on one side of the central tee unscrews while the opposite end tightens. A carriage bolt passing through the tee to a threaded knob allows putting a little squeeze on the threads to provide a tension adjustment. Although it requires a counterweight, this mount is nicely balanced and can be used successfully on a lightweight pedestal.

SHAFT EQUATORIALS. More conventional mounts are made with rotating shafts turning in sleeve bearings

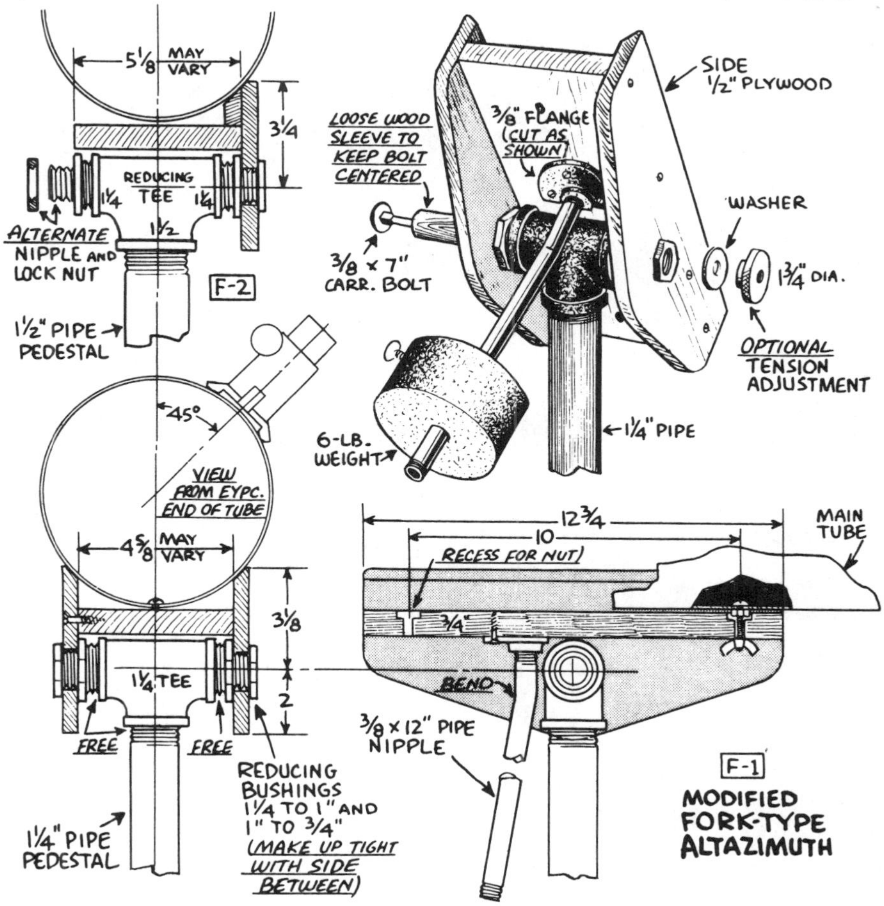

PIPE — STANDARD WEIGHT STEEL PIPE, BLACK OR GALVANIZED
BLACK PIPE AND FITTINGS ARE CHEAPEST, EASY TO PAINT... GALVANIZED LOOKS GOOD WITHOUT PAINT

NAME SIZE	OUTSIDE DIA.	INSIDE DIA.	WALL	THREADS PER INCH	WEIGHT PER FOOT
⅛"	.405"	.269"	.068"	27	.24 LBS.
¼"	.540	.364	.088	18	.42
⅜"	.675	.493	.091	18	.57
½"	.840	.622	.109	14	.85
¾"	1.050	.824	.113	14	1.13
1"	1.315"	1.049"	.133"	11½	1.68 LBS.
1¼"	1.66	1.380	.140	11½	2.27
1½"	1.90	1.610	.145	11½	2.72
2"	2.375	2.067	.154	11½	3.65
2½"	2.875	2.469	.203	8	5.79
3"	3.50	3.068	.216	8	7.58
3½"	4.00	3.548	.226	8	9.11
4"	4.50	4.026	.237	8	10.79

of metal, plastic or wood. Minimum specifications are about 1 in. dia shafts in bearings spanning at least 4 inches. The simplest construction makes use of solid wood bearing blocks, Fig. 6. For shafts, you have a choice of steel, brass or aluminum bar stock, which is truly round, but you must cut the pipe thread, or, you can use pipe shafts which are already threaded but you must machine round. Fig. 6 design shows both. The holes through the bearing blocks can be worked on the drill press, preferably followed by hand reaming to exact size. You can improve a wood bearing block considerably by using bearings of Devcon plastic steel or aluminum, Fig. 7.

The most common type of construction makes use of pipe and pipe fittings. Dozens of variations are possible. Tees are used as bearing housings and reducing bushings make the bearings. These parts are assembled permanently and bored straight through from one end to the other. Ream or grind if you have the equipment; otherwise you can get a fair fit by lapping with 400 or 600-grit abrasive grains with oil lubricant. Fig. 8 design departs from the construction mentioned by using a pipe cap as the end bearing for the polar shaft. This is fitted permanently in place and then bored. Fig. 9 is a medium-duty design recommended if you want fair stability without too much weight.

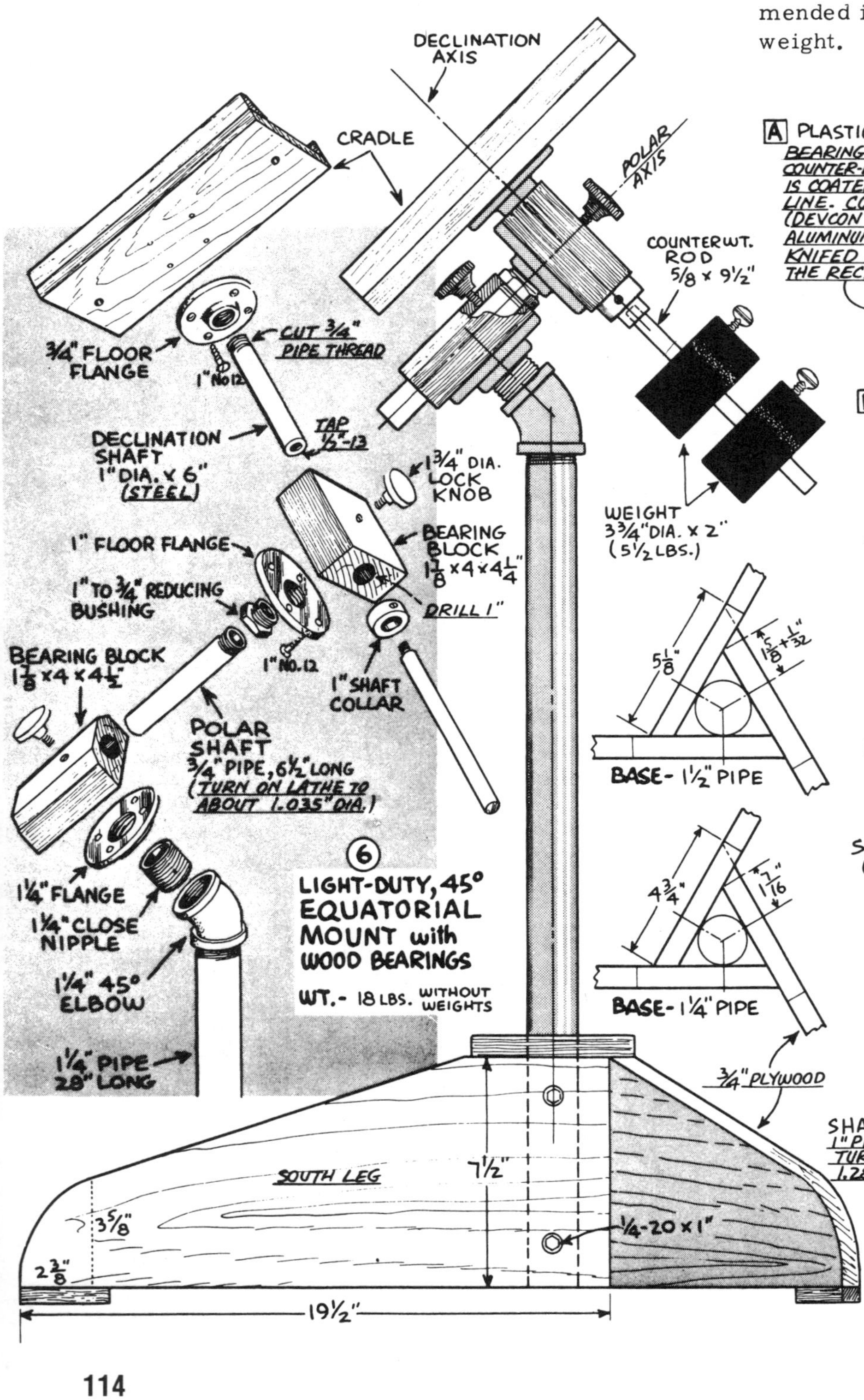

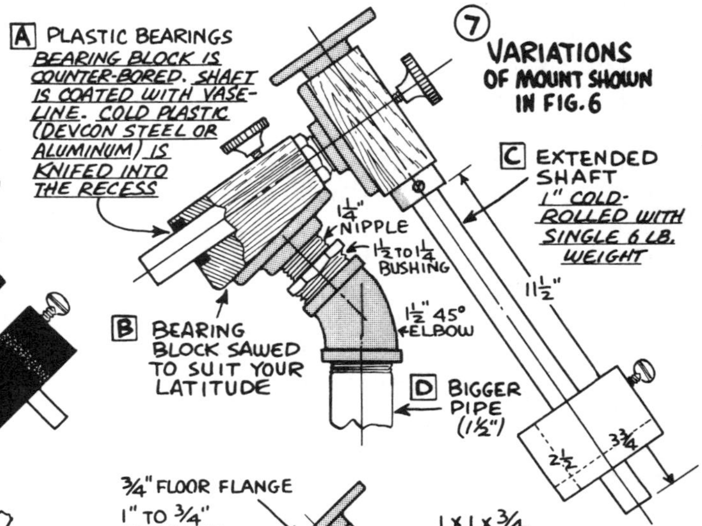

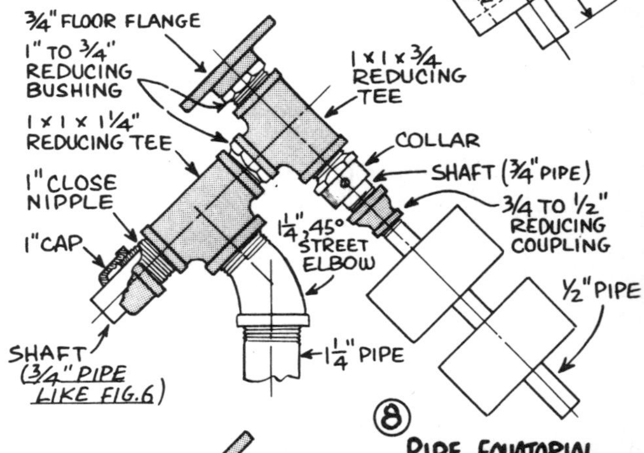

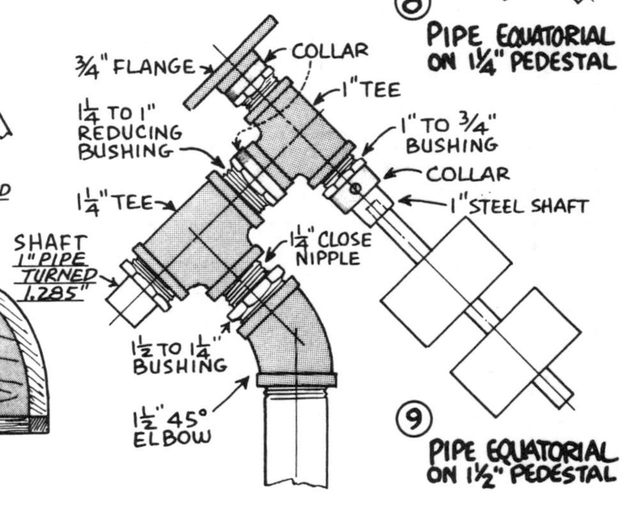

The only difficult part of making a pipe equatorial is the boring job, the 4 or 5-inch depth being near maximum for average home workshop equipment. An alternate is to rough-bore the bearings separately about .015 inch undersize, after which the assembly is made and the bearings hand reamed straight through to exact fit. Here again the required depth of cut is near the limit for ordinary reamers.

The use of pipe as bearings, as in Fig. 10, makes the boring job a little easier because the one long extension makes chucking easy and solid. Getting away from the deep boring job, many amateurs use cemented bushings, as shown in Fig. 11. In this construction, the threads are partially removed from both the bushings and the pipe tee, using straight cuts to produce a loose joint. The bushings are bored and fitted separately to the shaft; the final assembly is done with adhesive cement, such as DEVCON plastic steel or aluminum. There will be enough thread left on the joining parts to allow a screw-in assembly, which is done with the bushings mounted on the shaft. Meanwhile, you keep turning the shaft in the bushings. It may bind, but keep on turning because another half-turn may free it again. What you want is a moderate amount of bind -- a good firm fit--because it will loosen-up considerably after a slight workout. Experiment with a dry assembly before you apply the cement. It is wise to grease the shaft with vaseline to prevent sticking should you get some cement on it. A plastic

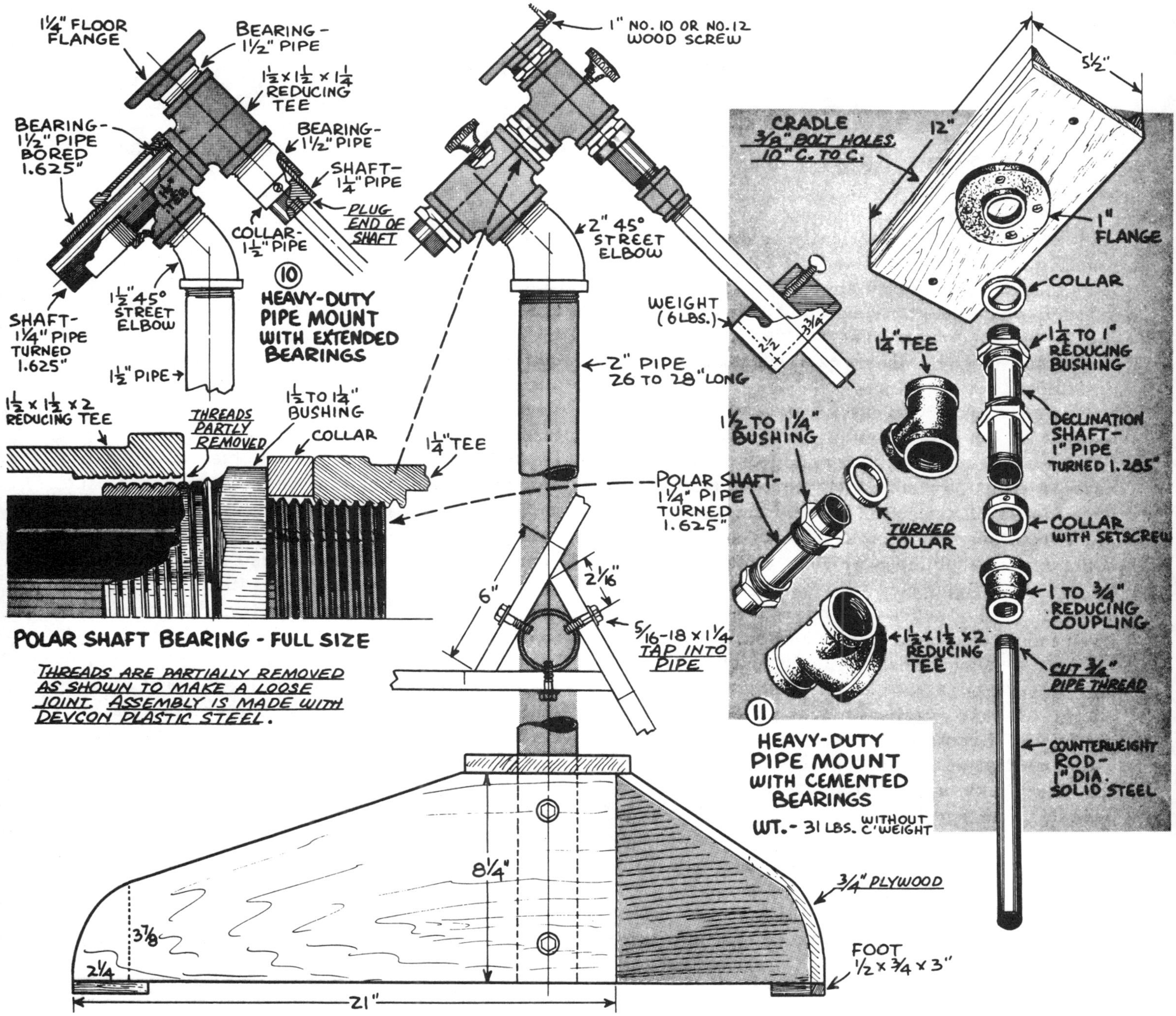

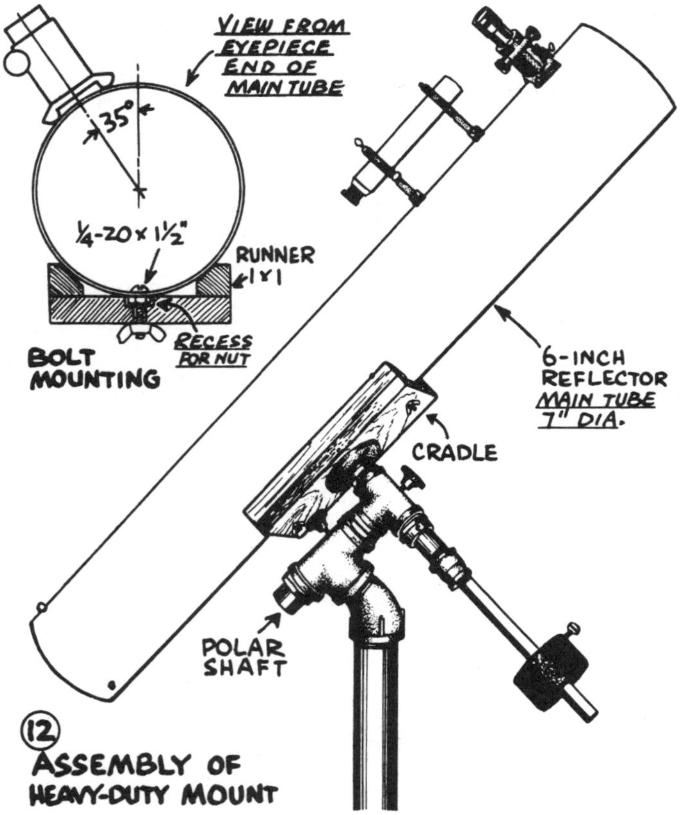

(12) ASSEMBLY OF HEAVY-DUTY MOUNT

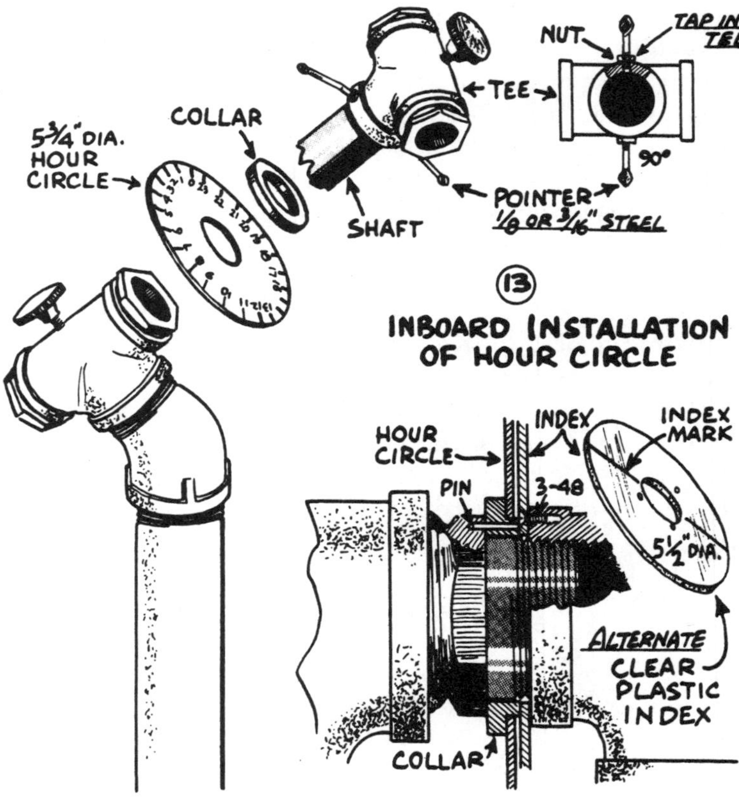

(13) INBOARD INSTALLATION OF HOUR CIRCLE

joint of this kind is practically as strong as the metal itself. However, should it ever be necessary to take such a joint apart, heat applied with a torch will make the cement friable, destroying its bonding power. The cementing technique can also be used with regular brass, bronze, plastic or steel bushings.

Fig. 11 equatorial is a heavy-duty mount using a 2-inch pipe pedestal. In portable mounts this is as heavy as you should go; bigger and heavier pipe may contribute to the stability of your mount but the job of transporting the equipment from basement to backyard is an item to be considered--anything over 50 lbs. total including the counterweights may kill your interest in stargazing. It is, of course, a different story with a fixed mount in a permanent location.

Fig. 12 shows the heavy-duty mount assembled and carrying a 6-inch reflector. Bolt mounting is shown, with the eyepiece mount 35 degrees from a position directly opposite the mounting bolts. This orientation is at its best when observing the south sky with telescope on east side of pedestal, the eyepiece pointing up and at about eye level. For north sky objects the telescope should be positioned west of pedestal.

INSTALLING SETTING CIRCLES. On most mounts you can install setting circles at any time without undue work. Mostly it is just the matter of boring a hole through the circle of such size as to fit the shafts of your mount. Usually the circle will be plastic, which should be mounted on plywood for added rigidity, the plywood base also serving to hold the plastic disk for boring the central hole. Installation can be either at outboard or inboard end of the shaft. Fig. 13 shows the inboard installation of the hour circle. A direct-reading hour circle is preferable--this is one with the hour numbers increasing counterclockwise, as shown (see also Fig. 15). Mount a disk of wood in the lathe in 3-jaw chuck and make a thin facing cut to true the outer surface, at the same time raising a boss at the center of the diameter of the original hole in the setting circle. Fasten the plastic disk to the wood with three small brads, then bore out the center to the required size to fit over the polar shaft. In a setup like Fig. 13, the circle is merely slipped over the shaft; in use, it will stay put since the collar above it provides the actual contact on which the telescope turns. However, if you want to make sure the circle does not move, mount it with a firm slip fit to the collar, which is then permanently pinned to the polar shaft housing, as shown in Fig. 13 alternate detail. So installed, nothing rubs against the circle at all and it will not move except when you set it by hand. The detail also shows an alternate index in the form of a clear plastic disk. In any case, a double-ended index is required to provide for reading the hour circle with the telescope either

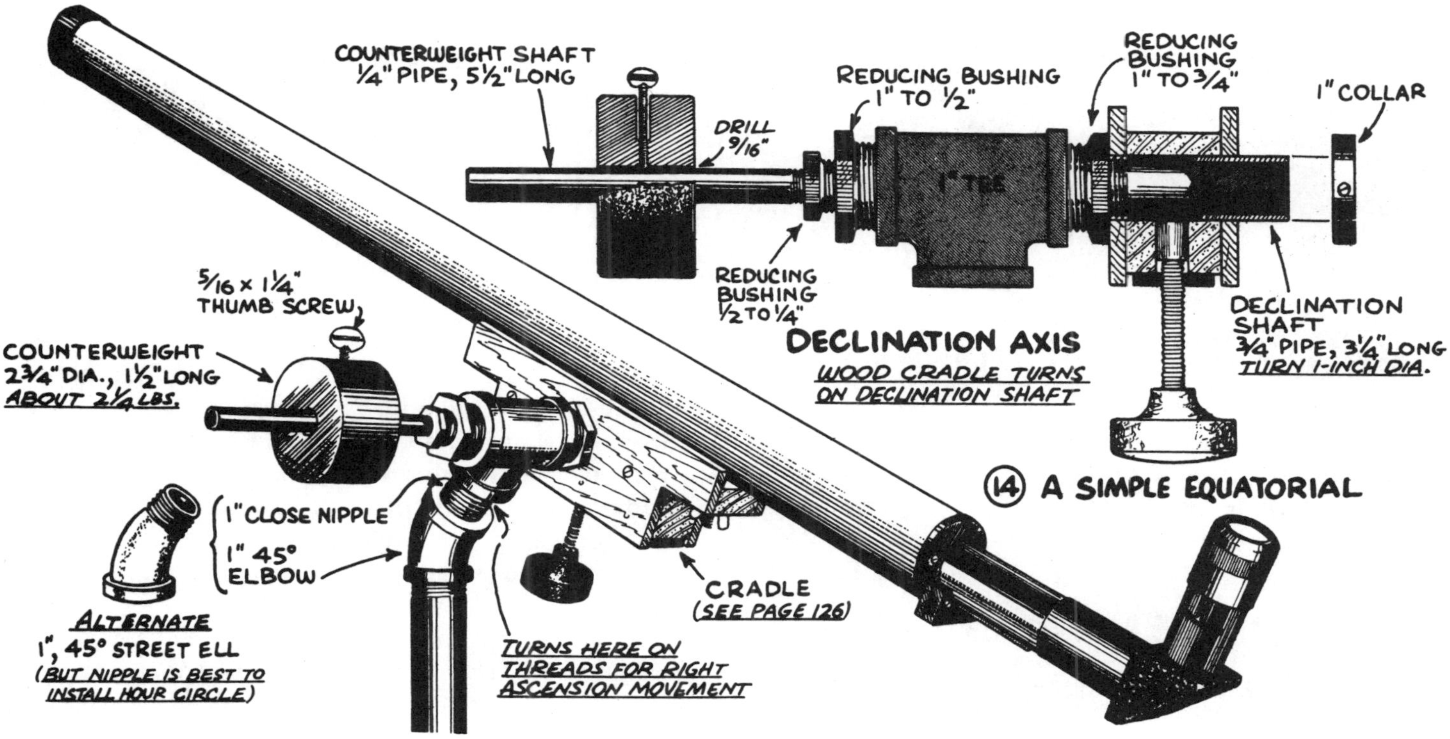

(SEE PAGE 126)

(14) A SIMPLE EQUATORIAL

east or west of pedestal.

MOUNTING A SMALL REFRACTOR. A small refractor will perform nicely on a light-duty mount, such as shown above and on page 126. The "on top" position of the telescope requires a counterweight to balance the telescope around the declination axis, this being provided by a metal plate on the underside of the cradle and a heavy declination lock knob. The wood cradle turns on the fixed declination shaft; the polar axis movement is the familiar turn-on-threads variety.

Setting circles are easy to fit and this work can be done at any time. In the construction shown below, both circles are stationary in use while the index pointers move when the telescope is moved. This is direct indexing in both R. A. and declination. Both circles should be a snug fit but free to turn - - they are "fixed" circles when in use, but they must permit setting by hand as required. There should be a little clearance between the hour circle and the pipe tee, this being controlled by the thickness of the mounting block. The clearance of about 1/8 inch is needed for the one-thread unscrewing action of the turn-on-threads polar axis.

In use, the telescope is west of pedestal, as shown in drawing above, for all south sky objects to your zenith. You have free movement through the meridian east or west. The north sky is similarly covered with the telescope east of pedestal. If you remember to always go north around by west, and always return the same way, the turn-on-threads polar movement will always work perfectly without jamming or unwinding.

(15)

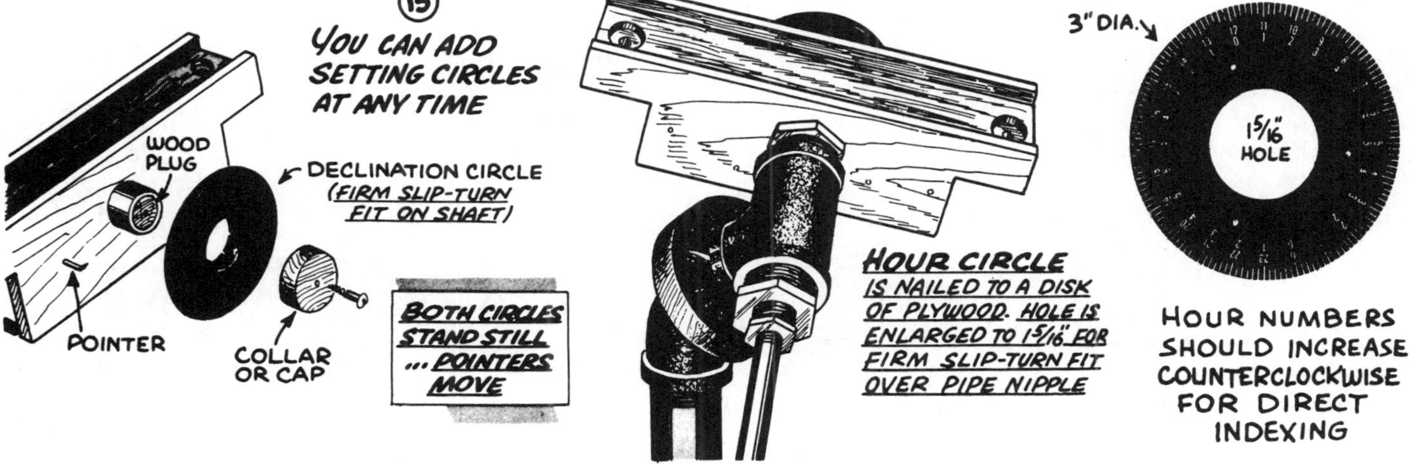

The CLOCK DRIVE

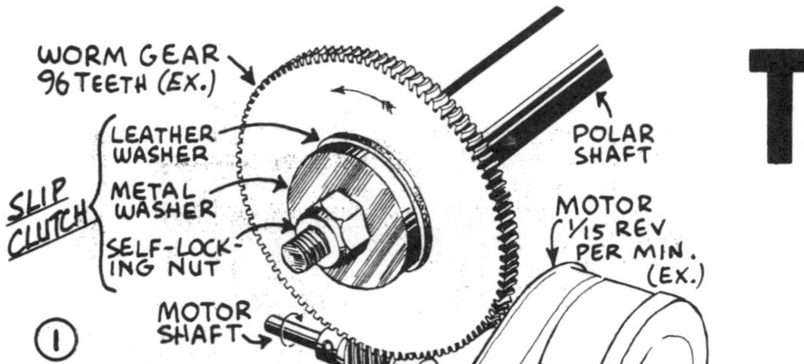

① TYPICAL MOTOR-GEAR COMBINATION FOR ONE REV PER DAY

1/15 REV OF WORM IN 1 MIN. TURNS 1/15 TOOTH = 1/4°
1 REV OF WORM IN 15 MIN. TURNS 1 TOOTH = 3¾°
4 REVS OF WORM IN 1 HOUR TURNS 4 TEETH = 15°
96 REVS OF WORM IN 1 DAY (1440 MIN.) TURNS 96 TEETH = 360° ... IS 1 REV OF GEAR

② VARIOUS MOTOR-GEAR COMBINATIONS FOR 1 REV PER DAY (ALL APPROX. 4 MINUTES SLOWER THAN SIDEREAL RATE)

MOTOR REVS PER MINUTE	TIME TO TURN WORM 1 REV	NUMBER OF TEETH TURNED IN 1440 MIN.	ANGULAR MOVEMENT PER TOOTH	NEAREST FAST DRIVE	
				NUMBER OF TEETH	DRIVE RATE VARIATION
1/2	2 MIN.	1440/2 = 720	1/2°	718	EXACT
1/4	4 MIN.	1440/4 = 360	1°	359	EXACT
1/5	5 MIN.	1440/5 = 288	1¼°	287	1 M. FAST
1/6	6 MIN.	1440/6 = 240	1½°	239	2 M. FAST
2/15	7½ MIN.	1440/7.5 = 192	1⅞°	191	3½ M. FAST
1/8	8 MIN.	1440/8 = 180	2°	179	4 M. FAST
1/10	10 MIN.	1440/10 = 144	2½°	143	6 M. FAST
1/12	12 MIN.	1440/12 = 120	3°	119	8 M. FAST
EXAMPLE ABOVE 1/15	15 MIN.	1440/15 = 96	3¾°	95	11 M. FAST
1/20	20 MIN.	1440/20 = 72	5°	71	16 M. FAST
1/24	24 MIN.	1440/24 = 60	6°	59	20 M. FAST
1/30	30 MIN.	1440/30 = 48	7½°	47	26 M. FAST
1/40	40 MIN.	1440/40 = 36	10°	35	36 M. FAST
1/45	45 MIN.	1440/45 = 32	11¼°	31	41 M. FAST
1/48	48 MIN.	1440/48 = 30	12°	29	44 M. FAST PER DAY

VARIOUS motor-gear combinations can be used to maintain the one-rev-per-day pace of the stars. In general, the more teeth in the worm gear, the more accurate and smoother the drive will be. At the same time, a big gear is out of place on a small telescope, making the 96-tooth worm gear of about 3 inches diameter a practical choice. This automatically fixes the motor speed, which can be none other than 1/15 r.p.m. to get the desired 1 rev per day.

Synchronous motors run on standard solar time. As you may know, the stars travel a little faster, making one revolution in 23 hours, 56 minutes of solar time. In other words, the motor runs slow--the telescope lags the stars by about 4 minutes per day. In actual practice, this is not an inconvenience. Even with an exact sidereal clock, the drive can never be truly exact if for no other reason than refraction by the earth's atmosphere which changes the apparent position of stars, and hence the driving rate. If an exact drive rate is required--as in photography--it is necessary to add hand guiding to the clock drive. Obviously, as long as you have to guide, it is just as easy to guide the little bit extra to correct the lag in the conventional drive. If you are curious about a drive faster than the sidereal rate, the right hand side of Fig. 2 table gives the nearest approach. Of course most of the odd-tooth gears are not available.

The same gearing used for a clock drive is often used for a manual drive, Fig. 3. The motorized drive is just a matter of adding the motor, Fig. 4. Many amateurs get the happy idea of a

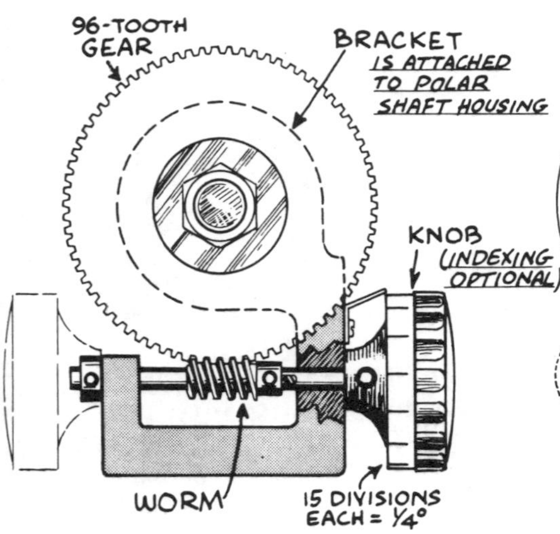

③ MANUAL SLOW-MOTION

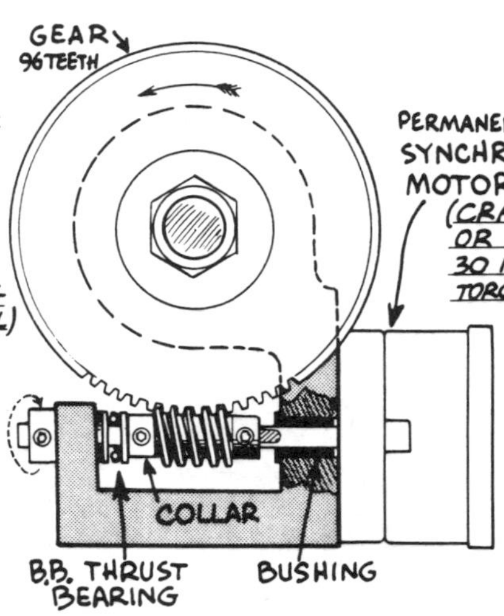

④ CLOCK DRIVE

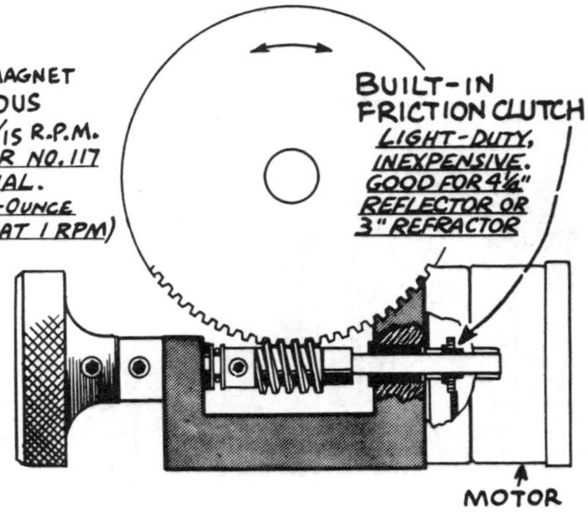

⑤ COMBINATION MANUAL SLOW-MOTION and CLOCK DRIVE

manual drive and a clock drive, all on the same shaft, Fig. 5. This calls for a clutch between worm and motor, and an apparent solution is the built-in friction drive which motor manufacturers supply for about 25¢ extra. However, your luck may run out at this point because many such mechanisms are strictly for no-load applications, such as setting the hands of a timing device. But some built-in clutches are husky enough to drive a small telescope, assuming it is in good balance. The better two-way drive has the clutch as a separate unit outside the motor where it is accessible and adjustable.

CLOCK SLOW-MOTION. Among amateur builders, favorite way of getting a slow-motion on the clock itself is to remove the mounting bolts from the clock motor and let it "float," controlled by a threaded rod, Fig. 7. In another system, Fig. 8, the motor and worm are moved as a unit. 1/8 inch travel is plenty for the purpose, which is only to correct the driving rate. Fig. 6 is a refinement of this idea, with the clock and worm again as a unit, but rotating around the polar axis. The slow-motion in this example is supplied by an eccentric cam operated by a lever.

SLIP CLUTCH. A slip or friction clutch must supply enough tension to drive the telescope, and yet must be loose enough to allow easy manipulation of the scope by hand. A common solution is simply a self-locking nut, Fig. 1. Better but more expensive is the keyed washer method shown in Fig. 7. The alternate detail shows a clutch with a backstop ring fastened to the shaft with a dog point set screw. The actual friction pressure comes from two thumb screws. This is a good slip clutch, offering any desired degree of friction with quick, easy release. The same idea is shown in Fig. 6 except the fixed backstop consists of a self-locking nut screwed tightly against a threaded washer. In all cases the tension knob or nut must turn with the polar shaft to keep from "winding-up" when the telescope is moved. If the nut stands still--its natural tendency--the friction will tighten when the scope is moved to the west, and loosen when scope is moved to the east.

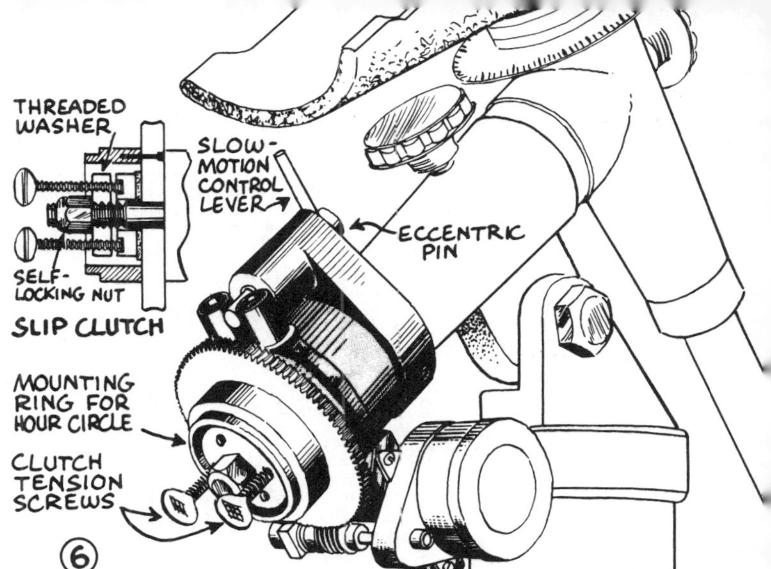

⑥ CLOCK DRIVE ON HEAVY-DUTY MOUNT

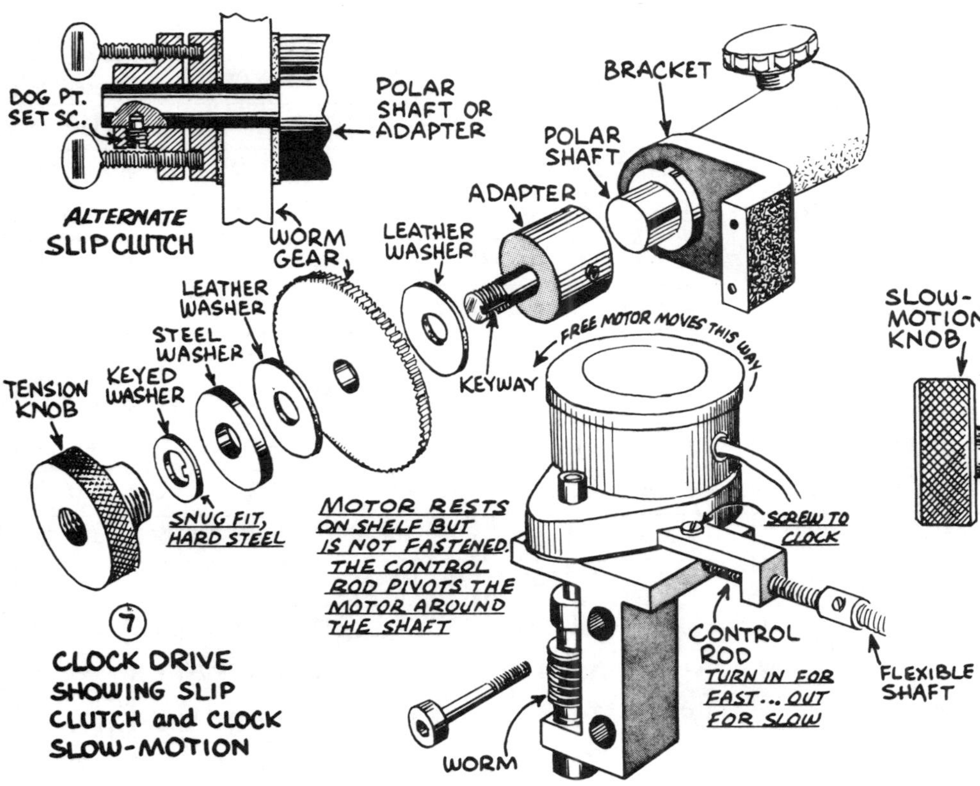

⑦ CLOCK DRIVE SHOWING SLIP CLUTCH and CLOCK SLOW-MOTION

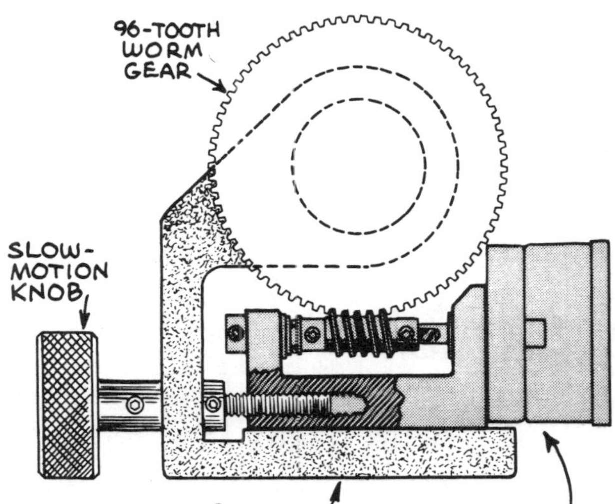

⑧ CLOCK SLOW-MOTION WITH SLIDING WORM

The BRIGHTEST STARS

STAR NAME		POSITION (1966)			VIS. MAG.	ON MERIDIAN AT 9p.m., L.M.T.
BAYER	POPULAR	R.A.	S.H.A.	DEC.		
α And	ALPHERATZ	0^h 07^m	358°	+28° 54′	2.1	Nov. 7
β Cas	CAPH	0 07	358	+58° 58′	2.3	Nov. 7
α Phe	ANKAA	0 25	354	−42° 29′	2.4	Nov. 11
α Cas	SCHEDAR	0 39	350	+56° 21′	2.2	Nov. 15
β Cet	DIPHDA	0 42	350	−18° 10′	2.0	Nov. 16
β And	MIRACH	1^h 08^m	343	+35° 26′	2.1	Nov. 22
α Eri	ACHERNAR	1 36	336	−57° 25′	0.5	Nov. 29
α UMi	POLARIS	2^h 00^m	330	+89° 06′	2.0	Dec. 5
γ And	ALMAK	2 02	330	+42° 10′	2.1	Dec. 6
α Ari	HAMAL	2 05	329	+23° 18′	2.0	Dec. 7
β Per	ALGOL	3^h 06^m	314	+40° 50′	2.2 to 3.5 Var.	Dec. 22
α Per	MIRFAK	3 22	310	+49° 45′	1.8	Dec. 26
α Tau	ALDEBARAN	4^h 34^m	292	+16° 27′	0.9	Jan. 13
β Ori	RIGEL	5^h 13^m	282	−8° 14′	0.1	Jan. 23
α Aur	CAPELLA	5 14	281	+45° 58′	0.1	Jan. 24
γ Ori	BELLATRIX	5 23	279	+6° 19′	1.6	Jan. 26
β Tau	ELNATH	5 24	279	+28° 35′	1.7	Jan. 26
ε Ori	ALNILAM	5 34	276	−1° 13′	1.7	Jan. 28
ζ Ori	ALNITAK	5 39	275	−1° 58′	1.7	Jan. 30
κ Ori	SAIPH	5 46	273	−9° 41′	2.2	Feb. 1
α Ori	BETELGEUSE	5 53	272	+7° 24′	0.4 to 1.3 Var.	Feb. 2
β Aur	MENKALINAN	5 57	271	+44° 57′	1.9	Feb. 3
β CMa	MIRZAM	6^h 21^m	265	−17° 56′	2.0	Feb. 12
α Car	CANOPUS	6 23	264	−52° 41′	−0.7	Feb. 12
γ Gem	ALHENA	6 36	261	+16° 26′	1.9	Feb. 13
α CMa	SIRIUS	6 44	259	−16° 40′	−1.5	Feb. 15
ε CMa	ADHARA	6 57	256	−28° 55′	1.5	Feb. 18
δ CMa	WEZEN	7^h 07^m	253	−26° 20′	1.8	Feb. 21
η CMa	Eta, CANIS MAJOR	7 23	249	−29° 14′	2.4	Feb. 25
α Gem	CASTOR	7 32	247	+31° 58′	1.6	Feb. 27
α CMi	PROCYON	7 38	246	+5° 19′	0.4	Mar. 1
β Gem	POLLUX	7 43	244	+28° 07′	1.1	Mar. 2
ζ Pup	Zeta, PUPPIS	8^h 02^m	239	−39° 54′	2.3	MAR. 7
γ Vel	Gamma, VELA	8 08	238	−47° 14′	1.8	MAR. 8
ε Car	AVIOR	8 22	235	−59° 24′	1.9	MAR. 12
δ Vel	Delta, VELA	8 44	229	−54° 35′	1.9	Mar. 18
λ Vel	SUHAIL	9^h 07^m	223	−43° 18′	2.2	Mar. 23
β Car	MIAPLACIDUS	9 13	222	−69° 35′	1.7	Mar. 25

STAR NAME		POSITION (1966)			VIS. MAG.	ON MERIDIAN AT 9 p.m., L.M.T.
BAYER	POPULAR	R.A.	S.H.A.	DEC.		
ι Car	Iota, CARINA	9^h 16^m	221°	−59° 08′	2.2	Mar. 26
α Hya	ALPHARD	9 26	219	−8° 31′	2.0	Mar. 28
α Leo	REGULUS	10^h 07^m	208	+12° 08′	1.4	Apr. 8
γ Leo	ALGEIBA	10 18	205	+20° 01′	2.0	Apr. 11
β UMa	MERAK	11^h 00^m	195	+56° 34′	2.4	Apr. 21
α UMa	DUBHE	11 02	195	+61° 56′	1.8	Apr. 22
β Leo	DENEBOLA	11 47	183	+14° 46′	2.1	May 3
α Cru	ACRUX	12^h 25^m	174	−62° 55′	0.8	May 13
γ Cru	GACRUX	12 29	173	−56° 55′	1.7	May 14
γ Cen	MUHLIFAIN	12 40	170	−48° 46′	2.2	May 17
β Cru	MIMOSA	12 46	169	−59° 30′	1.3	May 18
ε UMa	ALIOTH	12 53	167	+56° 09′	1.8	May 20
ζ UMa	MIZAR	13^h 23^m	159	+55° 06′	2.1	May 27
α Vir	SPICA	13 23	159	−10° 59′	1.0	May 27
η UMa	ALKAID	13 46	153	+49° 29′	1.9	June 2
β Cen	HADAR	14^h 01^m	150	−60° 13′	0.6	June 6
θ Cen	MENKENT	14 05	149	−36° 12′	2.1	June 7
α Boo	ARCTURUS	14 14	146	+19° 22′	−0.1	June 9
α Cen	RIGEL KENT	14 37	141	−60° 42′	0.1	June 15
β UMi	KOCHAB	14 51	137	+74 18′	2.1	June 19
α CrB	ALPHECCA	15^h 33^m	127	+26° 50′	2.2	June 29
α Sco	ANTARES	16^h 27^m	113	−26° 22′	1.0	July 13
α TrA	ATRIA	16 45	109	−68° 58′	1.9	July 18
ε Sco	Epsilon, SCORPIUS	16 48	108	−34 14′	2.3	July 19
λ Sco	SHAULA	17^h 31^m	97	−37° 05′	1.6	July 29
α Oph	RASALHAGUE	17 33	97	+12° 35′	2.1	July 30
θ Sco	Theta, SCORPIUS	17 35	96	−42° 59′	1.9	July 31
γ Dra	ELTANIN	17 56	91	+51° 30′	2.2	Aug. 5
ε Sag	KAUS AUSTRALIS	18^h 22^m	85	−34° 24′	1.9	Aug. 11
α Lyr	VEGA	18 36	81	+38° 45′	0.0	Aug. 15
σ Sgr	NUNKI	18 53	77	−26° 20′	2.0	Aug. 19
α Aql	ALTAIR	19^h 49^m	63	+8° 47′	0.8	Sept. 2
γ Cyg	SADR	20^h 21^m	55	+40 09′	2.2	Sept. 10
α Pav	PEACOCK	20 23	54	−56° 51′	1.9	Sept. 11
α Cyg	DENEB	20 40	50	+45 09′	1.3	Sept. 15
α Gru	AL NAIR	22^h 06^m	28	−47° 08′	1.7	Oct. 7
β Gru	Beta, GRUS	22 41	20	−47° 04′	2.1	Oct. 16
α PsA	FOMALHAUT	22 56	16	−29° 48′	1.2	Oct. 20

Collimation and Adjustments

A GOOD telescope must have good optics, and the good optics must be properly aligned or collimated. So there is a little problem here: If a telescope performs poorly, is it because of poor optics or poor collimation? Invariably, you can blame the optics since it is nearly impossible to align the average f/8 reflector so poorly as to cause marked deterioration of image quality. The "good" field of any f/8 telescope is over three times as large as the image field of an average low-power eyepiece. This point is illustrated in the drawing; it should be apparent that just ordinary care will assure a workable alignment.

Coma is difficult to see in an f/8 reflector because it is so far off-center as to be out of the field entirely. Coma is an off-axis aberration; by its nature it cannot occur on the optical centerline. Hence, any time you see coma at the center of the eyepiece field, you will know you are not aligned with the optical axis.

Spherical aberration in a telescope can be detected by the appearance of a star image inside and outside of best focus. A test star should be 2nd or 3rd magnitude, near the zenith for best seeing, or toward the north for least movement. At best focus the star image should be a small disk of light, quite nearly perfectly round, slightly radiant around the edge with perhaps a faint outer ring of light around the central image. Usually you will not see rings around star images, and if you want to see this effect you should use high power of 40x or 50x per inch of aperture in order to obtain a star image large enough to reveal its physical structure.

Rack the eyepiece slightly in from best focus and then out. If the expanded star image looks the same in both positions, it is a sign there is no spherical aberration. If, however, you see a bright ring inside focus and no bright ring outside focus, you are looking at S.A., in this case under-corrected (see drawing). The reverse case--bright ring outside focus and no ring inside focus--means over-corrected S.A.

Star tests of this kind work best with a refractor. With a reflector, you will pick up a silhouette of the diagonal which more or less obscures the central part of the expanded star.

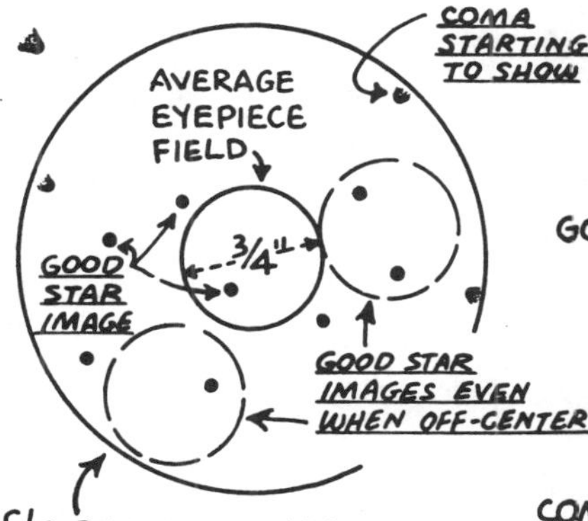

f/8 REFLECTOR HAS GOOD RESOLUTION OVER LARGE FIELD (2 9/16" DIA. - SEE TABLE BELOW)

GOOD STAR IMAGE

COMATIC STAR IMAGE

Permissible "GOOD FIELD" for PARABOLIC MIRRORS

MIRROR DIA.		f/4	f/5	f/6	f/7	f/8
3"	LINEAR→	.64"	1.00"	1.44"	1.96"	2.56"
	ANGULAR→	183'	229'	275'	321'	366'
4 1/4"	LINEAR→	.64"	1.00"	1.44"	1.96"	2.56"
	ANGULAR→	130'	164'	200'	224'	259'
6"	LINEAR→	.64"	1.00"	1.44"	1.96"	2.56"
	ANGULAR→	91'	115'	138'	160'	183'
8"	LINEAR→	.64"	1.00"	1.44"	1.96"	2.56"
	ANGULAR→	69'	86'	103'	120'	138'
10"	LINEAR→	.64"	1.00"	1.44"	1.96"	2.56"
	ANGULAR→	55'	69'	82'	96'	110'

LINEAR: *INCHES* ANGULAR: *MINUTES OF ARC*

A SHORT TELESCOPE LETS YOU SEE RESULTS OF MIRROR MOVEMENT

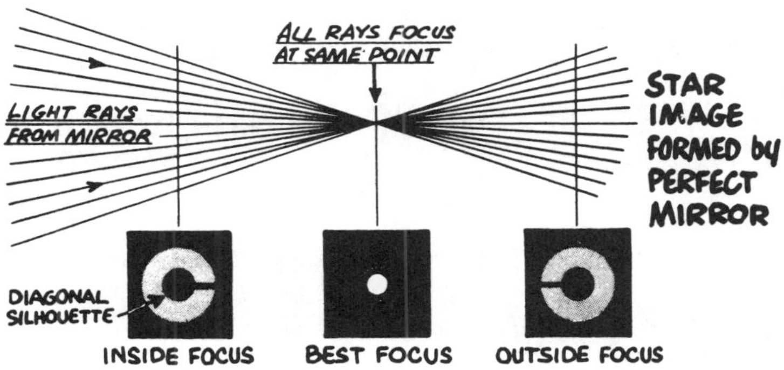

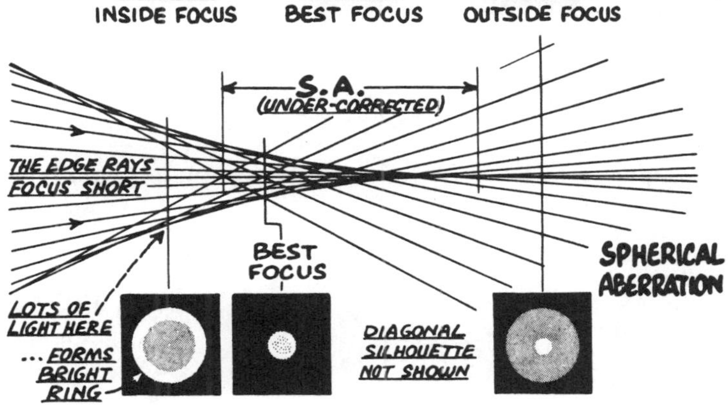

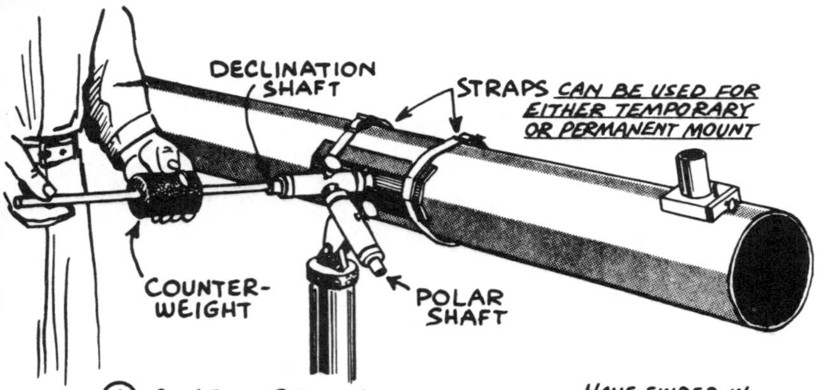

① SIMPLE BALANCING

BALANCING a TELESCOPE

ORDINARY balancing of a telescope is a simple matter. You put the main tube in a horizontal position, Fig. 1, checking to see that both lock knobs are loose. The telescope is then balanced by moving the counterweight in or out as needed. Any slight imbalance can be remedied by applying moderate tension with the lock knobs; too much imbalance will cause an erratic action, particularly noticeable with a clock drive.

The balance obtained in Fig. 1 operation is not true standing or static balance, as you will find by moving the main tube to a vertical position, Fig. 3. Now, in the form of eyepiece, finder, camera or other equipment, you have more weight at one side of the tube, with the result the telescope swings in that direction.

An approximate correction is often all you need--that is, you get some kind of weight in some position nearly opposite the finder or eyepiece. If you want to be more exact, the weight and position of each part must be determined. Then, as shown in Fig. 4 example, a simple drawing to scale will show what weight is needed to balance. If there are more than two weights involved, the solution for two of the weights, as in Fig. 4, is combined with the third weight.

The balancing weight can be placed anywhere along the main tube. Attachment to the cradle is often practical. Usually the weight is arranged to slide along the tube so that it can be used to maintain lengthwise balance as well as balance around the mechanical centerline.

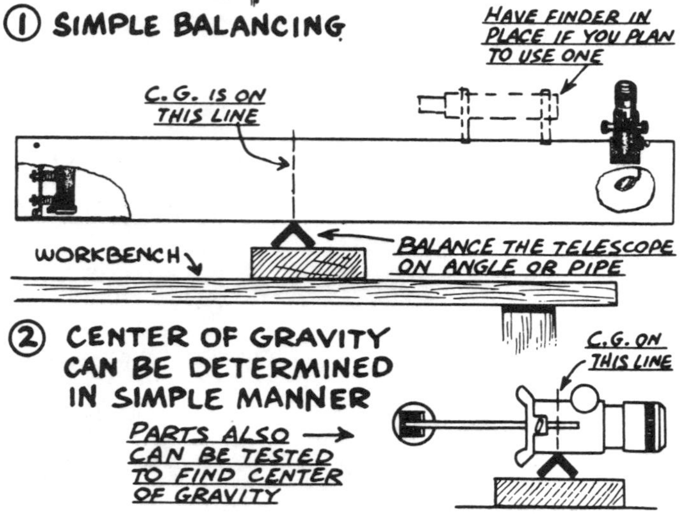

② CENTER OF GRAVITY CAN BE DETERMINED IN SIMPLE MANNER

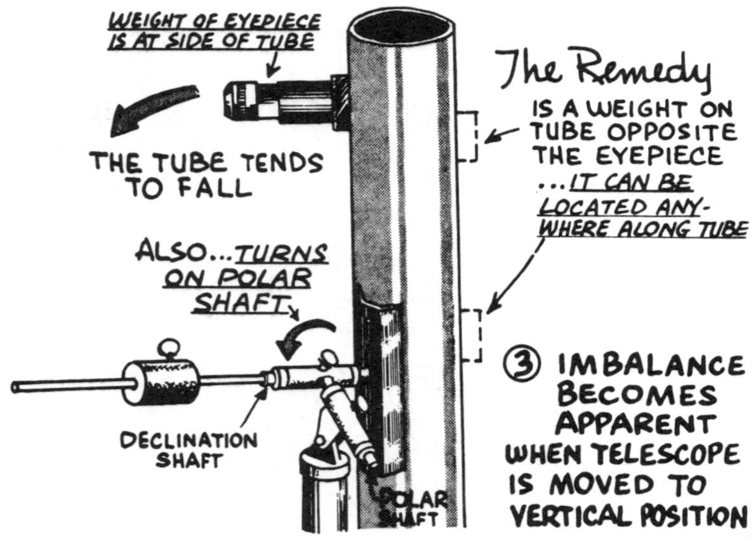

③ IMBALANCE BECOMES APPARENT WHEN TELESCOPE IS MOVED TO VERTICAL POSITION

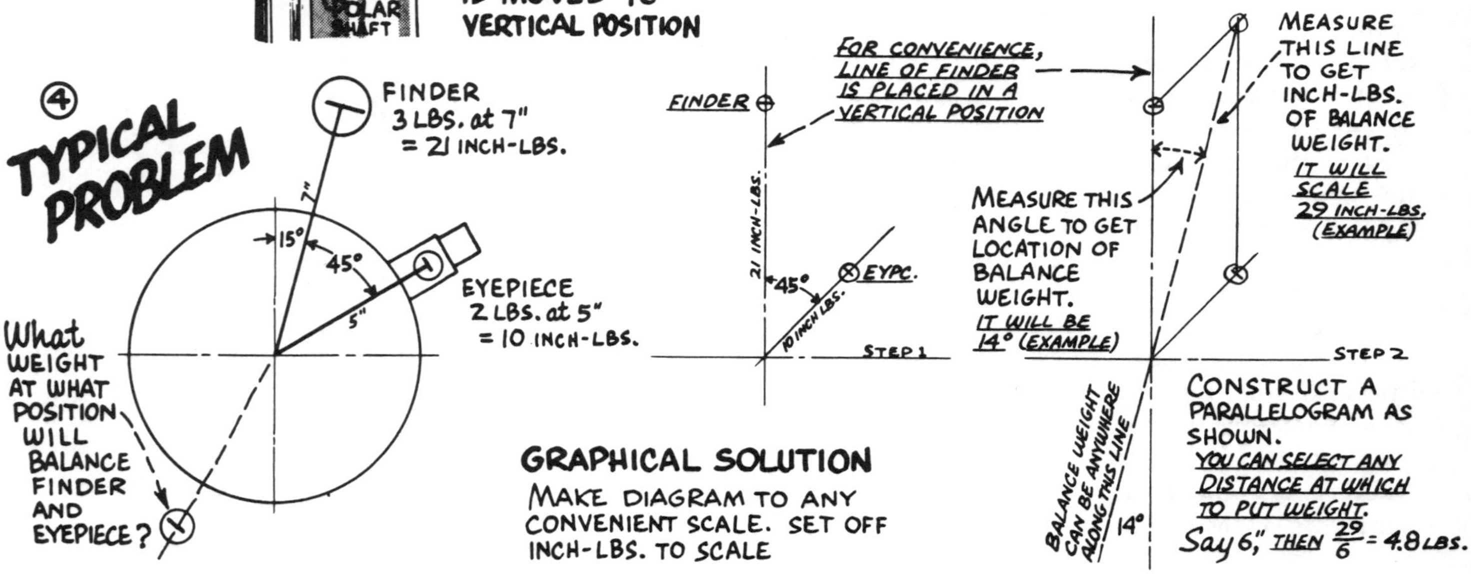

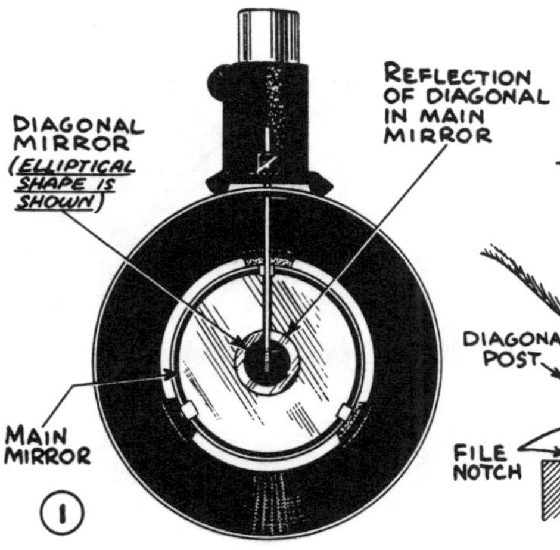

LOOKING DOWN THE MAIN TUBE

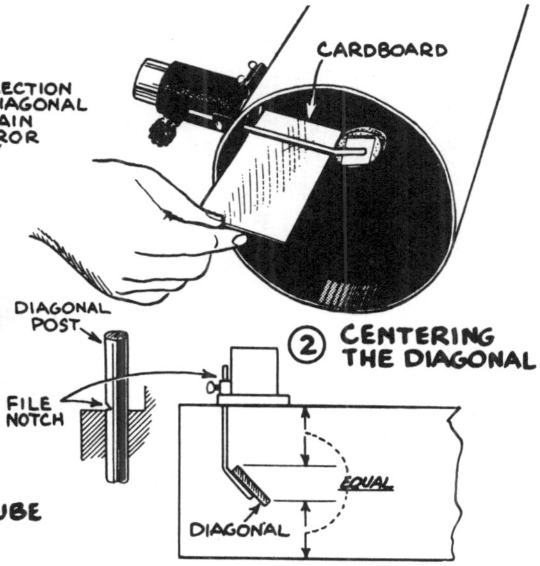

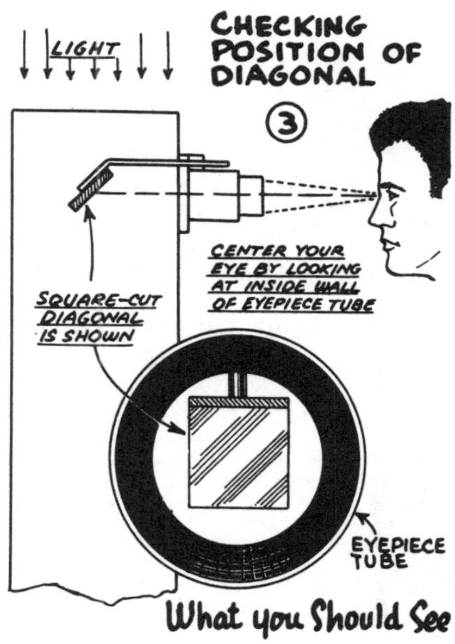

Collimating a Reflector

COLLIMATION is the process of aligning optical elements. If you have a reflecting telescope, collimation is a "must" because the instrument will not work well if the mirrors are out of alignment. As a basic starting point, both mirrors should be mechanically centered. This is just a matter of ordinary measuring, Fig. 2. The diagonal must also be centered mechanically under the eyepiece tube. Check this by looking through the eyepiece tube (without eyepiece) from a distance of several inches, Fig. 3.

Diagonal rotation comes next. Look through the eyepiece tube from a close position, Fig. 4. It helps in this operation to use a 1/8 inch sighting hole or a 1/4 inch eyepiece from which the lenses have been removed. What you look at is the reflection of the main mirror in the diagonal. If the reflection shows off-center, as at A, Fig. 4, turn the diagonal post slightly to center the reflection, as at B. A reflection at top or bottom of diagonal, C, usually indicates an angle error although a slight displacement toward the bottom, C, is normal, as will be shown later. Angle errors can be corrected by bending the diagonal post. This requires careful work--don't rush it until you are sure a correction must be made.

The final step is the adjustment of the main mirror. Look into the sighting ring as before, Fig. 5, but this time direct your critical attention to the small reflection of the diagonal. It will probably be off-center but is easily centered by turning the mirror cell adjusting nuts. It is convenient to have a helper turn the adjusting nuts

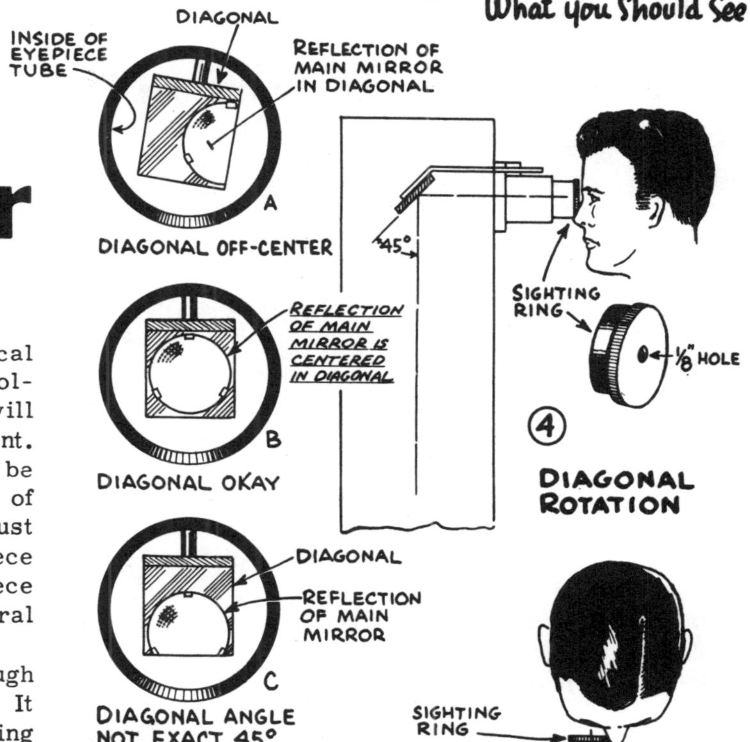

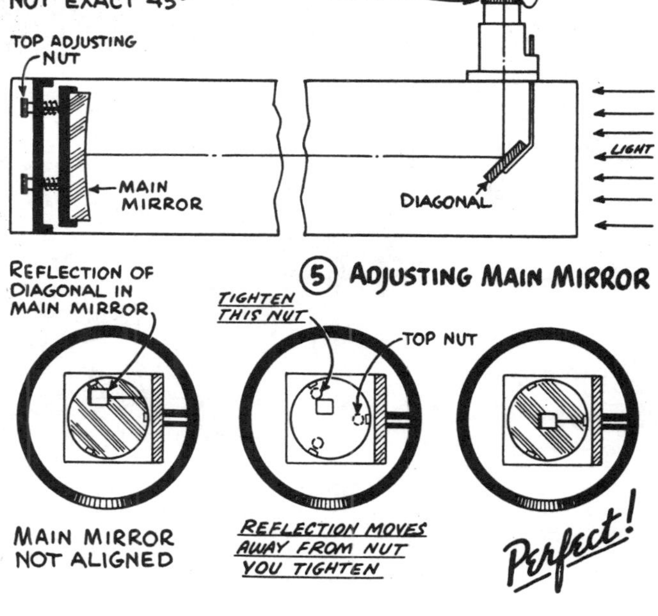

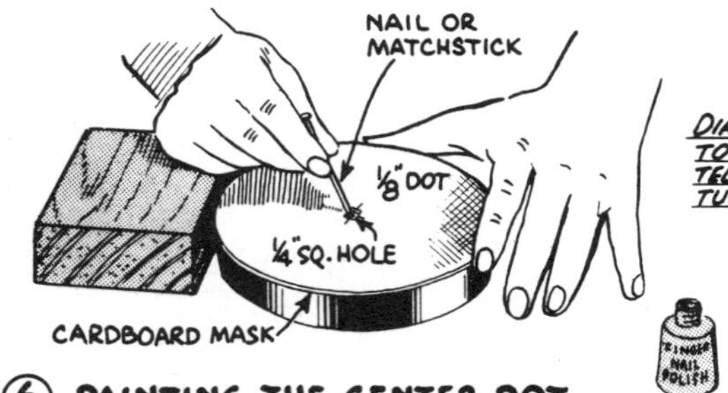

⑥ PAINTING THE CENTER DOT

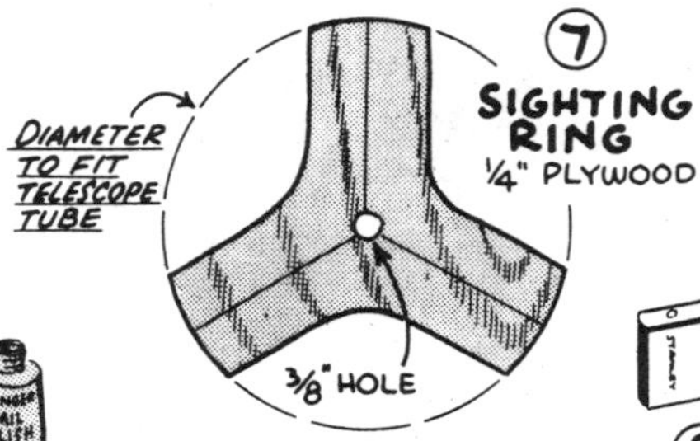

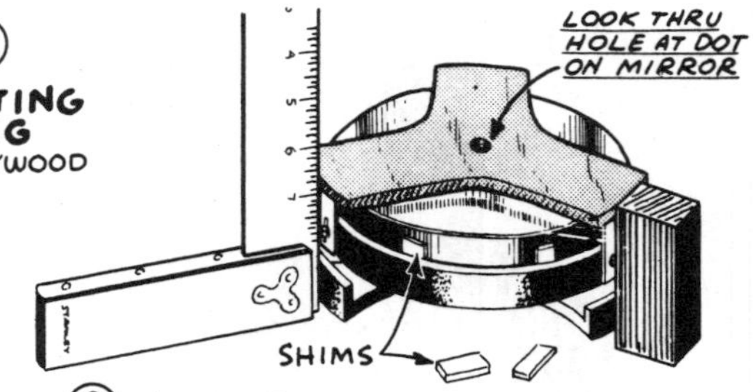

⑧ CENTERING THE MIRROR

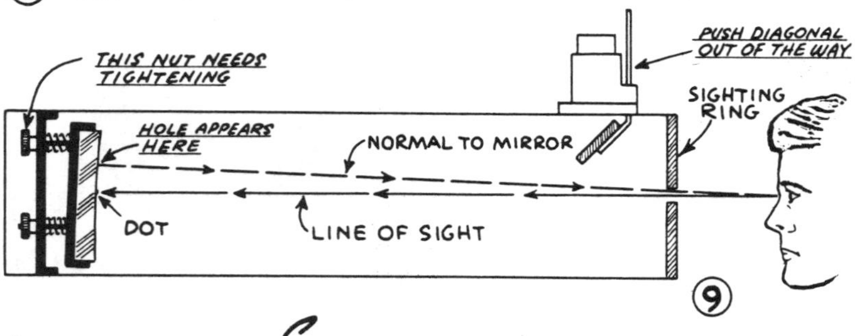

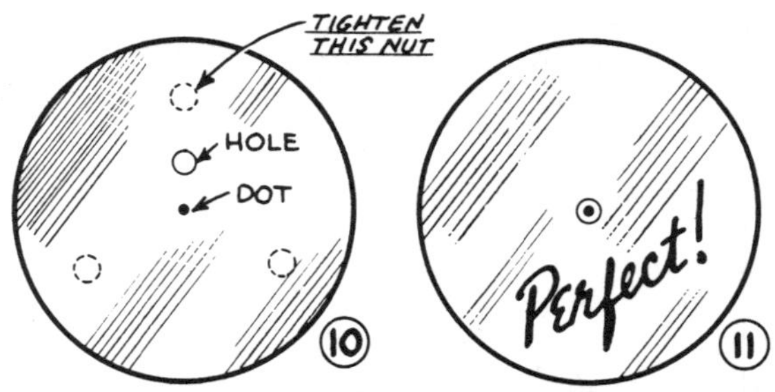

COLLIMATION IS Easy WITH CENTER-DOT MIRROR

THE HOLE MOVES AWAY FROM NUT YOU TIGHTEN... OR TOWARD NUT YOU LOOSEN

while you direct him. If you do the job single-handed, the simple rule is that the small reflection will move away from the adjusting nut you tighten. The top adjusting nut is the one in line with the reflection of the diagonal post, and the other two are right and left the same as your view of the mirror.

Regarding the whole process, keep in mind that the large reflection of the main mirror in the diagonal is entirely a diagonal adjustment--get this right before adjusting the main mirror screws which control only the small reflection. Strong illumination is helpful, such as open sky or a white ceiling, wall or cardboard screen flooded with light. Do not use direct illumination. As a final check, the alignment can be viewed at the exit pupil, using a magnifying glass held central with the optical centerline.

COLLIMATING WITH CENTER DOT. This method is fast and easy and is recommended for all reflectors where the diagonal can be readily pushed out of the way. The center dot is painted on the main mirror with any kind or color of paint applied through a cardboard mask, Fig. 6. You don't have to worry about the dot spoiling your mirror because this part of a telescope mirror is never used, being lost in the shadow of the diagonal. A needed accessory is a sighting ring, Fig. 7. The cutout shape allows light to enter the tube and the three-leg design corresponds to the three-point mounting of the mirror. The sighting ring also provides a convenient way of checking the mechanical centering of the main mirror, Fig. 8.

Fig. 9 shows how center-dot collimation works. In this method, you adjust the main mirror first and then the diagonal. Look through the

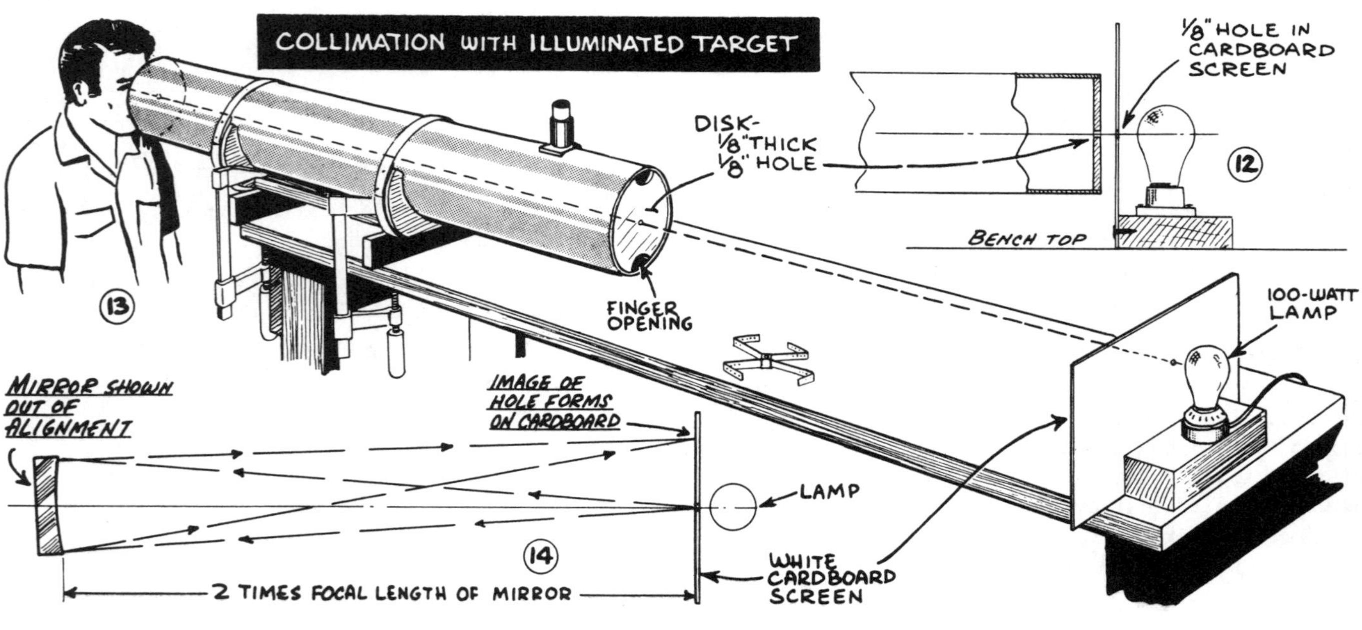

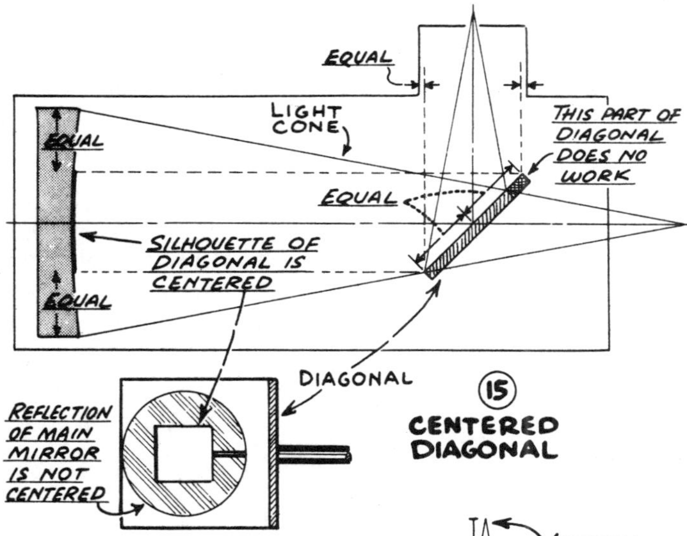

(15) CENTERED DIAGONAL

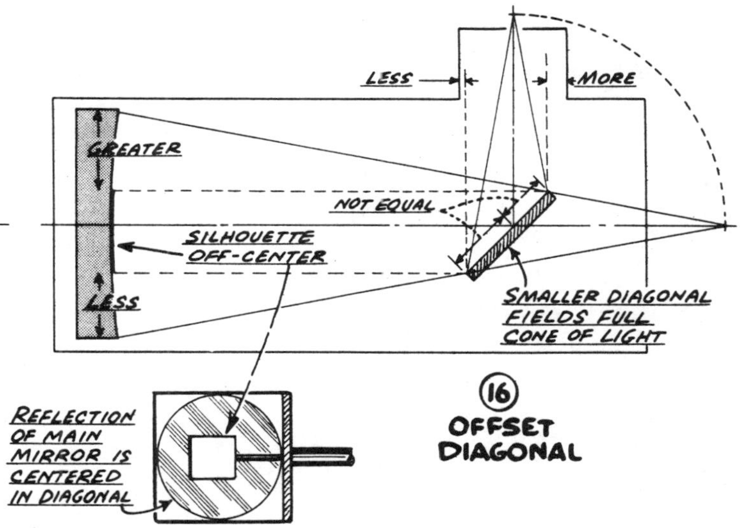

(16) OFFSET DIAGONAL

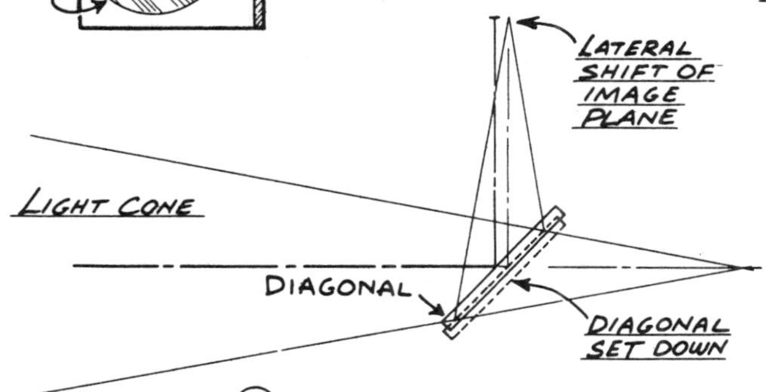

(17) DROPPED DIAGONAL

sighting hole at the painted dot. You will also see a reflection of the sighting hole. It is helpful to use a flashlight, shining the beam on the mirror if you want to see the dot plainly, and on the hole or your face if you want to see the hole. The idea of course is to center the hole around the painted dot. The required movement is obtained by tightening the adjusting nut on the side toward which the hole is displaced, Fig. 10. When you get the hole exactly centered around the dot, the main mirror is in perfect alignment. The diagonal adjustment follows and is the same as for ordinary collimation, Fig. 4. The main mirror being already collimated, the two axes are then in alignment and the small reflection automatically shows in the center of the main mirror.

COLLIMATING WITH TARGET. This method is accurate and convincing although it has a minor drawback in that the main mirror and cell must be removed. The setup is shown in Fig. 13. The telescope tube should be strapped or clamped securely to the workbench. The target is then made with hole at the same height as centerline of tube, Fig. 12. It is then fairly easy even without an assistant to move the target as needed until you can see the illuminated hole through the sighting disks fitted in both ends of the telescope tube.

Now, remove the sighting disks and mount the mirror. Use care to avoid disturbing the position of either tube or target. Turn out all lights except light behind target. If the target is approximately two times mirror focal length from the mirror, the mirror will reflect an image of the hole back onto the screen, as shown in Fig. 14. By turning adjusting nuts, it is easy to make the image coincide exactly with the target hole. You now have the mirror in perfect alignment with the mechanical centerline of tube. Fitting and adjustment of the diagonal follows, as previously described. When a spider diagonal mount with center hole is used, it can be fitted in place after the initial target sighting by removing the front disk only and sighting from rear disk through spider hole to target. The centering of a diagonal post can be similarly checked by dropping the diagonal below the light beam; the post will block the beam if accurately centered. Don't forget the trick of blowing smoke on the light beam if you want to see it clearly at any particular point.

OFFSET DIAGONAL. Properly, the diagonal should be off-center slightly. This becomes obvious in a high-speed system, Fig. 15, where it can be seen that the top part of a centered diagonal is wasted. With offset diagonal, Fig. 16, the same light cone can be fielded with a much smaller diagonal. Offsetting causes some complications since it can be seen, Fig. 16, that the diagonal is not mechanically centered with the eyepiece tube nor is its reflection centered on the main mirror. If you use an offset diagonal, the optical center should be marked. A rubber band snapped around the diagonal at this point provides a guide mark which can be centered in the main mirror.

Offsetting the diagonal is not commonly practiced if the mirror is f/8 or higher f/number. The offset for an f/8 mirror is a scant 1/16 inch--hardly worthwhile in view of the extra complications in collimating. However, it is worth noting that the centered diagonal will show the main mirror reflection slightly off-center, Fig. 15. This is correct for a centered diagonal. Do not try to obtain better centering by dropping the diagonal, Fig. 17, because this merely displaces the optical centerline and does not make the desired correction as viewed from the center of focusing tube.

Adjustment to POLE

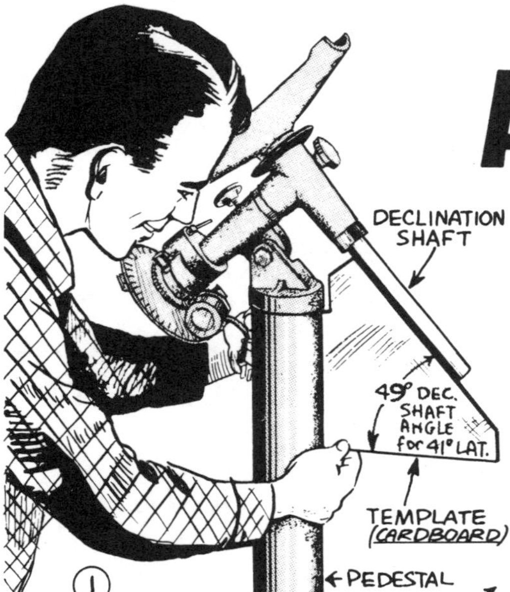

① POLAR SHAFT ANGLE...

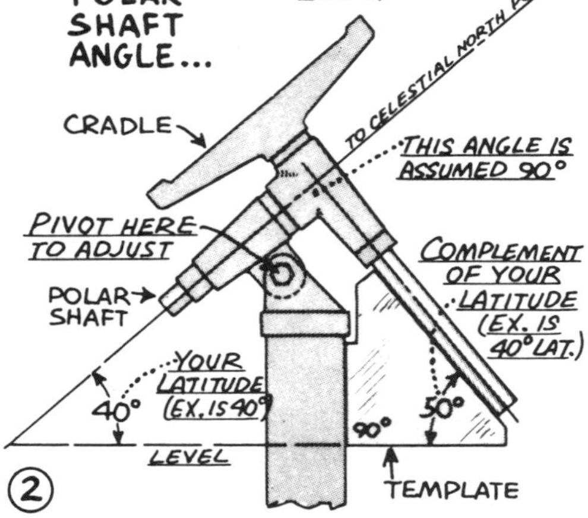

② ... IS SET IN A ROUNDABOUT WAY BY MEASURING THE TILT OF DECLINATION SHAFT

NORTH by guess is good enough for casual star-gazing, but when you get into more advanced observing and photography, the setting of the polar axis to the celestial north pole becomes an adjustment of considerable importance.

In any case, the first thing to do is to set the polar shaft at the same angle as the latitude of your location. When this is done, the polar shaft will point to the same height above the horizon as the celestial north pole itself. This adjustment can be made indoors, using such levels, protractors, etc. as may be available. If you have a pedestal mount, a roundabout yet practical method is to measure the angle between pedestal and declination shaft, as shown in Figs. 1 and 2.

BORE SIGHTING. The simple, direct way to get the polar shaft pointing exactly to the pole is to sight right through the bore of the polar shaft housing. This is not a practical method with many mounts because of the numerous parts that must be removed piecemeal. With some mounts, however, the

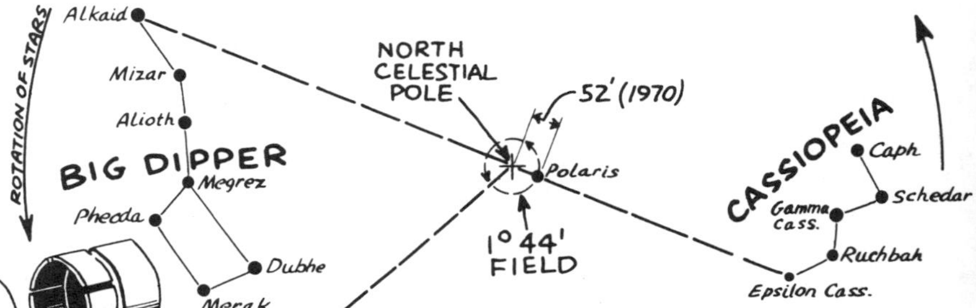

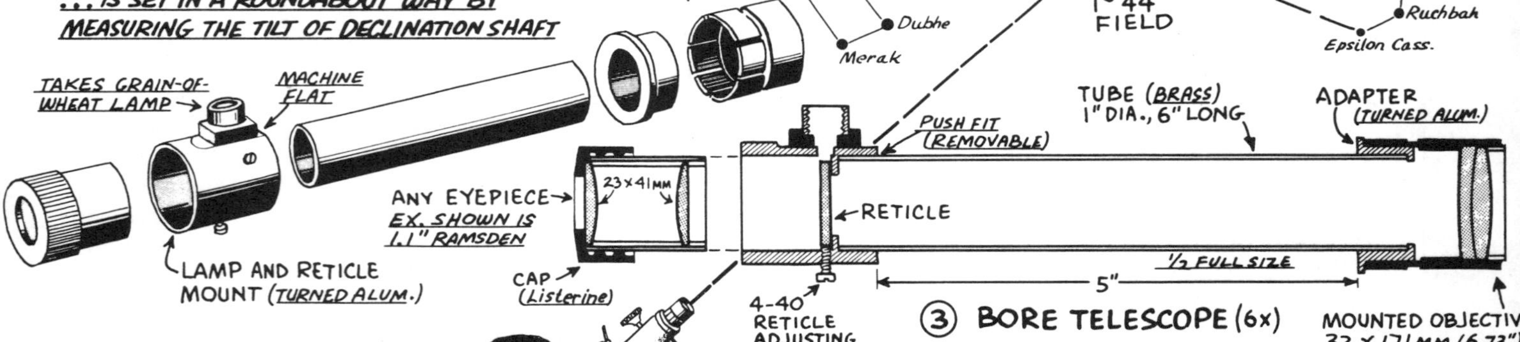

③ BORE TELESCOPE (6x)

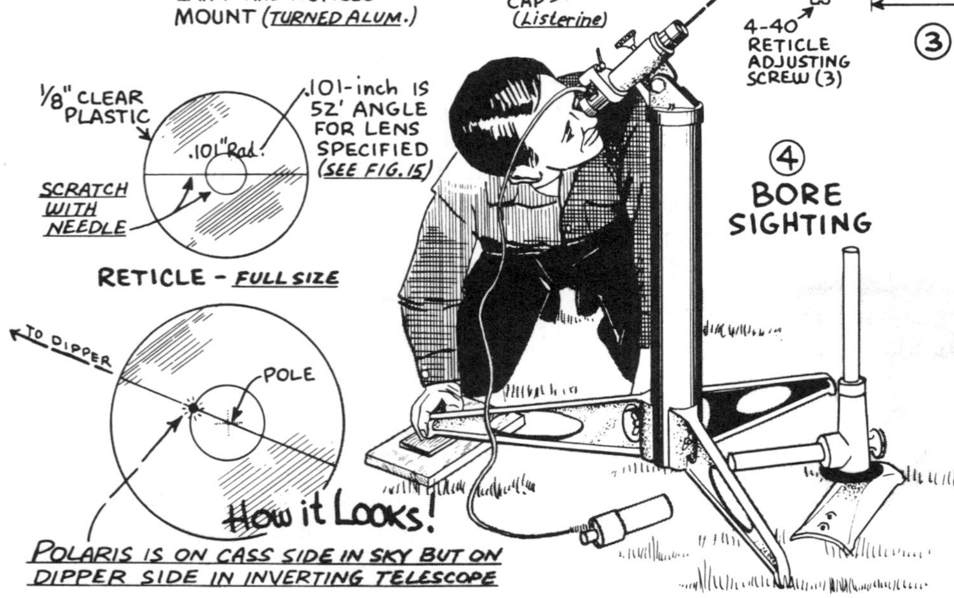

④ BORE SIGHTING

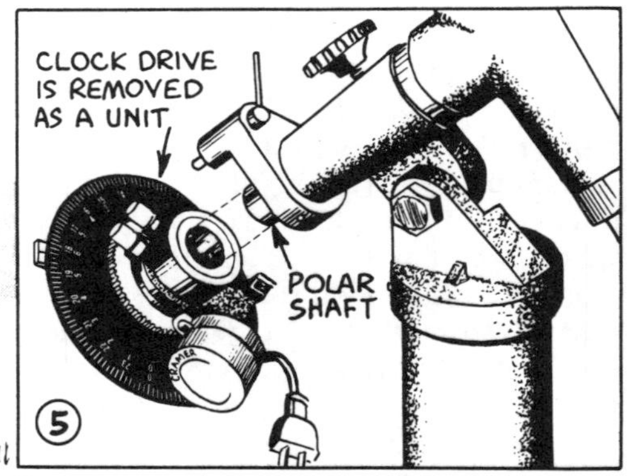

⑤ BORE SIGHTING IS PRACTICAL ONLY IF THE CLOCK DRIVE IS EASY TO REMOVE
DRIVE SHOWN IS SELF-CONTAINED, EASY TO REMOVE BY LOOSENING SET SCREWS

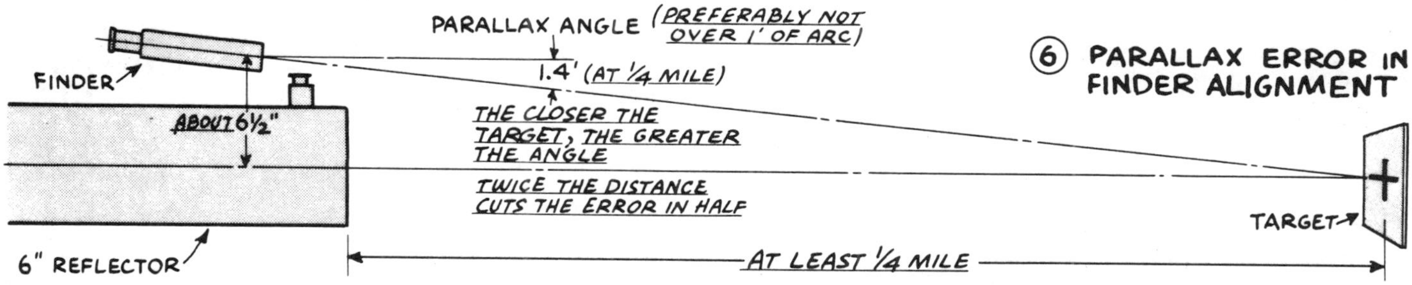

⑥ PARALLAX ERROR IN FINDER ALIGNMENT

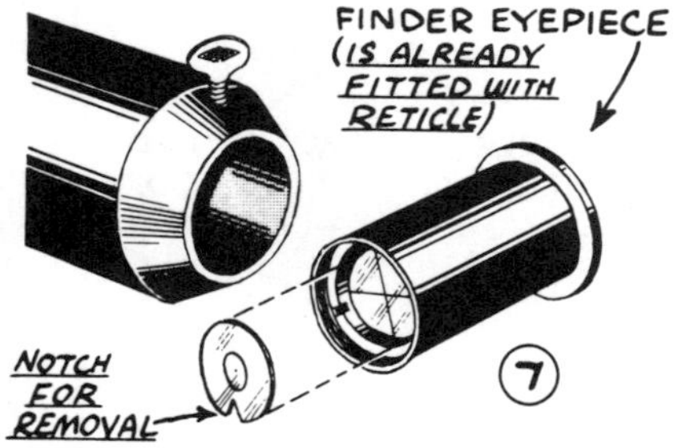

OVERLAY RETICLE CIRCLE

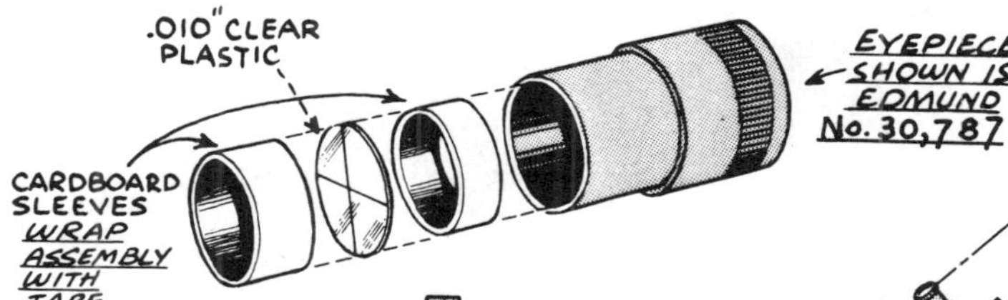

CROSSLINE RETICLE for MAIN TELESCOPE

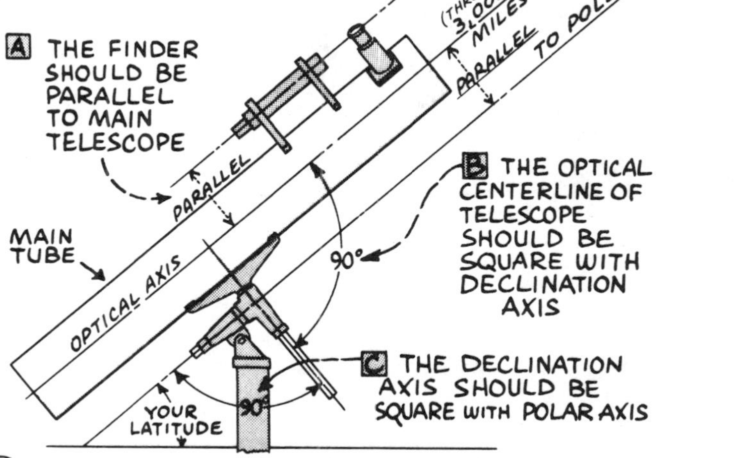

⑨ THE ANGLES OF THE POLAR ADJUSTMENT

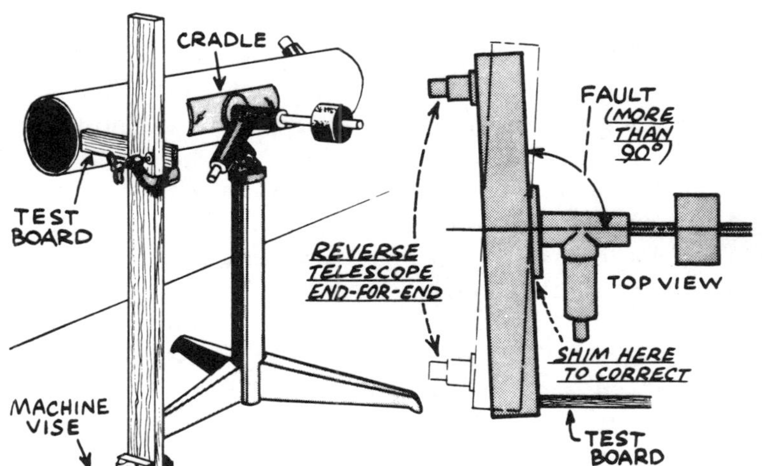

⑩ SQUARING THE CRADLE

whole drive can be removed by loosening two set screws. Of course, to sight through the bore, you need a bore telescope, Fig. 3. The optical centering of the reticle is checked by rotating the assembled telescope on vee blocks while looking at any target object--the center should stand still.

Fig. 4 shows the bore sight being made. The illuminated reticle is helpful but not essential since it is practical to work at dusk or in moonlight when an ordinary non-illuminated reticle is easily visible. Get Polaris in the field and then by shifting and shimming the base of the mount, put Polaris on the reticle circle at the same angle as your reference star, which can be either Alkaid in the Dipper or Epsilon in Cassiopeia. If desired you can turn either the bore telescope or the reticle holder to make the reticle line assume the same angle as the reference star, Fig. 4.

FINDER ALIGNMENT. Lacking a borescope, you can make the pole adjustment with the finderscope. This is considerably more complicated. To get started, you align the finder parallel with main telescope, Fig. 6. The finderscope eyepiece, Fig. 7, has the usual crossline reticle, as well as the polar circle which will be used later in setting to the pole position. The eyepiece of the main telescope has a plain crossline reticle Fig. 8. The actual sighting is usually done on a daytime object and is not difficult.

SQUARING THE CRADLE. This involves two 90 deg. angles, as can be seen in Fig. 9. Of these, the angle between polar shaft and declination shaft is fixed immovable by the manufacturer and nothing can be done about an error, which happily, is not often present.

There is usually a modest angle error between the declination shaft and the mounted telescope tube, seldom exceeding 1/4 degree. This can be corrected by placing cardboard shims between the cradle and main tube. A simple testing method is shown in Fig. 10. Reversing the main tube end-for-end, you will be able to detect a contact difference as small as 1/32 inch, which means about 5 min. of arc accuracy. This is close enough for a preliminary adjustment.

POLE ADJUSTMENT. Sighting the pole with a

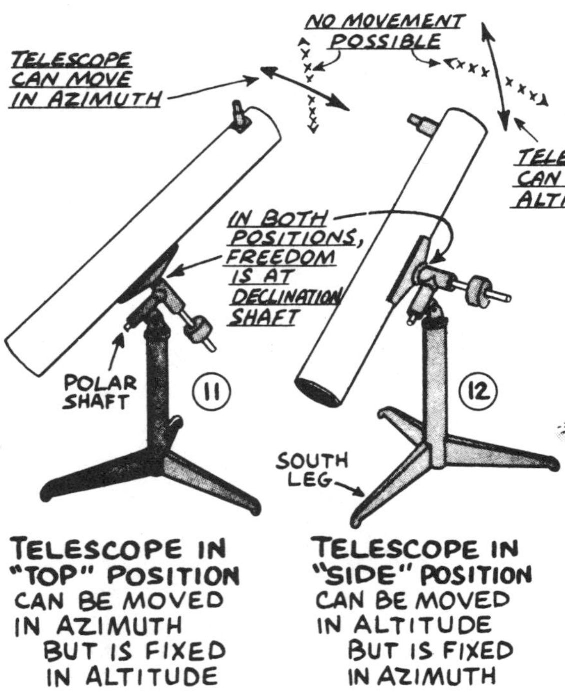

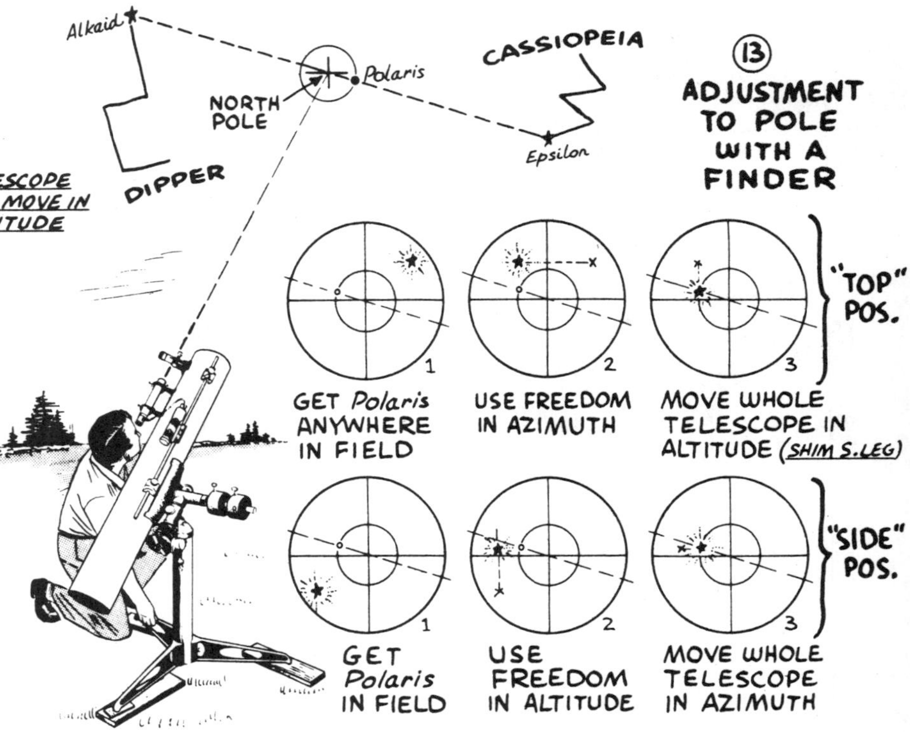

finderscope is much the same as with a bore-scope, except you have to do the sighting twice in order to eliminate the movement of the telescope itself.

Fig. 11 shows the main telescope in "top" position with main tube directly over the polar shaft. If you study this drawing, or, better, an actual equatorial mount in this position, you will see the telescope can be moved from side to side. This is a movement in azimuth. But you can't move the telescope up and down at all--you have no freedom in altitude.

The other basic position in making the polar adjustment is the "side" position, Fig. 12. Here you can move up and down in altitude, but you have no freedom in azimuth.

It is best to start with the telescope in top position. This position is used to set the altitude, which is not so apt to be disturbed by later manipulations. In top position, a slight movement in azimuth around the declination axis is permissible if needed to put Polaris at the desired position. Similarly, when in side position, Fig. 13, it is permissible to use a slight movement in altitude if needed. In both cases, these movements are not in the plane you are adjusting at the time. The various positions of Polaris as seen in the finderscope are shown in Fig. 13 diagrams. Of course, Polaris may be at any position in the sky. The whole operation can be done in about five minutes--but not the first time!

After completing the sighting to the pole, you can recheck the squareness of the cradle by slowly moving the telescope from side position west of pedestal to side position east of pedestal, all the while watching Polaris in the finderscope eyepiece. The star should walk around the reticle circle.

REFERENCE STARS. 25 years ago, Alkaid in the Dipper, and Epsilon, Cassiopeia, were nicely in line with Polaris and the pole. Currently, Thuban in Draco and No. 50 Cass are better aligned altho less bright, Fig. 14. Alkaid and Epsilon Cass are still quite reliable.

If you can catch Polaris at transit or elongation, the job of making the adjustment to pole is somewhat easier, more accurate. A fixed telescope can be wholly adjusted by using a transit position for the azimuth setting, and an elongation for altitude. Although this means a 6-hour wait, the procedure does not require a reticle circle and can be done with eyepiece with cross-line reticle used with main telescope. 10 or 15 minutes before or after the exact transit or elongation can be tolerated--Polaris moves about 2 min. of arc in 10 min. time.

TRANSIT TIME FROM PLANISPHERE. In ordinary use, a planisphere is used by putting the time opposite the date--the chart then shows the stars on the meridian. This procedure can be reversed --if you put a certain star on the meridian, you can read its time of transit opposite any date. Almak has the same R.A. as Polaris, and it is Almak you set on the meridian to get the transit time of Polaris, Fig. 16. Accuracy within 5 minutes is possible.

(14)

REFERENCE STARS for TRANSIT and ELONGATION OF POLARIS

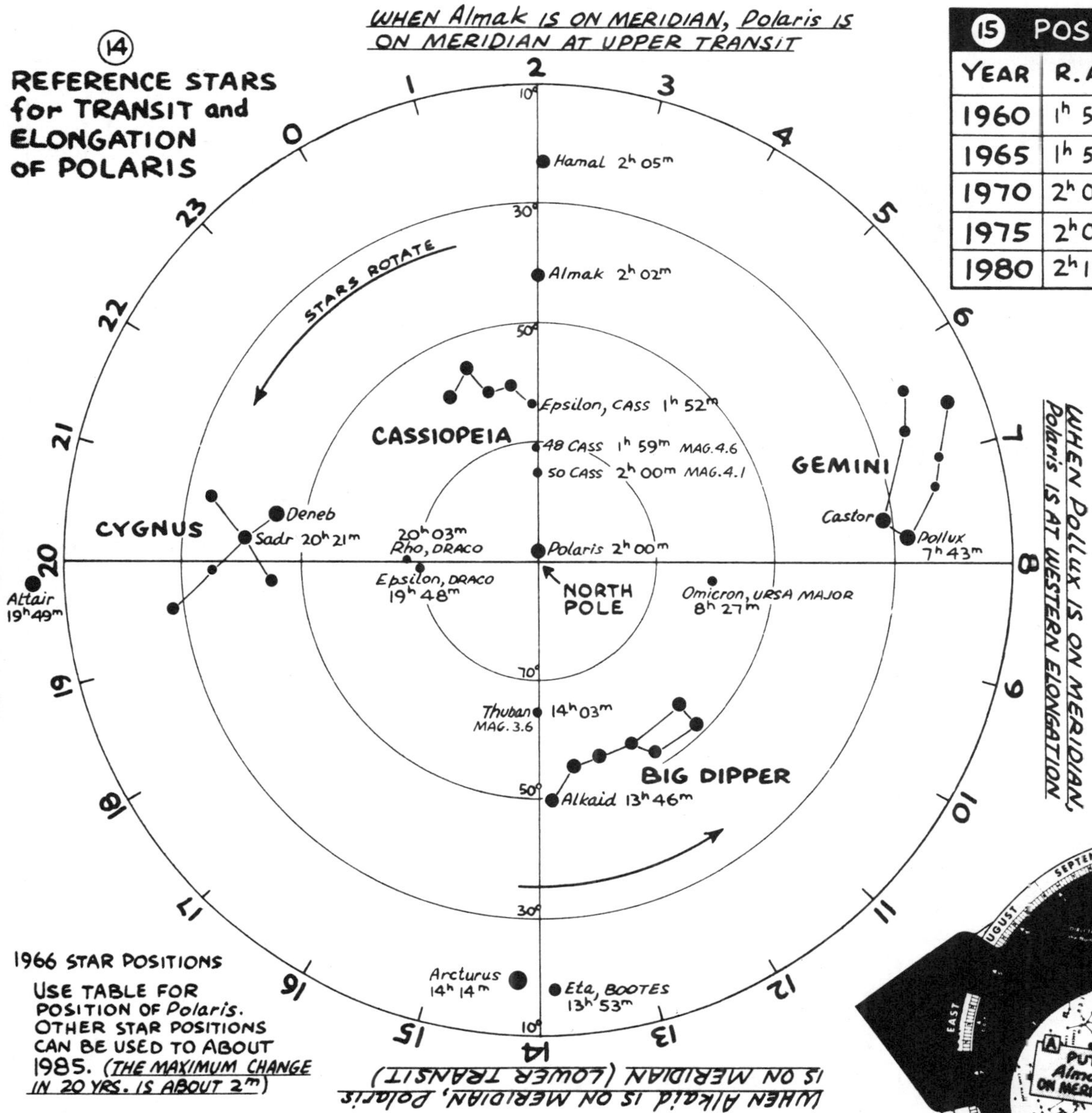

1966 STAR POSITIONS

USE TABLE FOR POSITION OF Polaris. OTHER STAR POSITIONS CAN BE USED TO ABOUT 1985. (THE MAXIMUM CHANGE IN 20 YRS. IS ABOUT 2m)

(15) POSITION OF POLARIS

YEAR	R.A.	DEC.	from POLE	RETICLE FACTOR*
1960	1h 56m	89° 5′	55′	.016
1965	1h 59m	89° 6′	54′	.016
1970	2h 03m	89° 8′	52′	.015
1975	2h 07m	89° 9′	51′	.015
1980	2h 11m	89° 10′	50′	.015

*MULTIPLY BY F.L. OF FINDERSCOPE OBJECTIVE TO GET RADIUS OF RETICLE CIRCLE

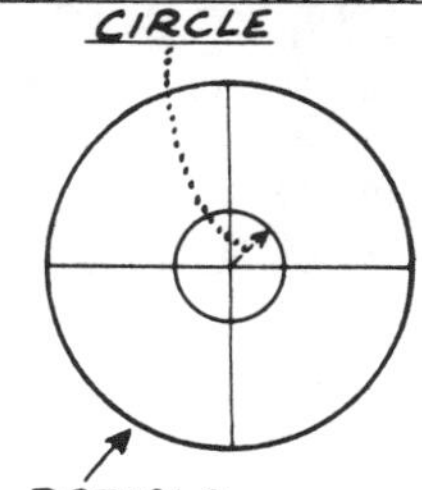

RETICLE (TO SUIT YOUR FINDER. .010-inch CLEAR PLASTIC. MARK CIRCLE WITH DIVIDERS)

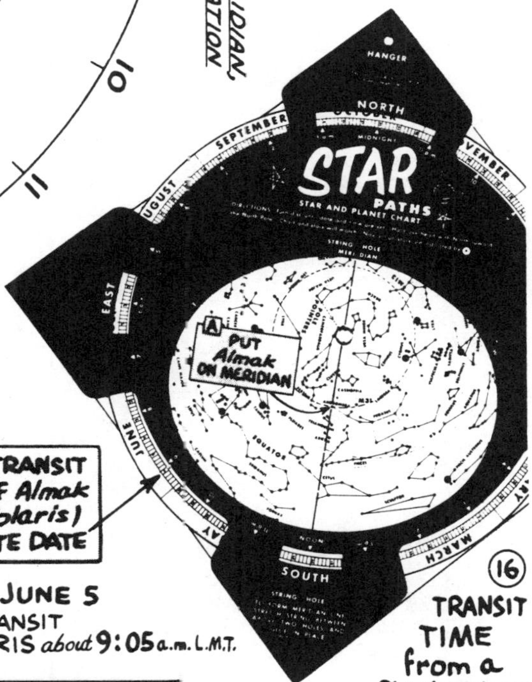

Example: JUNE 5
UPPER TRANSIT OF POLARIS about 9:05 a.m. L.M.T.

(16) TRANSIT TIME from a PLANISPHERE

TRANSIT BY CALCULATION. The transit time of Polaris or any star is simply a conversion of sidereal time to its local mean time equivalent. A specific example dealing with Polaris itself is shown in Fig. 17. You should surely know by now that sidereal time is the R.A. on the meridian at that instant. Currently the R.A. of Polaris is an even 2 hours. When it is 2 hr. sidereal time, Polaris will be on your meridian above the pole. The problem is simply a conversion of 2 hr. sidereal time to your local mean time. After you get the L.M.T. of the transit, you have to convert to your standard time. This is just a matter of adding or subtracting your time correction.

After you know the time of upper transit, you know that western elongation will be 6 sidereal hours later. In solar time, it will be 6 hours later, minus 1 minute. The minus 1 minute is the correction for converting a 6-hr. sidereal interval to solar time.

(17) TRANSIT TIME BY ARITHMETIC

(THIS PART OF CALCULATION IS SIDEREAL INTERVAL)

FORMULA: L.M.T. = R.A.* − G.S.T. (AT 0h) − CORRECTION

Example: Polaris, JULY 1, ZONE 6

R.A. --- 2h 00m + 24h = 25h 60m

Minus G.S.T. --- 18h 35m (0h JULY 1) (from TABLE, p.28) 18h 35m

.... 7h 25m

Minus CORRECTION 02m

L.M.T. 7h 23m UPPER TRANSIT IS MORNING

Plus 12 00

19h 23m

Minus 02m CORRECTION FOR 12h SID. TIME

JULY 1 --- L.M.T. 19h 21m LOWER TRANSIT

(SID. INTERVAL PLUS TIME ZONE NO. 6 (MAKE DIVISION AND CALL IT MINUTES))

CORR. 7h SID. INT. plus 6 ZONE; 6 | 13 = 2 MIN.

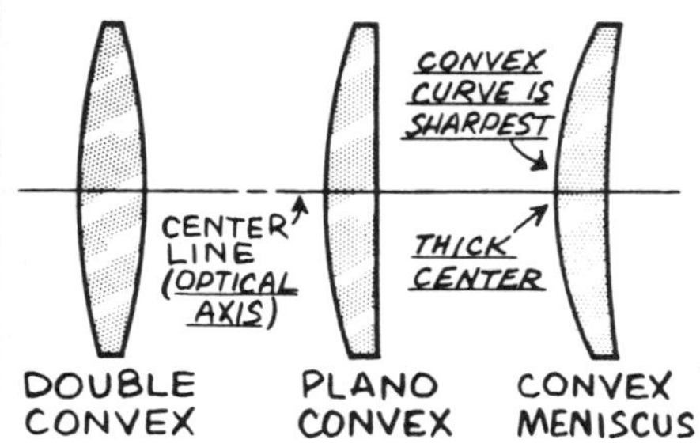

POSITIVE LENSES

A LENS IS POSITIVE IF IT CAN CONVERGE PARALLEL LIGHT TO FORM AN IMAGE. IF THE LENS IS A SINGLE PIECE OF GLASS, IT IS A SIMPLE LENS

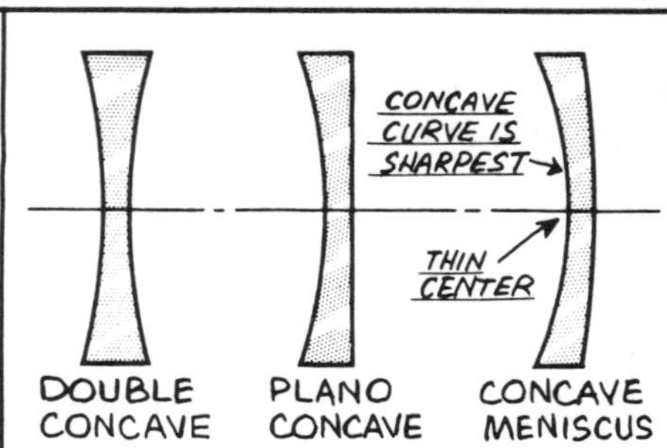

NEGATIVE LENSES

NEGATIVE LENSES CAUSE PARALLEL LIGHT RAYS TO DIVERGE — NO IMAGE IS FORMED. USED ALONE AT EYE, A NEGATIVE LENS REDUCES

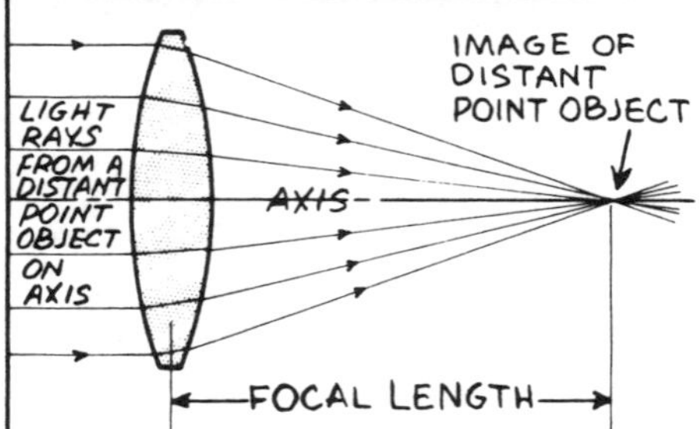

FOCAL LENGTH

DISTANCE FROM A LENS TO POINT WHERE IT FORMS AN IMAGE OF A DISTANT OBJECT. USUALLY MEASURED FROM CENTER OF LENS ALTHO NOT EXACT

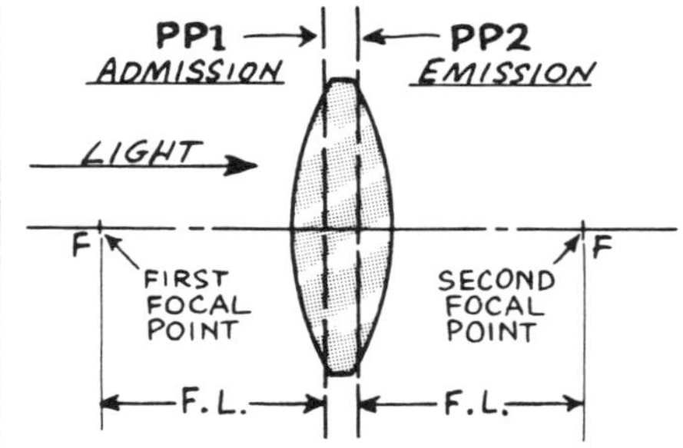

PRINCIPAL PLANES

IMAGINARY PLANES FROM WHICH EXACT FOCAL LENGTH MEASUREMENTS ARE TAKEN. IF THE LENS IS SYMMETRICAL, THE PP's ARE SYMMETRICAL

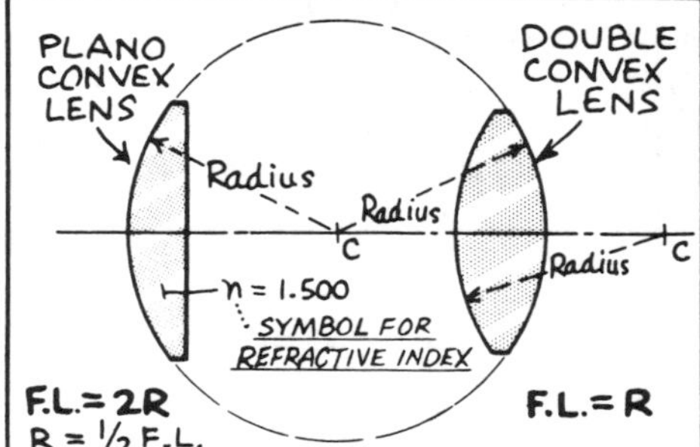

RADIUS of CURVATURE

ALL SIMPLE LENSES HAVE SPHERICAL CURVES. THE RADII DETERMINE THE FOCAL LENGTH. FORMULAS ARE APPROX. ...EXACT ONLY FOR THIN LENS, INDEX 1.5

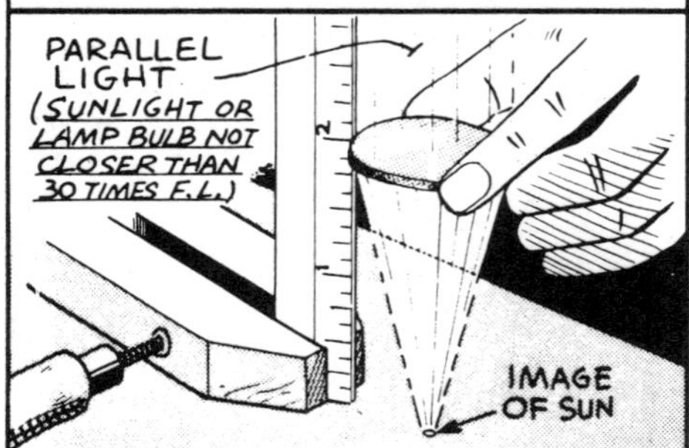

FINDING F.L. of a LENS

HOLD THE LENS IN SUNLIGHT AND ALONGSIDE A RULER. MOVE LENS UP AND DOWN TO FORM SMALLEST IMAGE OF SUN. READ THE F.L. ON RULER

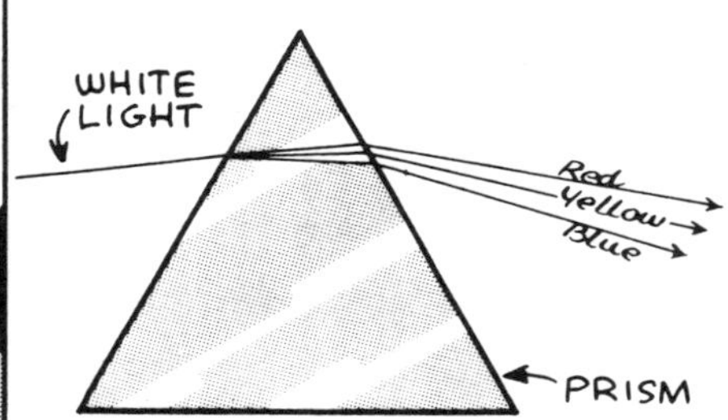

WHITE LIGHT

WHITE LIGHT IS COMPOSED OF ALL THE COLORS. A NARROW BEAM OF SUNLIGHT DIRECTED THRU A PRISM WILL EMERGE AS A COLORED BAND - THE SPECTRUM

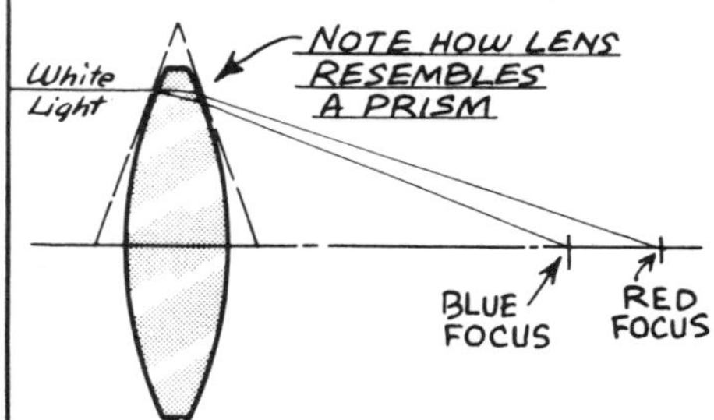

LONGITUDINAL COLOR

SIMPLE LENS BREAKS UP WHITE LIGHT JUST LIKE A PRISM. BLUE RAYS FOCUS CLOSER THAN RED. THE FAULT IS CONSTANT OVER WHOLE FIELD

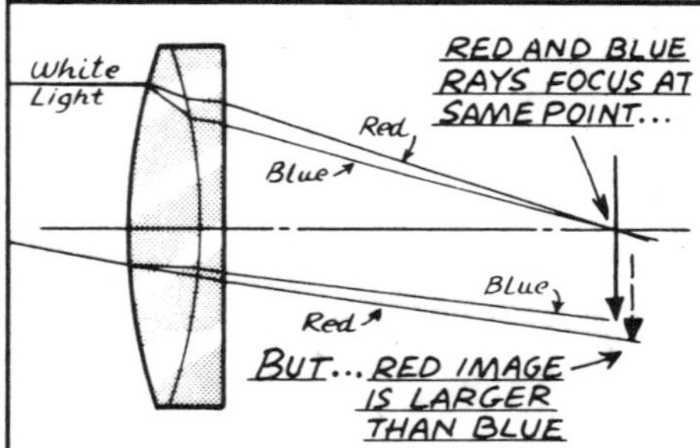

LATERAL COLOR

FAILURE OF THE LENS TO FORM SAME IMAGE SIZE IN ALL COLORS. THIS FAULT INCREASES WITH FIELD ANGLE... IS NOT PRESENT ON AXIS

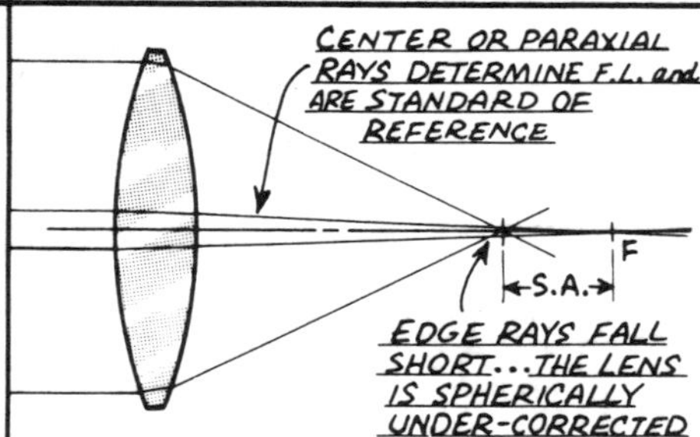

SPHERICAL ABERRATION

AS SHOWN, A POSITIVE LENS IS SPHERICALLY UNDER-CORRECTED. **S.A.** VARIES WITH f/VALUE OF LENS AND IS LESS FOR A SMALL APERTURE

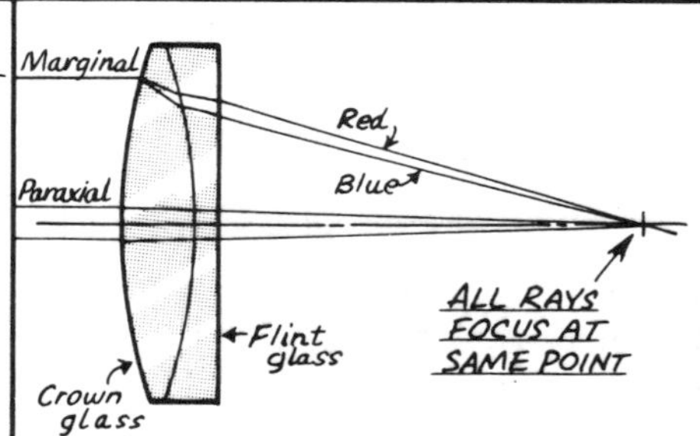

ACHROMATIC LENS

ACHROMATS ARE TWO-ELEMENT LENSES CORRECTED FOR LONGITUDINAL COLOR AND SPHERICAL ABERRATION FOR AN AXIAL OBJECT

Telescope Optics

OPTICS is a big subject. Most beginners avoid lens design, knowing very well that such work is a mathematical jungle. The simpler approach is to buy lenses, prisms and mirrors readymade and take it from there. Then the math work reduces to simple lens equations. If you use stock optics of good quality, you can be assured of good imagery. The problem of designing a telescope is just a matter of getting light through the instrument to form an image of a certain size at a certain location.

The astronomical telescope is a narrow-field instrument of 1 degree or less. As a result the objective is nearly immune to the off-axis aberrations--coma, distortion, field curvature and astigmatism. The two axial faults are chromatic and spherical aberration. As you may already know, chromatic aberration is non-existent for reflected light. In brief, if your interest is in reflecting telescopes, your only problem in image quality is that of spherical aberration. Even when you make your own mirror, the job of fashioning and correcting a single surface is well within the capabilities of the average person.

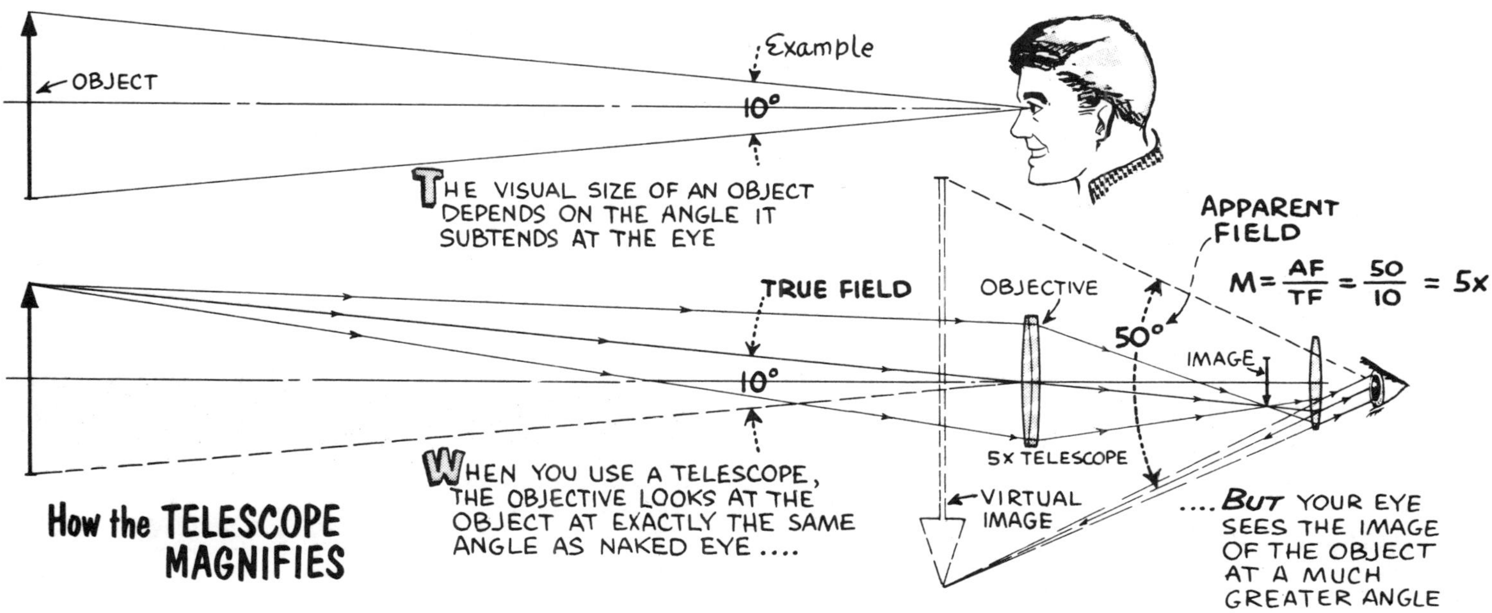

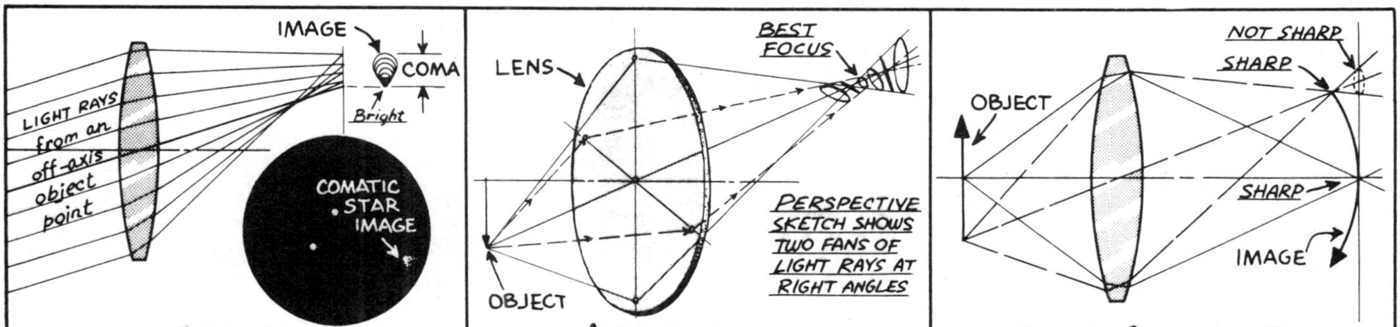

COMA

COMA IS OBLIQUE SPHERICAL ABERRATION. IT OCCURS ONLY OFF-AXIS WHERE THE UNSYMMETRIC DEFORMATION PRODUCES A COMET-SHAPE IMAGE

ASTIGMATISM

THIS FAULT SHOWS AN IMAGE POINT AS A SHORT LINE, WHICH CHANGES FROM HORIZONTAL TO VERTICAL POSITION ON EITHER SIDE OF BEST FOCUS

FIELD CURVATURE

THE NATURAL TENDENCY OF A LENS IS TO FORM A SHARP IMAGE ON A CURVED SURFACE BENT TOWARD THE LENS

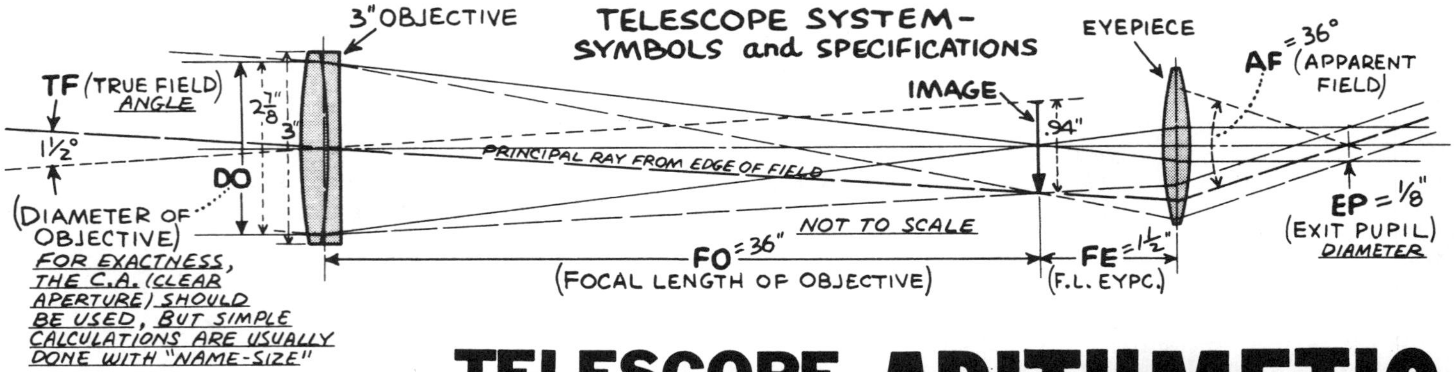

TELESCOPE ARITHMETIC

THREE WAYS TO CALCULATE MAGNIFICATION		EXAMPLE
1	$M = \frac{FO}{FE}$	$\frac{36}{1.5} = 24\times$
2	$M = \frac{DO}{EP}$	$\frac{3}{1/8} = 3 \times \frac{8}{1} = 24\times$
3	$M = \frac{AF}{TF}$	$\frac{36}{1.5} = 24\times$
	M. MUST BE KNOWN FOR OTHER CALCULATIONS	
4	$FO = M \times FE$	$24 \times 1\frac{1}{2} = 36''$
5	$FE = \frac{FO}{M}$	$\frac{36}{24} = \frac{3}{2} = 1\frac{1}{2}''$
6	$DO = M \times EP$	$24 \times \frac{1}{8} = 3''$
7	$EP = \frac{DO}{M}$	$\frac{3}{24} = \frac{1}{8}''$
8	$TF = \frac{AF}{M}$	$\frac{36}{24} = \frac{3}{2} = 1\frac{1}{2}°$
9	$AF = M \times TF$	$24 \times 1\frac{1}{2} = 36°$

ONE OF the first things you have to know about any telescope is its magnification. This is easily calculated by the No. 1 equation in box at left, a formula well-known to even the beginner.

Equations involving the field angle--Nos. 3, 8 and 9--can be applied to a part of the field as well as all of it. For example, if the separation of a double star is 6 seconds of arc (its true field angle) and you want to see it at 6 minutes of arc apparent field angle, you use equation No. 3, first changing the 6 minutes to seconds. Then by equation No. 3, M equals 360/6, equals 60x. Of course, you then have to find out what f.l. eyepiece will give 60x, or, in other words, you have to find **FE** when M and FO are known. The solution is equation No. 5.

The image diameter is easy to calculate if you know the apparent field, equation No. 10, and a simple transposition of this, No. 11, reveals the apparent field angle when the image size is known. The linear field diameter for any normal focal length can also be obtained directly from the Image Table in the chapter on eyepieces.

CALCULATION OF IMAGE DIA. and APPARENT FIELD by RADIANS*	EXAMPLE (AS ABOVE)
10 IMAGE (DIA.) = AF (RADIANS) × FE	AF = 36° 36° = .628 Radians (FROM TABLE) IMAGE = .628 × 1.5 = **.94"**
11 $AF\ (\text{RADIANS}) = \frac{\text{IMAGE}}{\text{FE}}$	$AF\ (\text{RADIANS}) = \frac{.94}{1.5} = .63^R$ .63 Radians = **36°** (FROM TABLE)

*THE UNIT OF CIRCULAR MEASURE. ONE RADIAN EMBRACES AN ARC OF A CIRCLE EQUAL TO ITS RADIUS. 1 RADIAN = 57.3°

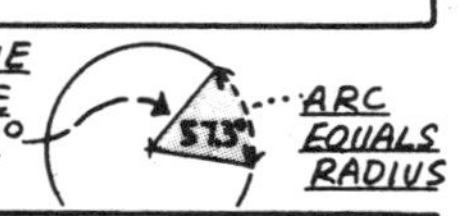

f/VALUE ... IS THE RATIO OF FOCAL LENGTH TO APERTURE

12	13	14
$f/\text{VALUE} = \frac{FO}{DO}$	$DO = \frac{FO}{f/\text{VALUE}}$	$FO = f/ \times DO$
EXAMPLE AS ABOVE: $f/ = \frac{36}{3} = f/12$	$DO = \frac{36}{12} = 3''$	$FO = 12 \times 3 = 36''$

TABLE - DEGREES TO RADIANS

DEG.	RAD.	DEG.	RAD.	DEG.	RAD.	DEG.	RAD.
5°	.087R	25°	.436R	41°	.716R	57°	.995R
10	.175	26	.454	42	.733	58	1.012
11	.192	27	.471	43	.751	59	1.030
12	.209	28	.489	44	.768	60	1.047
13	.227	29	.506	45	.785	61	1.065
14	.244	30	.524	46	.803	62	1.082
15	.262	31	.541	47	.820	63	1.100
16	.279	32	.559	48	.838	64	1.117
17	.297	33	.576	49	.855	65	1.135
18	.314	34	.593	50	.873	66	1.152
19	.332	35	.611	51	.890	67	1.169
20	.349	36	.628	52	.908	68	1.187
21	.367	37	.646	53	.925	69	1.204
22	.384	38	.663	54	.942	70	1.222
23	.401	39	.681	55	.960	71	1.239
24	.419	40	.698	56	.977	72	1.257

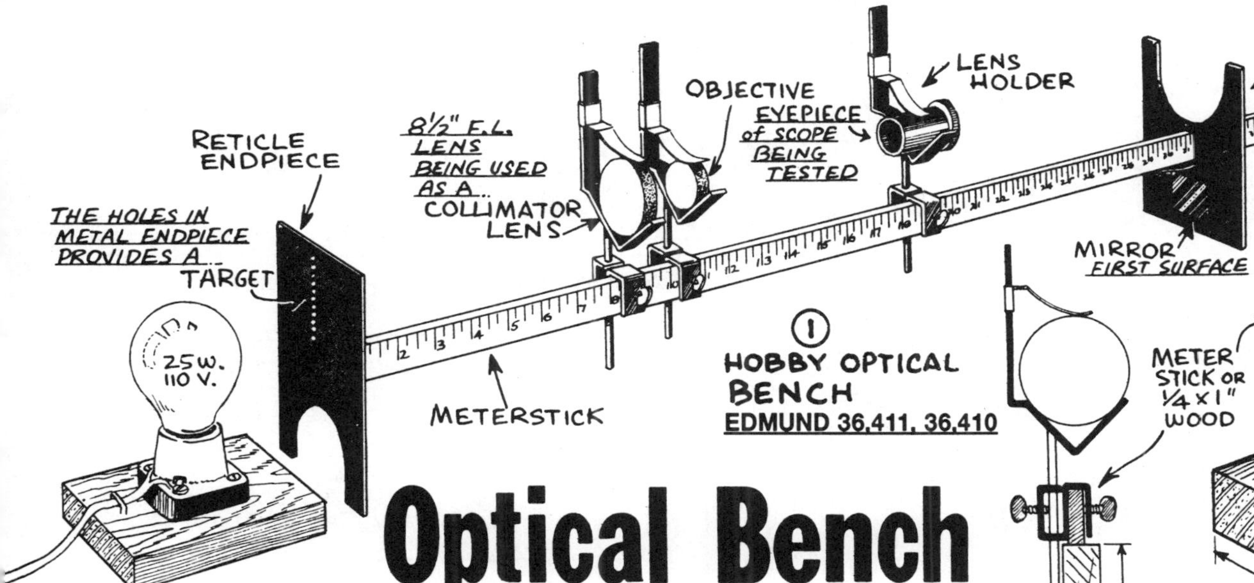

① HOBBY OPTICAL BENCH
EDMUND 36,411, 36,410

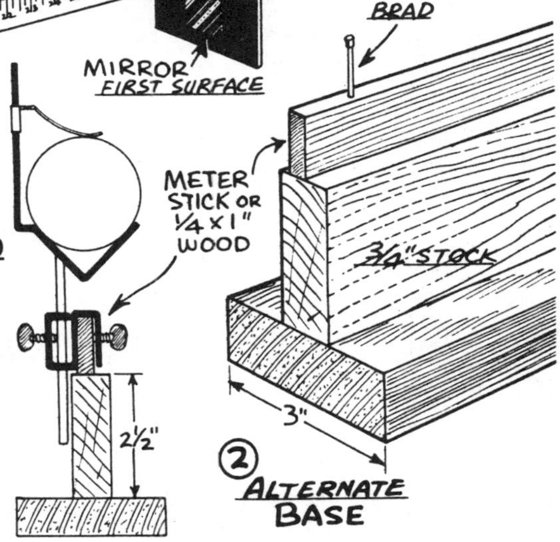

② ALTERNATE BASE

Optical Bench and Collimator

AN OPTICAL bench is the kind of equipment which may cost $5 or $5000. You can buy or build. Fig. 1 shows an inexpensive hobby optical bench you can buy. It is mounted on a wood meterstick. If you need a stronger or longer base, the construction shown in Fig. 2 can be used.

A collimator consists of some kind of illuminated reticle or target in the focal plane of an achromatic lens. Such an arrangement provides the equivalent of a distant target. A collimator can be built right on the optical bench as needed. In the equipment shown, the end plate is perforated with a vertical line of small holes. This is your "target." The collimator lens can be any good-quality achromat of 5 in. or more focal length. It is mounted at exactly one focal length from the reticle plate, a setting which is easily checked by auto-collimation as described on a following page. Fig. 1 setup shows a small finderscope being tested, the bench providing a means of holding the lenses while the collimator supplies the equivalent of a distant target.

Fig. 3 shows a simple homemade optical bench. The adjustable lens holders can handle lenses to 2-1/8 inch diameter, and sizes over this can be mounted in individual holders. The sliding vee blocks which clamp the lens in the grooved frame should be made of hardwood plywood. The collimator is a military gunsight, which requires only a simple conversion to 110-volt lighting, details of which are supplied with the merchandise.

The obvious weakness of the optical bench and collimator is that the equipment should be somewhere near the physical size of the largest telescope you plan to test. Small equipment works

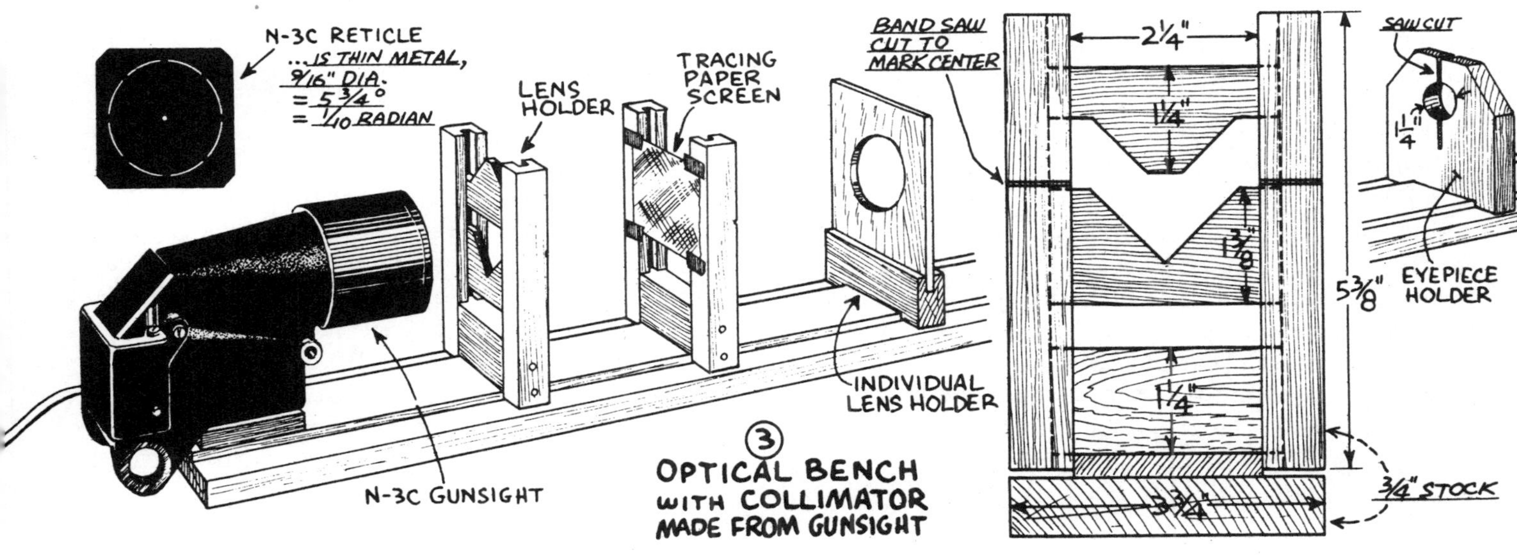

③ OPTICAL BENCH WITH COLLIMATOR MADE FROM GUNSIGHT

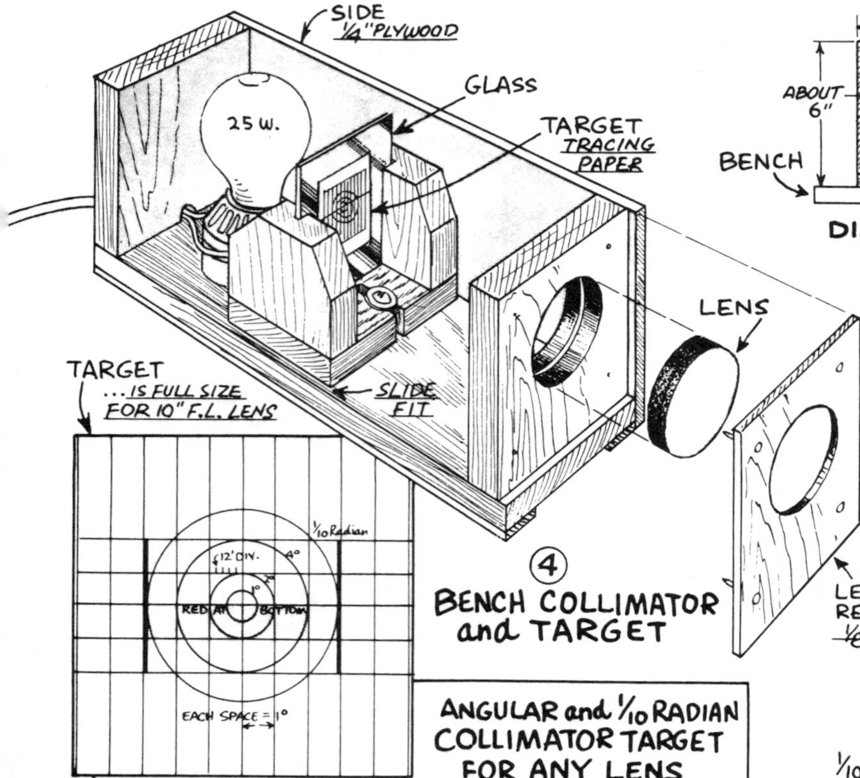

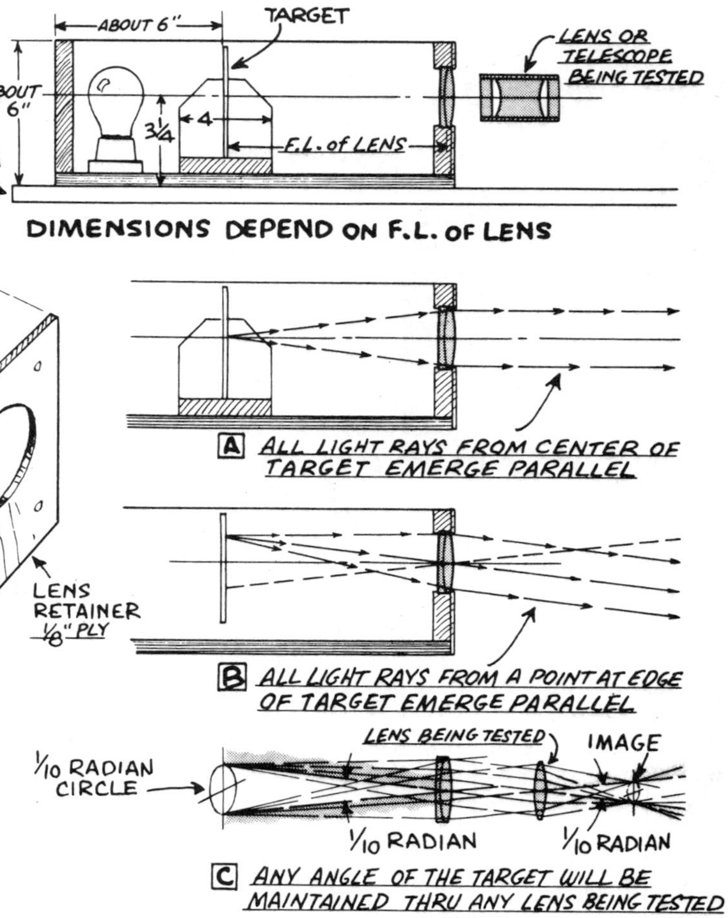

ANGULAR and 1/10 RADIAN COLLIMATOR TARGET FOR ANY LENS

	You want to find	FORMULA	Ex: 20" F.L. COLLI. LENS
MAKING THE TARGET	DIA. of 1/10 RADIAN CIRCLE	.1 Radian CIRCLE (DIA.) = F × .1	20 × .1 = 2.0"
	DISTANCE EQUAL TO 1°	1° SPACE = F × .0175	20 × .0175 = .35"
READING THE TARGET	TRUE FIELD of a TELESCOPE	COUNT THE DEGREE SPACES Ex: is about 4½°	Example: 2" F.L. YOU DO NOT KNOW F.L. IMAGE WILL READ .200"
	F.L. of a LENS BEING TESTED	F = IMAGE DIA. of .1 Radian CIRCLE / .1	F = .200 / .1 = 2.0"
	ALTERNATE: FOR LENSES OVER 5" F.L.	F = IMAGE OF 1° SPACE / .017	EX: 20" F.L. LENS (IT WILL READ .350") F = .350 / .017 = 20"

fine for riflescopes, finderscopes and small terrestrial and astro telescopes. Suitable equipment to test a 6-inch reflector is somewhat of an oversize luxury. However, you can do many tests and operations with a small collimator.

HOMEMADE COLLIMATOR. You can house a collimator in either a box or a tube. Fig. 4 shows a typical box job. The collimator lens should be a good quality achromat of fair size and focal length--3 inches diameter and 24 inches f.l. is a good size, suitable for some tests with telescopes as large as 6-inch aperture. Much smaller equipment is perfectly satisfactory for some operations. The collimator target is drawn with ink on tracing paper. The target is taped or cemented to a piece of glass, as shown. Simple rules for scaling the target to suit any focal length collimator lens are given in the drawing, Fig. 4.

Light from any distant object reaches your eye in parallel bundles. In the same manner, light emerges from the collimator in parallel bundles. That is, a point at the center of the target will send out a beam like A in Fig. 4; a point at the edge of target will send out a beam at some specific angle, as at B. All of the light is in parallel bundles, but the whole light cone is spreading, diverging. In other words, parallel light does not mean quite the same thing as a parallel "beam" of light.

Any angle that the target makes with the collimator lens will be reproduced exactly by any lens or telescope placed in front of the collimator. Fig. 4C shows the situation as it applies to the 1/10 radian circle. This particular unit is used for the determination of focal length. The image of the 1/10 radian circle produced by any lens, eyepiece or telescope will be 1/10 the focal length of said lens, eyepiece or telescope. In other words, if you measure the image diameter formed by any lens, you will know immediately its focal length, which is simply 10 times the image diameter. For short focal lengths under 5 in., a pocket comparator (measuring magnifier) is ideal for measuring the image diameter.

Optical Bench PROCEDURES

FOCAL LENGTH OF A LENS. When you focus a camera or telescope on a distant object, the image forms at one focal length behind the lens. So, if you have a lens of unknown f.l., you simply focus on a distant object and then measure the distance from lens to image, which is the focal length. Indoors, this is done with any bench collimator, as shown in Fig. 1.

FIELD OF A TELESCOPE. If your collimator target includes degree marks, the field of any telescope or binocular can be seen directly on the target--just count the number of degrees you can see. This is the true angular field. The Apparent Field is TF times M.

F.L. FROM IMAGE SIZE. This method is especially useful for eyepieces and other combinations of two or more lenses. You need a 1/10 radian reticle target, as described on a previous page. Set up the lens or eyepiece to be tested in the usual manner, and then measure the image it forms with a fine scale or with a direct-reading magnifier, Fig. 2. The focal length is 10 times the image diameter.

AUTO-COLLIMATION. The target for this is an opaque material in which is cut a small hole. The target is also a screen and should be white on the side facing away from the light. An ordinary flat mirror is held behind the lens being tested. When properly focused the lens will form an image of the target hole on the target itself, as shown in Fig. 3. The distance from target to lens is the focal length.

PRINCIPAL PLANES. Make some kind of setup similar to Fig. 4. The general idea is to juggle the eyepiece and screen back and forth until the image on screen is exactly in focus and exactly 1/2 inch long, the same as the target. This is 1x spacing, indicating the PP's are located two focal lengths from the object and image.

Alternately, PP2 can be located with a collimator, Fig. 5. Reversing the eyepiece would locate PP1, but usually the image will not be accessible. If the eyepiece is symmetrical, PP2 is all you need.

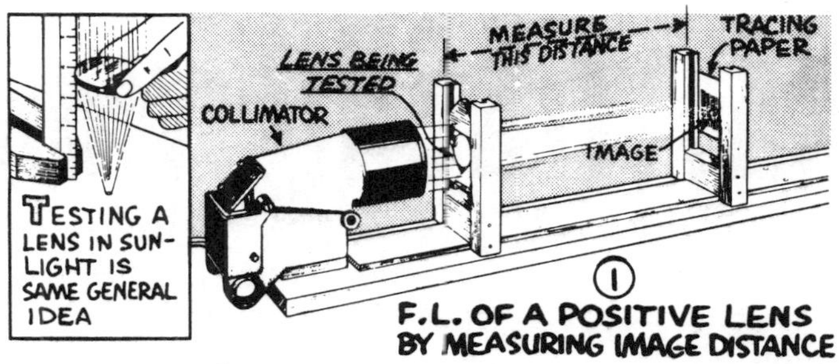

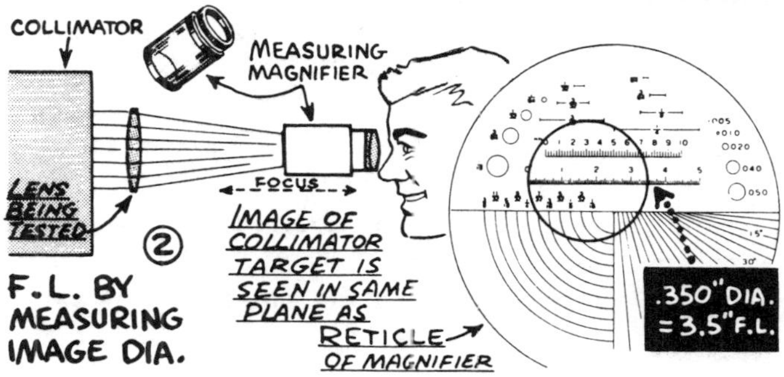

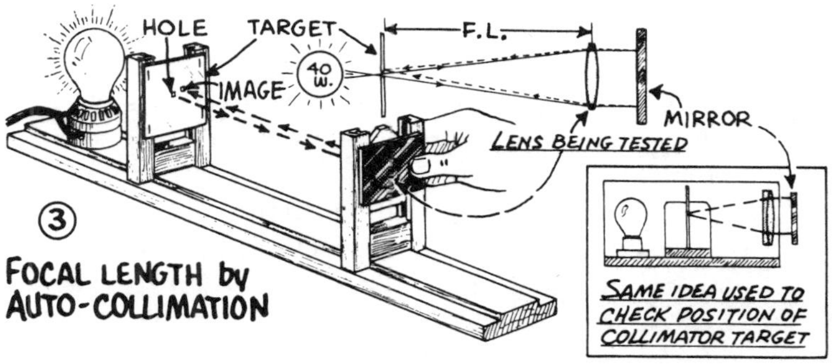

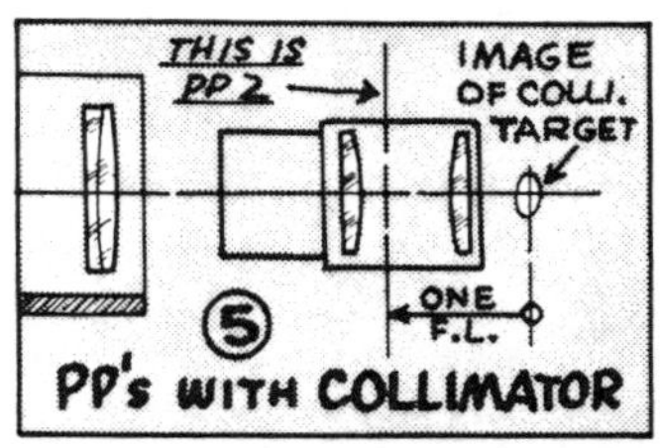

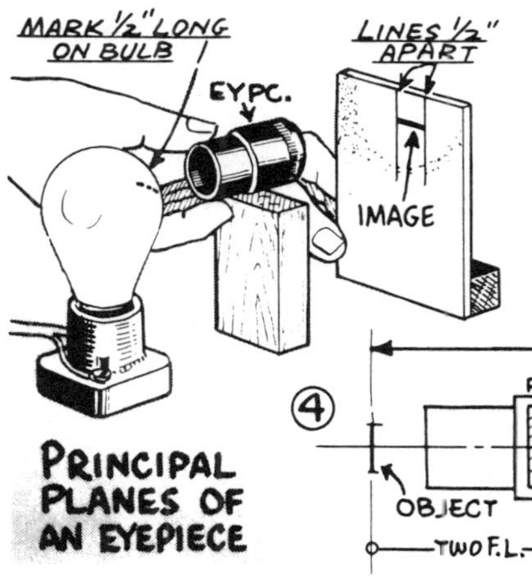

PRINCIPAL PLANES of SIMPLE POSITIVE LENSES

Usually it is close enough to set off the focal length or other measurement from the center of a lens, but if you want to be exact, the proper measuring points are the principal planes, as shown. When you make light ray diagrams, the light goes first to PP1, then parallel with axis to PP2 and then to the image point.

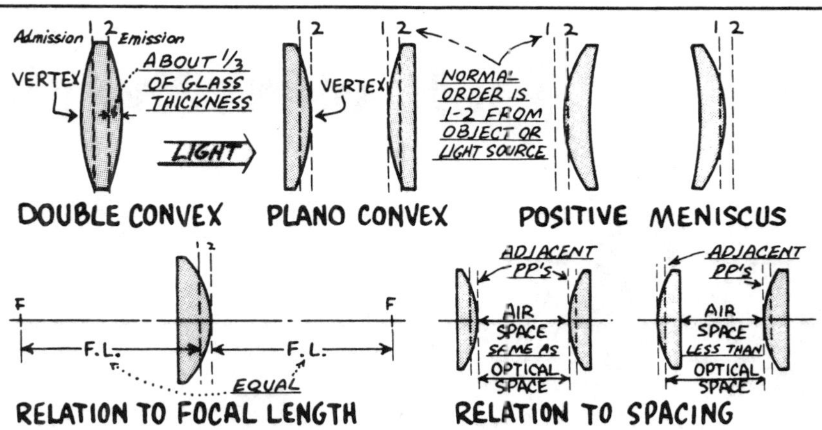

Ray Tracing from bench setups

IN designing a telescope, the first step is to select some suitable lenses for the objective and eyepiece. Next, you do some paperwork to get basic data. The third step is to "test" the design in some fashion, usually with optical bench and collimator target if the telescope is not too large. The end product of optical designing is some kind of plan drawing, showing how light gets through the system. The light path diagram, plus the actual "look and see" test on the optical bench, gives ample assurance that the telescope is AOK.

PAPERWORK. An example of a small refractor is shown on the opposite page. Instruments of this size and power are commonly used as finderscopes on larger telescopes. The preliminary data is obtained by applying the formulas given on page 2; information about the field angle and linear image diameter is given in the chapter on eyepieces. The preliminary paperwork reveals on overly large exit pupil of 1/2 inch diameter. Even in the dark, the pupil of your eye is not more than about 5/16 inch diameter. In brief, the design wastes a lot of light. So, if you were actually building this telescope, you would probably substitute a 1-1/4-inch diameter objective of the same focal length. This would assure better optical performance all around while retaining the maximum useful diameter of exit pupil.

BENCH TESTING. The first operation on the optical bench is to set up the objective and locate the image plane, Fig. 2. You can work in ordinary room light. The image is not confined in any manner at this stage and it will spread over a considerable area. Next, the eyepiece is mounted behind the image and moved back and forth until you see the image in sharp focus, Fig. 3. In other words, you focus the telescope--you make the focal plane of the eyepiece coincide with the focal plane of the objective. Under such circumstance, the emergent light is in parallel bundles. The final bench operation is to locate the exit pupil, Fig. 4.

LIGHT RAY DIAGRAM. From the bench setup, it is easy to pick off spacing dimensions and diameters. You can also check the angular field and linear diameter of image. The first stage in the light path diagram is the light cone from a point object at the center of the field. This funnels down to a corresponding point at the center of the image, and then emerges from the eyepiece as a parallel bundle of light rays, Fig. 5. The light cone for an edge-of-field object point is drawn next, Fig. 6. The most important light ray for edge-of-field object point is the ray that passes through the center of the objective. It is not deviated by the objective and goes directly from the point object at edge of field to the point image at edge of image, straight through to the eyepiece. This light ray through the center of the objective is called a principal or chief ray--if you can get it through to the exit pupil you are assured of no less than 50% lighting at the edge of field. You can see that for this particular telescope, the principal ray does get through, but the marginal ray criss-crossing the axis fails to strike the eyepiece. This instrument has a little better than 50% lighting at the edge of field, and such lighting is generally satisfactory, based on the fact the eye is self-compensating for such

Field of an Eyepiece

You can measure the linear image field of any positive eyepiece by introducing a folded-over strip of tracing paper into the open end. Get the folded-over end sharply in focus. Can you see the whole width of the paper? More? Less? Don't crowd--your eye should be at about the exit pupil position. Once you know the linear field, a simple diagram will reveal the Apparent Field.

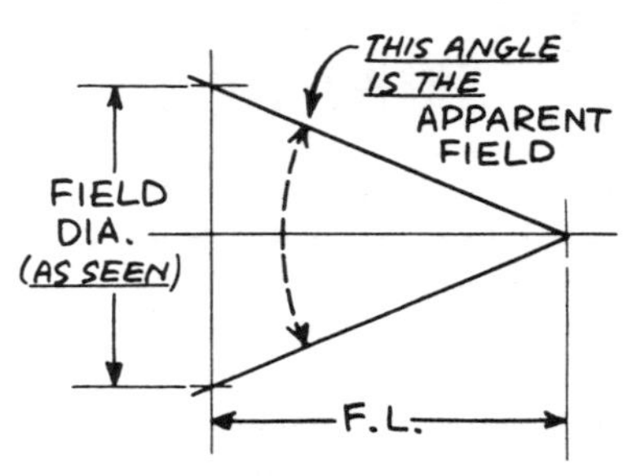

MEASURE THE ANGLE WITH A PROTRACTOR OR ADJUSTABLE TRIANGLE

lighting. Your eye sees sharpest at the center of the field, but it detects light and movement more readily at the edge of the field. Hence, if you lose a little light at the edge of field, it will not be noticed. As a matter of fact, you can look as closely as you like and you will not be able to see a 50% light loss at edge of field unless the overall illumination is very dim.

TRACE THROUGH TWO-LENS EYEPIECE. A two-lens eyepiece, Fig. 8, is set up in the same manner as a single lens eyepiece. You can locate the exit pupil. You can check the angular and linear field. From this incomplete data, it is possible to draw light rays on either side of the

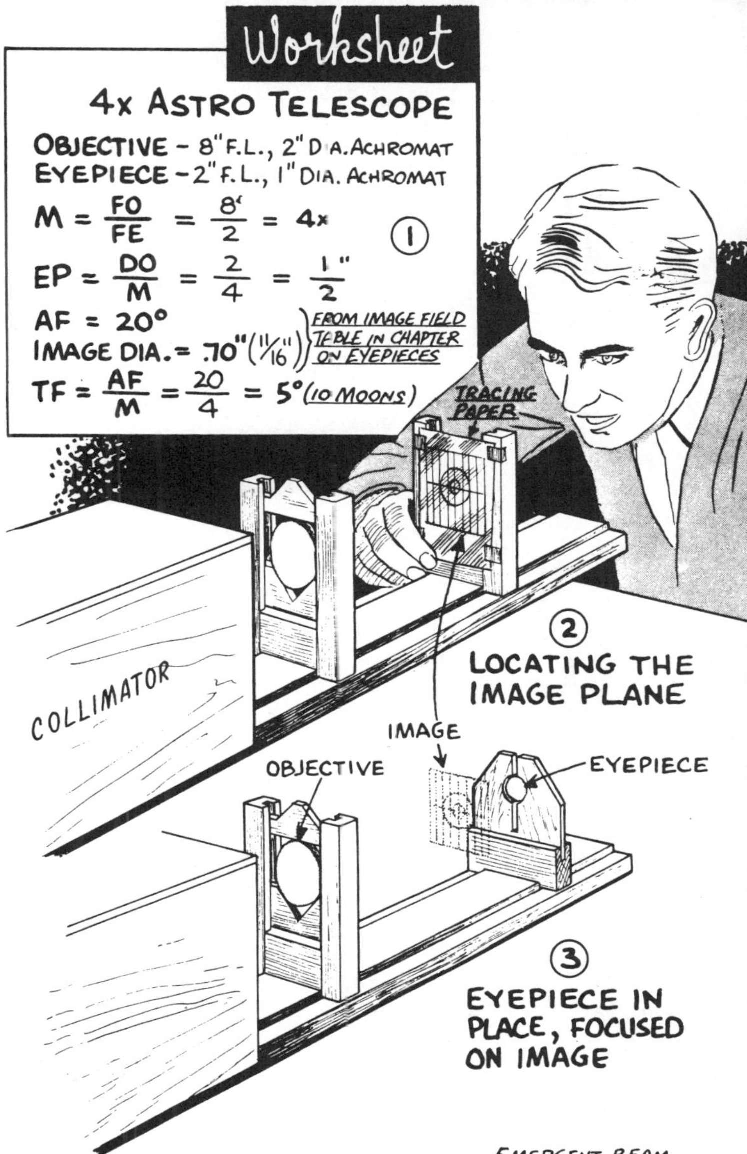

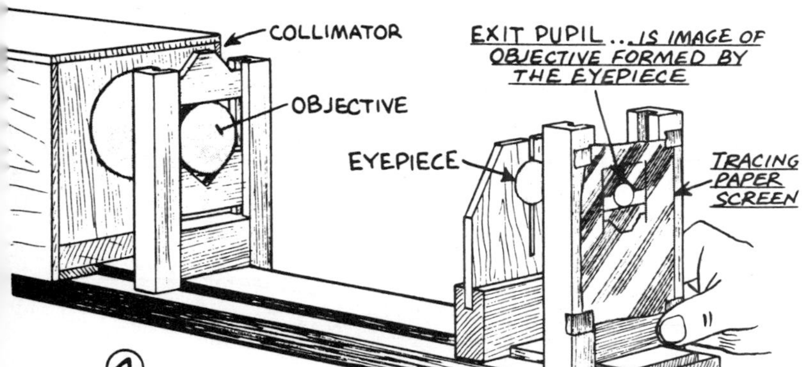

④ LOCATING THE EXIT PUPIL

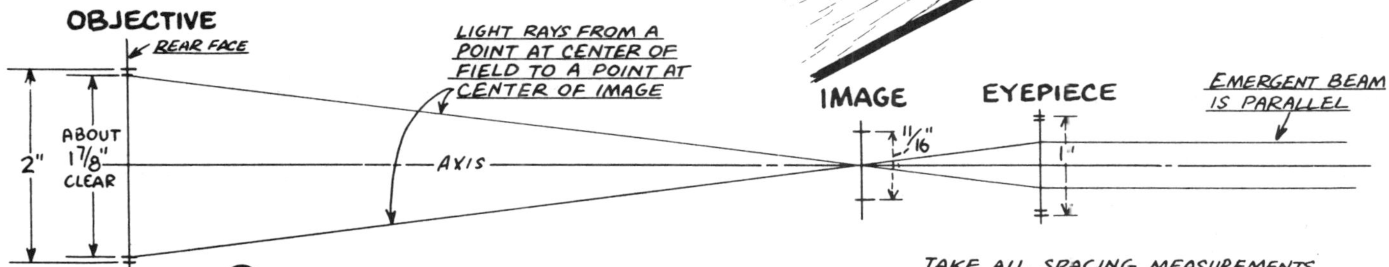

⑤ LIGHT CONE FOR A POINT AT CENTER OF IMAGE

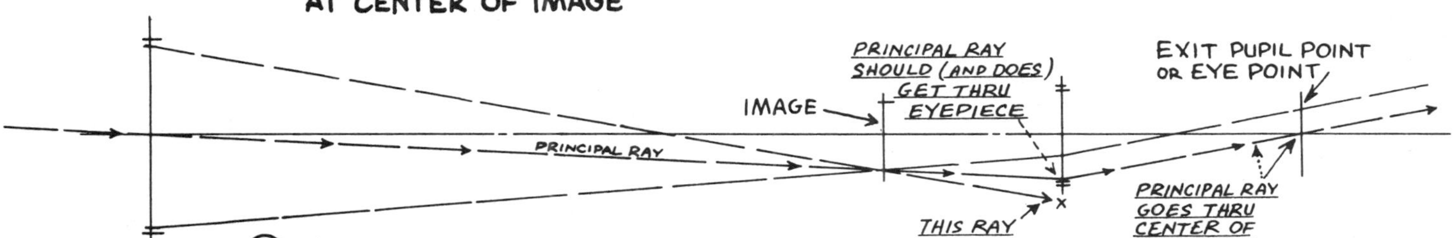

⑥ LIGHT CONE FOR A POINT AT EDGE OF IMAGE

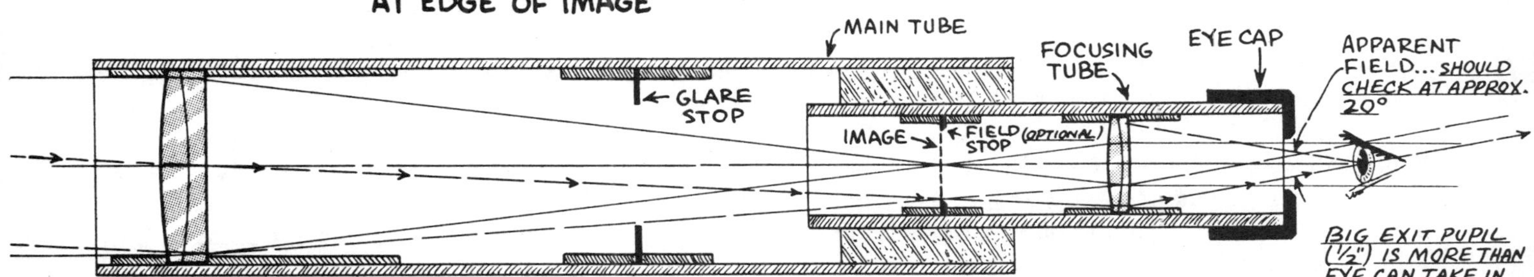

⑦ COMPLETE TELESCOPE (THIS DESIGN FOR DEMONSTRATION ONLY... NOT RECOMMENDED FOR ACTUAL USE)

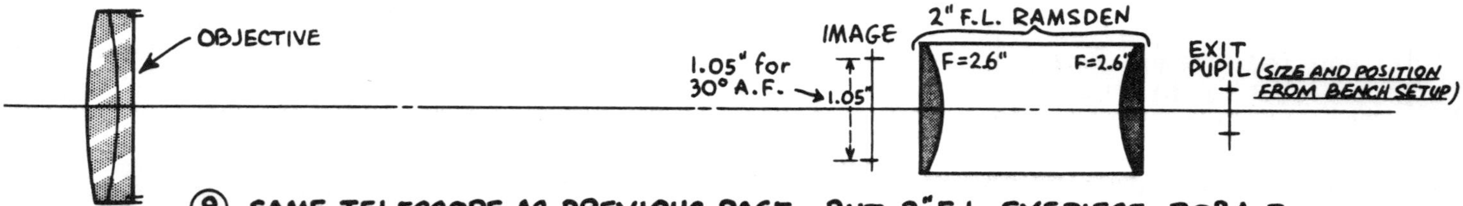

⑧ SAME TELESCOPE AS PREVIOUS PAGE, BUT 2" F.L. EYEPIECE, 30° A.F.

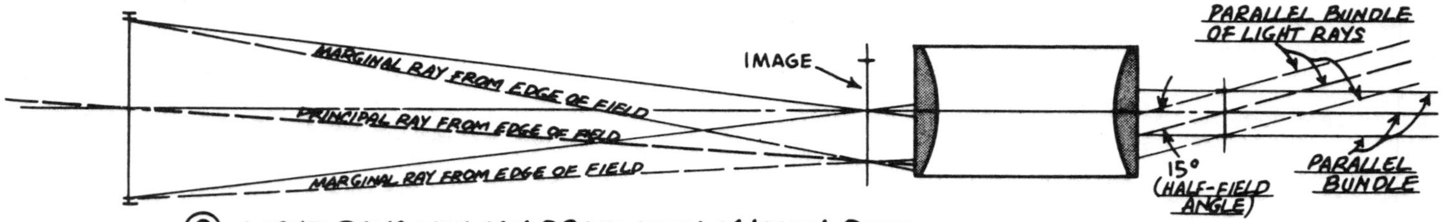

⑨ LIGHT RAYS YOU CAN DRAW WITH KNOWN DATA

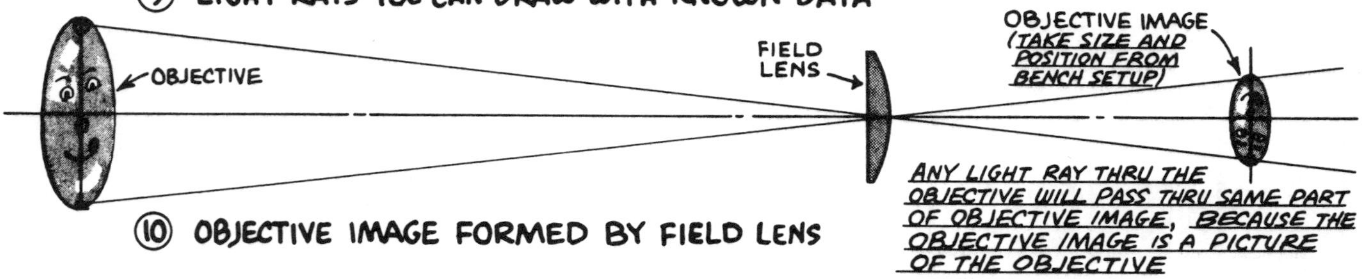

⑩ OBJECTIVE IMAGE FORMED BY FIELD LENS

eyepiece, Fig. 9. This is actually all you need know since it can be assumed the light rays get through, but just in case you want to trace the rays in approved fashion, you can do it easy enough if you do one lens at a time. Remove the eye lens but keep the field lens in its normal position. As before, you will pick up an image of the objective some distance behind the lens, Fig. 10. This is a little picture of the objective as formed by the field lens. If you put a mark on the objective, it will appear on the objective image because the objective image is a picture of the objective. If you can imagine light rays making visible tracks through the objective, they would make exactly the same tracks through the objective image. In brief, if a light ray passes through the center of the objective, it will also pass through the center of the objective image. Likewise, rays through the margins of the objective will go through the margins of the objective image. Thus, you have a simple and accurate guide to put the light rays through the field lens, Fig. 11. Then, putting the eye lens in place, you repeat the operation with the whole eyepiece, the final objective image being of course, the exit pupil, Fig. 12. This diagram also shows the manner of drawing light rays if you are using principal planes--you go first to PP1, then parallel with the axis to PP2.

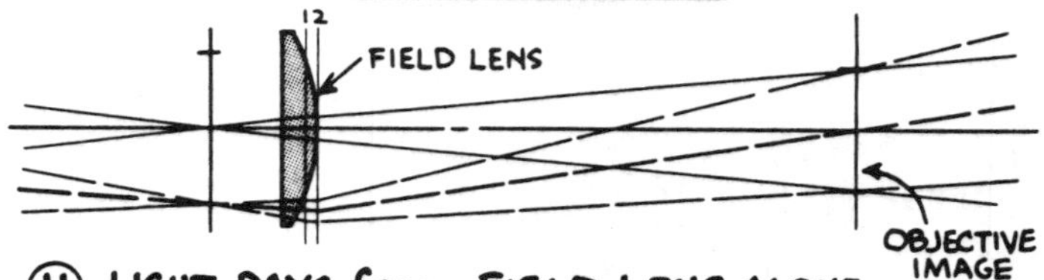

⑪ LIGHT RAYS from FIELD LENS ALONE

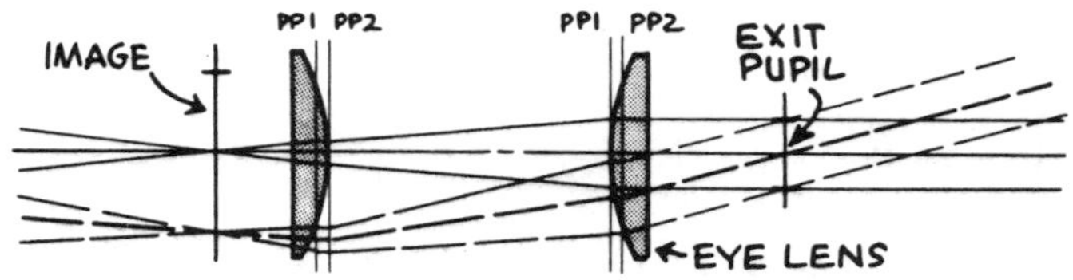

⑫ LIGHT PATH THRU EYEPIECE COMPLETED

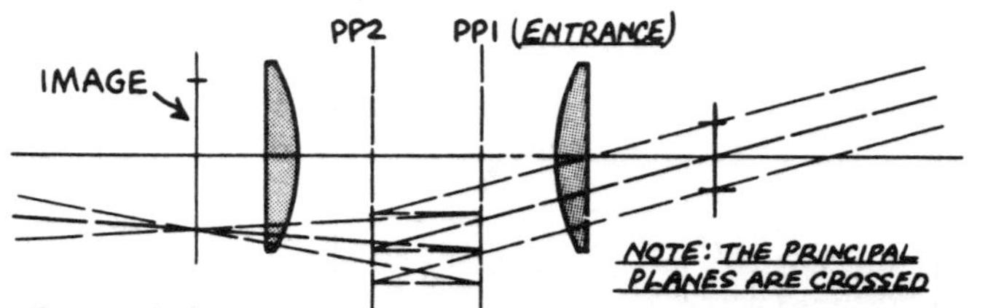

⑬ ALTERNATE RAY TRACE USING PRINCIPAL PLANES

RAY TRACING TO PRINCIPAL PLANES. When you are using a purchased eyepiece, it is inconvenient to take the eyepiece apart for the lens-by-lens trace just described. Instead, you find the principal planes of the eyepiece as a whole and then draw light rays to the PP's, ignoring the lenses entirely except for the single item of diameter. The method of finding the PP's has already been described; the manner of making the drawing is as shown in Fig. 13. You draw a light ray to the image and keep right on going until you strike PP1. From PP1 to PP2, the ray is parallel to the axis. From PP2, the light ray goes to the corresponding part of the objective image, i.e., the exit pupil.

OBJECT-IMAGE MATH

WHERE is the image? This is a basic problem in all optical designing--you can't even get started until you know the answer. The common textbook solution is the classical equation:

$$\frac{1}{F} = \frac{1}{L} + \frac{1}{L'}$$

The F, of course, means focal length. L means "length," the L without a prime mark being the length to the object while the primed L is the distance to the image. Other symbols are often used, notably S (Space) instead of L (Length). Some writers use the letter A for the object distance, and B for the image distance, which is the system used in this book. It is easily memorized from the fact that you must first have an object to look at before you can have an image; hence, A (first letter of alphabet), is the object distance. B is the image distance.

The numeral 1 above each quantity means that the reciprocal of each quantity is used in the calculation. The reciprocal of any number is the number divided into 1. If you have a mirror or lens of 50 inches focal length, its reciprocal is 1/50. Some simple problems can be worked with the reciprocal in this fractional form, but usually you have to convert to the decimal equivalent, which, for this example, can be calculated mentally: 1/50 is 2/100 or .02. In actual work you must refer to a Table of Reciprocals, which can be found in many math and engineering handbooks. For the benefit of beginners who may not have such a table, the basic Gaussian (Karl Friedrich Gauss, 1777-1855, German astronomer and physicist) formula is converted to other

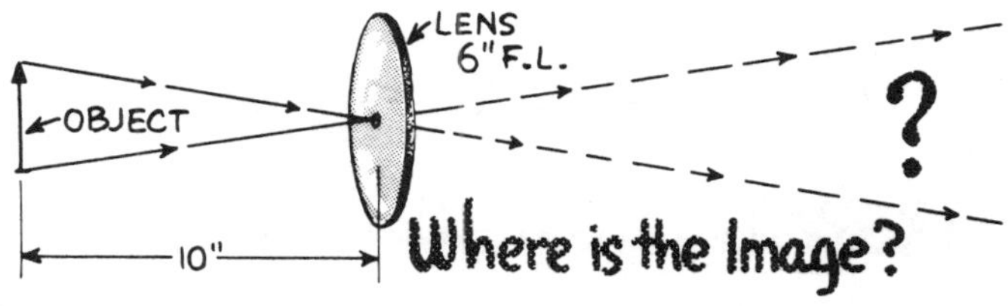

Where is the Image?

SOLUTION: *THIS IS A CASE 1 PROBLEM (See next page)*

YOU KNOW: F = 6"
A = 10"

then **CASE 1 [2]** $B = \frac{F \times A}{A - F} = \frac{6 \times 10}{10 - 6} = \frac{60}{4} = 15"$

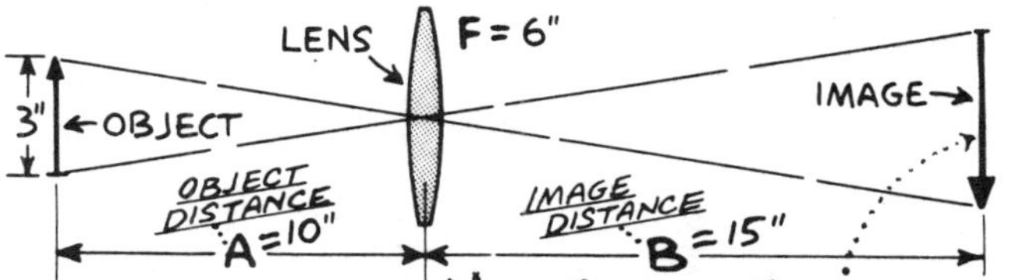

How BIG is the Image?

LINEAR M. **ALL CASES** $= \frac{B}{A} = \frac{15}{10} = 1.5\times$

① Example OF A CASE 1 PROBLEM

HENCE: IF THE OBJECT IS 3" HIGH AND M. IS 1½×:

IMAGE SIZE $= 3 \times 1\frac{1}{2} = 4\frac{1}{2}"$

② **IMAGE FORMED BY TWO OR MORE LENSES OR MIRRORS**

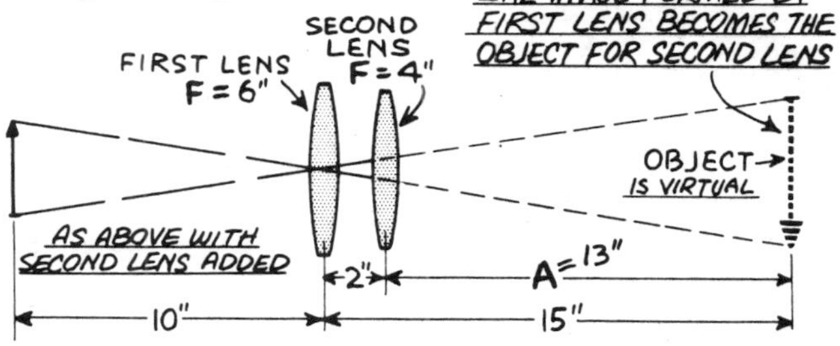

SOLUTION: *SECOND LENS IS A CASE 4 PROBLEM*

YOU KNOW: F = 4"
A = 13"

THEN **CASE 4 [2]** $B = \frac{F \times A}{F + A} = \frac{4 \times 13}{4 + 13} = \frac{52}{17} = 3.06"$

M. FOR SECOND LENS:

$M = \frac{B}{A} = \frac{3.06}{13} = .24\times$

IMAGE SIZE $= 4.5 \times .24 = 1.08"$

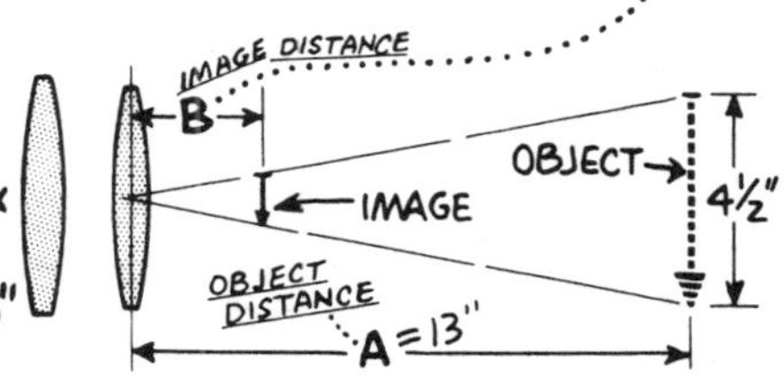

③ **VIRTUAL IMAGE** *FORMED BY FIRST LENS IS TREATED LIKE A* **REAL OBJECT** *FOR SECOND LENS... IS CASE 1*

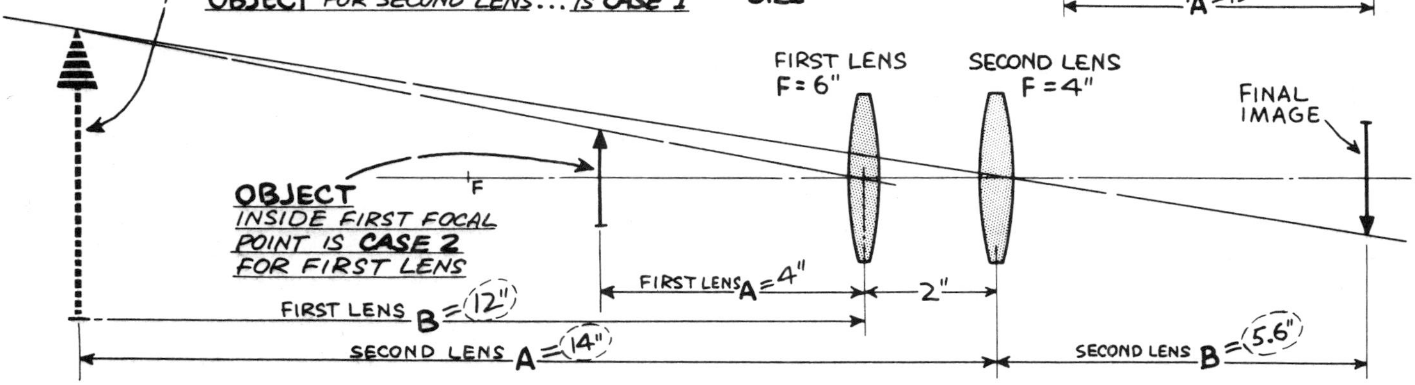

OBJECT-IMAGE MATH for a SINGLE POSITIVE LENS or CONCAVE MIRROR

CASE 1 — OBJECT AT MORE THAN ONE FOCAL LENGTH FROM LENS

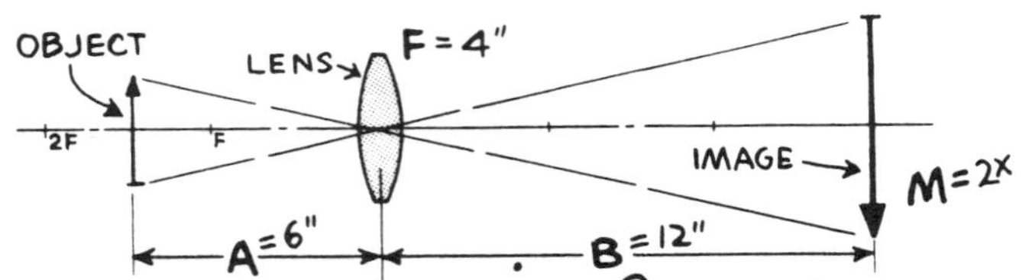

Example

1	$B=(M+1)\times F$	$B=(2+1)\times 4=3\times 4=12''$
2	$B=\frac{F\times A}{A-F}$	$B=\frac{4\times 6}{6-4}=\frac{24}{2}=12''$
3	$B=A\times M$	$B=6\times 2=12''$
4	$A=\frac{B}{M}$	$A=\frac{12}{2}=6''$
5	$A=\frac{F}{M}+F$	$A=\frac{4}{2}+4=2+4=6''$
6	$A=\frac{F\times B}{B-F}$	$A=\frac{4\times 12}{12-4}=\frac{48}{8}=6''$
7	$M=\frac{B}{A}$	$M=\frac{12}{6}=2\times$
8	$M=\frac{F}{A-F}$	$M=\frac{4}{6-4}=\frac{4}{2}=2\times$
9	$M=\frac{B-F}{F}$	$M=\frac{12-4}{4}=\frac{8}{4}=2\times$
10	$F=\frac{A\times M}{M+1}$	$F=\frac{6\times 2}{2+1}=\frac{12}{3}=4''$
11	$F=\frac{B}{M+1}$	$F=\frac{12}{2+1}=\frac{12}{3}=4''$
12	$F=\frac{A\times B}{A+B}$	$F=\frac{6\times 12}{6+12}=\frac{72}{18}=4''$
13	$\frac{1}{F}=\frac{1}{A}+\frac{1}{B}$	$\frac{1}{F}=\frac{1}{6}+\frac{1}{12}$; $\frac{1}{F}=\frac{2}{12}+\frac{1}{12}=\frac{3}{12}$; $F=\frac{12}{3}=4''$
14	$\frac{1}{A}=\frac{1}{F}-\frac{1}{B}$	$\frac{1}{A}=\frac{1}{4}-\frac{1}{12}$; $\frac{1}{A}=\frac{3}{12}-\frac{1}{12}=\frac{2}{12}$; $A=\frac{12}{2}=6''$
15	$\frac{1}{B}=\frac{1}{F}-\frac{1}{A}$	$\frac{1}{B}=\frac{1}{4}-\frac{1}{6}$; $\frac{1}{B}=\frac{3}{12}-\frac{2}{12}=\frac{1}{12}$; $B=\frac{12}{1}=12''$

ALTERNATE EQUATIONS USING RECIPROCALS (rows 13–15)

THE SAME EQUATIONS ARE USED FOR A POSITIVE (CONCAVE) MIRROR

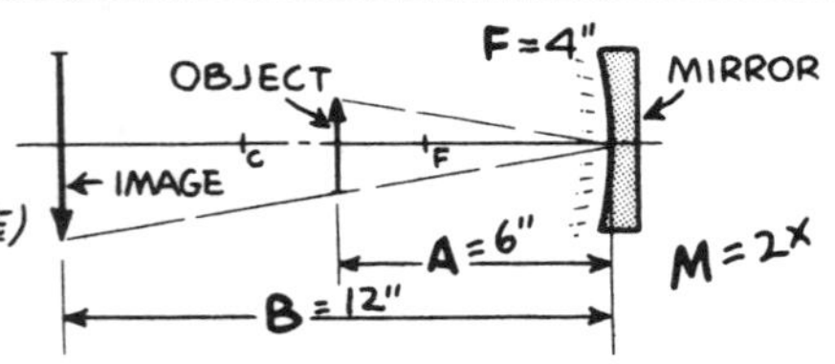

TWO CASES: WHEN THE OBJECT IS MORE THAN **F** BUT LESS THAN **2F** FROM LENS (*SHOWN*), THE SYSTEM IS PROJECTION-- THE IMAGE IS LARGER THAN OBJECT.

WHEN **A** IS MORE THAN **2F**, **B** WILL BE LESS THAN **2F** AND THE IMAGE WILL BE SMALLER THAN OBJECT

CASE 2 — OBJECT AT LESS THAN ONE FOCAL LENGTH FROM LENS

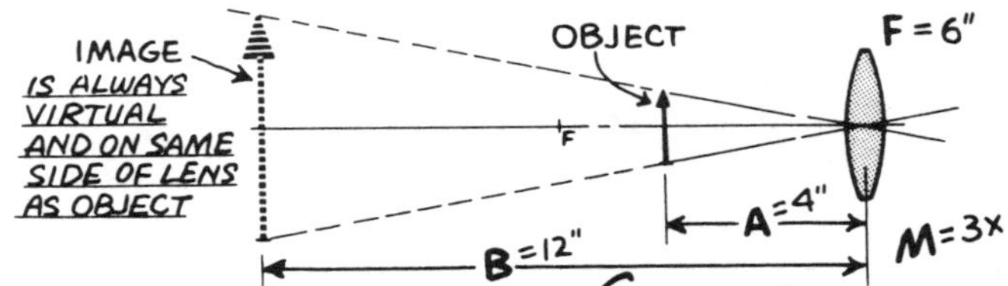

Example

1	$B=(M-1)\times F$	$B=(3-1)\times 6=2\times 6=12''$
2	$B=\frac{F\times A}{F-A}$	$B=\frac{6\times 4}{6-4}=\frac{24}{2}=12''$
3	$B=A\times M$	$B=4\times 3=12''$
4	$A=\frac{B}{M}$	$A=\frac{12}{3}=4''$
5	$A=F-\frac{F}{M}$	$A=6-\frac{6}{3}=6-2=4''$
6	$A=\frac{F\times B}{F+B}$	$A=\frac{6\times 12}{6+12}=\frac{72}{18}=4''$
7	$M=\frac{B}{A}$	$M=\frac{12}{4}=3\times$
8	$M=\frac{F}{F-A}$	$M=\frac{6}{6-4}=\frac{6}{2}=3\times$
9	$M=\frac{B+F}{F}$	$M=\frac{12+6}{6}=\frac{18}{6}=3\times$
10	$F=\frac{A\times M}{M-1}$	$F=\frac{4\times 3}{3-1}=\frac{12}{2}=6''$
11	$F=\frac{B}{M-1}$	$F=\frac{12}{3-1}=\frac{12}{2}=6''$
12	$F=\frac{A\times B}{B-A}$	$F=\frac{4\times 12}{12-4}=\frac{48}{8}=6''$
13	$\frac{1}{F}=\frac{1}{A}-\frac{1}{B}$	$\frac{1}{F}=\frac{1}{4}-\frac{1}{12}$; $\frac{1}{F}=\frac{3}{12}-\frac{1}{12}=\frac{2}{12}$; $F=\frac{12}{2}=6''$
14	$\frac{1}{A}=\frac{1}{F}+\frac{1}{B}$	$\frac{1}{A}=\frac{1}{6}+\frac{1}{12}$; $\frac{1}{A}=\frac{2}{12}+\frac{1}{12}=\frac{3}{12}$; $A=\frac{12}{3}=4''$
15	$\frac{1}{B}=\frac{1}{A}-\frac{1}{F}$	$\frac{1}{B}=\frac{1}{4}-\frac{1}{6}$; $\frac{1}{B}=\frac{3}{12}-\frac{2}{12}=\frac{1}{12}$; $B=\frac{12}{1}=12''$

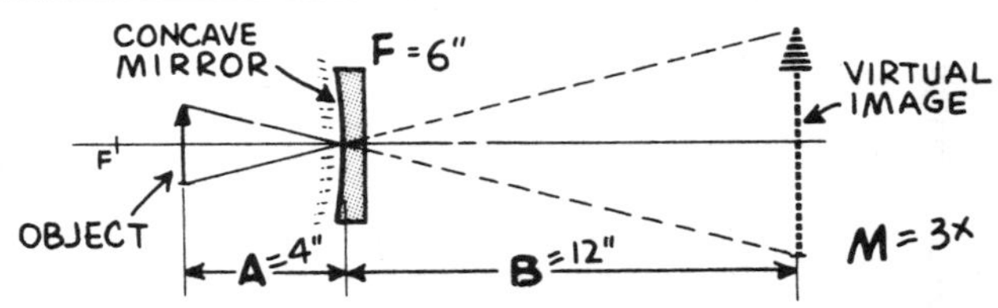

WHEN THE OBJECT-TO-LENS DISTANCE IS LESS THAN **F**, THE IMAGE IS VIRTUAL, ERECT AND MAGNIFIED. **M** IS NEVER LESS THAN **1X** ... IS GREATEST WHEN OBJECT IS JUST INSIDE **F**.

THE SAME ACTION IS OBTAINED WITH A CONCAVE MIRROR EXCEPT VIRTUAL IMAGE APPEARS BEHIND THE MIRROR

form suited to simple arithmetical solution, as shown in the Tables on the following pages.

FIVE OBJECT-IMAGE CASES. A Case 1 problem concerns a single positive lens (or mirror) with the object located at more than one focal length from the lens. Fig. 1 is an example. You know the focal length (F) of the lens and the distance to the object (A). The problem is to find the image distance, B, and this is easily calculated with the second equation in the CASE 1 Table.

If the problem involves two or more lenses, the image formed by the first lens becomes the object for the second lens, as shown in Fig. 2. This example is a continuation of Fig. 1 example, with a second lens added. The problem is to calculate the image position as formed by the second lens, using the image formed by the first lens as the object. In this particular instance, the second lens is a Case 4 problem, i.e., it is concerned with a virtual object to the right of a positive lens. You use Case 4 Table. Again you know F and A, so B is found by using Equation No. 2.

Sometimes the first lens looks at an object at less than one focal length, and this is Case 2, of which Fig. 3 is an example. Fig 3 also shows the situation where the virtual image formed to the left of the first lens becomes the real object for the second lens.

Case 5 covers the situation of a negative lens or mirror inside the focus of a primary lens or mirror, a situation which many readers will immediately identify as the Barlow Case, because this is the way a Barlow amplifying lens works in a telescope. The five cases covered by the tables will handle practically any kind of object-image problem. An important exception is the common telescope situation where the object is at infinity. For such a target, the equations are useless, but

SYMBOLS:

M... IS **LINEAR MAGNIFICATION**... IS ACTUAL RATIO OF IMAGE SIZE TO OBJECT SIZE. *DO NOT CONFUSE WITH ANGULAR MAGNIFICATION USED TO SPECIFY THE POWER OF A TELESCOPE*

F... IS **FOCAL LENGTH**

A... IS DISTANCE FROM LENS TO **OBJECT**

B... IS DISTANCE FROM LENS TO **IMAGE**

FOR TECHNICAL EXACTNESS, SPACING SHOULD BE SET OFF FROM THE ADJACENT PRINCIPAL PLANE OF THE LENS OR VERTEX OF MIRROR

OBJECT-IMAGE MATH INDEX

IF YOU KNOW	FIND, WITH
F and A	M....8, B.....2
F and B	M....9, A....6
F and M	A....5, B....1
M and A	F....10, B....3
M and B	F....11, A....4
A and B	M....7, F....12

USE THIS INDEX FOR ALL CASES. WORK ALL PROBLEMS WITH SIMPLE ARITHMETIC

CASE 3 OBJECT-IMAGE MATH

SINGLE NEGATIVE LENS OR CONVEX MIRROR. OBJECT AT ANY DISTANCE FROM LENS

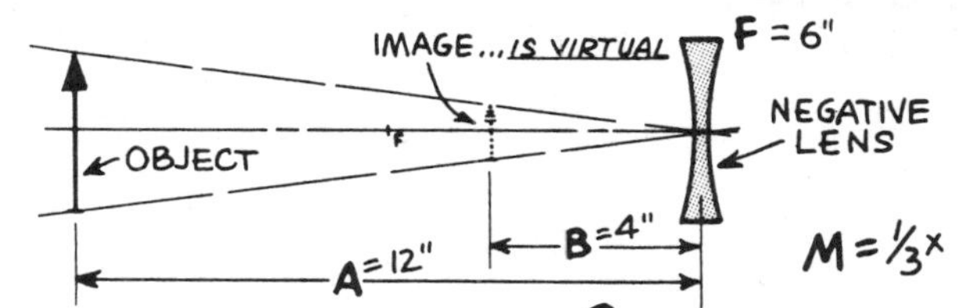

		Example
1	$B = (1-M) \times F$	$B = (1-\frac{1}{3}) \times 6 = \frac{2}{3} \times 6 = 4''$
2	$B = \frac{F \times A}{F + A}$	$B = \frac{6 \times 12}{6 + 12} = \frac{72}{18} = 4''$
3	$B = A \times M$	$B = 12 \times \frac{1}{3} = \frac{12}{3} = 4''$
4	$A = \frac{B}{M}$	$A = \frac{4}{\frac{1}{3}} = 4 \times \frac{3}{1} = 12''$
5	$A = \frac{F}{M} - F$	$A = \frac{6}{\frac{1}{3}} - 6 = 6 \times \frac{3}{1} - 6 = 18 - 6 = 12''$
6	$A = \frac{F \times B}{F - B}$	$A = \frac{6 \times 4}{6 - 4} = \frac{24}{2} = 12''$
7	$M = \frac{B}{A}$	$M = \sqrt[4]{\frac{4}{12}} = \frac{1}{3}x$ (OR .33x)
8	$M = \frac{F}{A + F}$	$M = \frac{6}{12 + 6} = \frac{6}{18} = \frac{1}{3}x$
9	$M = \frac{F - B}{F}$	$M = \frac{6 - 4}{6} = \frac{2}{6} = \frac{1}{3}x$
10	$F = \frac{A \times M}{1 - M}$	$F = \frac{12 \times \frac{1}{3}}{1 - \frac{1}{3}} = \frac{4}{\frac{2}{3}} = 4 \times \frac{3}{2} = 6''$
11	$F = \frac{B}{1 - M}$	$F = \frac{4}{1 - \frac{1}{3}} = \frac{4}{\frac{2}{3}} = 4 \times \frac{3}{2} = 6''$
12	$F = \frac{A \times B}{A - B}$	$F = \frac{12 \times 4}{12 - 4} = \frac{48}{8} = 6''$
13	$\frac{1}{F} = \frac{1}{B} - \frac{1}{A}$	$\frac{1}{F} = \frac{1}{4} - \frac{1}{12}$; $\frac{1}{F} = \frac{3}{12} - \frac{1}{12} = \frac{2}{12}$; $F = \frac{12}{2} = 6''$
14	$\frac{1}{A} = \frac{1}{B} - \frac{1}{F}$	$\frac{1}{A} = \frac{1}{4} - \frac{1}{6}$; $\frac{1}{A} = \frac{3}{12} - \frac{2}{12} = \frac{1}{12}$; $A = \frac{12}{1} = 12''$
15	$\frac{1}{B} = \frac{1}{A} + \frac{1}{F}$	$\frac{1}{B} = \frac{1}{12} + \frac{1}{6}$; $\frac{1}{B} = \frac{1}{12} + \frac{2}{12} = \frac{3}{12}$; $B = \frac{12}{3} = 4''$

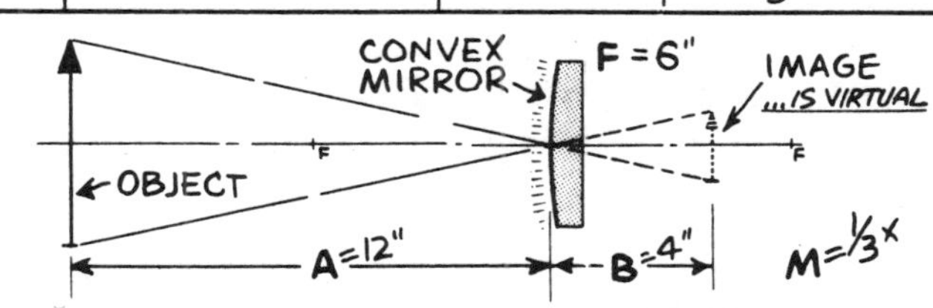

THE IMAGE IS ALWAYS VIRTUAL, ERECT AND REDUCED. **B** IS ALWAYS LESS THAN **A**, AND ALSO LESS THAN **F**. **M** IS ALWAYS UNDER 1X. WHEN THE OBJECT IS AT INFINITY, THE EQUATIONS BECOME USELESS AND UNNECESSARY. IN SUCH CASE:

$A = \infty$ (INFINITY) $B = F$ $M = 0$ (*NEARLY*)

OBJECT-IMAGE MATH for Second of Two Lenses with Virtual Object

CASE 4

SECOND OF TWO POSITIVE LENSES OR MIRRORS. VIRTUAL OBJECT AT ANY DISTANCE

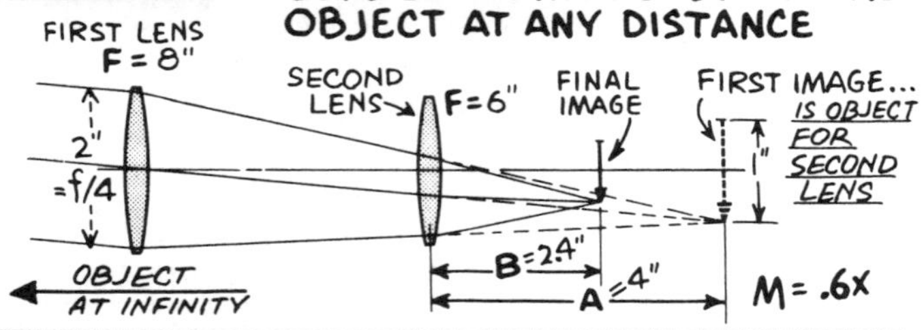

1	$B = (1 - M) \times F$	$B = (1 - .6) \times 6 = .4 \times 6 = 2.4''$
2	$B = \frac{F \times A}{F + A}$	$B = \frac{6 \times 4}{6 + 4} = \frac{24}{10} = 2.4''$
3	$B = A \times M$	$B = 4 \times .6 = 2.4''$
4	$A = \frac{B}{M}$	$A = \frac{2.4}{.6} = 4''$
5	$A = \frac{F}{M} - F$	$A = \frac{6}{.6} - 6 = 10 - 6 = 4''$
6	$A = \frac{F \times B}{F - B}$	$A = \frac{6 \times 2.4}{6 - 2.4} = \frac{14.4}{3.6} = 4''$
7	$M = \frac{B}{A}$	$M = \frac{2.4}{4} = .6X$
8	$M = \frac{F}{F + A}$	$M = \frac{6}{6+4} = \frac{6}{10} = .6X$
9	$M = \frac{F - B}{F}$	$M = \frac{6 - 2.4}{6} = \frac{3.6}{6} = .6X$
10	$F = \frac{A \times M}{1 - M}$	$F = \frac{4 \times .6}{1 - .6} = \frac{2.4}{.4} = 6''$
11	$F = \frac{B}{1 - M}$	$F = \frac{2.4}{1 - .6} = \frac{2.4}{.4} = 6''$
12	$F = \frac{A \times B}{A - B}$	$F = \frac{4 \times 2.4}{4 - 2.4} = \frac{9.6}{1.6} = 6''$
13	$\frac{1}{F} = \frac{1}{B} - \frac{1}{A}$	$\frac{1}{F} = \frac{1}{2.4} - \frac{1}{4}$; $\frac{1}{F} = .4167 - .25 = .1667$; $F = \frac{1}{.1667} = 6''$
14	$\frac{1}{A} = \frac{1}{B} - \frac{1}{F}$	$\frac{1}{A} = \frac{1}{2.4} - \frac{1}{6}$; $\frac{1}{A} = .4167 - .1667 = .25$; $A = \frac{1}{.25} = 4''$
15	$\frac{1}{B} = \frac{1}{A} + \frac{1}{F}$	$\frac{1}{B} = \frac{1}{4} + \frac{1}{6}$; $\frac{1}{B} = .25 + .1667 = .4167$; $B = \frac{1}{.4167} = 2.4''$

CALCULATIONS USING A TABLE OF RECIPROCALS

WHOLE SYSTEM: *(ORIGINAL OBJECT MUST BE AT INFINITY)*

FINAL IMAGE DIA. = FIRST IMAGE × **M** = 1 × .6 = **.6"**

E.F.L. = F.L. OF FIRST LENS × **M** = 8 × .6 = **4.8"**

f/VALUE = f/VALUE OF FIRST LENS × **M** = f/4 × .6 = **f/2.4**

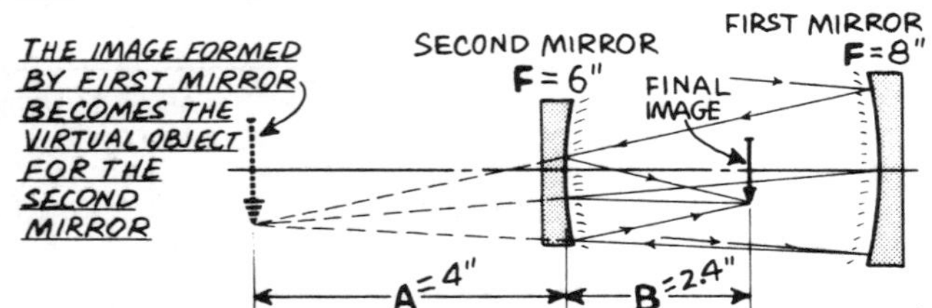

SAME SYSTEM WITH MIRRORS M = .6X

CASE 5

THE NEGATIVE LENS OR MIRROR OF A POS-NEG COMBINATION. VIRTUAL OBJECT LESS THAN F

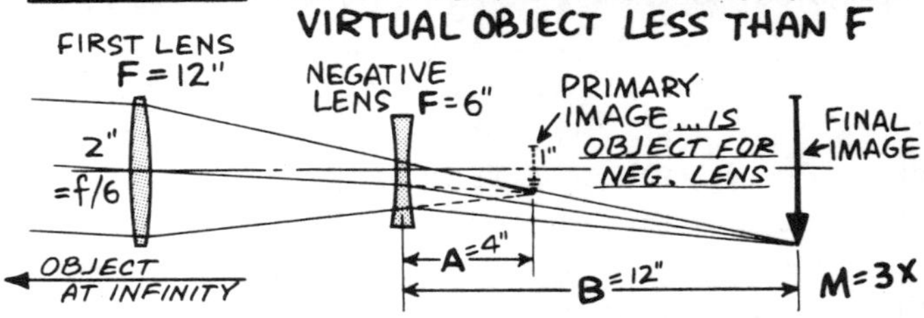

1	$B = (M - 1) \times F$	$B = (3 - 1) \times 6 = 2 \times 6 = 12''$
2	$B = \frac{F \times A}{F - A}$	$B = \frac{6 \times 4}{6 - 4} = \frac{24}{2} = 12''$
3	$B = A \times M$	$B = 4 \times 3 = 12''$
4	$A = \frac{B}{M}$	$A = \frac{12}{3} = 4''$
5	$A = F - \frac{F}{M}$	$A = 6 - \frac{6}{3} = 6 - 2 = 4''$
6	$A = \frac{F \times B}{F + B}$	$A = \frac{6 \times 12}{6 + 12} = \frac{72}{18} = 4''$
7	$M = \frac{B}{A}$	$M = \frac{12}{4} = 3X$
8	$M = \frac{F}{F - A}$	$M = \frac{6}{6 - 4} = \frac{6}{2} = 3X$
9	$M = \frac{F + B}{F}$	$M = \frac{6 + 12}{6} = \frac{18}{6} = 3X$
10	$F = \frac{A \times M}{M - 1}$	$F = \frac{4 \times 3}{3 - 1} = \frac{12}{2} = 6''$
11	$F = \frac{B}{M - 1}$	$F = \frac{12}{3 - 1} = \frac{12}{2} = 6''$
12	$F = \frac{A \times B}{B - A}$	$F = \frac{4 \times 12}{12 - 4} = \frac{48}{8} = 6''$
13	$\frac{1}{F} = \frac{1}{A} - \frac{1}{B}$	$\frac{1}{F} = \frac{1}{4} - \frac{1}{12}$; $\frac{1}{F} = \frac{3}{12} - \frac{1}{12} = \frac{2}{12}$; $F = \frac{12}{2} = 6''$
14	$\frac{1}{A} = \frac{1}{F} + \frac{1}{B}$	$\frac{1}{A} = \frac{1}{6} + \frac{1}{12}$; $\frac{1}{A} = \frac{2}{12} + \frac{1}{12} = \frac{3}{12}$; $A = \frac{12}{3} = 4''$
15	$\frac{1}{B} = \frac{1}{A} - \frac{1}{F}$	$\frac{1}{B} = \frac{1}{4} - \frac{1}{6}$; $\frac{1}{B} = \frac{3}{12} - \frac{2}{12} = \frac{1}{12}$; $B = \frac{12}{1} = 12''$

WHOLE SYSTEM:

FINAL IMAGE DIA. = FIRST IMAGE × **M** = 1 × 3 = **3"**

E.F.L. = F.L. OF FIRST LENS × **M** = 12 × 3 = **36"**

f/VALUE = f/VALUE OF FIRST LENS × **M** = f/6 × 3 = **f/18**

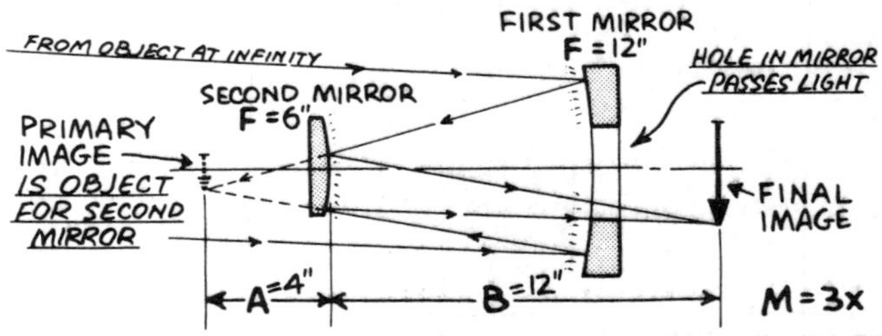

also unnecessary. For example, with an object at infinity, the image distance (B) is always one focal length, and there is no need to make a calculation. Also note that an object at infinity does not permit the calculation of linear magnification. For example, the moon at the focus of a telescope of 50 in. focal length will show an image about 1/2 inch diameter. Comparing this to the actual size of the moon (about 2000 miles) is useless. The kind of magnification used in such cases is angular magnification, which compares the apparent angular size of the object seen through the telescope with the angular size of the object as seen with the unaided eye. Differing from this, linear magnification is the exact ratio of image size to object size--it is B/A for all cases. Usually, magnification is thought of in the sense of being bigger, but linear magnification can indicate same size (1x) or even minification, such as 1/2x, as well as actual enlargement, like 2x.

OTHER CALCULATIONS. Quite often you will have to figure the equivalent or effective focal length of two positive lenses spaced a certain distance apart. Eyepieces are common examples. The E.F.L. calculation is easily made with the equation given in Fig. 4, which also shows an example. The E.F.L. is shortest when the lenses are close together, Fig. 6, increasing as the spacing distance is increased, until "d" is equal to the combined focal lengths, at which spacing the system becomes an astro telescope with virtual image and infinite focal length, Fig. 7.

The calculation for the location of principal planes of two combined positive lenses is shown in Fig. 5, and will be found useful if you design your own Ramsden and Huygens eyepieces. After finding the E.F.L. and principal planes, the doublet is treated very much like a single thick lens. Fig. 8 duplicates the example shown in Fig. 3 except the two lenses are now treated as a single unit. The object at same distance as before is now more than one f.l. from PP1, making the calculation a Case 1 problem.

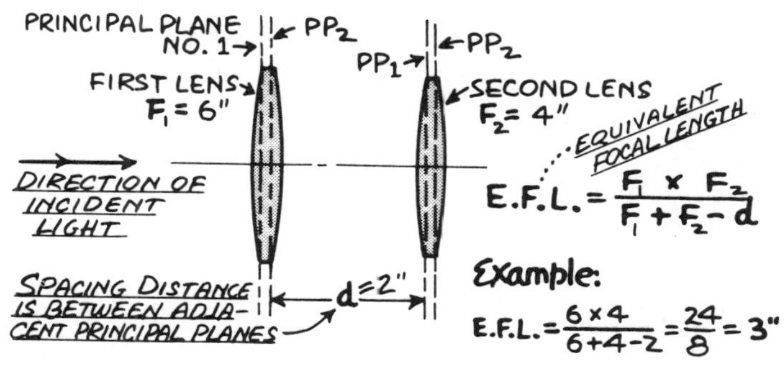

④ E.F.L. OF TWO POSITIVE LENSES

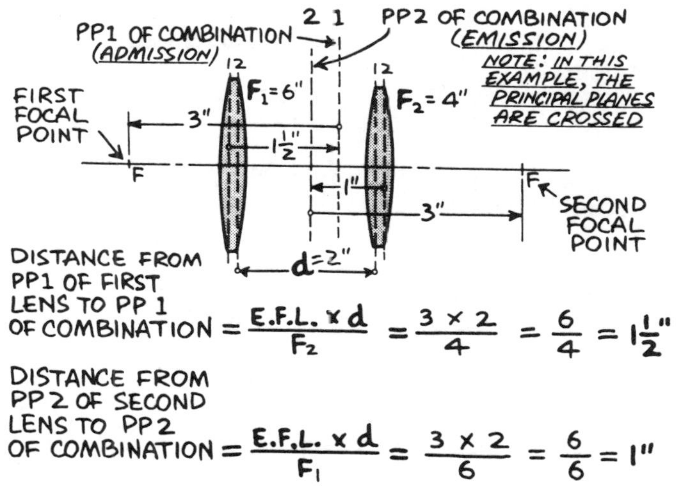

⑤ PRINCIPAL PLANES OF COMBINED LENSES

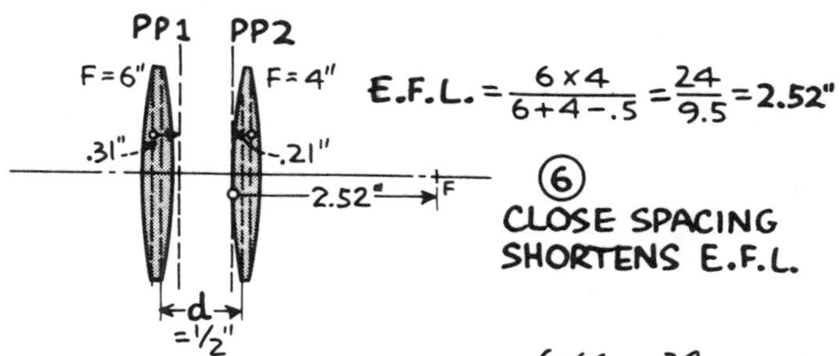

⑥ CLOSE SPACING SHORTENS E.F.L.

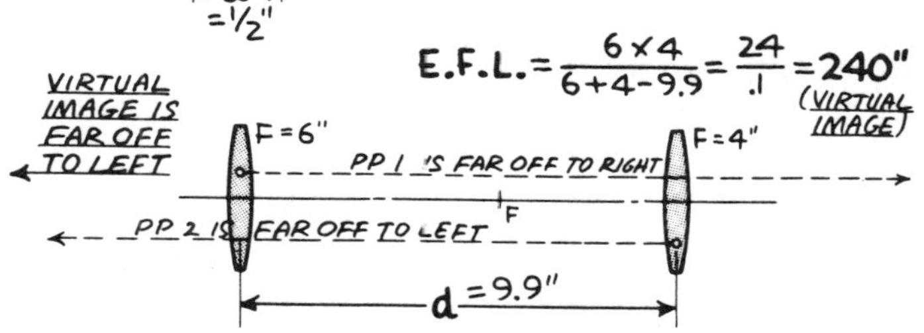

⑦ WIDE SPACING INCREASES E.F.L.
(THIS EXTREME EXAMPLE IS PRACTICALLY A TELESCOPE)

⑧

COMBINED LENSES ARE TREATED LIKE A SINGLE THICK LENS
Example is same as Fig. 3

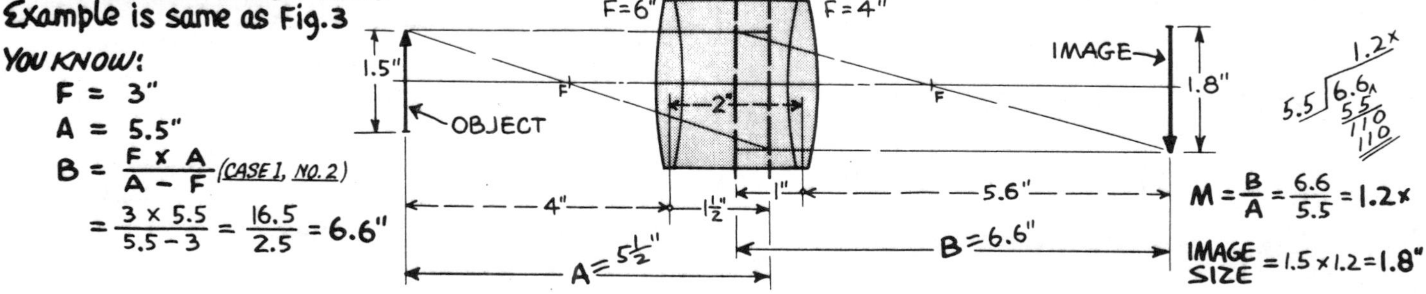

graphical RAY TRACING

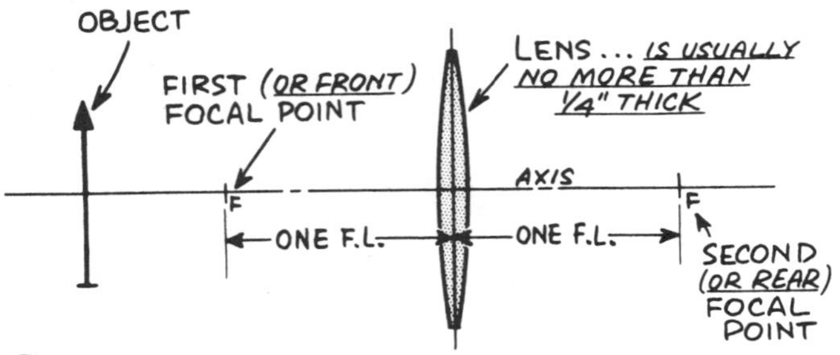

① **LENSES ARE ASSUMED TO BE THIN.** *IF THE LENS IS THIN, A LINE THROUGH THE CENTER PROVIDES A REASONABLY ACCURATE REFERENCE PLANE FOR REFRACTION AND MEASUREMENTS. THE PRINCIPAL PLANES SHOULD BE REFERENCE LINES IF LENS IS THICK*

ONE WAY of solving the various object-image problems in optical design is to make an accurate drawing of the system to scale, after which you can run in the needed light rays graphically by following a few simple rules.

PARALLEL RAY METHOD. You start with the basic drawing shown in Fig. 1. Then, by following the three simple rules shown, Fig. 2, you can trace three light rays through the lens, any two of which will locate the image position. The use

② PARALLEL RAY METHOD - POSITIVE LENS

Rule 1: A LIGHT RAY PASSING THROUGH THE CENTER OF LENS IS NOT DEVIATED

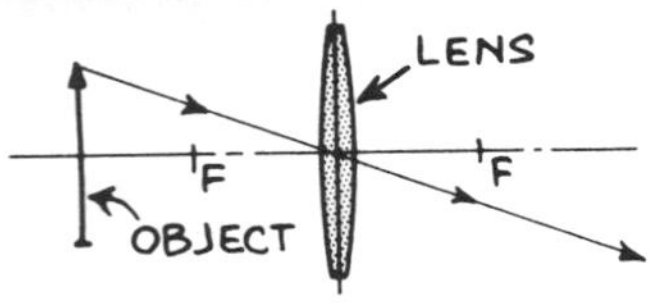

Rule 2: A LIGHT RAY PARALLEL WITH AXIS WILL, AFTER REFRACTION, PASS THRU THE REAR FOCAL POINT

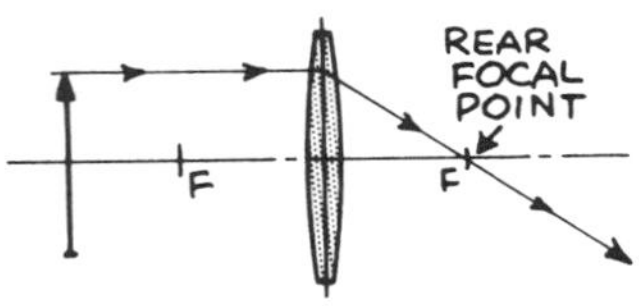

Rule 3: A LIGHT RAY THROUGH THE FIRST FOCAL POINT WILL BE REFRACTED PARALLEL WITH THE AXIS

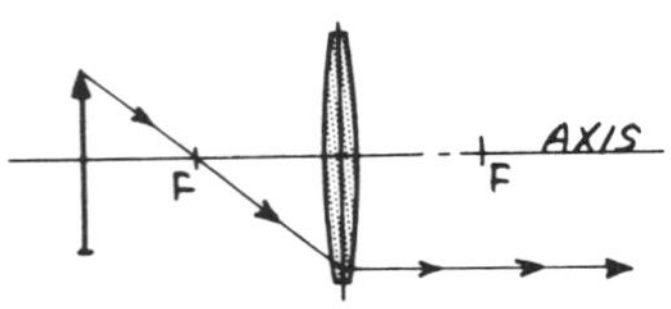

THE INTERSECTION OF ANY TWO OF THE THREE LIGHT RAYS SHOWN WILL LOCATE THE POSITION OF THE IMAGE

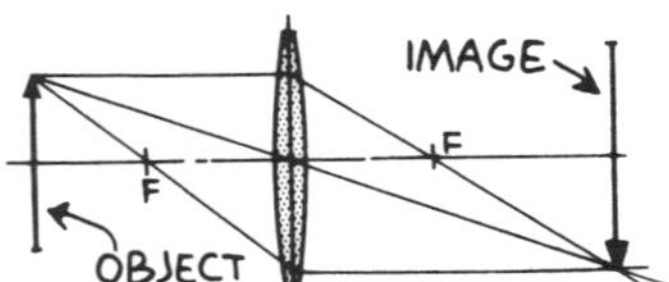

AFTER LOCATING THE IMAGE, YOU CAN "HOOK-UP" ANY NUMBER OF EXTRA RAYS *THE KEY RAYS FOR A SECOND LENS ARE EASILY LOCATED IN THIS MANNER, AFTER WHICH THE FINAL IMAGE IS LOCATED AS BEFORE*

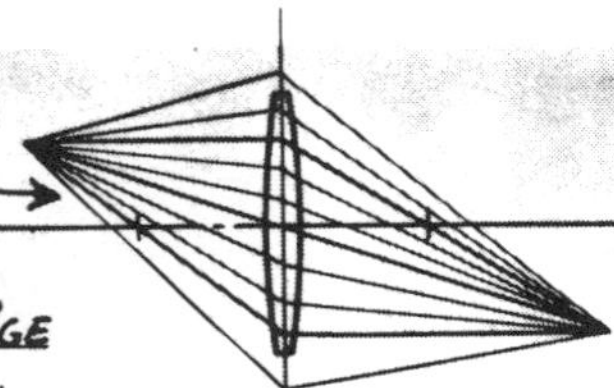

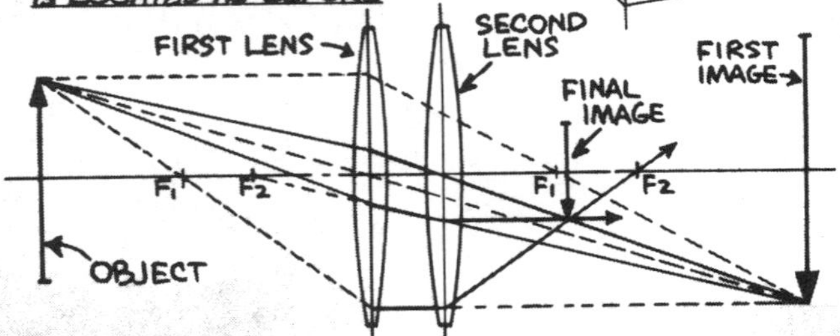

③ **PARALLEL RAY METHOD for TWO LENSES**

④ PARALLEL RAY METHOD-NEGATIVE LENS

Rule 1: A LIGHT RAY PASSING THRU THE CENTER OF LENS IS NOT DEVIATED

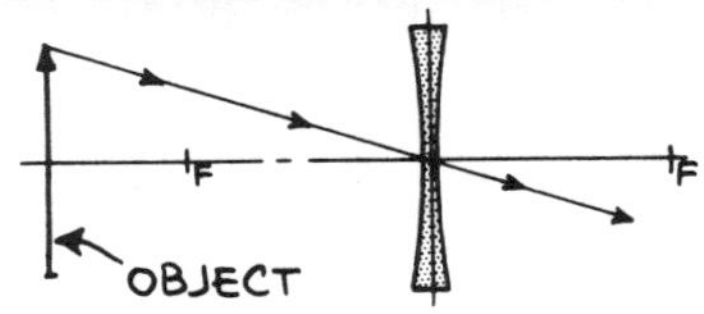

Rule 2: A LIGHT RAY PARALLEL TO THE AXIS WILL, AFTER REFRACTION, APPEAR TO COME FROM THE FRONT FOCAL POINT

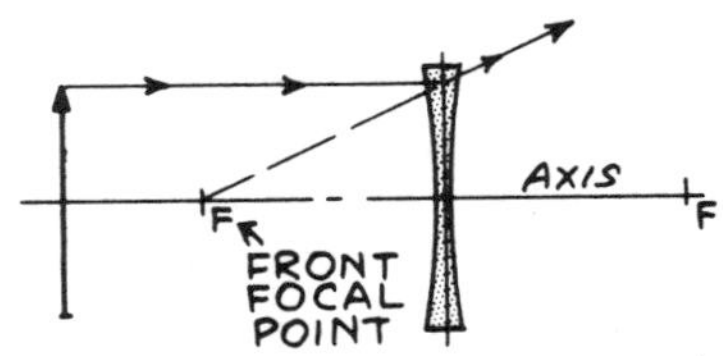

Rule 3: A RAY DIRECTED TOWARD THE REAR FOCAL POINT WILL BE REFRACTED PARALLEL TO THE AXIS. *A BACKWARD EXTENSION OF THIS RAY WILL PASS THRU THE IMAGE POINT*

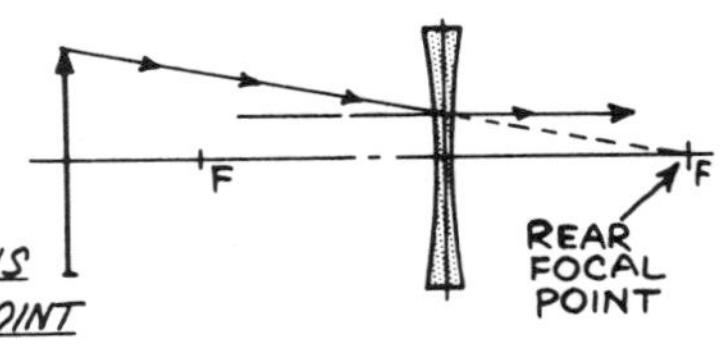

THE INTERSECTION OF ANY TWO OF THE THREE RAYS SHOWN WILL LOCATE THE POSITION OF THE IMAGE

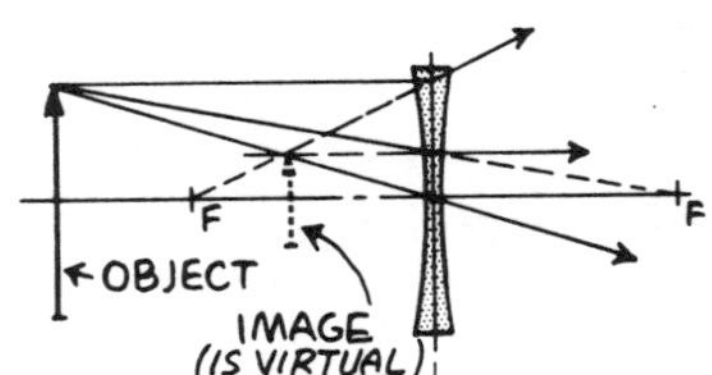

⑤ **NEGATIVE-AND-POSITIVE**

THE VIRTUAL IMAGE FORMED BY THE NEGATIVE LENS BECOMES A REAL OBJECT FOR THE POSITIVE LENS

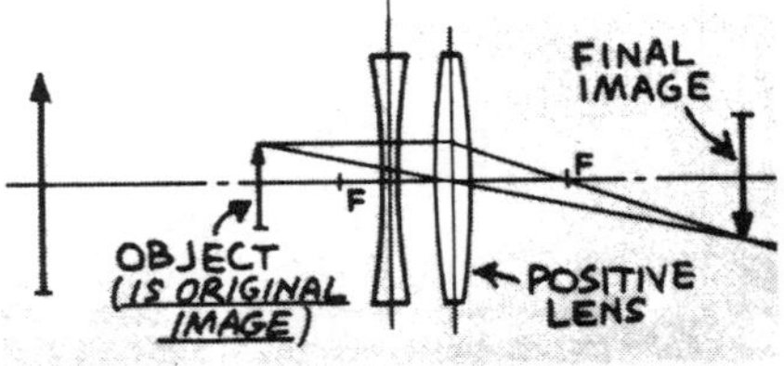

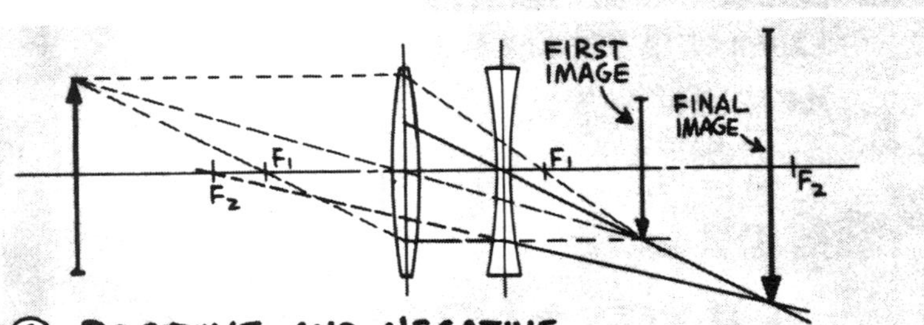

⑥ **POSITIVE-AND-NEGATIVE**

⑦ POSITIVE (CONCAVE) MIRROR

Rule 1: A LIGHT RAY THRU THE CENTER OF CURVATURE IS NOT DEVIATED - *IT IS REFLECTED BACK OVER THE SAME PATH*

Rule 2: A RAY PARALLED TO THE AXIS WILL BE REFLECTED THROUGH THE FOCAL POINT

Rule 3: A RAY PASSING THRU THE FOCAL POINT WILL BE REFLECTED PARALLEL TO AXIS

THE INTERSECTION OF ANY TWO OF THE THREE RAYS SHOWN WILL LOCATE THE POSITION OF IMAGE

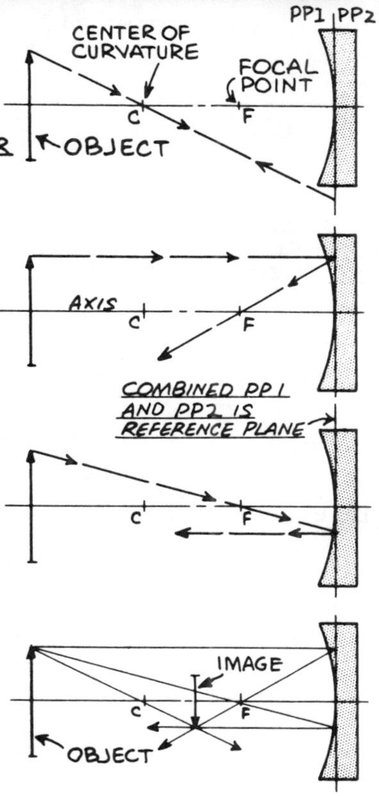

⑧ NEGATIVE (CONVEX) MIRROR

Rule 1: A RAY AIMED AT THE CENTER OF CURVATURE IS NOT DEVIATED

Rule 2: A RAY PARALLEL TO AXIS WILL, AFTER REFLECTION, APPEAR TO COME FROM THE FOCAL POINT

Rule 3: A RAY AIMED AT THE FOCAL POINT WILL BE REFLECTED PARALLEL TO AXIS. *A BACKWARD EXTENSION OF THIS RAY WILL PASS THRU THE IMAGE POINT*

THE INTERSECTION OF ANY TWO RAYS WILL LOCATE THE IMAGE POSITION

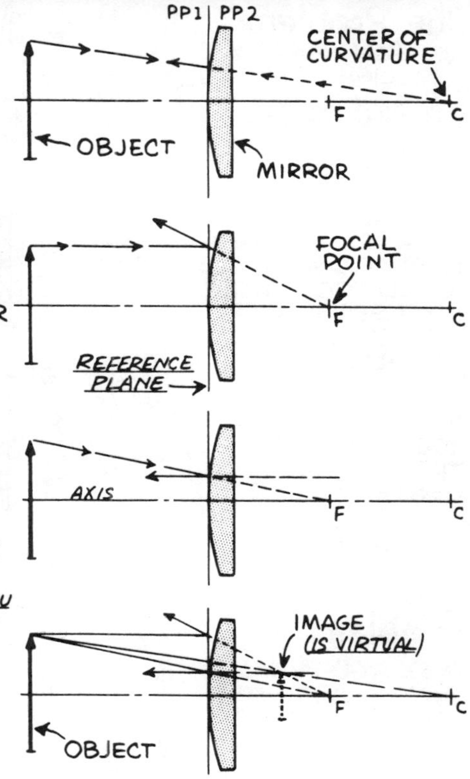

of an entering and emergent parallel ray gives this method its name. Fig. 4 explains the parallel ray method as applied to a negative lens, while Figs. 7 and 8 cover similar situations where a mirror is the optical element.

Once you have located the image, you can run in any number of additional light rays connecting the object point to the image point, Fig. 3. This is the basis for tracing light rays through two or more lenses. The general idea is to locate in the light cone of the first lens, that ray which passes through the center of the second lens. Figs. 3 and 6 are examples. With one parallel light ray already available, you can then plot the image position as formed by the second lens. If the first lens of a pair forms a virtual image to the left, as in Fig. 5 example, the image itself immediately becomes the object for the second lens. Fig. 9 is another example where a virtual image to the left would be immediately available as the object for a second lens.

When two lenses are involved, it is often simpler to calculate the e.f.l. and principal planes of the combo, after which it can be treated very much like a single lens. Between principal planes, all light rays are drawn parallel with the axis. You always draw to PP1 first, since this is the plane of admission. If

⑨ OBJECT AT LESS THAN ONE F.L. FROM A POSITIVE LENS OR MIRROR

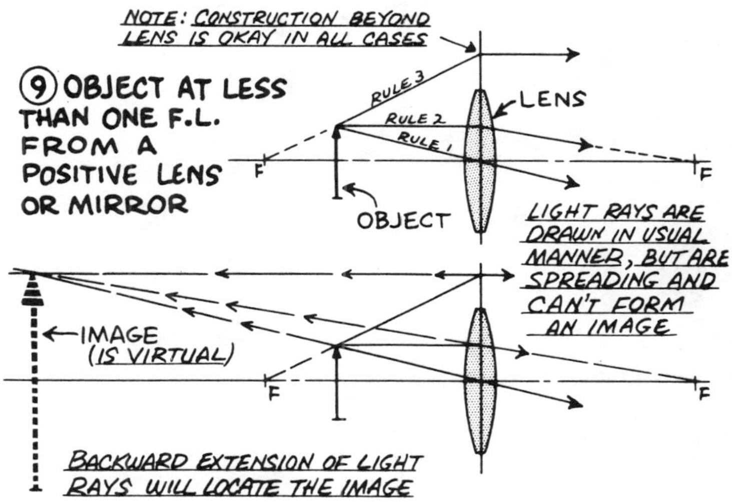

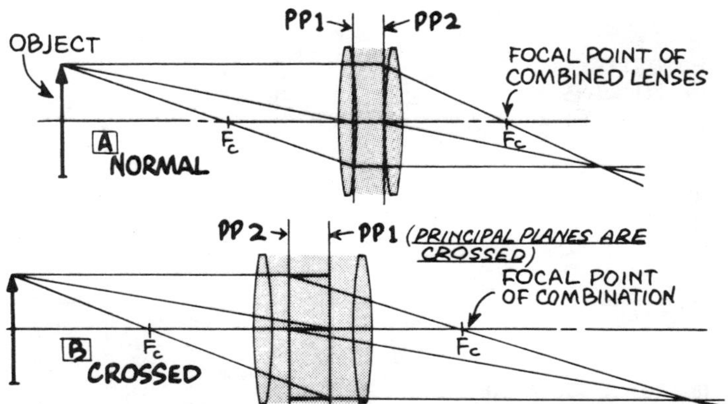

⑩ CONSTRUCTION WITH PRINCIPAL PLANES

OBLIQUE RAY METHOD (11) with a POSITIVE LENS (12) with a POSITIVE (Concave) MIRROR

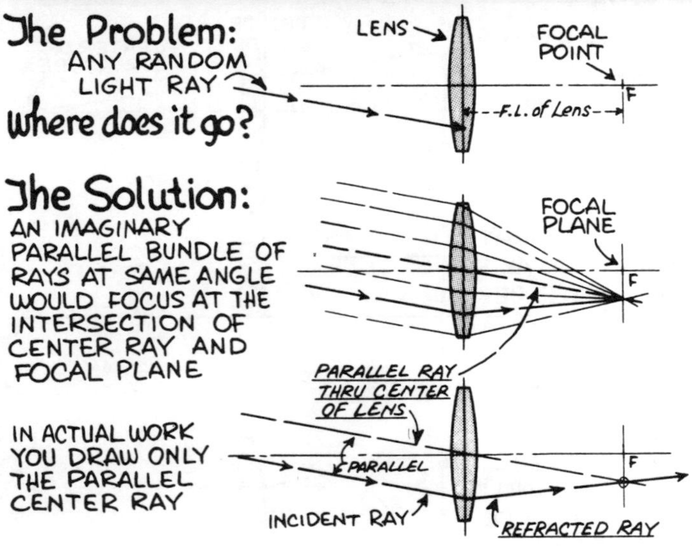

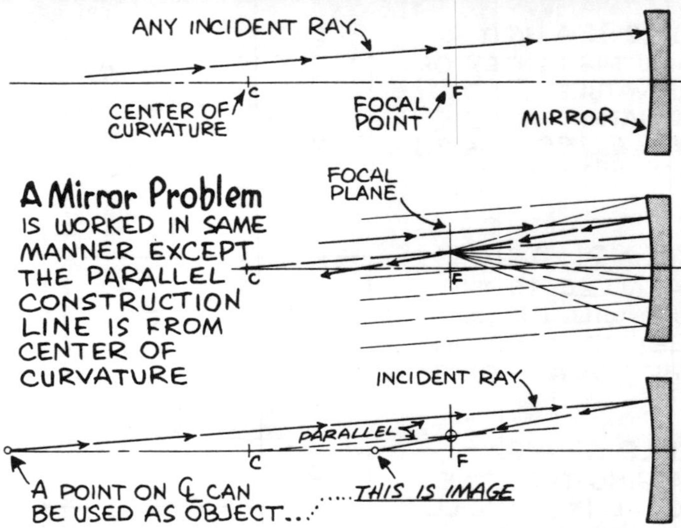

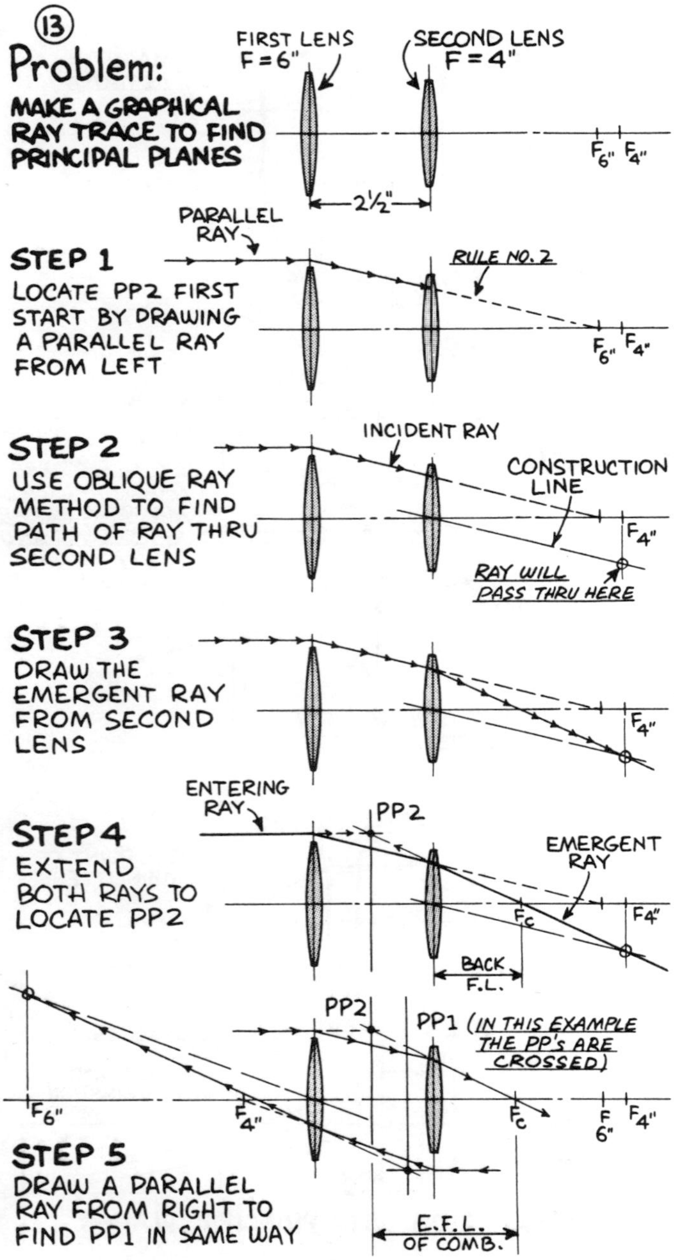

the principal planes are crossed, this means you must then backtrack to PP2, as can be seen in Fig. 10B example.

OBLIQUE RAY METHOD. This is probably the fastest way to trace a single ray through successive lenses. The general idea can be seen in Fig. 11--you don't know where the light ray is going, but you do know that an imaginary ray through the center of the lens would come to a focus at the focal plane. By making this ray parallel to the incident ray, you establish a point through which the incident ray must pass.

Fig. 12 shows the oblique ray method applied to a mirror. This is worked very much like a lens, the main difference being that the construction ray parallel to incident ray is drawn through the center of curvature, as shown. It should be noted in both cases that the intersection of the construction lines marks a point through which the light ray must pass--it is not the location of an image.

PRINCIPAL PLANES. By tracing a single ray from the left, you can locate the plane of emergence (PP2) of a simple lens duplet, as shown in successive steps in Fig. 13. It can be seen that this graphical trace also gives the back focal length and e.f.l., both of which can be scaled from the drawing. The graphical trace itself makes use of the parallel ray method to trace the light ray through the first lens, Step 1, followed by the oblique ray method for the second lens, as shown in Steps 2 and 3.

The plane of admission, PP1, is traced in the same manner by running in a parallel ray from the right, as in Step 5. If the two lenses have the same f.l., the PP's are symmetrical.

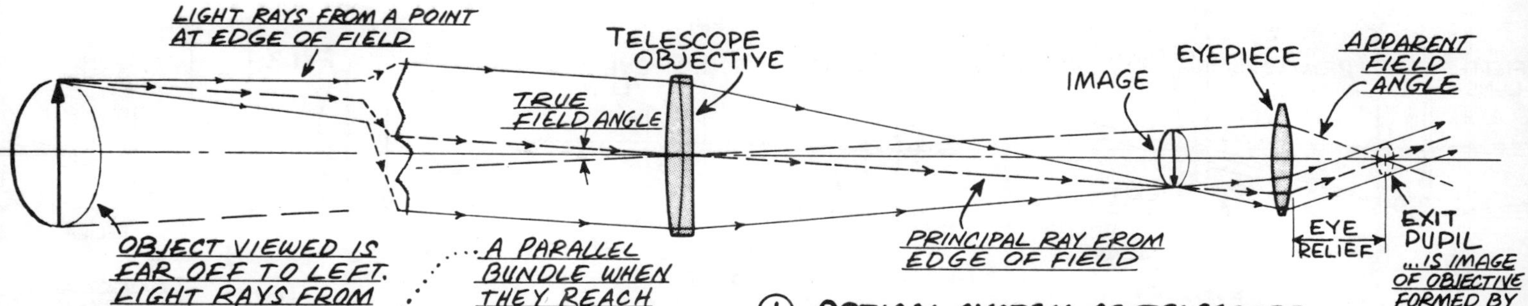

① OPTICAL SYSTEM OF TELESCOPE

EYEPIECES
you can make

WHILE a few hardy amateurs grind and polish their own simple lenses to make eyepieces, the usual "homemade" eyepiece is one assembled from stock lenses. Some common types you can make are shown in Fig. 2. Except for the modest field of less than 40 degrees, homemade eyepieces perform practically as well as complex, expensive oculars.

RAMSDEN. The Ramsden eyepiece is often made of two identical lenses, Fig. 2, since this simplifies construction. However, the eye lens can be made much smaller and still pick up all useful light rays. The original Ramsden is a 1-1-1 design, fully corrected for lateral color, Fig. 4.

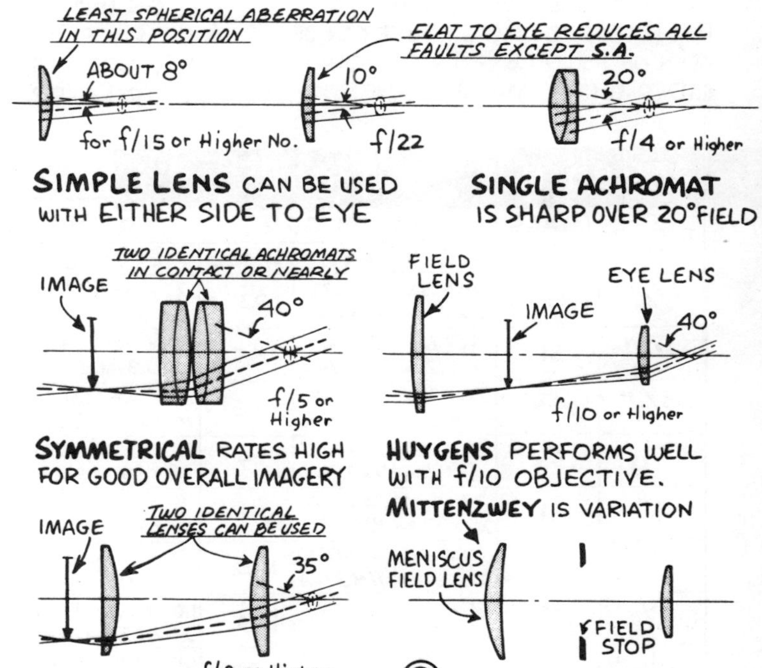

② EYEPIECES YOU CAN MAKE

③ LINEAR FIELD OF TELESCOPE IMAGE WITH VARIOUS EYEPIECES

APPARENT FIELD OF EYEPIECE		FOCAL LENGTH OF EYEPIECE 1/6"	1/5"	1/4"	3/8"	1/2"	5/8"	3/4"	7/8"	1"	1 1/16"	1 1/8"	1 1/4"	1 3/8"	1 1/2"	1 3/4"	2"
	65°	.19"	.23"	.28"	.42"	.57"	.71"	.85"	.99"	1.13"	1.20"	1.27"	1.42"	1.55"	1.70"	1.98"	2.27"
AVERAGE ERFLE	60°	.17	.21	.26	.39	.52	.66'	.79	.92	1.05	1.11	1.18	1.32	1.44	1.57	1.84	2.09
	55°	.16	.19	.24	.36	.48	.60	.72	.84	.96	1.02	1.08	1.20	1.32	1.44	1.68	1.92
AVERAGE KELLNER OR ORTHOSCOPIC	50°	.14	.17	.22	.32	.44	.54	.65	.76	.87	.92	.98	1.08	1.20	1.31	1.52	1.74
AV. SYMMETRICAL OR PLOSSL	45°	.13	.16	.20	.29	.39	.49	.59	.68	.78	.83	.88	.98	1.07	1.17	1.36	1.57
LIMIT for HUYGENS	40°	.12	.14	.17	.26	.35	.44	.52	.61	.70	.74	.79	.88	.96	1.05	1.22	1.40
LIMIT for RAMSDEN	35°	.10	.12	.15	.23	.31	.38	.46	.53	.61	.65	.69	.76	.84	.92	1.07	1.22
USUAL FIELD OF CHEAP TERRESTRIAL SCOPES	30°	.09	.10	.13	.20	.26	.33	.39	.46	.52	.55	.59	.65	.72	.78	.91	1.05
	25°	.07	.09	.11	.16	.22	.27	.33	.39	.44	.47	.49	.55	.60	.66	.77	.87
SINGLE ACHROMAT	20°	.06	.07	.09	.13	.17	.22	.26	.31	.35	.37	.39	.44	.48	.52	.61	.70
SINGLE SIMPLE LENS	10°	.03	.03	.04	.07	.09	.11	.13	.15	.17	.18	.19	.21	.23	.26	.30	.35

Example: YOU PLAN TO USE A 50° EYEPIECE OF 1" F.L. WHAT IS IMAGE DIA.? **SOLUTION:** IN LEFT COL., LOCATE 50° A.F. ON SAME LINE UNDER 1" F.L, READ .87" (7/8") DIA. OF FIELD AT IMAGE PLANE

GRAY AREA: STANDARD 1¼" EYEPIECE HOLDER WILL NOT COVER. YOU MUST MAKE OVERSIZE FOCUSING EYEPIECE HOLDER

HUYGENS EYEPIECE ONLY: READ PRIMARY IMAGE DIA. FROM TABLE... BUT READ FINAL IMAGE (STOP DIA.) 2 LINES DOWN

Ex. 1" F.L. HUYGENS, 40° A.F.
PRIMARY IMAGE = .70"
FINAL IMAGE = .52" (APPROX.)

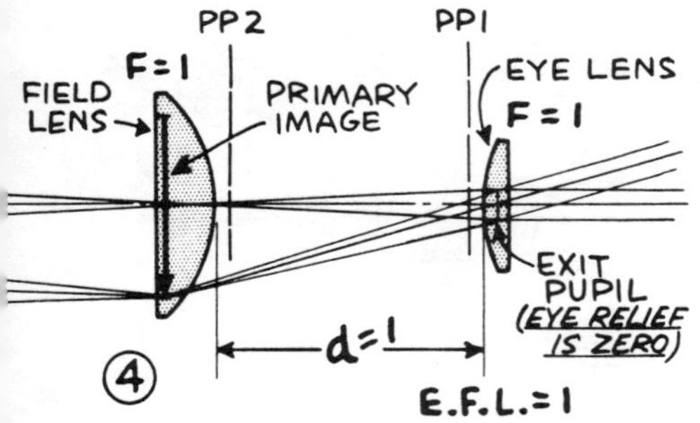

1-1-1 RAMSDEN IS CORRECTED FOR LATERAL COLOR BUT EYE RELIEF IS ZERO... DUST CAN BE SEEN ON FIELD LENS WHICH COINCIDES WITH IMAGE PLANE

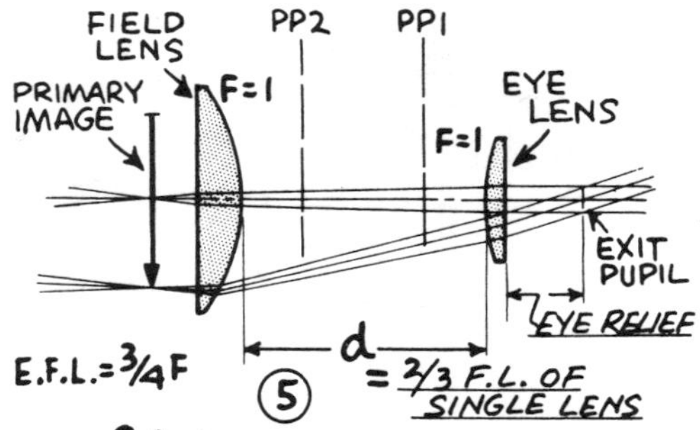

1-1-2/3 RAMSDEN IS STOCK DESIGN WITH LENSES SAME F.L. AND SPACE 2/3 F.L. OF SINGLE LENS. SPACING LESS THAN ½(F+F) CAUSES LATERAL COLOR

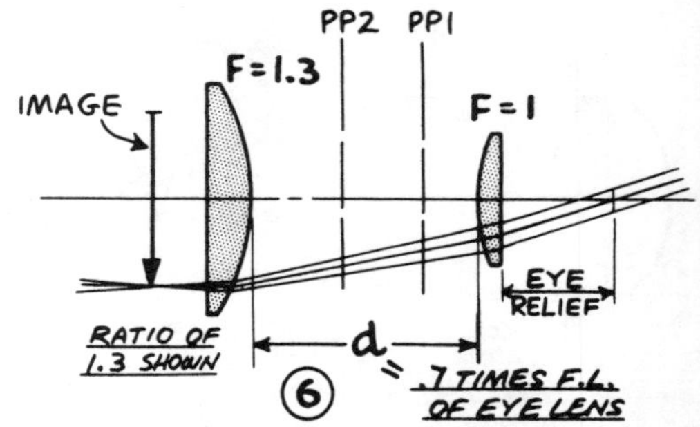

LONGER F.L. FIELD LENS IS BEST FOR POWERS OVER 10X ...GIVES BETTER COMA CORRECTION ...LONGER EYE RELIEF

POWER	RATIO
10X OR LESS	1 OR 1.1
10X TO 40X	1.1 OR 1.2
30X OR MORE	1.2 OR 1.3

⑦ RAMSDEN Specifications

E.F.L.	IMAGE 35° FIELD	RATIO OF 1			RATIO OF 1.1			RATIO OF 1.2			RATIO OF 1.3		
		FIELD	EYE	SPACE	FIELD	EYE	SPACE	FIELD	EYE	SPACE	FIELD	EYE	SPACE
1/4"	.15"	.24×.32" 6×8mm	.14×.32" 4×8mm	.23"	.25×.35" 6×9mm	.14×.32" 4×8mm	.22"	.25×.37" 6×10mm	.15×.31" 4×8mm	.22"	.26×.40" 7×10mm	.15×.31" 4×8mm	.22"
1/2"	.31	.42×.65" 11×17mm	.22×.65" 6×17mm	.45	.43×.70" 11×18mm	.23×.63" 6×16mm	.45	.44×.75" 11×19mm	.24×.62" 6×16mm	.44	.46×.80" 12×21mm	.25×.62" 7×16mm	.43
3/4"	.46	.60×.98" 15×25mm	.30×.98" 8×25mm	.68	.61×1.05" 16×27mm	.31×.95" 8×24mm	.67	.64×1.12" 16×29mm	.32×.94" 8×24mm	.66	.66×1.20" 17×31mm	.34×.92" 9×24mm	.64
7/8"	.53	.69×1.13" 18×29mm	.33×1.13" 9×29mm	.79	.71×1.22" 18×31mm	.35×1.10" 9×28mm	.77	.73×1.30" 19×33mm	.36×1.09" 9×28mm	.76	.76×1.39" 20×36mm	.38×1.07 10×27mm	.75
1"	.61	.78×1.30" 20×33mm	.32×1.30" 10×33mm	.91	.81×1.40" 21×36mm	.39×1.27" 10×33mm	.89	.83×1.50" 21×38mm	.41×1.25" 11×32mm	.87	.86×1.60" 22×41mm	.43×1.23" 11×31mm	.86
1 1/8"	.69	.82×1.46" 22×37mm	.41×1.46" 11×37mm	1.02	.90×1.57" 23×40mm	.43×1.42" 11×36mm	1.00	.92×1.68" 24×43mm	.45×1.40" 11×36mm	.98	.96×1.79" 25×46mm	.47×1.38" 12×35mm	.96
1 1/4"	.76	.96×1.63" 25×42mm	.45×1.63" 12×42mm	1.14	1.00×1.75" 26×45mm	.47×1.59 12×41mm	1.11	1.02×1.87" 26×47mm	.50×1.56" 13×39mm	1.09	1.06×2.00 27×51mm	.52×1.54" 14×39mm	1.08
1 1/2"	.92	1.14×1.95" 29×50mm	.53×1.95" 14×50mm	1.37	1.19×2.10" 30×54mm	.56×1.91" 14×49mm	1.34	1.22×2.25 31×57mm	.59×1.87" 15×47mm	1.31	1.26×2.40 32×61mm	.62×1.85" 16×47mm	1.29
1 3/4"	1.07	1.32×2.28" 34×58mm	.60×2.28" 16×58mm	1.59	1.37×2.45 35×62mm	.64×2.22 16×56mm	1.56	1.41×2.62 36×67mm	.67×2.19" 17×56mm	1.52	1.46×2.80 37×71mm	.71×2.15" 18×55mm	1.51

LENSES ARE LISTED BY DIA. AND F.L. IN INCHES AND MM. ALL DIAMETERS ARE APPROX. CL. APERTURE PLUS 1/16" FOR MOUNTING

***FOR F.L. NOT LISTED:** USE VALUES FOR 1" F.L. FROM TABLE AND MULTIPLY BY F.L. YOU WANT*

Worksheet ⑧

YOU WANT 1" (25MM) F.L. RAMSDEN FOR 48X. PREFERABLE FIELD-EYE LENS RATIO IS 1.2 OR 1.3 BUT YOU CAN USE WHOLE RANGE:

DATA FROM TABLE ABOVE	RATIO OF 1	RATIO OF 1.1	RATIO OF 1.2	RATIO OF 1.3
FIELD LENS	20×33MM	21×36MM	21×38MM	22×41MM
EYE LENS	10×33MM	10×33MM	11×32MM	11×31MM

AVAILABLE LENSES (EDMUND – 1965)

FIELD:
19.5 × 38MM
22 × 40
23 × 41 ← BEST
19 × 42
26 × 42

EYE:
13×13 × 31MM
11×15 × 32MM
12.5 × 35MM ← BEST

RATIO = $41 \div 35 = 1.2$ Approx. (OK)

SPACE = 35 × .7 = 24.5MM (SAY 25MM)

$$\text{E.F.L.} = \frac{41 \times 35}{41+35-25} = \frac{1435}{51} = 28\text{MM}$$

(IS LONG BUT COULD BE USED)

TRY LESS SPACE: 35 × .6 = 21MM

$$\text{E.F.L.} = \frac{41 \times 35}{41+35-21} = \frac{1435}{55} = 26\text{ MM}$$

(OK)

This construction is excellent for an erecting system or for a projecting eyepiece, but it is poor for ordinary telescope use in that there is no eye relief.

To provide eye relief and also move the image away from the field lens, a spacing of 2/3 the f.l. of the individual lens is often used, and this is the standard Ramsden, Fig. 5. The eyepiece is improved slightly by making the field lens a little longer focal length, Fig. 6. The specifications in Fig. 7 Table can be juggled a little to suit, an example being shown in Fig. 8. Spacing can be reduced to as little as 50% of the f.l. of the eye lens without introducing an excessive amount of lateral color.

HUYGENS. In Huygens' time, (Christian Huygens, 1629-1695), the refractor was the only telescope of note, and this eyepiece was designed

⑨ GENERAL FEATURES OF THE HUYGENS EYEPIECE

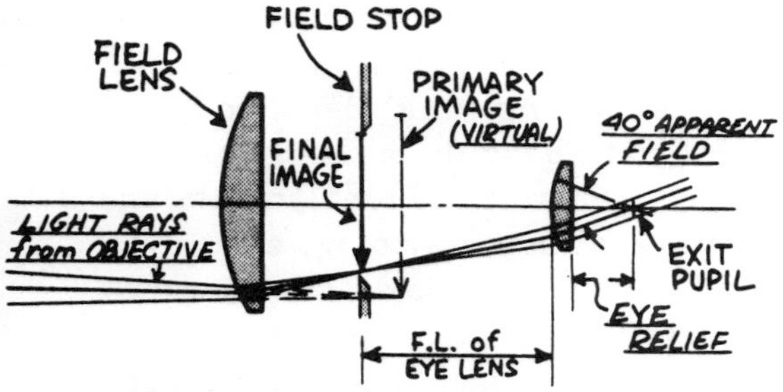

LIGHT PATH: WITHOUT EYEPIECE, THE TELESCOPE OBJECTIVE WOULD FORM PRIMARY IMAGE AS SHOWN. THE FIELD LENS INTERCEPTS THE LIGHT RAYS AND FORMS A REAL FINAL IMAGE WHICH IS VIEWED BY THE EYE LENS

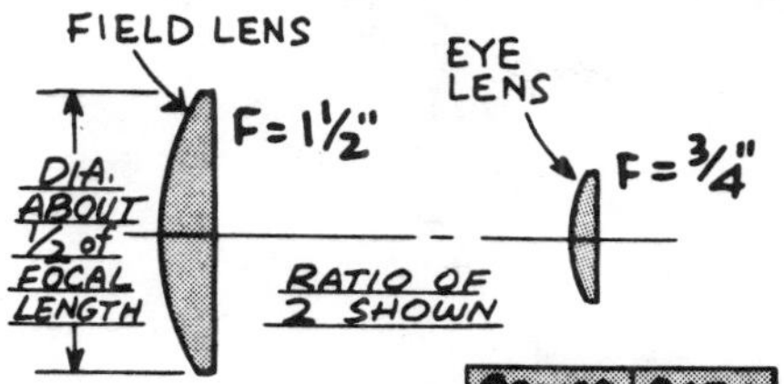

RATIO: FIELD LENS IS ALWAYS LONGER F.L. THAN EYE LENS. BEST RATIO DEPENDS ON POWER OF SCOPE AND SHOULD BE SELECTED FROM TABLE →

POWER	RATIO
3x	1.3
4x	1.35
5x	1.4
10x	1.7
20x	2.0
30x	2.3
50x	2.6
or More	3.0

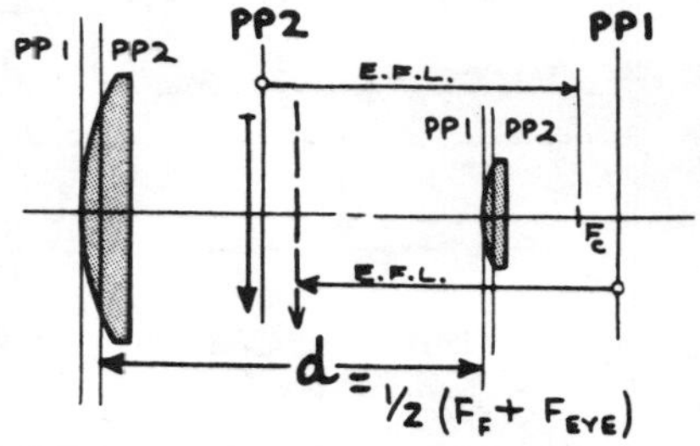

SPACING: LATERAL COLOR IN ANY SIMPLE EYEPIECE IS CORRECTED BY SPACING OF ½ THE COMBINED FOCAL LENGTHS. (*LONGITUDINAL COLOR IS NOT CORRECTED*). SPACING IS MEASURED BETWEEN ADJACENT PRINCIPAL PLANES

⑩ HUYGENS Specifications - 40° FIELD

E.F.L.	RATIO OF 2				RATIO OF 2.3				RATIO OF 2.6				RATIO OF 3			
	FIELD	EYE	SPACE	STOP	FIELD	EYE	SPACE	STOP	FIELD	EYE	SPACE	STOP	FIELD	EYE	SPACE	STOP
1/4"	.25 x .38" 7 x 10 MM	.13 x .19" 4 x 5 MM	.28"	.13" D.	.25 x .42" 6 x 11 MM	.12 x .18" 4 x 5 MM	.30"	.13" D.	.25 x .45" 7 x 12 MM	.14 x .17" 4 x 5 MM	.31"	.12" D.	.25 x .50" 7 x 13 MM	.14 x .17" 4 x 5 MM	.33"	.12" D.
1/2"	.44 x .75" 11 x 19 MM	.20 x .37" 5 x 10 MM	.56	.26	.44 x .83 11 x 21 MM	.20 x .36" 5 x 9 MM	.59	.25	.44 x .90 11 x 23 MM	.22 x .35" 6 x 9 MM	.62	.25	.45 x 1.00 12 x 26	.22 x .33" 6 x 9 MM	.66	.24
3/4"	.62 x 1.13" 16 x 29 MM	.27 x .56" 7 x 14 MM	.84	.39	.62 x 1.25 16 x 32 MM	.27 x .54" 7 x 14 MM	.89	.38	.62 x 1.35 16 x 34	.29 x .52" 8 x 13	.93	.37	.64 x 1.50 17 x 38	.29 x .50 7 x 13 MM	.99	.36
7/8"	.71 x 1.31" 18 x 34	.30 x .65" 8 x 17 MM	.98	.46	.71 x 1.44 18 x 37	.30 x .63 8 x 16 MM	1.04	.44	.71 x 1.57 18 x 40	.33 x .60" 9 x 15 MM	1.08	.43	.74 x 1.74" 19 x 44	.33 x .58" 9 x 15 MM	1.16	.41
1"	.81 x 1.50 21 x 38 MM	.34 x .75 9 x 19 MM	1.12	.53	.81 x 1.66" 21 x 42	.34 x .72 9 x 18	1.19	.51	.81 x 1.80 21 x 46	.37 x .69" 10 x 18	1.24	.49	.84 x 2.00 22 x 51	.37 x .67 10 x 17 MM	1.33	.47
1 1/8"	.90 x 1.68 23 x 43	.37 x .84" 10 x 22	1.25	.59	.90 x 1.86 23 x 47	.37 x .81" 10 x 21	1.33	.57	.90 x 2.02 23 x 52	.41 x .77" 11 x 20 MM	1.39	.55	.93 x 2.25 24 x 57	.41 x .75 11 x 19 MM	1.49	.53
1 1/4"	1.00 x 1.88 25 x 48	.41 x .94" 11 x 24	1.40	.66	1.00 x 2.08 25 x 53	.41 x .90" 11 x 23	1.49	.64	1.00 x 2.25 25 x 57	.45 x .86 11 x 22	1.55	.61	1.04 x 2.50 27 x 64	.45 x .84 11 x 22	1.66	.59
1 1/2"	1.19 x 2.25 30 x 57 MM	.48 x 1.12" 12 x 29 MM	1.69	.80	1.19 x 2.49 30 x 63	.48 x 1.08 13 x 28	1.78	.76	1.19 x 2.70 30 x 69	.53 x 1.04 14 x 27 MM	1.86	.74	1.23 x 3.00 31 x 76	.53 x 1.00 14 x 26	1.99	.71

FOR FOCAL LENGTHS NOT LISTED, RATIO OF 2 CAN BE CALCULATED BY THE FOLLOWING:

Example: 1" F.L.

F.L. EYE = ¾ E.F.L. = ¾ × 1 = ¾"

F.L. FIELD = 1½ E.F.L. = 1½ × 1 = 1½"

or 2 TIMES EYE = 2 × ¾ = 1½"

SPACE = 1⅛ E.F.L. = 1⅛ × 1 = 1⅛"

or 3/2 TIMES EYE = 3/2 × ¾ = 1⅛"

FOR FOCAL LENGTHS NOT LISTED, USE VALUES FOR 1" F.L. AS MULTIPLYING FACTORS

Example → E.F.L. to be .8" RATIO OF 2.3

FIELD LENS = 1.66 × .8 = 1.33" F.L.
EYE LENS = .72 × .8 = .58" F.L.
SPACE = 1.19 × .8 = .95"
STOP DIA. = .51 × .8 = .41"

You Can CHECK $\text{E.F.L.} = \frac{F \times F}{F + F - d} = \frac{1.33 \times .58}{1.33 + .58 - .95} = \frac{.7714}{.96} = .803$"

GENERAL RULES FOR RATIO OF 3:

Example: 1" F.L.

EYE = ⅔ E.F.L. = ⅔ × 1 = ⅔" F.L.

FIELD = 2 E.F.L. = 2 × 1 = 2" F.L.

or 3 TIMES EYE = 3 × ⅔ = 2"

SPACE = 4/3 EFL = 4/3 × 1 = 1⅓"

or 2 TIMES EYE = 2 × ⅔ = 1⅓"

STOP - *FOR RATIO OF 3... IS MIDWAY BETWEEN LENSES*

solely for the refractor with which it functions rather well. Specifically, the Huygens is intended for use with telescopes of narrow aperture--f/10 or higher f/number. The best feature of the Huygens is that it can be fully corrected for lateral color; in other respects it is slightly inferior to the Ramsden. The field lens is always longer focal length than the eye lens. The ratio to use is that which gives the best correction for coma, this being dependent on the power of the telescope, or, more exactly, the number of focal lengths of the eyepiece contained in the distance from eyepiece to objective. If you start out with the proper ratio for a certain power, Fig. 9, the eyepiece will be well-corrected for coma. Fig. 10 Table gives specifications for various focal lengths. These are general calculated values which may be modified slightly to suit such lenses as may be available. For astro use, the field lens-eye lens ratio is preferably 2.3 or more for best coma correction, longest eye relief.

ACHROMATIC DUPLETS. One of the more popular and practical homemade eyepieces is based on the general idea of using two achromats. The

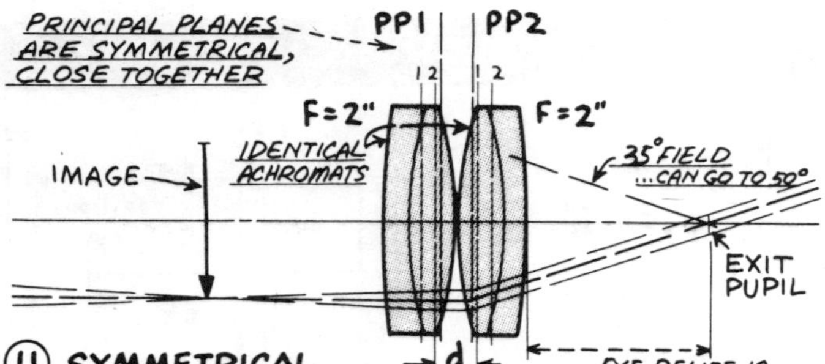

(11) SYMMETRICAL

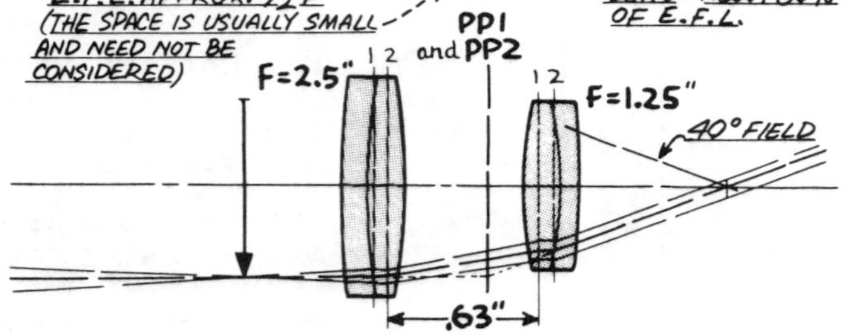

(12) UNSYMMETRICAL DUPLET (SPECIFICATIONS FOR 1" F.L.)

construction is usually symmetrical, Fig. 11, with the two lenses in contact or nearly so. Many excellent eyepieces can be made in this manner. Unsymmetrical construction, Fig. 12, allows you to use a big field lens for light pickup in combination with a smaller and stronger eye lens. The ratio of focal lengths and the spacing can be almost anything; specifications given in Fig. 12 are popular because they are simple and result in a single principal plane for the combination. In any case, the lenses used should be good-quality, conventional achromats, i.e., duplets designed for incident parallel light, corrected for spherical aberration and achromatized for the C (red) and F (blue) lines of the spectrum. Not all war surplus achromats comply with these basic specifications.

DESIGN PROCEDURE. For conventional Huygens or Ramsden eyepieces, the specific lenses needed are given in the tabular data. For other types, you can find diameter in a roundabout way from Fig. 3 Table, which gives the image size. For most eyepieces the field lens must be a little larger than the image diameter. Note that for the Huygens, the final image, which is also the field stop, is slightly smaller than the primary image.

Specifications given in other optical books may be followed if desired, and if radius only is given you can readily convert to f.l. by using the formula given in Fig. 14. When the f.l. specified is not what you want, it is easy to change to any desired f.l. by the method shown in Fig. 13. This can also be applied to any single lens, such as a telescope objective, or to any optical system.

CHANGE (13) Specifications

YOU WILL FIND EYEPIECE DESIGNS IN MANY OPTICAL BOOKS. *IF THE F.L. IS NOT SUITABLE, YOU CAN CHANGE TO ANY DESIRED F.L. BY MULTIPLYING ALL "SPECS" BY A FACTOR:*

FACTOR = F.L. YOU WANT / ORIGINAL F.L.

Example: Conrady,* p.498

RAMSDEN EYPC., RATIO 1.25
F.L. = 1.56"

Conrady's specifications:
FIELD LENS F.L. = 2.42"
EYE LENS F.L. = 1.93"
SPACE = 1.35"

YOU WANT, SAY, 1" F.L., SO MULTIPLYING FACTOR IS:

$$\frac{\text{F.L. YOU WANT} \ldots 1}{\text{ORIGINAL SPEC.} \ldots 1.56} = .64$$

THEN:
FIELD = 2.42 × .64 = 1.55"
EYE = 1.93 × .64 = 1.24"
SPACE = 1.35 × .64 = .86"

check

$$\text{E.F.L.} = \frac{1.55 \times 1.24}{1.55 + 1.24 - .86} = \frac{1.92}{1.93} = 1'' \text{ APPROX. OK}$$

*APPLIED OPTICS AND OPTICAL DESIGN – A. E. Conrady

(14)

Conversion Formulas

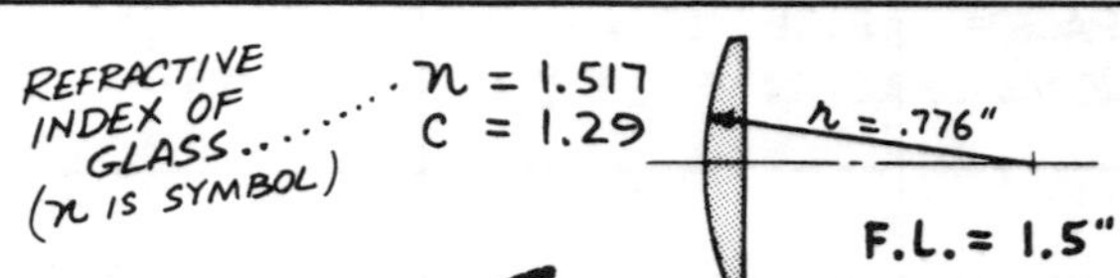

PLANO-CONVEX EYEPIECE LENSES ARE OFTEN SPECIFIED BY RADIUS OR CURVATURE, RATHER THAN F.L.

CONVERT AS NEEDED BY USING THESE FORMULAS

THE APPROX. CONVERSION OF RADIUS TO F.L. IS SIMPLY:

F.L. = 2r

THIS IS EXACT ONLY IF GLASS HAS INDEX OF 1.50. OTHERWISE, USE FORMULA NO. 2

	KNOWN	FIND	Formula	Example
1	radius	curvature	$c = \frac{1}{r}$	$c = \frac{1}{.776} = 1.29$ ..... 1.2887 IS EXACT
2	radius	F.L.	$\text{F.L.} = \frac{r}{n-1}$	$\text{F.L.} = \frac{.776}{1.517-1} = \frac{.776}{.517} = 1.5''$
3	curvature	radius	$r = \frac{1}{c}$	$r = \frac{1}{1.29} = .775''$ ... APPROX. OK should use four decimal places
4	curvature	F.L.	$\text{F.L.} = \frac{1}{(n-1) \times c}$	$\text{F.L.} = \frac{1}{.517 \times 1.29} = \frac{1}{.667} = 1.5''$
5	F.L.	curvature	$c = \frac{1}{(n-1) \times F}$	$c = \frac{1}{.517 \times 1.5} = \frac{1}{.776} = 1.29$
6	F.L.	radius	$r = F \times (n-1)$	$r = 1.5 \times .517 = .776''$

The REFRACTOR

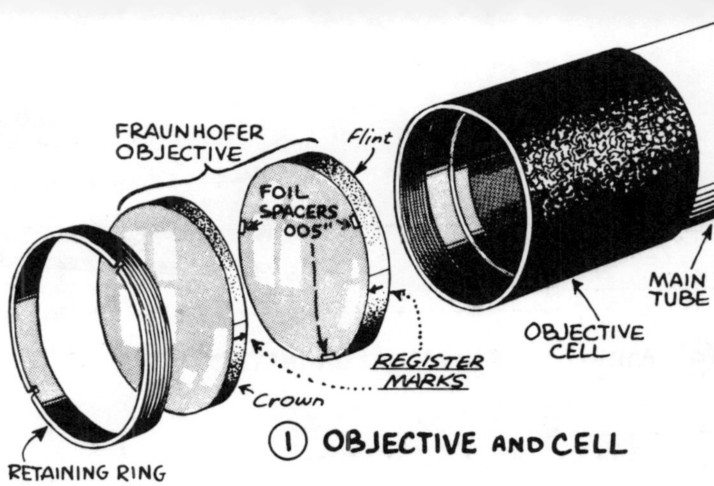

① OBJECTIVE AND CELL

THE TYPICAL refractor objective is a broken-contact or air-spaced style, usually purchased complete with cell, Fig. 1. Sold unmounted, the same lens has register marks to show how the two elements should be put together. Adhesive foil is supplied, from which you cut three pieces about 1/16 x 3/16-inch, attaching these to the concave side of the flint element, as shown. Practically all air-spaced objectives are now coated on all four surfaces, practically eliminating the 18 to 20 percent light loss from four uncoated surfaces.

Small achromats up to 2 in. diameter are usually cemented for convenience in handling and mounting. An equi-convex front lens, identical internal radii and a flat-back flint are conditions often specified for inexpensive achromats, Fig. 2. The same drawing shows the air-contact type used in many inexpensive telescopes. Over 3 in. diameter, air-spacing becomes almost necessary to eliminate the risk of cement failure caused by unequal glass expansion. In top-quality air-spaced objectives, the Fraunhofer-type is a favorite of long standing. The Clark is also excellent. Both of these are corrected for coma, which is usually neglected in a cemented achromat.

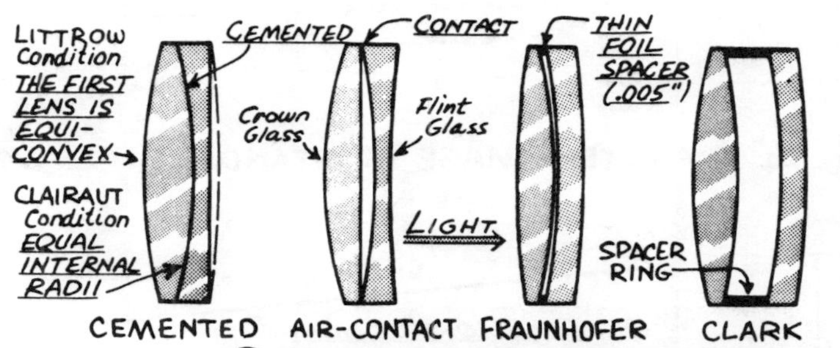

② ACHROMATIC TELESCOPE OBJECTIVES

COLOR CORRECTION. With two glass elements to work with, the lens designer can correct an achromat for any two colors. For a visual instrument like the telescope, the two best colors are the F-line in the blue part of the spectrum and the C-line in the red. These two lines bracket that portion of the spectrum to which the eye is most sensitive. There is, of course, a residual of uncorrected color, Fig. 3, both between and beyond the CF lines. This secondary color will put a hairline of purple light around a bright star, but unless you look you will rarely be aware of the color fringe.

CF achromatism is quite satisfactory for ordinary photography on pan film. A yellow filter is often specified, but this need is universal for long-range daytime photography with any lens. The ordinary photographic objective leans a bit more to the blue end of the spectrum to favor a wider range of film emulsions. An astrographic object glass is often corrected for blue only and can be used only with blue-sensitive, i.e., color-blind film. Some aerial camera lenses are achromatized for the high red end of the spectrum

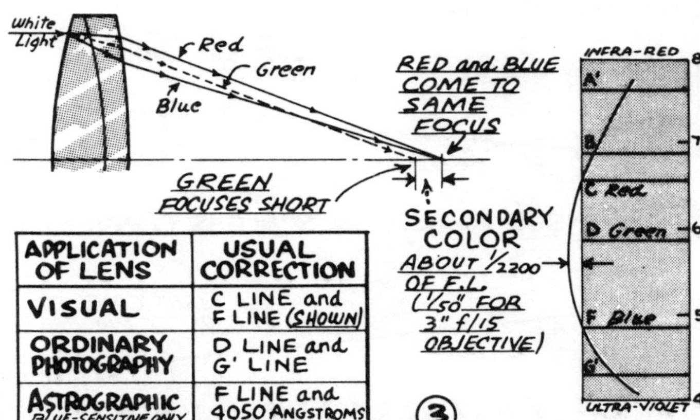

APPLICATION OF LENS	USUAL CORRECTION
VISUAL	C LINE and F LINE (SHOWN)
ORDINARY PHOTOGRAPHY	D LINE and G' LINE
ASTROGRAPHIC (BLUE-SENSITIVE ONLY)	F LINE and 4050 ANGSTROMS
AERIAL PHOTOGRAPHY	A' LINE and D LINE *

* DG' ACHROMATISM ALSO USED

③ COLOR CORRECTION

for use with infra-red and other red-sensitive emulsions. Such a glass is poor for visual use because the blue rays are badly out of focus.

THE LIGHT CONE. The moon is 1/2 degree in angular diameter. This is commonly taken as the minimum fully-illuminated field of any telescope. You can find what it amounts to in linear size by multiplying the f.l. of objective by .009. The situation is shown in Fig. 4. The field will be fully-illuminated if you can draw the three light rays shown without obstruction. As a matter of fact, the lowermost of the three rays is the only one that matters--if you get this one through, the others will clear automatically.

While keeping the minimum field in mind, most builders try for a bit more. A common standard

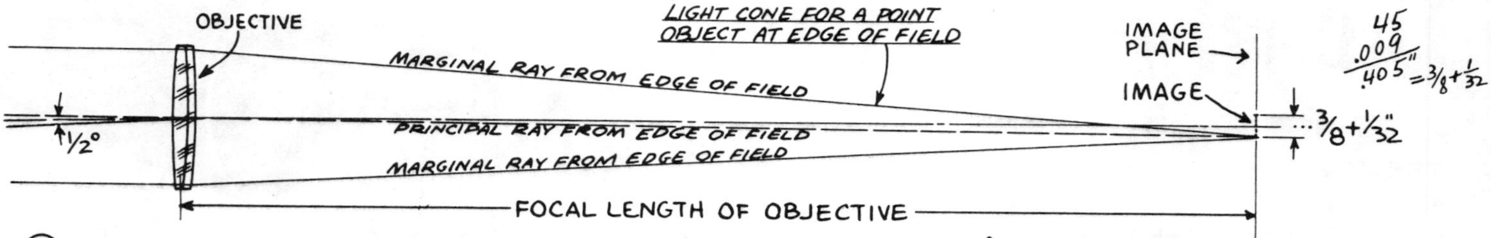

④ MINIMUM FULLY ILLUMINATED FIELD = 1/2° = F.L. x .009 Example: 3" f/15 REFRACTOR

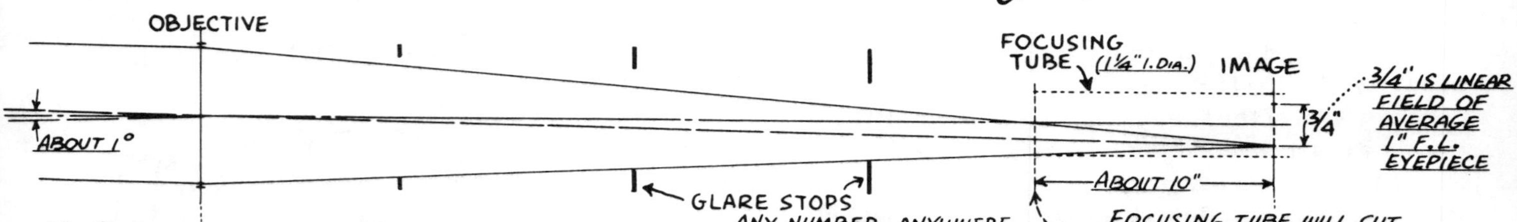

⑤ 3/4" DIAMETER IMAGE IS STANDARD

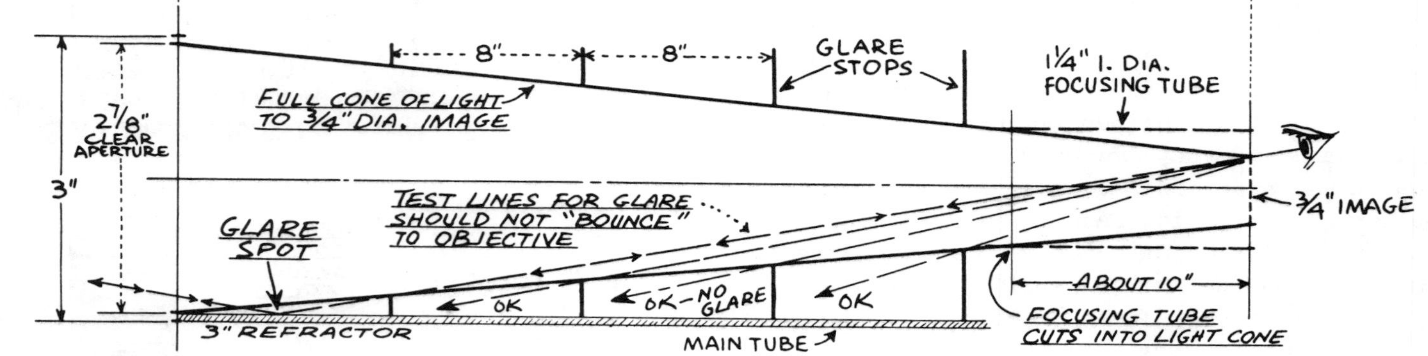

⑥ CONDENSED HORIZONTAL SCALE IS CONVENIENT WAY TO DRAW LIGHT CONE

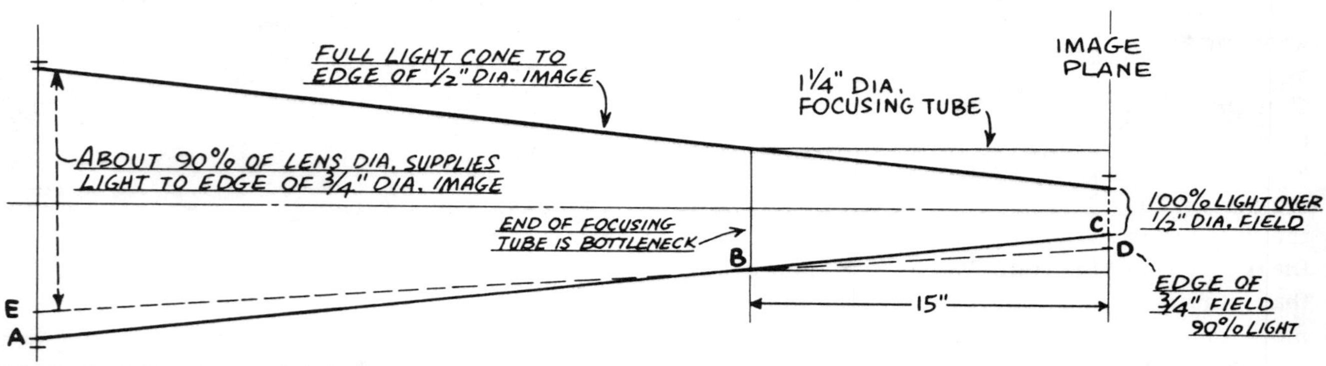

⑦ FOCUSING TUBE TOO FAR FORWARD WILL REDUCE THE SIZE OF FULLY-ILLUMINATED FIELD

is a 3/4 in. linear image field, regardless of objective focal length. This particular diameter is used because it is about the linear size of image seen through a 1-inch eyepiece. In most cases you will have no trouble in lighting a 3/4 inch image. Properly, the bottleneck should be the eyepiece itself, but more often it is the focusing tube that limits the light cone. Fig. 5 shows the situation with 3 in. refractor--the focusing tube can extend 10 in. forward from the image before encroaching on the light cone.

Any light outside the light cone is useless and should be blocked-off to eliminate glare. This is done with a set of 3 or 4 glare stops. You are assured the glare stops are really working if you can draw lines as in Fig. 6 without striking the objective. This diagram shows one glare spot at front of tube, which could be eliminated with a narrow glare stop at that point. Alternately, black flock paper offers fairly good protection and is especially useful when glare stops become too shallow.

Fig. 7 shows the case where the focusing tube encroaches on the light cone. How much light does this cut off? Make a condensed scale diagram like Fig. 7. Draw a line from the edge of

⑧ FOCUSING MOVEMENT from ZERO

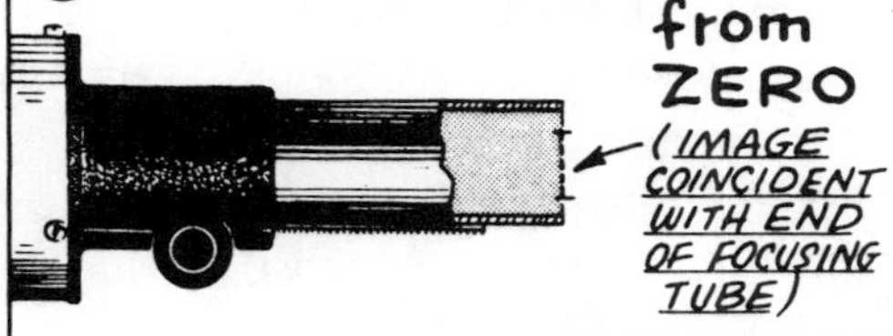

	IN	OUT
ORDINARY EYEPIECES	0"	1/2"
BARLOW LENS (1.83" F.L.)	3/8" 2x	1/2" TO 3x
STAR DIAGONAL	3"	0"
SUN DIAGONAL	3"	0"
ROOF PRISM ERECTOR	3"	0"
LENS ERECTOR	0"	1 1/2"
PORRO PRISM ERECTOR	7"	0"
35MM CAMERA (AT FOCUS)	2"	0"

(FOR DISTANT OBJECT ONLY)

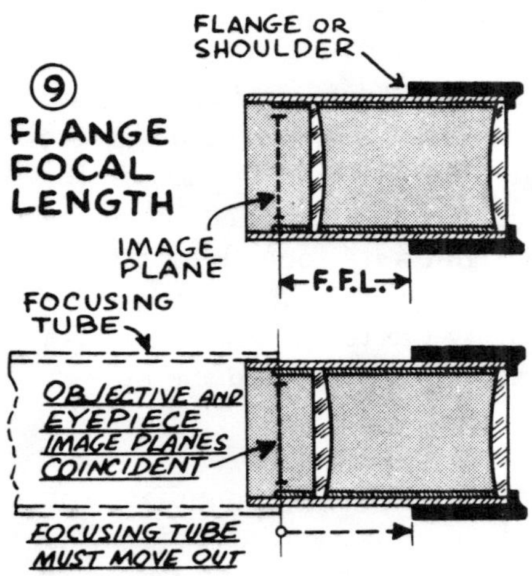

⑨ FLANGE FOCAL LENGTH

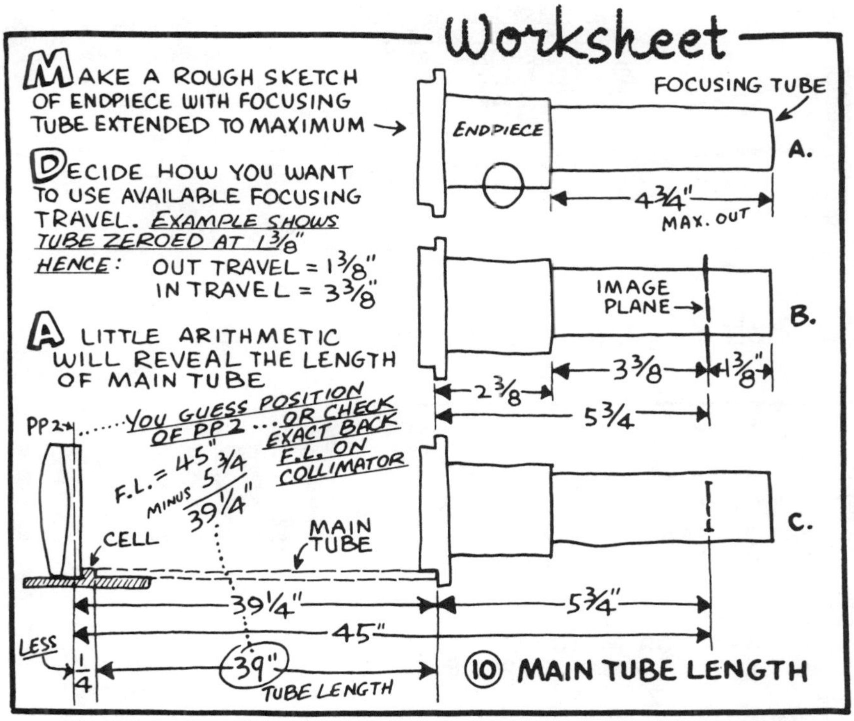

⑩ MAIN TUBE LENGTH

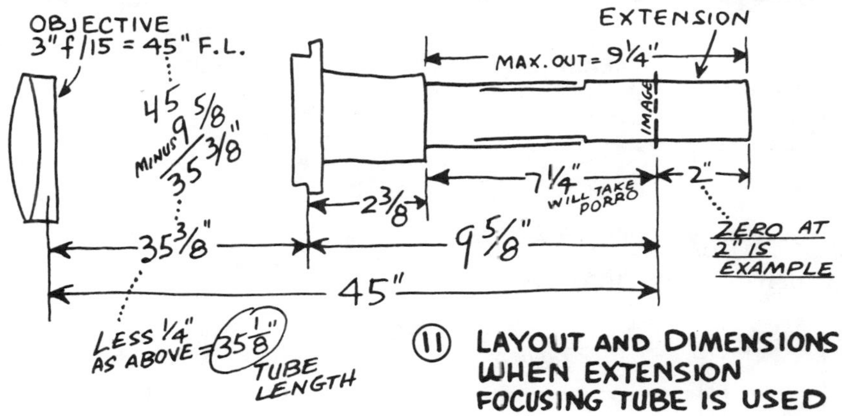

⑪ LAYOUT AND DIMENSIONS WHEN EXTENSION FOCUSING TUBE IS USED

objective at A, to edge of focusing tube at B, finally cutting the image plane at C. This shows the field of full illumination. Now, draw another line from D through B, cutting the objective at E. What you are drawing in each instance is the outermost ray of the light cone. All other rays of the cone will get through, meaning that all of the objective above point E will contribute light to point D at the edge of a 3/4 in. image. As can be seen, this is about 90% lighting and entirely practical--in an actual telescope you can't see this slight light loss even when you try.

LENGTH OF MAIN TUBE. The zero position of the focusing tube, Fig. 8, is a matter of choice, selected according to what accessories you plan to use. Eyepieces alone need only "out" focusing movement from the zero position, as can be seen in Fig. 9, the distance being the same as the flange focal length of the eyepiece.

Fig. 10 Worksheet shows the focusing tube zeroed at a distance of 1-3/8 in. "in" from the maximum "out" position. With a single focusing tube, this position will accommodate all ordinary eyepieces and all accessories except a big Porro erector. The rest of the problem takes a little thinking, but it is just a matter of simple arithmetic to find the proper length of the main tube. It is a good idea to check the exact back focal length of your objective. A 3 in. f/15 objective is supposed to be 45 in. f.l., but may be as much as 1/2 inch more. If you do not check the objective f.l., you can allow for this possible increase by cutting the main tube about 1/2 inch longer than calculated. An actual test of the telescope will then show the exact situation and you can take it from there. A Porro prism erector requires the use of either an extension tube or a draw tube. An extension tube is simplest and cheapest, but has the fault that the combined long length of focusing tube encroaches to some extent on the normal light cone.

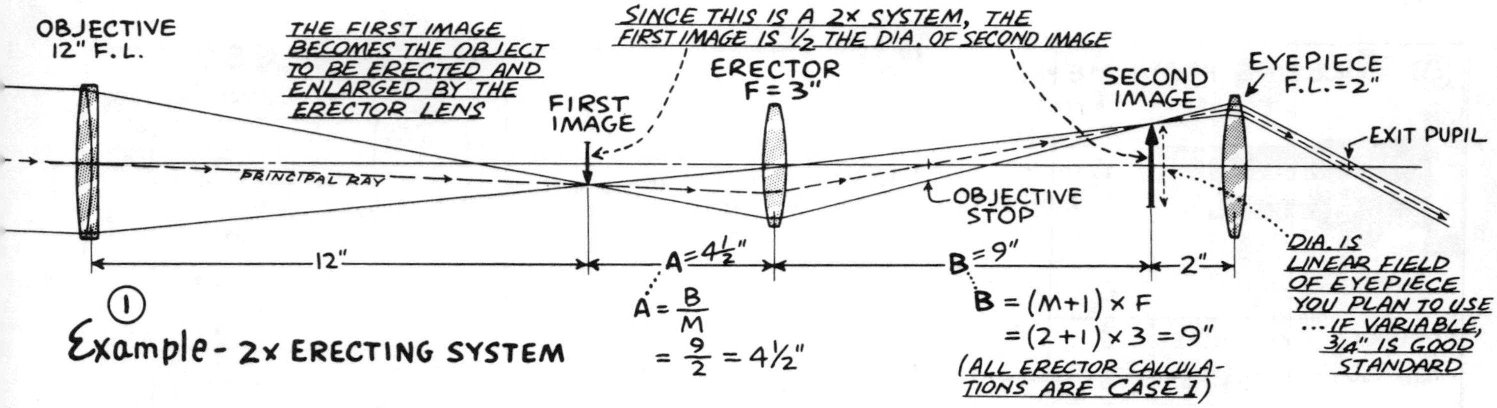

① Example - 2x ERECTING SYSTEM

② TWO WAYS TO APPLY ERECTOR MAGNIFICATION

1. MULTIPLY F.L. OF OBJECTIVE BY ERECTOR M. TO GET E.F.L.
   Example (ABOVE): 12 × 2 = **24"** E.F.L.

OR

2. DIVIDE F.L. OF EYEPIECE BY ERECTOR M. TO GET E.F.L. OF ERECTING EYEPIECE
   Example: $\frac{2}{2} = 1"$ E.F.L.

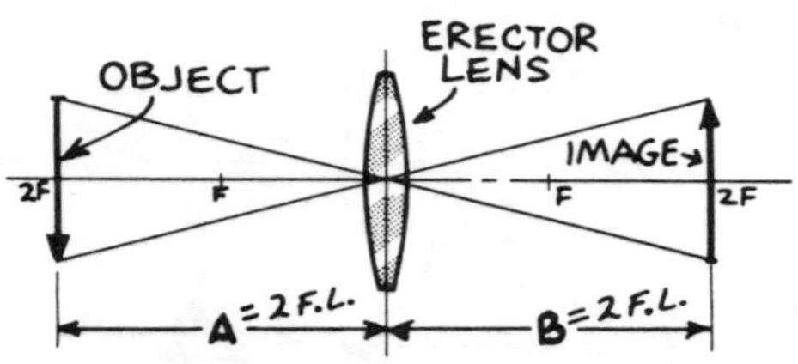

③ UNIT MAGNIFICATION (1x)

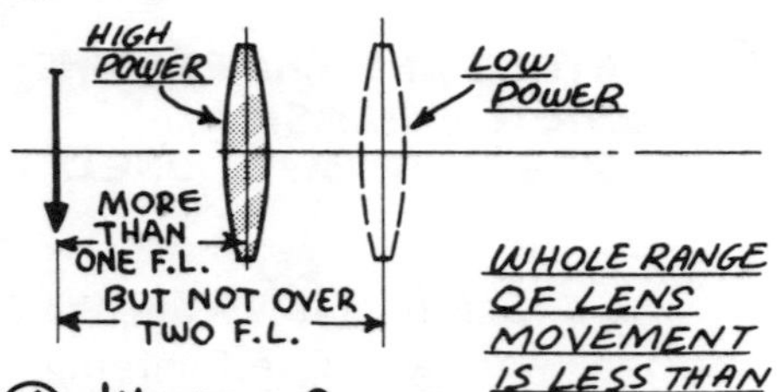

④ WHOLE RANGE

ERECTING Systems

AN ERECTOR lens erects the image. It can also magnify. Obviously it moves the image from one position to another position. From these varied functions, an erector lens is often called an amplifier lens, or a relay lens, or a transfer lens. Call it what you like, it is all one and the same optical system. The magnification factor can be applied to either the objective or eyepiece, Fig. 2, to get an equivalent focal length for the combination. Unit or 1x magnification is obtained when the erector lens is at a distance of two times its focal length from the object, Fig. 3. Of course, the "object" for an erector system is simply the image formed by the telescope objective. High M. is obtained by moving the lens closer to the object, Fig. 4.

ERECTOR ARITHMETIC. First, decide what magnification you want. It is then a simple matter to determine object and image positions, using Case 1 equations, which are repeated in Fig. 1 example. This is actually all you have to know. However, if you want to trace light rays through the system, you

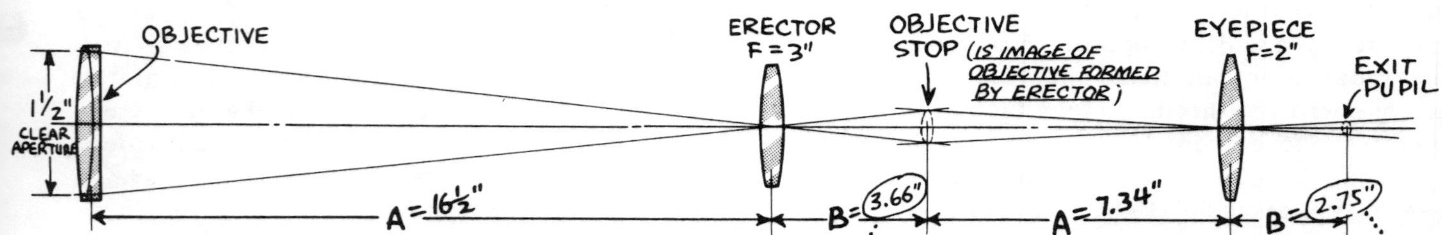

Calculation - OBJECTIVE STOP POSITION and DIAMETER

A = 16.5"

F = 3"

$$B = \frac{F \times A}{A - F} = \frac{3 \times 16.5}{16.5 - 3} = \frac{49.5}{13.5} = 3.66''$$

$$M = \frac{B}{A} = \frac{3.66}{16.5} = .22\times$$

DIAMETER OF OBJECTIVE STOP = 1.5 × .22 = .33"

⑤ SAME EXAMPLE AS FIG. 1

SAME Calculation FOR EXIT PUPIL

A = 7.34" F = 2"

$$B = \frac{F \times A}{A - F} = \frac{2 \times 7.34}{7.34 - 2} = \frac{14.68}{5.34} = 2.75''$$

$$M = \frac{B}{A} = \frac{2.75}{7.34} = .37\times$$

EXIT PUPIL DIAMETER = .33 × .37 = .122" = 1/8"

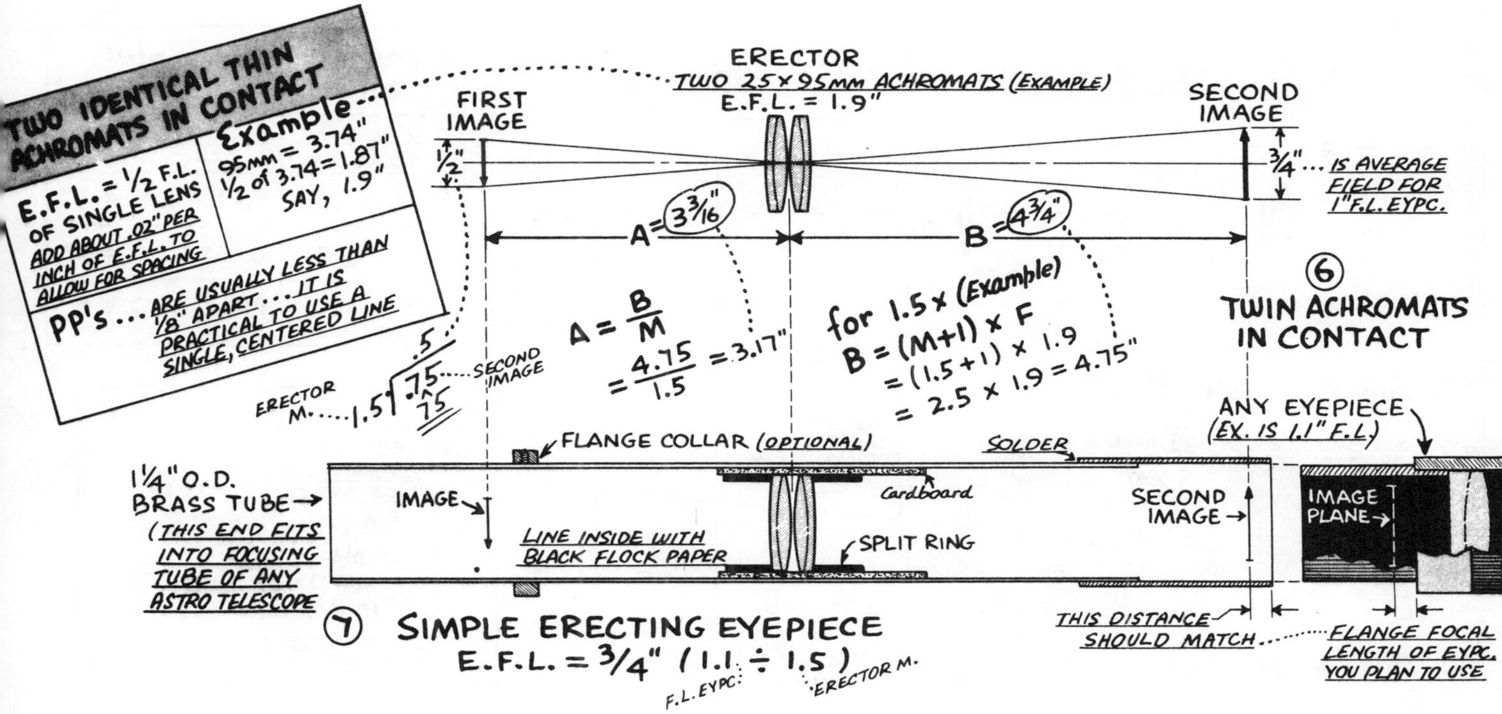

will have to locate the exit pupil. Looking at Fig. 5, you can see that objective and erector alone make up just like any ordinary astro telescope. What would be normally the exit pupil of such a combination now becomes the objective stop. This is a picture of the objective lens as seen by the erector lens. An actual physical stop is usually fitted at this point. The exit pupil of the whole instrument is the image of the objective stop as seen by the eyepiece. The simple arithmetical work is shown in Fig. 5. Note also in this diagram how a simple graphical trace can be used to determine diameters of both objective stop and exit pupil. If you are making a bench setup, no arithmetic is needed because all spacing and stop dimensions are picked off directly from the bench setup. However, it is always a good idea to run through the math work in order to become familiar with the procedure.

TWIN ACHROMATS. For good performance while retaining simple design and construction, a set of twin achromats in contact or nearly so is the most popular, practical erecting system, Fig. 6. The achromats range in diameter from 15 to 30 mm and in f.l. from 30 to 100mm. The odds are that a long focal length system will perform better than a short one, but at the same time it is desirable to keep the system as short as possible. 18x38mm is a nice size for erector lenses where maximum compactness is desired.

An erector tube for use with any astro telescope can be a very simple device, as shown in Fig. 7. You can improve on this by adding an objective stop, the location and size of which is determined as already described. A stop at the first image plane is often used. This is seen in sharp focus when you look through the whole erecting eyepiece.

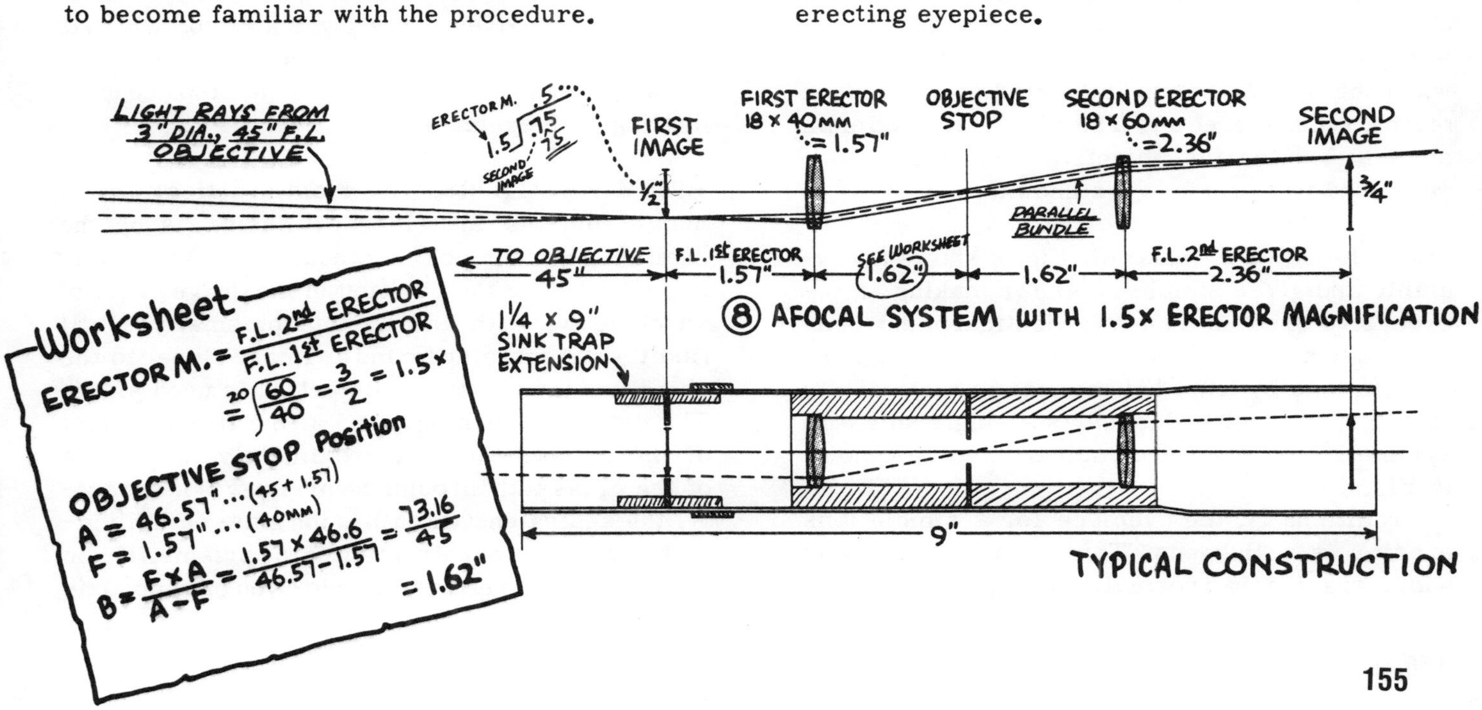

General RULES	Example: 4" E.F.L.
F = ½ E.F.L. INDIVIDUAL LENS	F = ½ × 4 = 2"
E.F.L. = 2F	E.F.L. = 2 × 2 = 4"
d = 1½ F	d = 1.5 × 2 = 3"
PP's = 3F ...ARE CROSSED	PP's = 3 × 2 = 6"

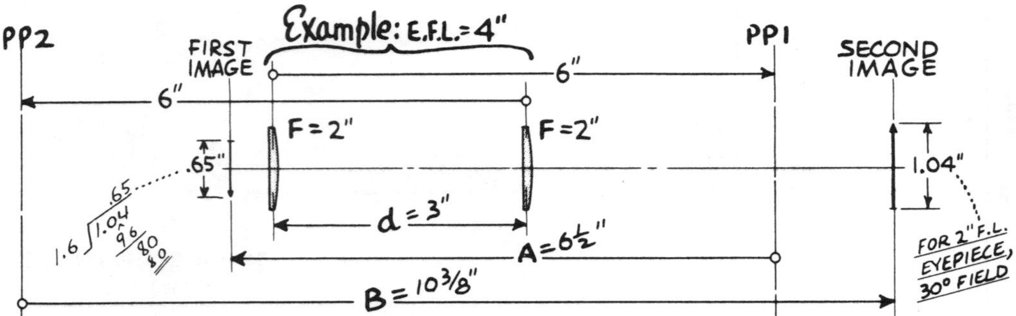

(9) SIMPLE LENS ERECTOR (EX. IS 1.6×)

ERECTOR M. SHOULD BE BETWEEN 1× AND 1.8×

HIGHER M. WILL PUT FIRST IMAGE TOO CLOSE TO LENS

Calculate M. AS USUAL

Example is 1.6×

B = (1.6+1) × 4 = 10.4"

A = 10.4 ÷ 1.6 = 6.5"

SET OFF THESE DISTANCES FROM PRINCIPAL PLANES, AS SHOWN

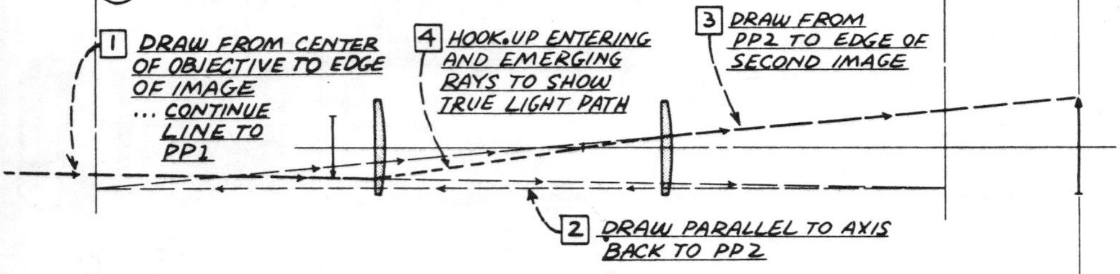

(10) DRAWING THE LIGHT RAYS (PRINCIPAL RAY ONLY IS SHOWN TO AVOID A CONFUSION OF LINES)

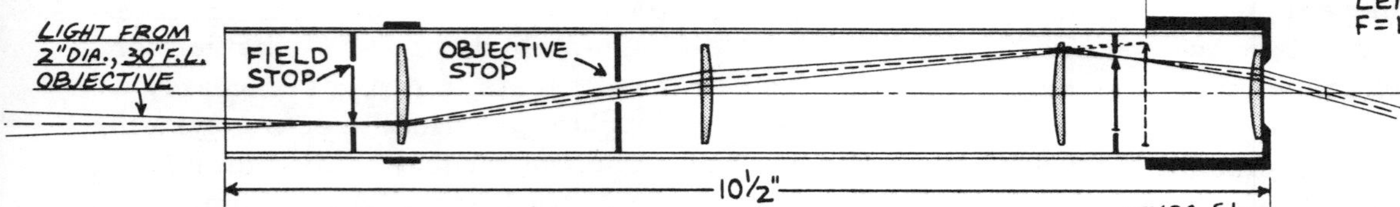

(11) WHOLE ERECTING EYEPIECE WITH 2" F.L. HUYGENS (E.F.L. = $\frac{2}{1.6}$ = 1.25") (2 ... EYPC. F.L.; 1.6 ... ERECTOR M.)

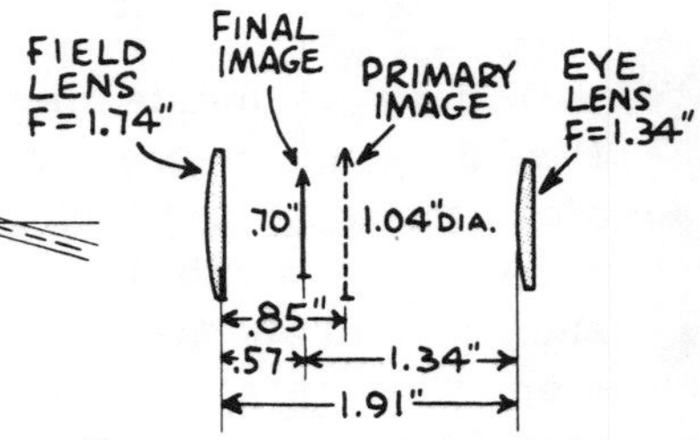

2" F.L. HUYGENS (1.3 RATIO) 1.3 TO 1

THE AFOCAL SYSTEM. This uses two achromats, but of different focal lengths. Each is used in the afocal position, that is, at its own focal length from the respective image. Magnification is obtained by the difference in focal lengths, Fig. 8. A system like this is usually wide-spaced. Since the light emerging from the first erector is parallel, the space between first and second erector is free optical space which can be varied as desired without changing the power of the system. A practical minimum spacing is obtained when the objective stop is located on the surface of the second erector. The practical maximum spacing locates the objective stop midway between the two erectors, as shown.

SIMPLE LENS ERECTOR. Fig. 9 shows an example and gives simple rules for making an inexpensive erector from two identical plano-convex lenses. Light rays can be traced through the system lens by lens, but the more practical way is to treat the two lenses as a single unit with symmetrical, crossed principal planes, as shown in Fig. 10.

Traditionally, the eyepiece for a simple lens erector is a Huygens. This is lower ratio and wider spacing than normal, both departures arising from the fact that the "object" for the eyepiece now becomes the objective stop instead of the objective itself. Huygens eyepieces made for microscope use are suitable since the working conditions are quite similar. Specifications for a suitable 2 inch f.l. Huygens for use with an erecting system are given in Fig. 11. You can scale this up or down as desired by dividing the desired new e.f.l. or lens f.l. by the similar specification given. This gives a factor which is then applied to all of the specifications given. This process is illustrated in the chapter dealing with eyepieces.

PRISM ERECTORS. The most practical way to handle prisms in telescope design is by the "equivalent air block" method. Fig. 13 explains. If you compare the position of the image plane of an objective with and without prisms, you will find the prism setup forms an image closer to the objective than when the objective is used alone. The difference can be taken as the air equivalent of the prisms--it is approximately two-thirds of the glass path through the prisms. More exactly, the equivalent air path is the glass path divided by the index of refraction of the prisms. If you don't know the refractive index you can approxi-

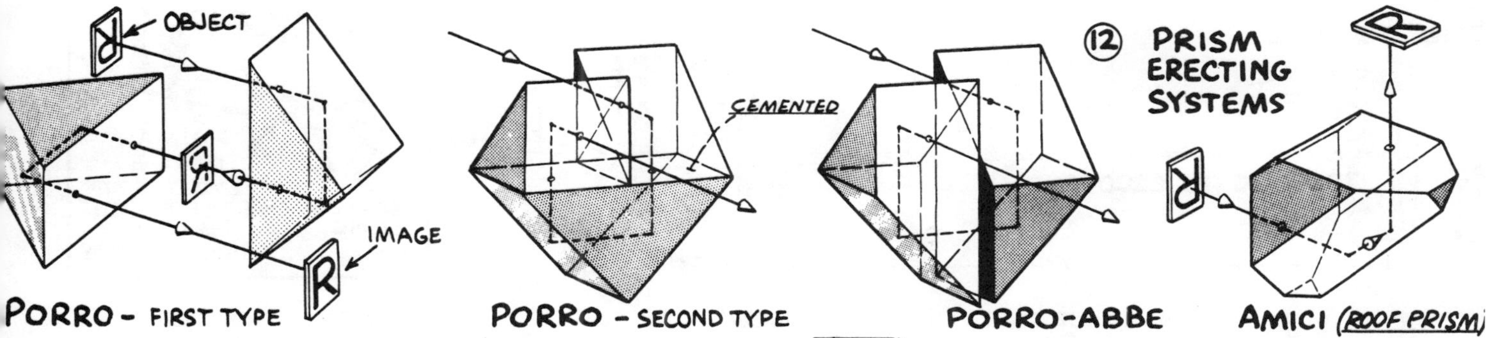

(12) PRISM ERECTING SYSTEMS

mate by the two-thirds rule, or you can determine air blocks exactly by making the simple bench test shown. The glass path itself of any prism is easily determined from a full-size drawing of the prism. Often the corners of a prism are cut off. When you are making a drawing, the corners should be restored to obtain the full side or hypotenuse from which the glass path is determined.

To design a prism erector attachment for an astro telescope, you start by making a full-size schematic diagram of the system, Fig. 14. This is then redrawn in stretchout form, substituting air blocks for the glass path. Measuring this final diagram will show the amount of "in" focusing movement needed. It is excessive for a Porro system of the first type, being a whopping 7-1/4 inches for the example shown, which is an actual commercial product. You can shorten the eyepiece tube on this to about 1 inch and otherwise "squeeze" the assembly to cut the focusing movement to about 6 inches. If you use a special eyepiece mounted in 7/8 in. diameter tube, it can be worked alongside the prism, eliminating the projecting eye tube completely. The second type of Porro prism is more compact--4-3/4 inches using the same-size prisms and mechanical dimensions. The Porro-Abbe is about the same. A roof prism is treated as a simple right angle prism of the same overall size; this takes less than 3 in. "in" movement, although some of the gain comes from the fact that a roof prism with overall size equal to a right angle prism will be about 25% less in face width.

To see how the prisms field the light cone, you simply draw them into the system as air blocks, Fig. 15. The width of the air block is the actual face width of the prism; the length is calculated as already described. For most applications, 50% edge-of-field lighting is satisfactory, meaning that if you can get the principal ray through the prisms, the lighting will be okay. Optically, the effect of prisms is the same as a thick piece of glass with parallel surfaces. Such a glass block has optical characteristics similar to a weak negative lens.

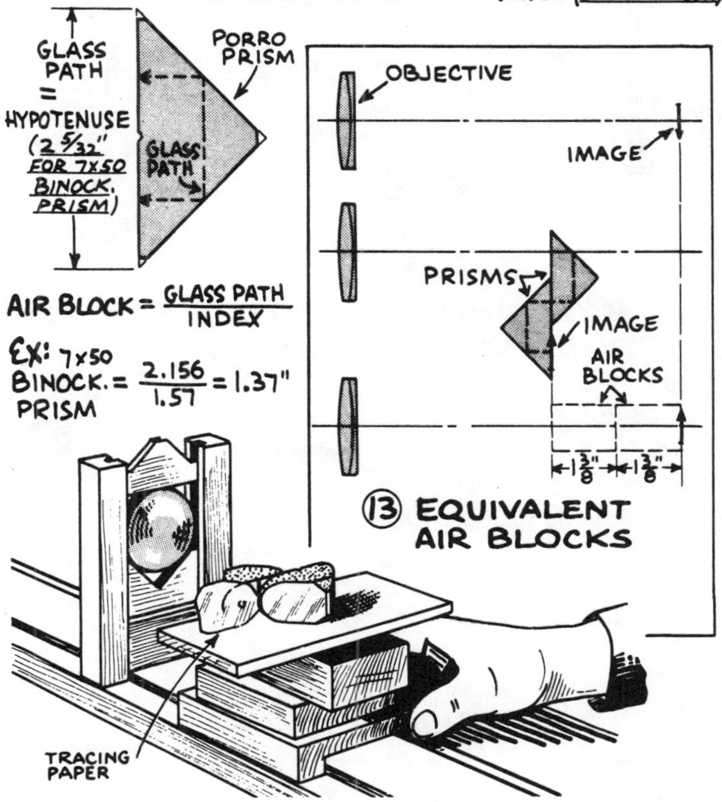

(13) EQUIVALENT AIR BLOCKS

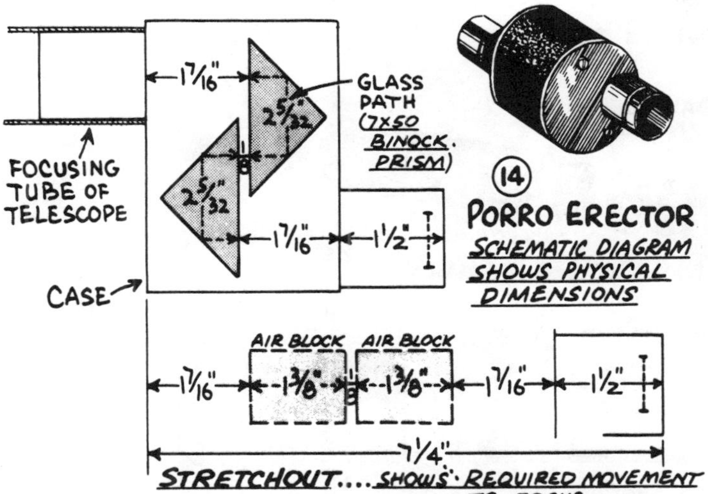

(14) PORRO ERECTOR
SCHEMATIC DIAGRAM SHOWS PHYSICAL DIMENSIONS

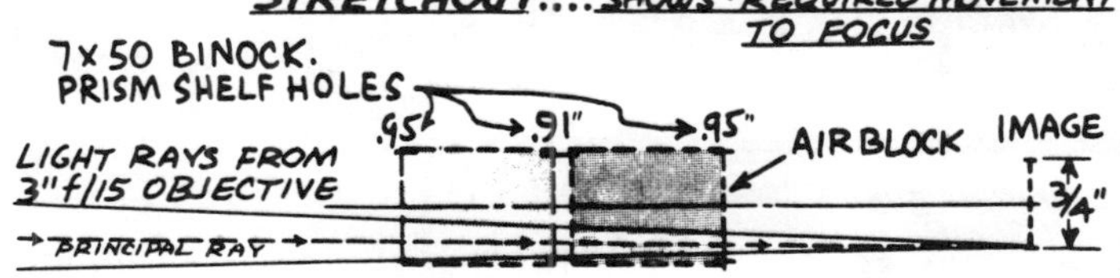

(15) LIGHT PATH - PRISMS SHOULD PASS AT LEAST THE PRINCIPAL RAY. Example SHOWS WHOLE CONE GETS THRU

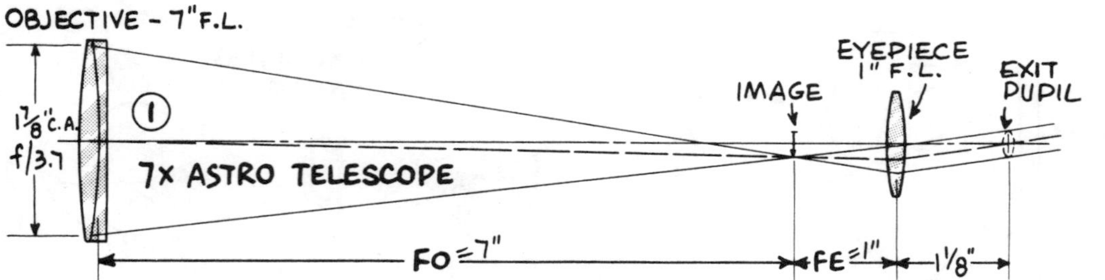

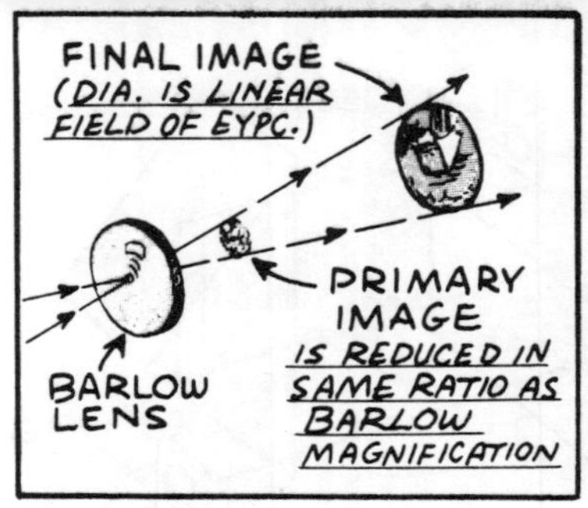

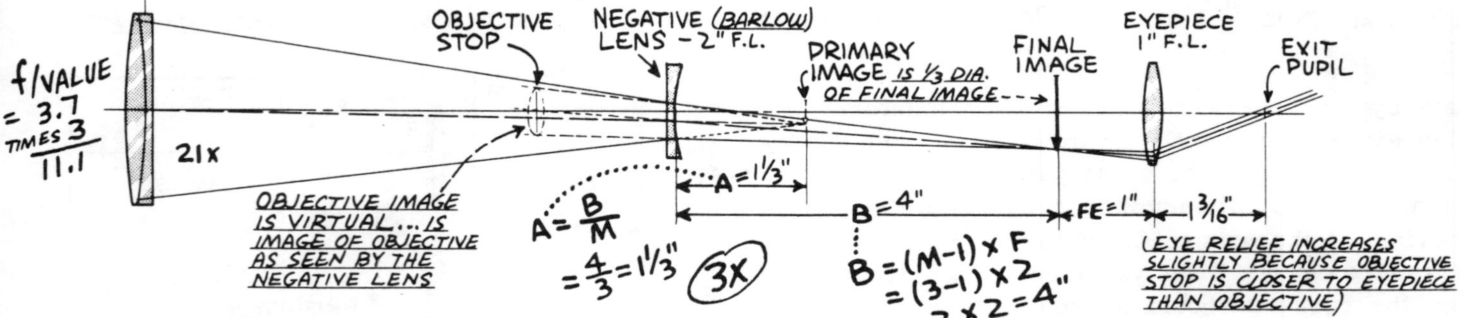

② ... AS ABOVE WITH 3X BARLOW (E.F.L. = 7 x 3 = 21")

the BARLOW Lens

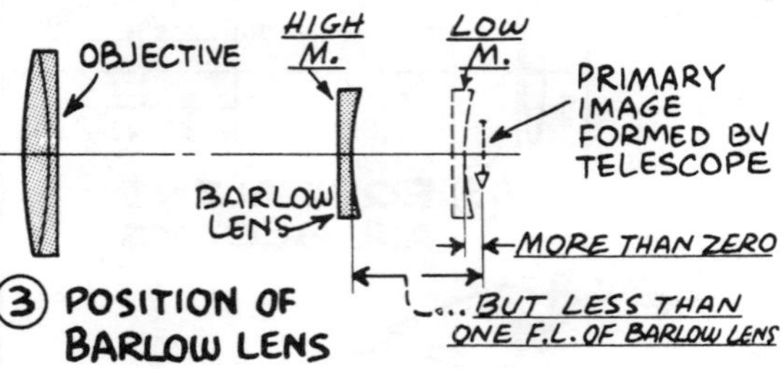

③ POSITION OF BARLOW LENS

A BARLOW lens is a negative lens used inside the focal plane of a telescope objective. Its normal diverging action reduces the convergence of the light cone, forming a larger image at a slightly greater distance. All Barlow lenses are designed for a certain magnification factor--usually 2x--but work well over a moderate range of powers.

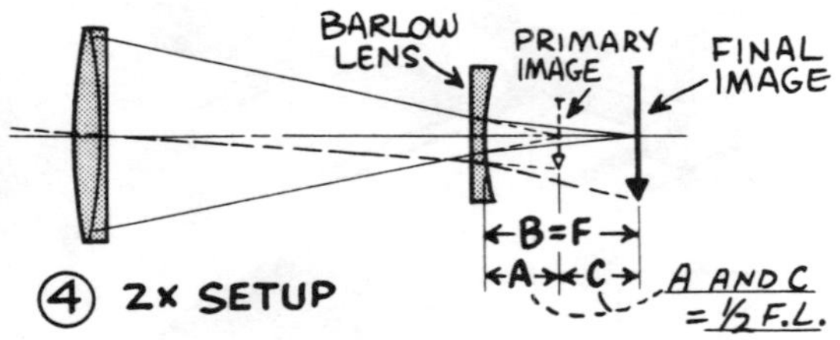

④ 2x SETUP

TYPICAL BARLOW SYSTEM. A drawing of a Barlow system begins with the usual light rays from objective to image, except, knowing the Barlow will enlarge the primary image, you make it just that much smaller, Fig. 1. A and B spacing distances are then calculated for the desired magnification, using Case 5 equations. The linear field of eyepiece is set off at the final image plane, and the light ray intercepts are extended from the Barlow lens to edge of final image, Fig. 2. If you want to locate the objective stop, it can be done graphically by extending the light rays backwards, as shown in Fig. 2. As can be seen, the objective stop is a virtual image; if calculated, you use Case 3 equations. The position of objective stop must be known if you want to calculate (Case 1) the exit pupil position. In most cases, only the A and B spacing distances are needed. Glare stops can be fitted anywhere along the light cone.

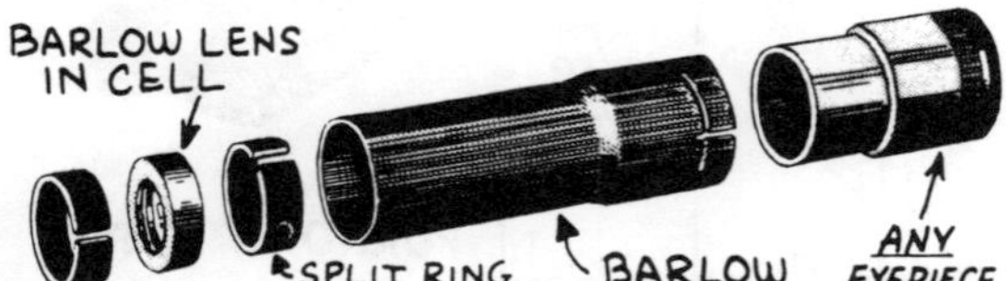

⑤ BARLOW ASSEMBLY

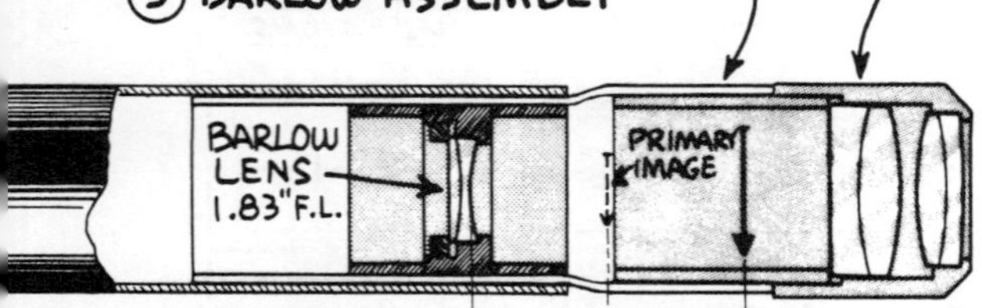

⑥ SPACING FOR 2X

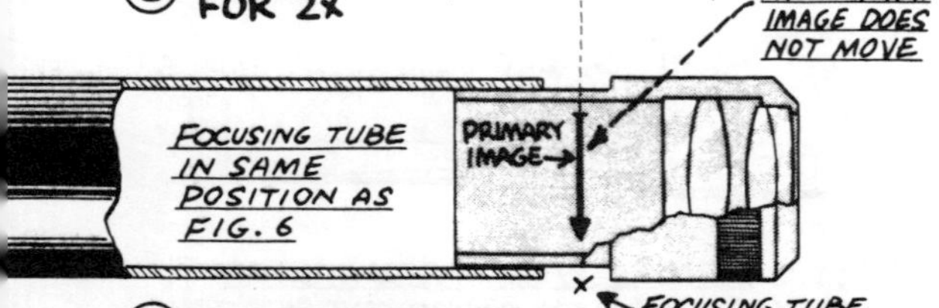

⑦ EYEPIECE ALONE

FOCUSING MOVEMENT. Normally a Barlow setup requires "out" movement of the focusing tube. A goodly amount of "out" movement is supplied by the Barlow tube itself, Fig. 5. The net result is that "in" movement of the focusing tube is needed for the popular 2x setup, Fig. 6. Fig. 7 illustrates in reverse fashion--with eyepiece alone

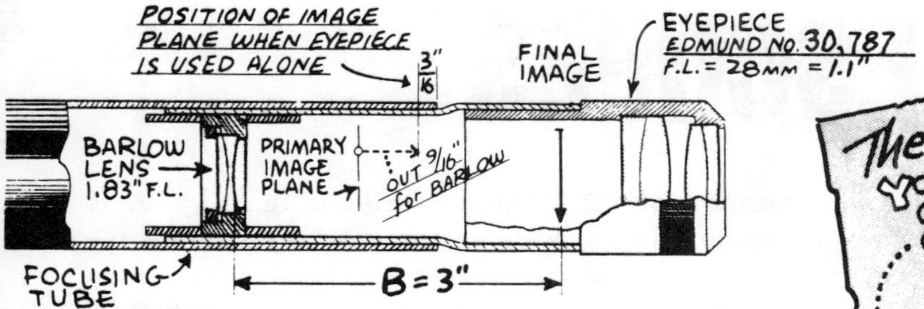

CASE 5 - NO. 9 $M = \frac{3 + 1.83}{1.83} = \frac{4.83}{1.83} = 2.64\times$

CASE 5 - NO. 4 $A = \frac{B}{M} = \frac{3}{2.64} = 1.14'' = 1\frac{1}{8}''+$

(8) MAXIMUM M. - *EXAMPLE IS EDMUND NO. 40-477 BARLOW and No. 30,787 EYEPIECE*

IF M. IS APPLIED TO EYEPIECE:

$$\frac{FE}{M} = \frac{1.1}{2.6} = .4''\text{ E.F.L.}$$

you have to focus "out" about 1/2 inch, indicating that the Barlow setup itself must focus "in."

Increased M. pushes the image back and requires more "out" movement. The maximum case for the equipment specified is shown in Fig. 8. This takes actual "out" movement of the focusing tube. To determine the focusing tube movement on a drawing like this, you measure from primary image plane to the position the focal plane of the eyepiece would normally occupy if used alone.

SPECIAL SETUPS. The parfocal Barlow, Fig. 9, has the same focus for either eyepiece alone or with Barlow--the focusing tube does not move. For this kind of setup, you simply space A and B distances from the primary image plane, which is made coincident with the focal plane of the eyepiece used alone.

A Barlow system is sometimes built-in as a permanent part of a telescope. Fig. 10 is an example. Because the diagonal is close to the primary image, which itself is reduced in size, it is possible to field the full light cone of an f/8 mirror with a 3/4 inch diagonal. For the arrangement shown, spacing distance B is fixed at about 8 in. You select some suitable magnification and then calculate F as shown. A good-quality simple plano-concave lens of the specified focal length will usually perform quite well.

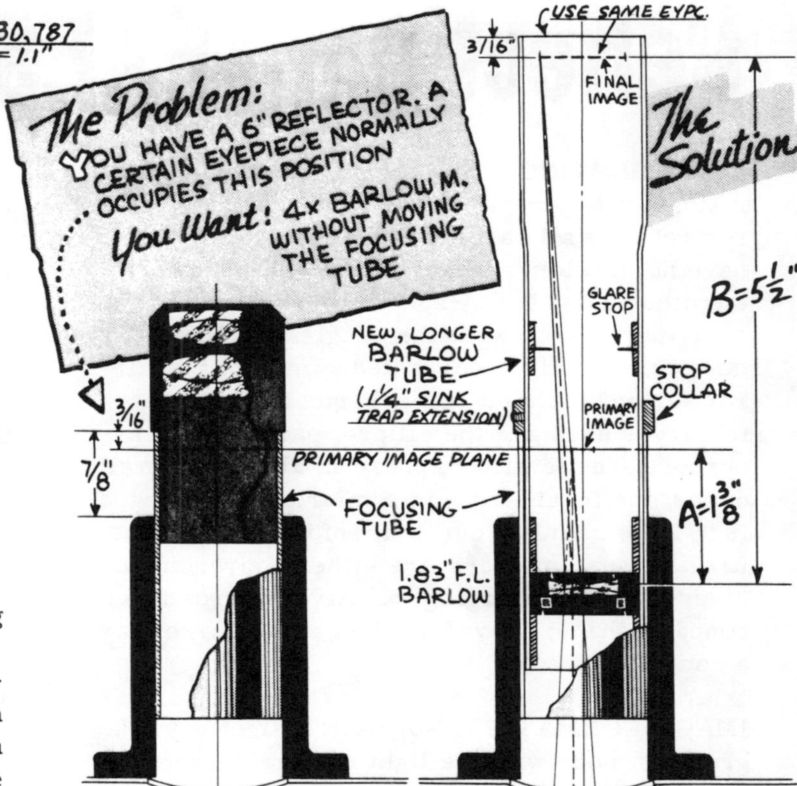

(9) 4x BARLOW, PARFOCAL

Calculations

$M = 4\times \qquad F = 1.83''$

CASE 5-1 $B = (M-1) \times F = (4-1) \times 1.83 = 5.49''$

CASE 5-4 $A = \frac{B}{M} = \frac{5.49}{4} = 1.37''$

WHOLE SYSTEM: E.F.L. = 48 × 4 = 192"

f/ = f/8 × 4 = f/32

(10) BUILT-IN BARLOW
M = 3x

WHOLE SYSTEM:

E.F.L. = 48 × 3 = 144"

f/ = f/8 × 3 = f/24

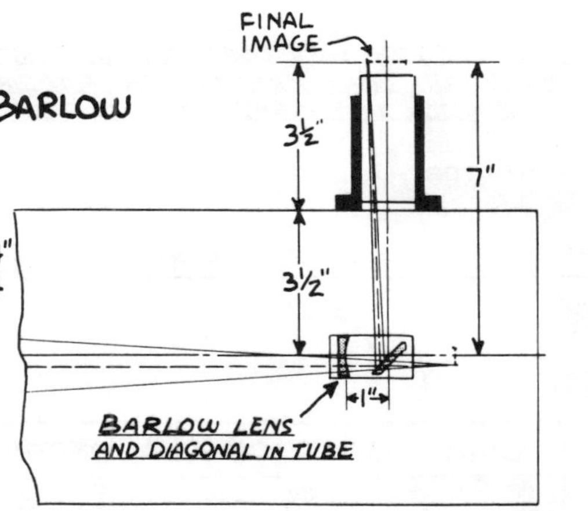

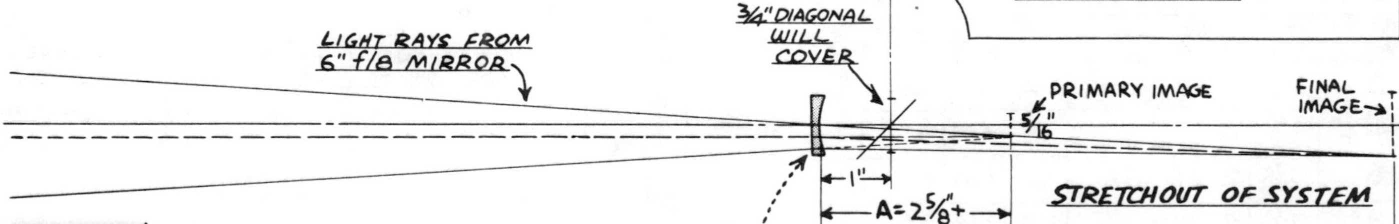

CASE 5 - 11 $F = \frac{B}{M-1} = \frac{8}{3-1} = \frac{8}{2} = 4''\text{ F.L. BARLOW}$

Reflecting TELESCOPES

A REFLECTING telescope uses mirror optics instead of lenses. A big advantage is that a mirror reflects all wavelengths of light equally--you have no problem at all with false color. However, the other axial fault--spherical aberration --is still there, and, with the single surface of the mirror, can only be corrected by making the mirror surface aspheric (not spherical). The aspheric curves used are the ellipse, parabola and hyperbola. The familiar sphere is also used. In all cases, the focal length of a mirror of any shape is 1/2 the radius of curvature of its central zone. Like lenses, mirrors are either converging or diverging. A converging (positive) mirror has a concave shape; a diverging (negative) mirror has a convex shape.

IMAGE FORMATION. Normally a light ray diagram is made with the light coming in from the left, but if this procedure is reversed, the net performance of a mirror is the same as a lens, Fig. 1. The light cone from a point object at edge of field is the only one you need draw. The outer ray of this cone defines the limit of the useful light. The principal ray (the one passing through the center of the objective) never actually gets through in a reflecting telescope, but it is no less useful as a guide line.

A common rule of thumb is that the secondary flat for a Newtonian reflector should not obstruct more than 6% of the incoming light. Some additional light is lost in a gray zone about twice the diameter of the diagonal silhouette, Fig. 2. In compound telescopes with secondary mirror, the obstruction may be as large as one-half the primary diameter, resulting in a light loss of 25%.

NEWTONIAN REFLECTOR. The manner of making a layout for a Newtonian reflector is shown in Fig. 3. Steps 1 to 5 are obvious. Step 6 requires the selection of some practical image size, which can be 3/4 in. for any telescope, although the example shown is somewhat larger. The edge-of-field light cone misses the diagonal slightly, Step 10. This is fairly standard practice, but the lighting should not fall below 50%. Step 11 shows that about 70% of the objective diameter contributes light to the edge of field. You will want to know what size image gets 100% lighting, and this is revealed by Step 12.

A final check is to see that the incoming light clears the main tube. Steps 15, 16 and 17 show that it does, even as far out as 6-3/4 in. from the eyepiece centerline. However, to assure adequate clearance, the front tube projection is usually trimmed to 3-1/2 or 4 in. as shown.

COMPOUND TELESCOPES. All astro telescopes are compound optical instruments in the sense that an enlarged image is formed by the objective and this image is further enlarged by the eyepiece. However, the accepted meaning of a compound telescope indicates an instrument with a built-in secondary optical system which en-

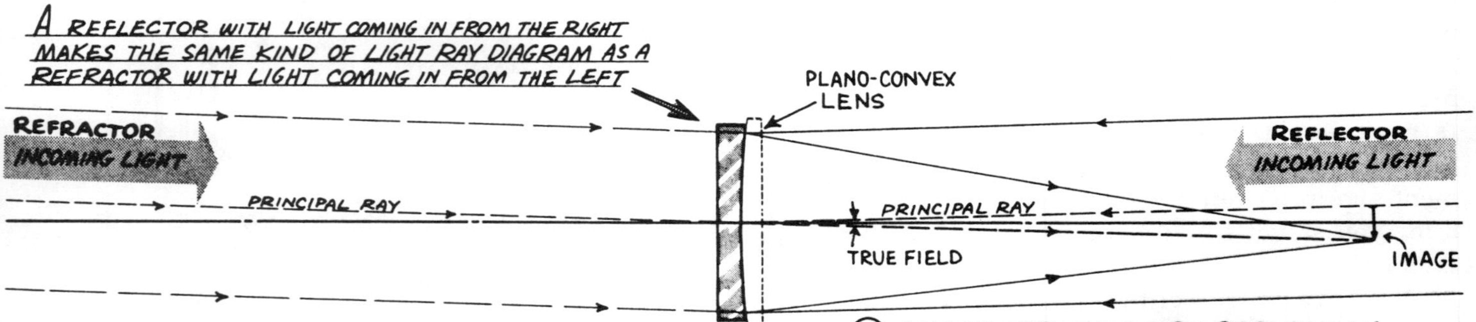

① IMAGE FORMATION BY REFLECTION

LIGHT LOSS BY DIAGONAL OBSTRUCTION							
MIRROR		MIRROR AREA	USUAL DIAG.	LOSS-SQ.IN.		LOSS-%	
NAME DIA.	C.A.			ELL.	RECT.	ELL.	RECT.
3"	2 7/8"	6.49 SQ.IN.	7/8" WIDE	.60 SQ.IN.	.87 SQ.IN.	8.8%	12%
4 1/4"	4 1/8	13.36	1 1/16	.87	1.30	6.6	9.9
6"	5 7/8	27.11	1 1/4	1.23	1.89	4.5	7.0
8"	7 7/8	48.71	1 1/2	1.77	2.64	3.6	5.4

ELL.-ELLIPTICAL SHAPE RECT.-RECTANGULAR. LOSS INCLUDES EDGE

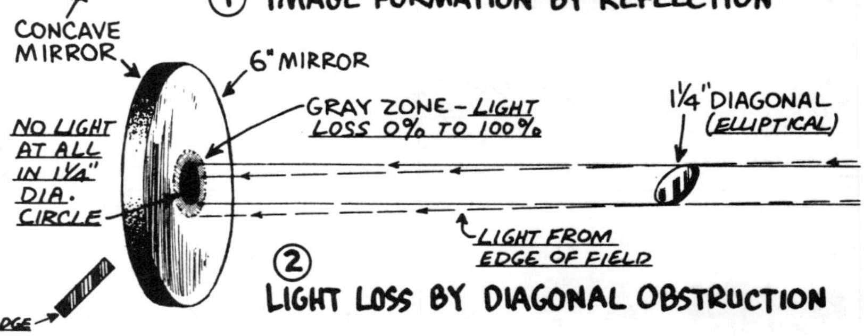

② LIGHT LOSS BY DIAGONAL OBSTRUCTION

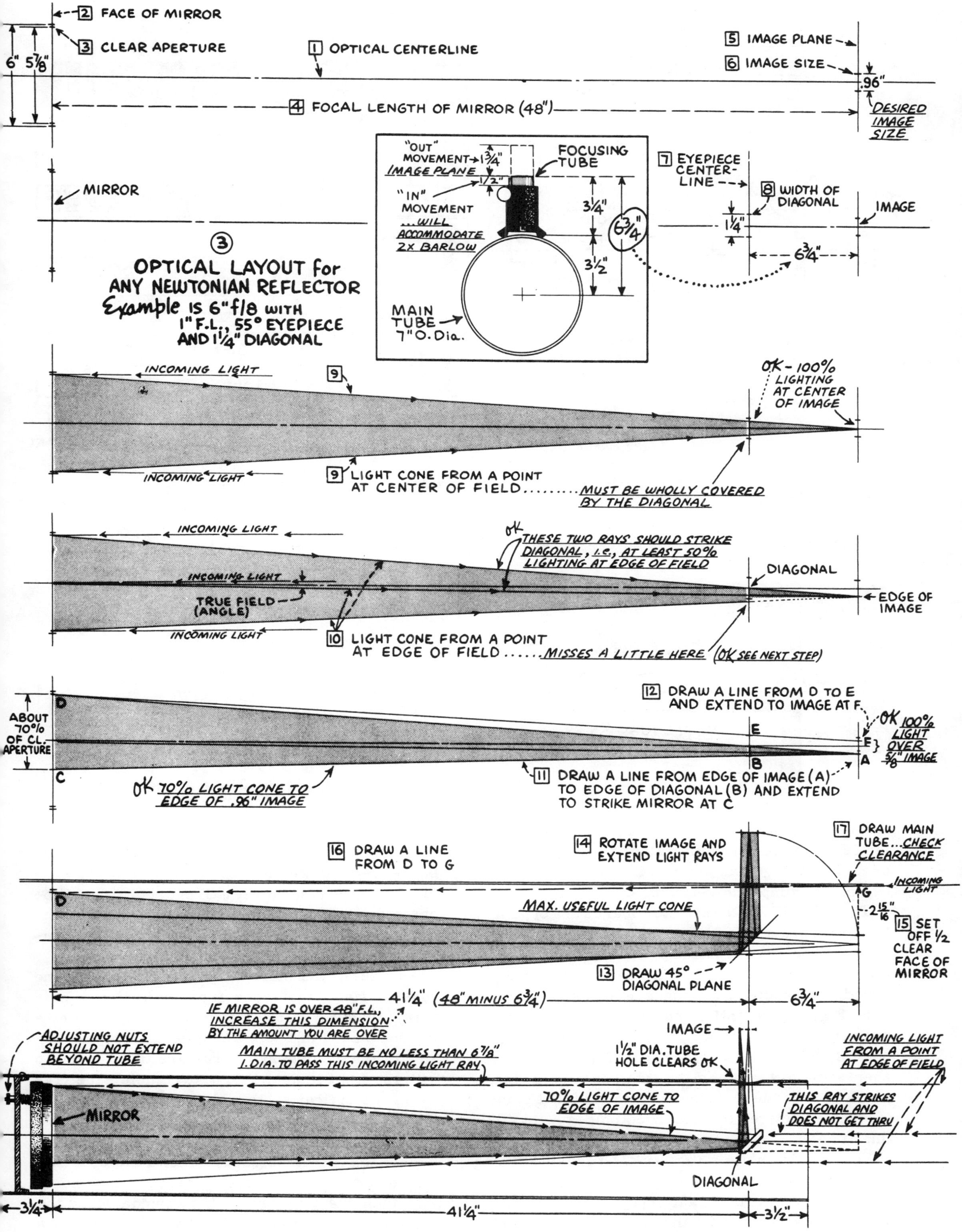

2 FACE OF MIRROR
3 CLEAR APERTURE
1 OPTICAL CENTERLINE
5 IMAGE PLANE
6 IMAGE SIZE
6" 5⅞"
.96"
4 FOCAL LENGTH OF MIRROR (48")
DESIRED IMAGE SIZE
"OUT" MOVEMENT 1¾"
IMAGE PLANE
½"
FOCUSING TUBE
"IN" MOVEMENT ...WILL ACCOMMODATE 2X BARLOW
3¼"
3½"
6¾"
MAIN TUBE 7" O. Dia.
7 EYEPIECE CENTER-LINE
8 WIDTH OF DIAGONAL
IMAGE
1¼"
6¾"
MIRROR
3
OPTICAL LAYOUT for ANY NEWTONIAN REFLECTOR
Example is 6" f/8 with 1" F.L., 55° EYEPIECE AND 1¼" DIAGONAL
INCOMING LIGHT
9
OK - 100% LIGHTING AT CENTER OF IMAGE
9 LIGHT CONE FROM A POINT AT CENTER OF FIELD......... MUST BE WHOLLY COVERED BY THE DIAGONAL
OK THESE TWO RAYS SHOULD STRIKE DIAGONAL, i.e., AT LEAST 50% LIGHTING AT EDGE OF FIELD
DIAGONAL
EDGE OF IMAGE
TRUE FIELD (ANGLE)
10 LIGHT CONE FROM A POINT AT EDGE OF FIELD...... MISSES A LITTLE HERE (OK SEE NEXT STEP)
12 DRAW A LINE FROM D TO E AND EXTEND TO IMAGE AT F.
OK 100% LIGHT OVER ⅝" IMAGE
ABOUT 70% OF CL. APERTURE
D
C
E
B
F
A
OK 70% LIGHT CONE TO EDGE OF .96" IMAGE
11 DRAW A LINE FROM EDGE OF IMAGE (A) TO EDGE OF DIAGONAL (B) AND EXTEND TO STRIKE MIRROR AT C
16 DRAW A LINE FROM D TO G
14 ROTATE IMAGE AND EXTEND LIGHT RAYS
17 DRAW MAIN TUBE...CHECK CLEARANCE
INCOMING LIGHT
G
2 15/16"
15 SET OFF ½ CLEAR FACE OF MIRROR
MAX. USEFUL LIGHT CONE
13 DRAW 45° DIAGONAL PLANE
41¼" (48" MINUS 6¾")
6¾"
IF MIRROR IS OVER 48" F.L., INCREASE THIS DIMENSION BY THE AMOUNT YOU ARE OVER
ADJUSTING NUTS SHOULD NOT EXTEND BEYOND TUBE
MAIN TUBE MUST BE NO LESS THAN 6⅞" I.DIA. TO PASS THIS INCOMING LIGHT RAY
IMAGE
1½" DIA. TUBE HOLE CLEARS OK
INCOMING LIGHT FROM A POINT AT EDGE OF FIELD
70% LIGHT CONE TO EDGE OF IMAGE
THIS RAY STRIKES DIAGONAL AND DOES NOT GET THRU
MIRROR
DIAGONAL
3¼"
41¼"
3½"

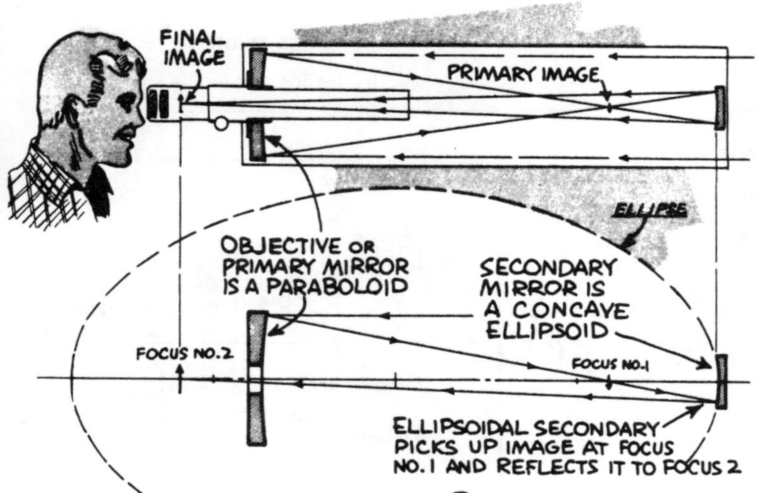

④ GREGORIAN OPTICAL SYSTEM

⑤ HOW TO *Calculate* A GREGORIAN		
You Specify	OBJECTIVE MIRROR — ANY DIA., 3" AND UP USUALLY LOW f/ - f/2 TO f/8	Examble (DRAWING BELOW) 6" DIA., 30" F.L. = f/5
	M. — M. OF SECONDARY USUALLY 4X TO 6X	5x
	ℓ — IMAGE DISTANCE BEHIND OBJECTIVE	6" THIS ALLOWS "IN" TRAVEL FOR A STAR DIAGONAL
You Calculate	GREGORIAN $A = \frac{F+\ell}{M-1}$	$\frac{30+6}{5-1} = \frac{36}{4} = 9"$
	CASE 1, NO.3 $B = A \times M$	$9 \times 5 = 45"$ B WILL BE SELF-EVIDENT FROM DRAWING
	CASE 1, NO.11 SECONDARY $F = \frac{B}{M+1}$	$\frac{45}{5+1} = \frac{45}{6} = 7\frac{1}{2}"$ (=15" radius)
WHOLE TELESCOPE	E.F.L. = FO × M	E.F.L. = 30 × 5 = 150"
	f/VALUE = f/OBJ. × M	f/VALUE = f/5 × 5 = f/25

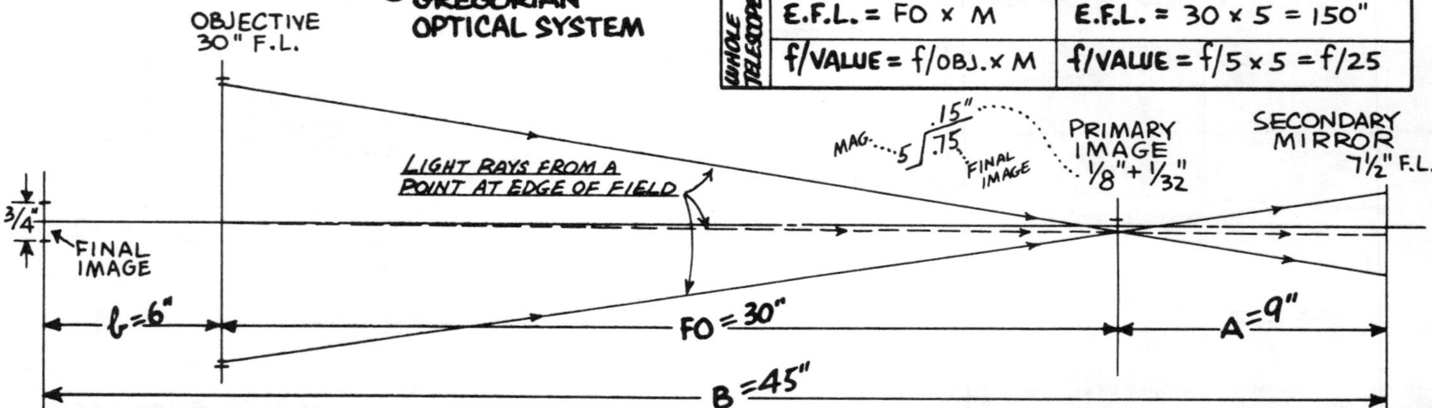

⑥ GREGORIAN LAYOUT (START)

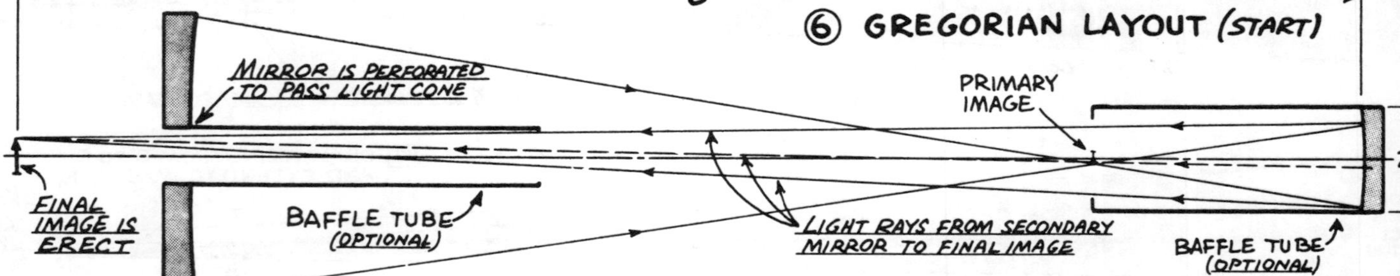

⑦ COMPLETE GREGORIAN LAYOUT

larges the primary image. So defined, a Newtonian is a simple reflector rather than a compound one. The Gregorian and the Cassegrain are the two basic compound reflectors.

The GREGORIAN reflector makes use of positive projection to erect and enlarge the primary image. The primary image itself is formed by a paraboloidal mirror. This image is then magnified and erected by a concave ellipsoidal mirror, Fig. 4. The optical nature of an ellipsoidal mirror is such that a light ray passing through one of the conjugate foci is reflected without aberration to the other. As shown in the drawing, one foci is made to coincide with the focal plane of the objective; the other coincides with the final image.

You can calculate a variety of Gregorian telescopes by the simple rules given in Fig. 5. However, if you try to make a low-power design, such as 2x, you will find that the long throw and large secondary mirror will make the design impractical. A performance fault is that stray light can barrel right down the main tube and into the eyepiece. If you want to use a Gregorian for daytime observing, you must use baffle tubes to limit the light rays to exactly that cone of light which contributes to the image.

As usual, you start the layout by drawing the edge-of-field light cone from objective to primary image, Fig. 6, extending the rays to strike the secondary mirror. The intercepts at the surface of the secondary mirror are then hooked up with the final image, Fig. 7, to complete the layout of the optical system. As usual, only one side of the light cone need be drawn; the outer ray of this is your guide for glare stops or baffle tubes.

A CASSEGRAIN telescope shows the usual inverted astro image. This system is more compact than the Gregorian. If you understand the Barlow lens, it is easy to visualize the Cassegrain

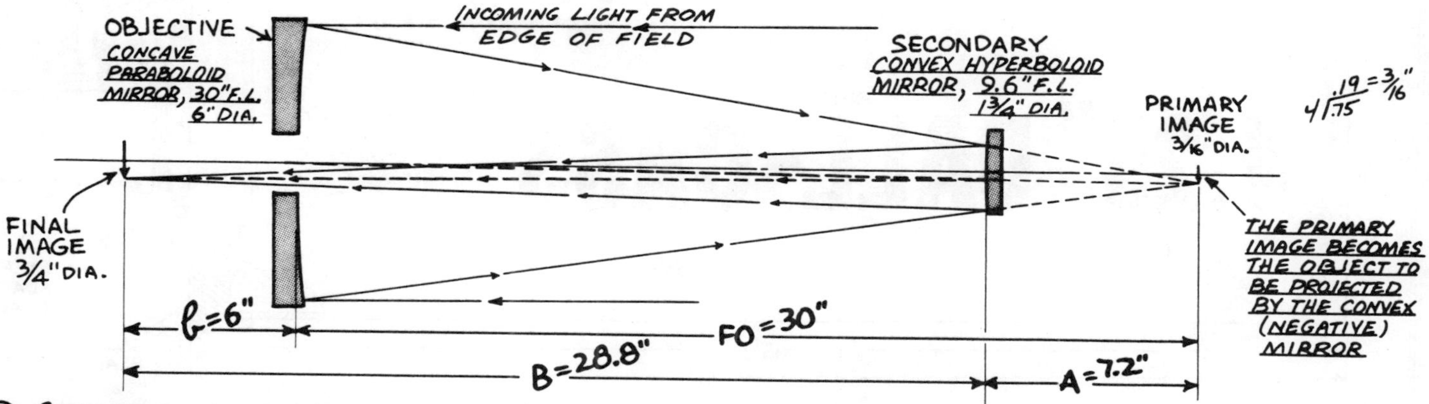

⑧ **CASSEGRAIN OPTICAL SYSTEM – 4x EXAMPLE**

as being a similar system done with mirrors. All calculations are the same as for a Barlow (Case 5) except you need one preliminary equation to include distance b, which is image distance behind the primary mirror. This is given in Fig. 10, together with the two Case 5 equations needed to complete the math work.

The drawing of a Cassegrain system is quite simple after you have determined the spacing. As usual, a cone of light is drawn to the edge of the primary image. At the points where these rays cut the surface of the secondary mirror, the rays are reversed and drawn to the edge of the final image, Fig. 8. The outer ray of the light cone is your guide for all diameters along the light cone; it tells you how big to make the secondary mirror, hole in primary, glare stops, etc. Like the Gregorian, the Cass needs glare stops or a baffle tube to stop stray light.

Like any mirror, the hyperbolic secondary may be used at various object-image positions, but it has one specific set of conjugate foci that give perfect imagery without spherical aberration. This pair of stigmatic foci are made to coincide with the positions of the primary and final images, as can be seen in Figs. 8 and 9. The foci should not be confused with the focal length of the mirror, which is, as usual, one-half the radius of the center zone.

Both the Gregorian and Cass can be built with unperforated primary mirrors. Amateurs often favor this construction to get around the sometimes difficult job of cutting a hole in the primary. Fig. 11 shows a typical unperforated Cassegrain. The right-angle bend in the light cone can be handled with the diagonal and mechanical parts used for a Newtonian reflector. The bend in the light cone serves also as a light baffle. It also erects the image but leaves it reversed left to right. The Cass requires good optics and the beginner should not tackle this work until he has completed the simpler Newtonian reflector.

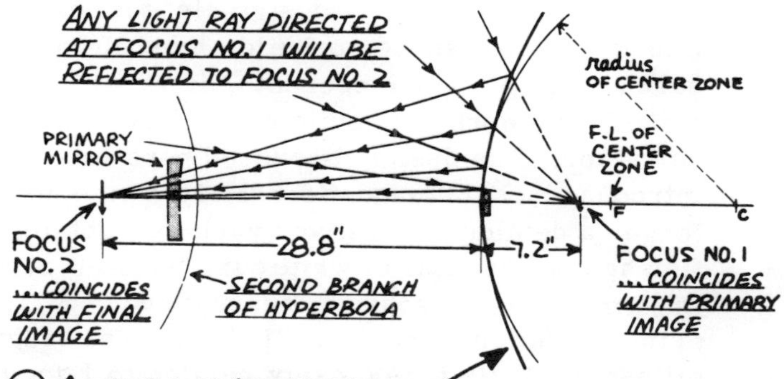

⑨ **CONVEX HYPERBOLOID**

⑩ **HOW TO Calculate A CASSEGRAIN**

You Specify	OBJECTIVE MIRROR ANY DIA., 3" and UP USUALLY LOW f/-f/2 TO f/6	Example: (DRAWING ABOVE) 6" DIA., 30" F.L. = f/5
	M. M. OF SECONDARY 2X TO 6X	4x
	b IMAGE DISTANCE BEHIND OBJECTIVE	6"
You Calculate	CASS $A = \frac{F+b}{M+1}$	$\frac{30+6}{4+1} = \frac{36}{5} = 7.2"$
	$B = A \times M$	$7.2 \times 4 = 28.8"$
	SECONDARY $F = \frac{B}{M-1}$	$\frac{28.8}{4-1} = \frac{28.8}{3} = 9.6"$ F.L.
WHOLE TELESCOPE	E.F.L. = FO × M	E.F.L. = 30 × 4 = 120"
	f/VALUE = f/OBJ. × M	f/VALUE = f/5 × 4 = f/20

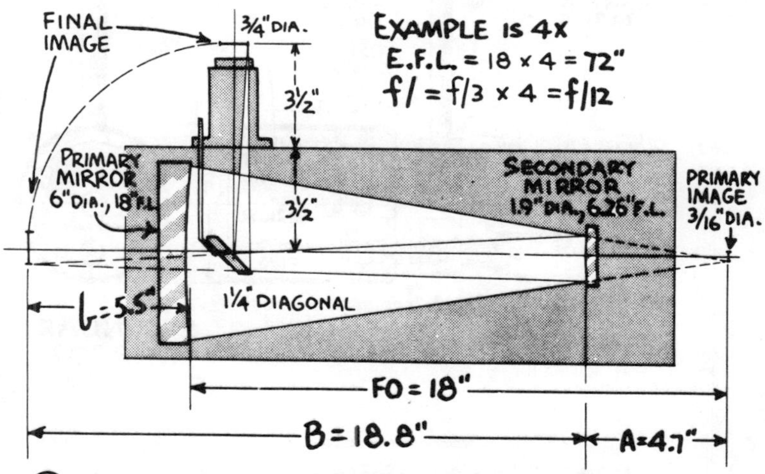

⑪ **CASSEGRAIN with UNPERFORATED PRIMARY**

4.25" DALL-CASS

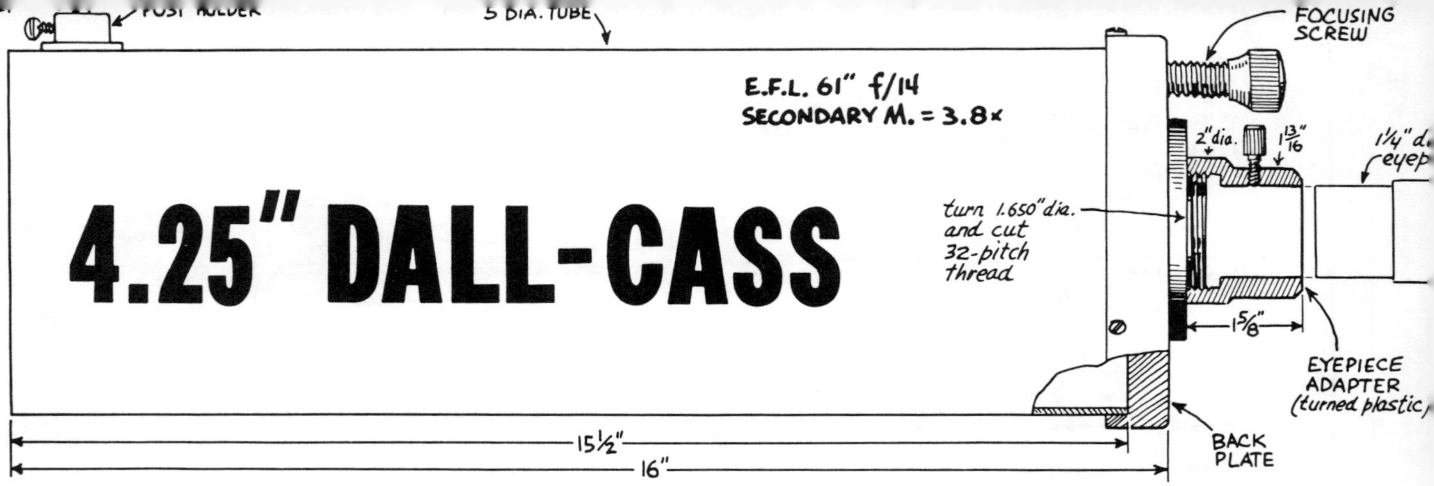

The Dall-Kirkham design is the easiest Cassegrain telescope to build. Even if you use purchased optical components, it is still about twice the work of building a Newtonian or refractor. The design discussed here was first introduced as Edmund No. 71,306 which is no longer a catalog item. If you wish help obtaining finished optics, call or write our National Sales Department for sources nearest you. The primary shown here is a 16 inch focal length ellipsoid. While it is closely related to Edmund No. 42,451 (17-1/2 inch focal length parabola) it differs not because it has a hole in the middle, but it has less aspheric figure. If a parabolic primary is used, then a hyperbolic secondary is required and testing is more difficult. In this design the secondary is simply a 1-1/4 inch diameter convex mirror with a negative 5.6" focal length. Used together the combination of ellipsoidal primary and spherical secondary eliminates spherical aberration.

Focusing is done internally by sliding the main mirror back and forth on the hollow spindle. The range is from about 30 ft. to infinity. The moving mirror also allows you to obtain various magnifications, the idea being much the same as a Barlow lens in a refractor. In performance, the field is sharp and clear at any distance. Brightness is only about half of an f/8 reflector, but still brilliant on daytime objects. The double set of front and rear baffle tubes kills almost all of the glare light which is so annoying in similar instruments without baffles. In weight and size, mounting can be as simple as an ordinary camera tripod. The short main tube of a Cassegrain with most of the weight at the rear makes fork mounting a first choice for astro use.

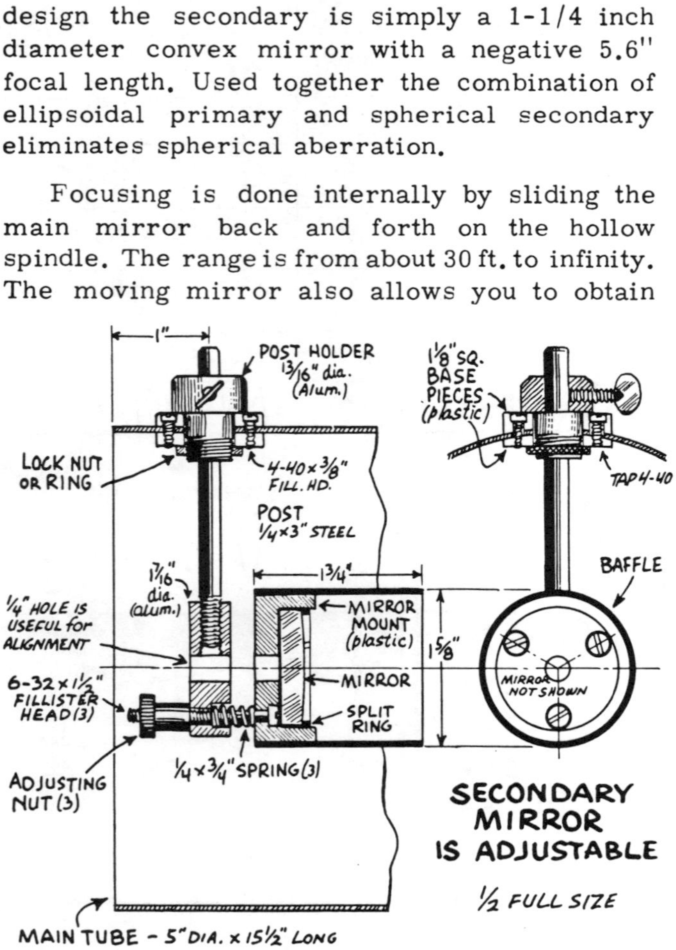

SECONDARY MIRROR IS ADJUSTABLE

½ FULL SIZE

MOUNTING THE SECONDARY. This is about the same problem as a Newtonian, and the best solution seems to be the single post, although some builders prefer three or four-arm spiders. A base block is needed for the post mounting, this being made of inside and outside plastic parts to be bolted through the tube, as shown. Then drill a 1/2 in. hole through the mounted base to take the post holder of turned aluminum. As can be seen in the drawing, the secondary mirror mount is the familiar spring mounting worked by three screws. When you are collimating the instrument, you can manipulate the adjusting nuts with your hand while your eye is peering through the hollow spindle. The mirror is held in place in the mount with a split ring of steel or fiber. It is a good idea to drill a 1/4 in. hole at the center of the mirror mount and its supporting base; if you use any kind of 3-arm aligning jig, perfect centering is assured when you have the hole in the mirror mount directly under the hole in the centering jig.

MOUNTING THE PRIMARY. The first thing you need here is the hollow spindle on which the mirror slides. The hole in the mirror is 1-1/16-in. diameter. The spindle bushing made of plastic must fit inside this hole. Allowing 1/16-in. wall thickness for the bushing, you get 15/16-in. for the outside diameter of the spindle itself. Again allowing 1/16-in. wall, you get 13/16-in. as the inside diameter of the spindle. You may be able

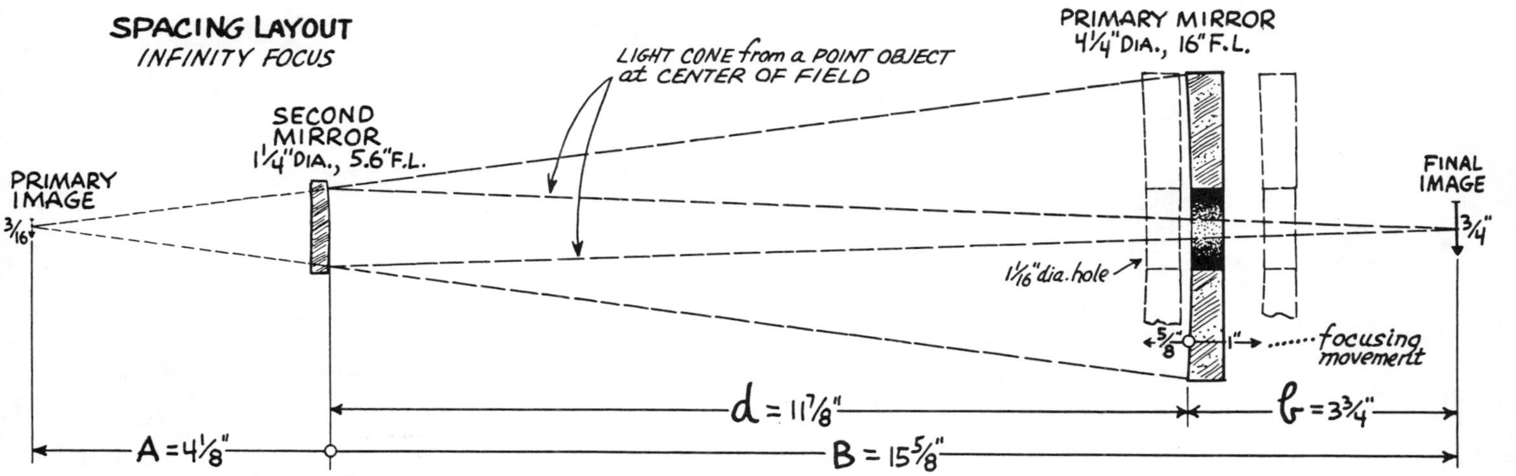

to find steel or brass tubing this size; if not the job of turning the 5-in. length is not excessive, even if you have to hog it out from a solid bar. On the inside of the spindle you should run a fine thread of about 48-pitch, the purpose of this being to minimize glare. You will note that although the whole telescope is completely shielded from light outside the field of view, the light from the field of view still comes down the spindle, where the glare protection is no better than a refractor without glare stops. But the internal thread acts like hundreds of tiny glare stops and is very effective in stopping unwanted reflections.

The spindle bushing which supports the mirror is turned from plastic. We used Tenite plastic for the test model and that is one nice plastic to machine. Some plastics used for pipe and pipe fittings are excellent machine stock, but others have a rubbery texture and are difficult to work.

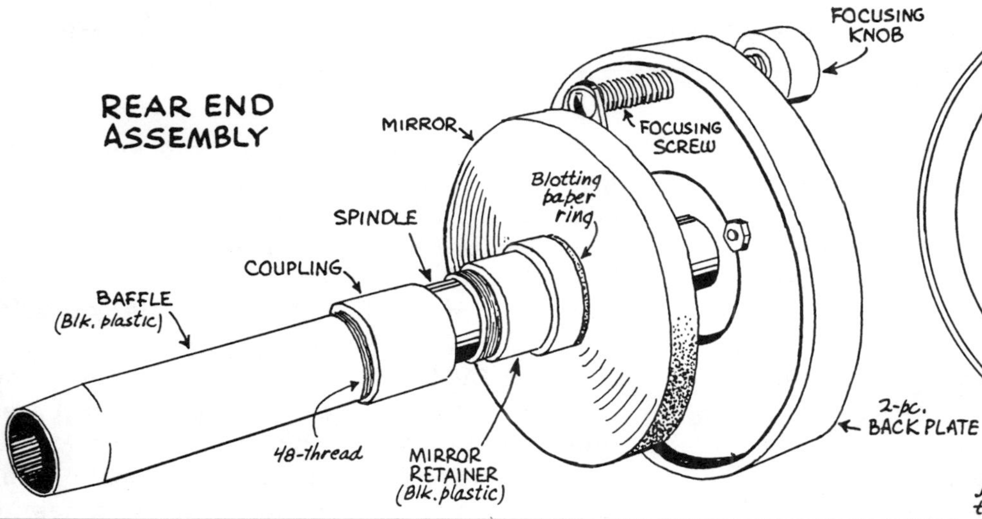

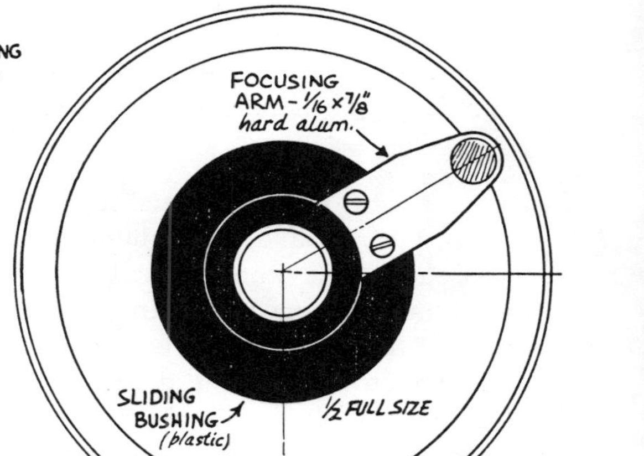

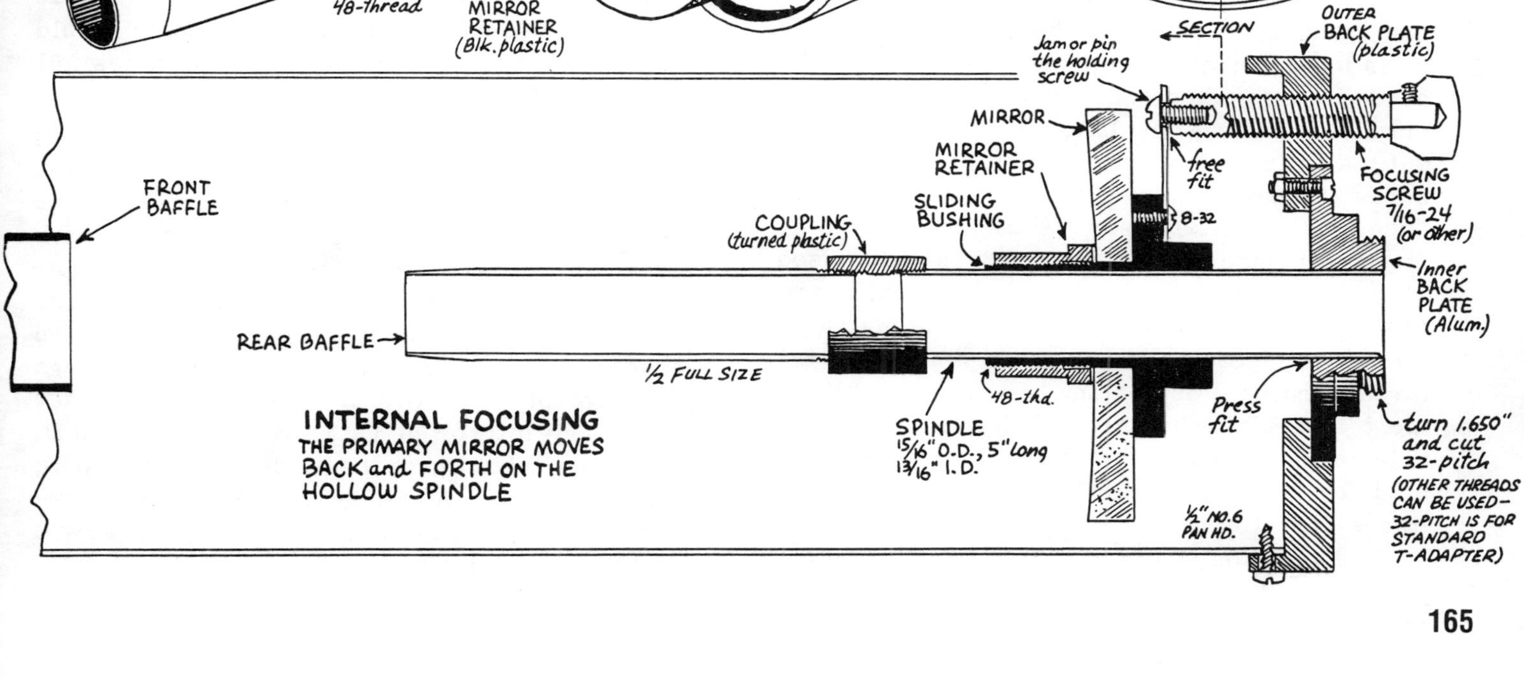

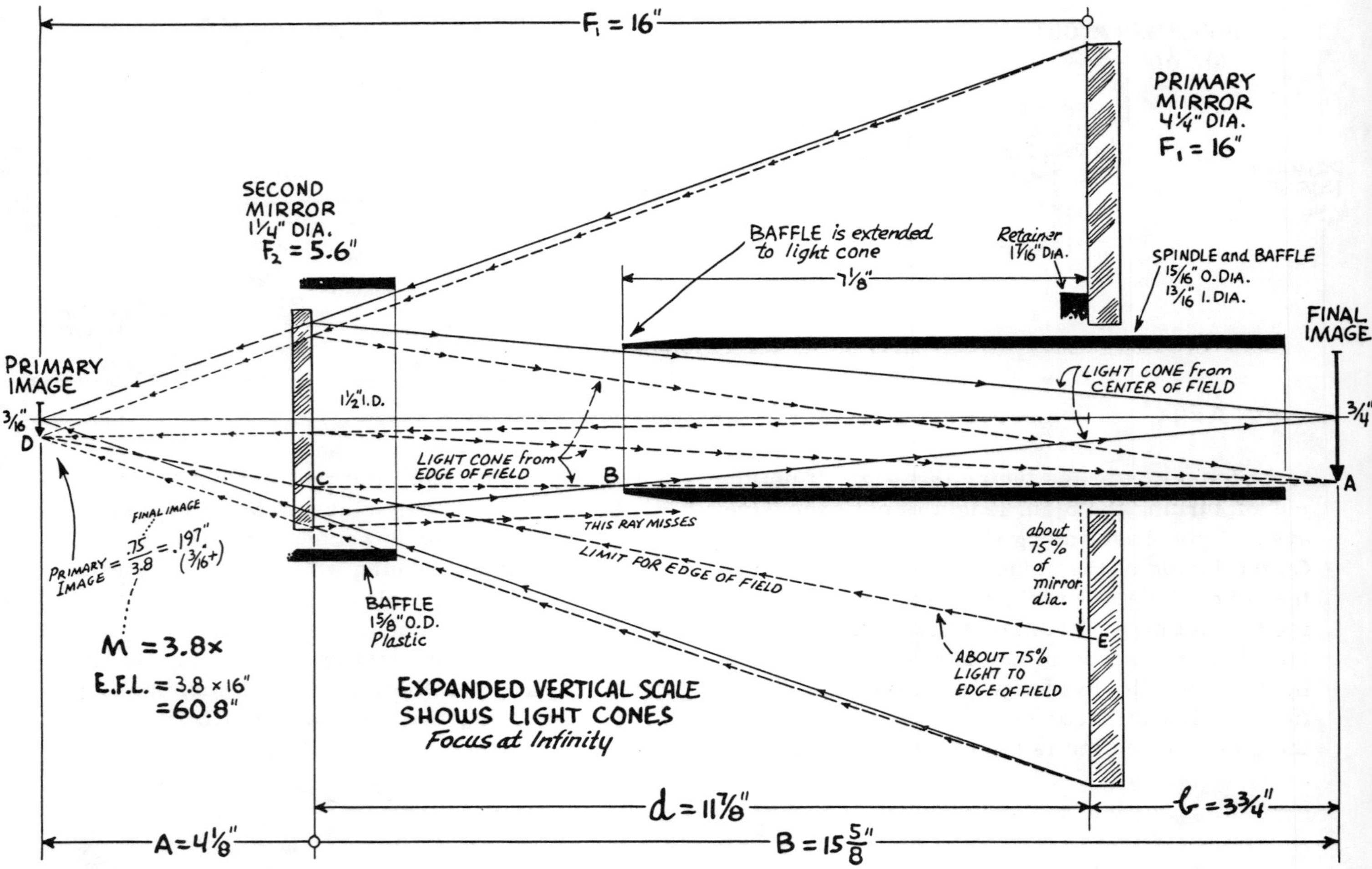

Turn the big end of the bushing first with the work mounted in a three-jaw chuck; then reverse it and you can do all other operations including boring with this second chucking. There is a 48-pitch thread to be cut on the small end of the bushing, as can be seen in the drawing. The mating part for this is a turned plastic retainer, which is used with a blotting paper washer to fasten the mirror to the bushing.

The back plate is made in two parts, the inner one being aluminum and the outer plastic. The main feature of the two-piece construction is the needed turning stock is more readily available in smaller pieces. A combination of red plastic and satin aluminum looks nice. A 7/16-in., 24-pitch screw thread is used for the focusing screw. Other sizes can be used: 3/8-24 is a more common thread and just as good; 3/8-16 can be used although it's a bit too coarse for the job. In any case, you must have a tap to cut the matching thread in the back plate--the tap you have is the thread you use. Important here is exact alignment of the hole in the focusing arm with the hole in the back plate--have both parts assembled and then drill a pilot hole through both parts in one operation.

The focusing screw allows 3/4-in. backward movement and 3/8-in. forward movement of the mirror, plus 1/4-in. each way to take care of possible misfits. You turn the screw out for near objects; you can get down to about 30 ft. You turn the screw forward or in if you want or need to form the image at a greater distance.

LIGHT DIAGRAM. The light diagram above shows how light passes through the Cassegrain telescope. In making your own diagram for this or other telescopes, the horizontal scale should be full size and the vertical scale two times full size.

This particular telescope does not field a full cone of light from edge of field, but the pickup is a comfortable 75%. As you may know, when you get 50% or better lighting at edge of field, your eyes will not detect any light loss for the simple reason the eye is more sensitive to light in this area. To determine the widest light cone from edge of field, draw a line from point A through the edge of the baffle at B, extending it to the secondary mirror at C. Then connect C and D, extending this line to the primary mirror at E, which marks the maximum cone--all of the mirror above this point not actually silhouetted will contribute to the edge-of-field light.

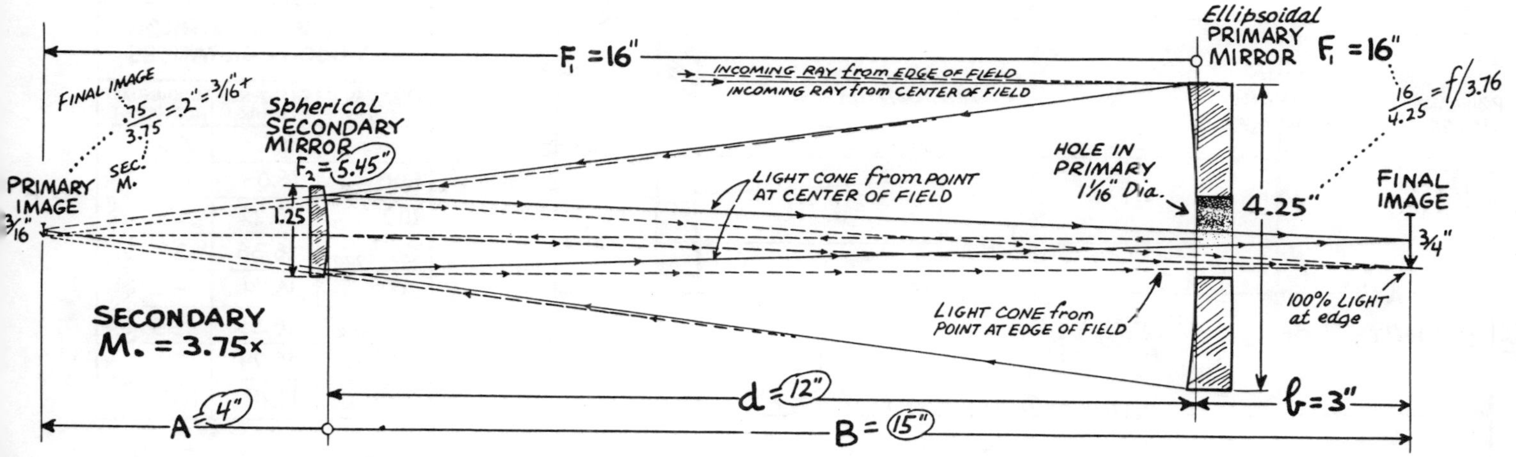

TYPICAL CASS SYSTEM

designing a Dall-Kirkham Cassegrain

THE DALL-CASS is popular with telescope builders who grind their own because the primary is a simple ellipsoid of a certain percent of a similar paraboloid, while the secondary is a plain sphere. The general idea is to balance the spherical aberration of the positive primary with the same amount of s.a. in the opposite direction produced by the negative second mirror. Technically, any design is made for a certain amount of secondary magnification, but up to twice the basic M. can be used with no apparent loss of image quality.

The simple thin-lens equations can be used for calculating any design. Our example is much the same as the current Edmund kit but with a little shorter f.l. secondary to exactly match the arithmetic of the equations. You start by specifying a few of the unknown quantities and from this starting data you can get all needed dimensions, as shown in the drawing.

A feature of the Cass system which may surprise you is that a small change in the spacing distance (d), can make a big difference in the focal plane distance (b). You can get the idea from the table at bottom of page; this is equally true in the opposite direction. Because the image plane changes so radically with even a small change of spacing, you will not see many Cass telescopes with the spring-mounted primary which is almost standard for the Newtonian design. Of course you can see why: You move the spring-mounted primary a little when you collimate, and this little movement can put the image plane an inch or more out of position. A related construction feature is that the usual rack-and-pinion focusing is almost useless if you want to view land objects at short range.

Most commercial Cass and Maksutov tele-

you Specify:

	GUIDELINES	EXAMPLE
DIAMETER OF PRIMARY MIRROR	SAME MIRROR BLANKS AS ORDINARY REFLECTORS—3, 4¼, 6, 8 and 10." 4¼" and 6" MOST POPULAR	4¼"
FOCAL LENGTH of PRIMARY	USUALLY LOW f/NUMBER, f/3 to f/5, i.e., the F.L. WILL BE 3 to 5 times the DIAMETER	16"
MAGNIFICATION of SECONDARY	USUALLY 2.5x to 6X. BOTH F.L. and f/value of PRIMARY ARE MULTIPLIED BY SECONDARY M.	3.75×
b IS IMAGE DISTANCE behind PRIMARY	SHOULD BE SHORT for a COMPACT TELESCOPE – 3" to 4". Extra b-distance is easily obtained with slightly less space between mirrors	3"

then Calculate:

	EXAMPLE as above
CASS $A = \frac{F_1 + b}{M + 1}$	$\frac{16+3}{3.75+1} = \frac{19}{4.75} = 4"$
$B = A \times M$	$4 \times 3.75 = 15"$
$d = B - b$	$15 - 3 = 12"$
$F_2 = \frac{B}{M-1}$	$\frac{15}{3.75-1} = \frac{15}{2.75} = 5.45"$ (EXACT 5.4545")

Whole TELESCOPE (with OBJECT at INFINITY) — SLIGHT DIFFERENCES MAY OCCUR FROM USING ONLY TWO DECIMAL PLACES

$f/ = f/_{\text{OF OBJ.}} \times M$	$f/ = f/3.76 \times 3.75 = f/14.1$
$\text{E.F.L.} = F_1 \times M$	$\text{E.F.L.} = 16 \times 3.75 = 60"$ (EXACT)
ALTERNATE $\text{E.F.L.} = \frac{F_1 \times F_2}{F_2 + d - F_1}$	$\text{E.F.L.} = \frac{16 \times 5.45}{5.45 + 12 - 16} = \frac{87.2}{1.45} = 60.13$

USE SIMPLE ARITHMETIC. ALL QUANTITIES ARE POSITIVE and the MINUS SIGN MEANS ONLY SUBTRACTION

Note: A LITTLE LESS 'd' MEANS A LOT MORE 'b'

	d	b	A	B	M	EFL
EXAMPLE as above →	12"	3"	4"	15"	3.75×	60"
	11.9	4.61"	4.1"	16.51"	4.03×	64.5"
JUST 1/10-inch SHORTER at 'd' MAKES NEARLY 1 5/8" LONGER at 'b'	11.8	6.47	4.2	18.27	4.35	69.6
	11.7	8.62	4.3	20.32	4.73	75.7
	11.6	10.4	4.4	22.77	5.18	82.9
Formulas →		$B - d$	$F_1 - d$	$\frac{F_2 \times A}{F_2 - A}$	$\frac{B}{A}$	$F_1 \times M$

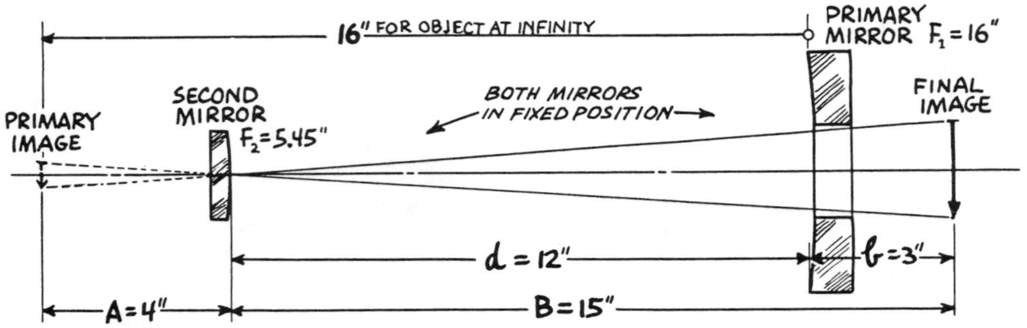

A INFINITY FOCUS

E.F.L. of 16" F.L. MIRROR at VARIOUS DISTANCES

DISTANCE TO OBJECT	DISTANCE TO IMAGE	Amount over NORMAL F.L.
∞	16"	.00"
500 FT. 6000"	16.04	.04
100 FT. 1200"	16.22	.22
60 FT. 720"	16.36	.36
50 FT. 600"	16.44	.44
40 FT. 480"	16.55	.55
30 FT. 360"	16.74	.74
20 FT. 240"	17.14	1.14

ADD THIS AMOUNT TO d and SUBTRACT from b

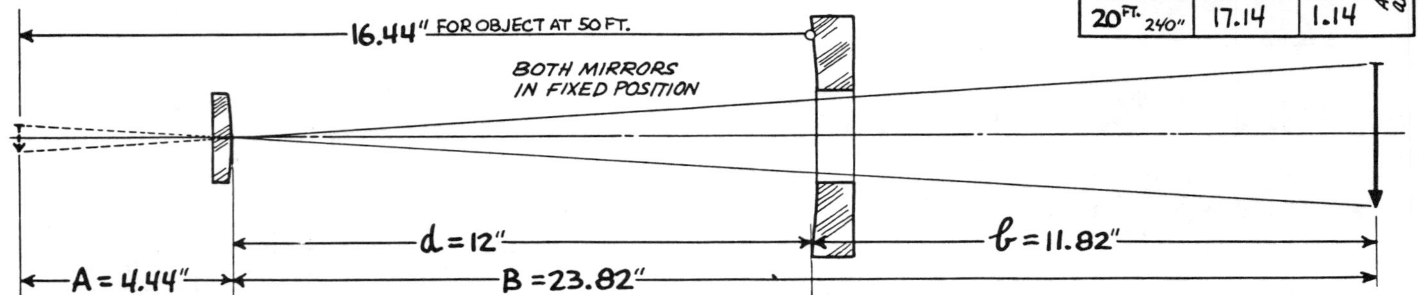

B FOCUS at 50 FT. with FIXED MIRRORS THROWS IMAGE FAR TO REAR... *may be beyond focusing range*

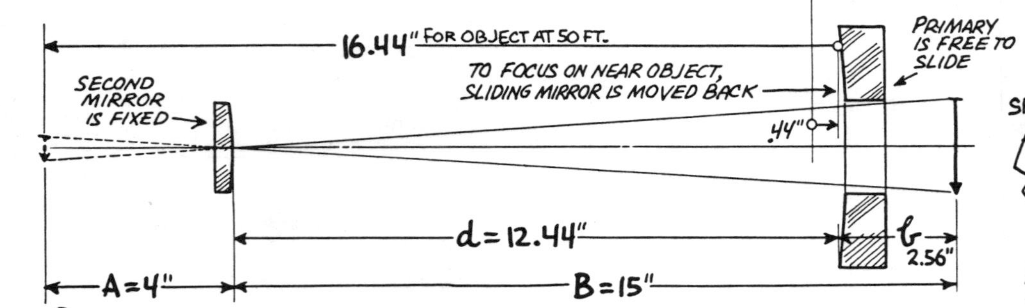

C **Constant IMAGE PLANE** IS OBTAINED BY SLIDING MIRROR... *image always in same position for any distance*

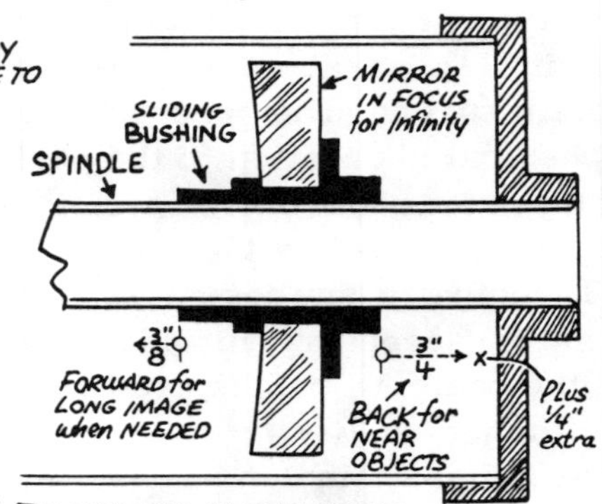

D FOCUSING MOVEMENT

scopes use internal focusing accomplished by moving the primary mirror. Using this focusing method you have a constant image plane for either distant or near objects. In addition, you can extend the image plane with a short extension to get higher M. with the same constant image plane at the new position.

The whole subject of focusing is shown in the diagrams on this page. Fig. A shows the mirror spacing and image position for an object at infinity. Now, if you focus on a near object, say at 50 ft., the effective focal length of the primary will be 16.44" and the primary image will be located at this greater distance. Then the secondary augments the final image distance, pushing it back several inches, as in diagram B.

But if you are using a movable mirror to focus, you simply take up the extra .44 inch image distance by increasing the mirror spacing by this amount. The result is that you have the same A and B-distances as before and the image remains in the same position, Fig. C. The idea is that you increase the mirror spacing for near objects; in the opposite direction, you make the spacing shorter if you want to extend the image plane. Originally distance "b" is made the least practical distance you are likely to use for the image plane. It is then quite simple to extend this with a small forward movement of the primary.

Since it is likely you will use the scope for land views as well as celestial, most of the focusing movement should be a movement of the mirror to the rear, which is what you need for near objects. In exact figures, the mirror should be allowed about 1 inch of backward movement for a 16 in. f.l. primary, as shown in diagram D. This will get you down to about 30 ft. with some to spare to cover a possible error in the initial spacing between mirrors. In the opposite direction, 3/8 in. less spacing will put the image plane about 10 inches behind the main mirror, which is plenty for a star diagonal and other attachments requiring a long image position. Here again about 1/4 in. extra should be allowed. So the whole mirror movement for a 16 in. f.l. objective is about 1-5/8 inches. When fully extended the focusing knob may interfere with mounting a camera with a short T-adapter. The cure is to use a short extension tube to make the telescope-to-camera connection.

GRINDING AND TESTING. If you have made one or more parabolic mirrors, you will have no trouble with the Dall-Kirkham primary. You work this just like a paraboloid except the correction is less--somewhere between 60 and 80-percent of a parabola of the same diameter and focal length. Dall's formula for this, slightly modified, is shown in the drawing with our sample design worked out, the correction being 70.6% of a similar paraboloid. You then calculate the normal correction for a paraboloid of the same diameter and focal length. Multiply these figures by .706 and you have the mirror correction for the ellipsoid. Knife-edge testing then proceeds in the same manner as for the paraboloid. The pattern you see on the mirror is the familiar doughnut--the same as for a paraboloid--but the shadows are not quite as contrasty. When you get a smooth figure 10% more or less than the exact correction, you can call the mirror finished.

An alternate method of testing the ellipsoid is by using the conjugate foci. You need to know the eccentricity of the ellipse, which is simply the square root of the percentage correction, as shown. Putting this figure through a simple formula will give you the conjugate foci, i.e., that particular pair of object-image distances where an object placed at one focus will be imaged without spherical aberration at the other. This being the case, light from a small pinhole at the near focus should vanish instantly and completely when a knife-edge is pushed into the light cone at the long focus, as shown in the bottom diagram. Try various positions for the knife-edge up to several inches away from the calculated image plane--when you get a null test, you know the figure is an ellipsoid.

MAKING THE SECONDARY. Being spherical, this is the simplest of all shapes to grind, but being a negative (convex) curve, you have problems in finding the focal length and determining the "figure." An easy way to find the f.l. is simply by measuring the f.l. of the glass tool, giving it a superficial polish if needed.

One common way of testing the figure of the secondary is to make it up as a simple plano-convex lens. The convex surface becomes concave when tested through the glass, using an optical flat to return the light from the pinhole.

DALL-KIRKHAM FORMULA

% Correction

of a SIMILAR PARABOLOID

$$\% \text{ CORRECTION} = 1 - \left(\frac{A^2 \times (A+B)^2}{B^2 \times F_1 \times F_2}\right)$$

$$= 1 - \left(\frac{4^2 \times (4+15)^2}{15^2 \times 16 \times 5.4545}\right)$$

$$= 1 - \left(\frac{\not{16} \times 19^2}{225 \times \not{16} \times 5.4545}\right)$$

$$= 1 - \left(\frac{361}{1227.2625}\right)$$

$$= 1 - .29415 = .70585$$

CORRECTION = 70.6%

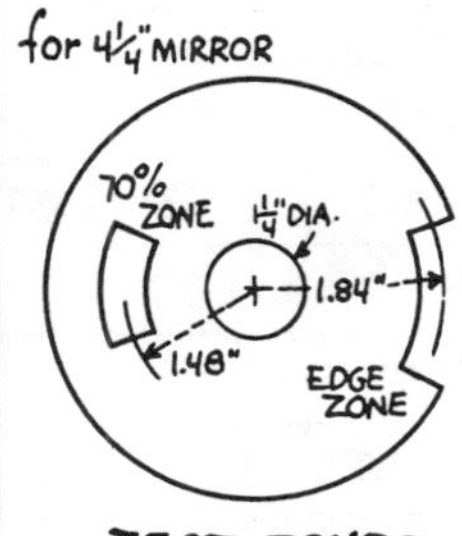

TEST ZONES

% Correction applied:

	PARABOLA	ELLIPSE
EDGE ZONE	$\frac{r^2}{R} = \frac{1.844^2}{32} = \frac{3.400}{32} = .106"$ times .706	.075"
70% ZONE	$\frac{r^2}{R} = \frac{1.484^2}{32} = \frac{2.202}{32} = .069"$ times .706	.049"

Eccentricity of ELLIPSE: Example

$$\text{ECCENTRICITY } (e) = \sqrt{\text{CORR.}} = \sqrt{.706} = .840$$

Conjugate FOCI:

$$1^{st} \text{ FOCUS} = \frac{R}{1+e} = \frac{32}{1+.84} = \frac{32}{1.84} = 17.391"$$

$$2^{nd} \text{ FOCUS} = \frac{R}{1-e} = \frac{32}{1-.84} = \frac{32}{.16} = 200"$$

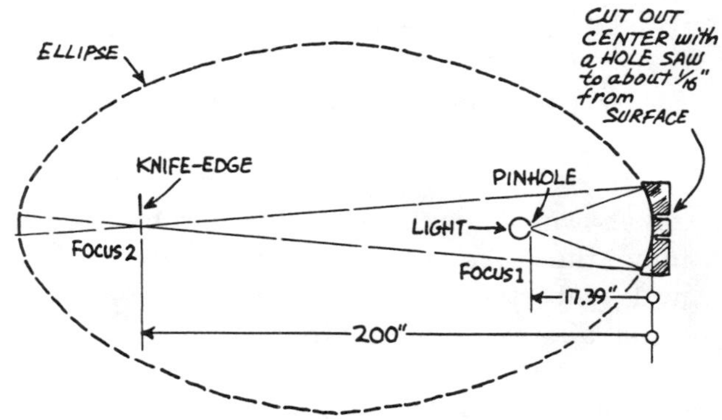

NULL TEST at CONJUGATE FOCI

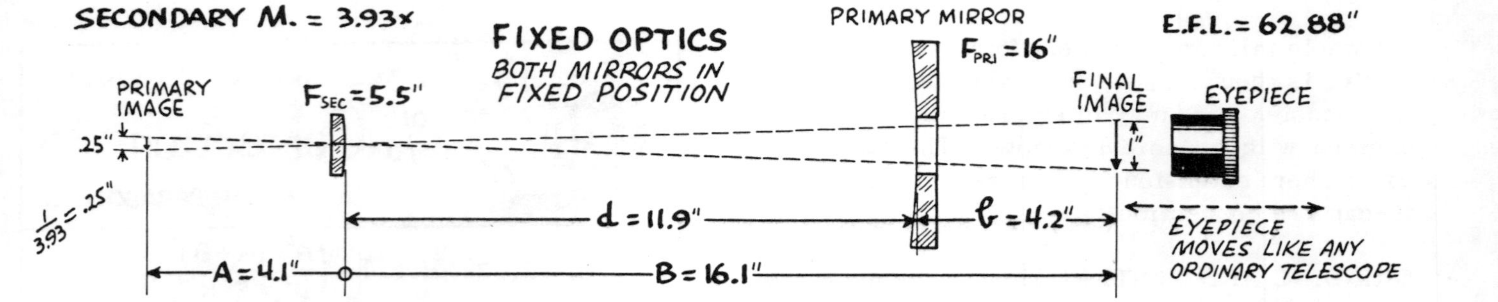

VARIOUS SPECIFICATIONS for 16" F.L. PRIMARY and 5.5" F.L. SECONDARY

USE ANY LINE... Diagram above is second line (Example below: line 4.2)

A	B	d	b	SEC. M.	E.F.L.
4.0"	14.67"	12.0"	2.67"	3.67×	58.72"
4.1	16.10	11.9	4.20	3.93×	62.88
4.2	17.77	11.8	5.97	4.30×	68.80
4.3	19.71	11.7	8.01	4.58×	73.28
4.4	22.00	11.6	10.40	5.00×	80.00
4.5	24.75	11.5	13.25	5.50×	88.00
4.6	28.11	11.4	16.71	6.11×	97.76
4.7	32.31	11.3	21.01	6.87×	109.92

Formulas: $A=\frac{F_{PRI}+b}{M+1}$ $B=\frac{F_{SEC}\times A}{F_{SEC}-A}$ $d=F_{PRI}-A$ $b=B-d$ $M_{SEC}=\frac{B}{A}$ $E.F.L.=M\times F_{PRI}$

"A" Less than 4" — **"b" INSIDE PRIMARY**

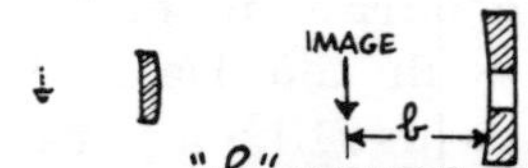

A	B	d	b	SEC. M.	E.F.L.
3.9"	13.41"	12.1"	+1.31"	3.44×	54.99"
3.8	12.29	12.2	+.09	3.24×	51.78
3.7	11.31	12.3	−.99	3.06×	48.90
3.6	10.42	12.4	−1.98	2.89×	46.32
3.5	9.62	12.5	−2.87	2.75×	44.00
3.4	8.90	12.6	−3.69	2.62×	41.90
3.3	8.25	12.7	−4.45	2.50×	40.00
3.2	7.65	12.8	−5.15	2.39×	38.26

for "b" INSIDE PRIMARY: $A=\frac{F_{PRI}-b}{M+1}$ $B=\frac{F_{SEC}\times A}{F_{SEC}-A}$ $d=F_{PRI}-A$ $b=B-d$ $M_{SEC}=\frac{B}{A}$ $E.F.L.=M\times F_{PRI}$

Caution: YOU CAN'T INTRODUCE VARIOUS VALUES OF "b" INTO THE FORMULAS FOR "A" BECAUSE "b" DEPENDS ON F_{SEC} WHICH IS ALREADY SPECIFIED

IF A CHANGE IN F_{SEC} IS PERMISSIBLE, YOU CAN USE THE "A" FORMULA FOR ANY SUITABLE VALUES FOR M_{SEC} AND b

like: $M_{SEC} = 4\times$ and $b = 6"$

then: $\frac{F_{PRI}+b}{M+1} = \frac{16+6}{4+1} = \frac{22}{5} = 4.4"$ (is A)

$B = A \times M = 4.4 \times 4 = 17.6$

$F_{SEC} = \frac{B}{M-1} = \frac{17.6}{4-1} = \frac{17.6}{3} = 5.87"$ (instead of 5.5")

but IF YOU WANT "b" = 6" WITH THE ORIGINAL SECONDARY OF 5.5" F.L., YOU WOULD USE THE NEAREST "b" IN THE TABLE, WHICH IS 5.97"

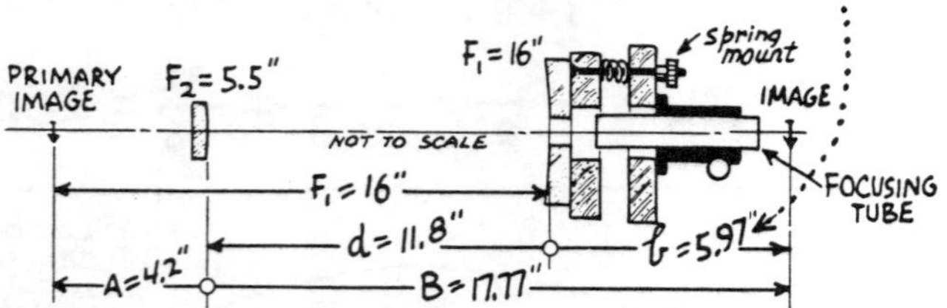

BIG "b" NEEDED FOR Rack-and-pinion FOCUSING

as shown. A dark red filter can be used to remove the false color. Unfortunately this is not a null test--the figure you see is the same as for a paraboloid but fainter. What you want is a smooth curve without zones. Some amateurs have obtained good results with small spherical secondaries without testing.

SPACING VARIATIONS. If you buy a finished set of Dall-Cass optics you will get a specification sheet. Usually this will list a minimal "b" dimension of about 3 inches, which you may have to increase or decrease to suit the construction and accessories you plan to use. If you want to use conventional rack-and-pinion focusing you will need a much greater "b" distance. The same is true for a star diagonal. On the other hand if you want a built-in erector, distance "b" is much less, even becoming negative with the image between the two mirrors.

If you are using a set of finished optics made by someone else, you will wonder if changes in the spacings are permissible. Look at the left-hand table above in the column giving the "b" distance. Notice the top line specifies about 3 in. while the bottom line is over 20 inches. Is this really practical? Yes it is practical because what you are working with is a simple projection system. Like with a slide projector you can get a sharp image at 2 ft. or 6 ft. or 12 ft. or whatever. Of course, for technical exactness, any variation in the spacing of a Dall-Cass would call for a change in the correction of the primary mirror. This is 70.6% for the first line in table. For the bottom line--if you were polishing your own mirror--the exact correction would be 67.1%. If you have ever ground and polished a mirror, you will know that either of

4¼" DALL-KIRKHAM with ERECTOR

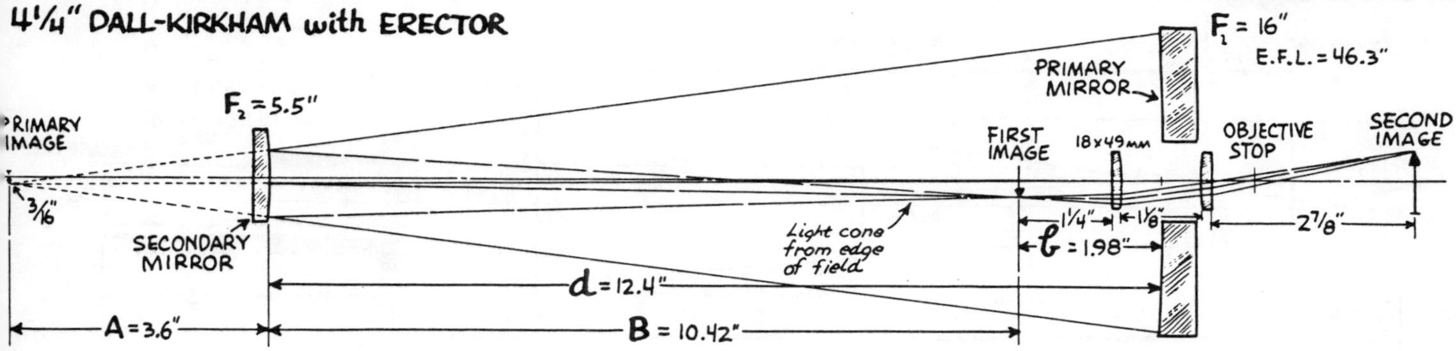

GENERAL SPECIFICATIONS:
Line 4 in right col. on opp. page
E.F.L. of ERECTOR – 1.43"
M. OF ERECTOR – 1.7×
M. OF SECONDARY – 2.89×
WHOLE M. – 1.7 × 2.89 = 4.91×
E F.L. TELESCOPE – 16×4.91 = 78.6"
f/PRIMARY – 16/4.25 = f/3.75
f/TELESCOPE – 3.75 × 4.91 = f/18
POWER with 1.1" EYEPIECE EDMUND No. 30,787 $\frac{78.6}{1.1}$ = 71×
IMAGE DIA. – .875
ERECTOR IMAGE – .875/1.7 = .51"
SECONDARY IMAGE – .51/2.89 = .18"
OBJECTIVE STOP – 1/8" DIA., 9/16" *behind second lens. Locate this visually. 5/32" dia. may be used to avoid possible light loss in focusing*

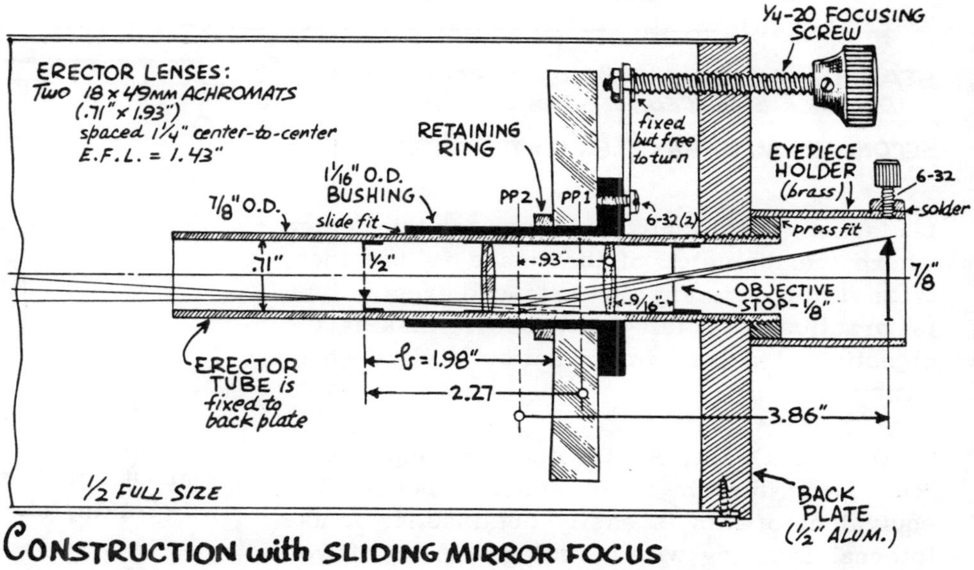

CONSTRUCTION with SLIDING MIRROR FOCUS

the figures mentioned is satisfactory, that is, a variation of 3 or 4-percent from a specified correction is permissible. As a matter of fact, any commercial Dall-Cass or Maksutov with sliding primary will use the full range of the table for focusing.

Less secondary magnification requires a smaller "b" distance. The tolerance here quickly gets beyond 3 or 4-percent, but the first five lines in the right-hand table on opposite page are within this limit. Line 7 departs 5% from normal; the last line is 6% off; if you were actually polishing a mirror to these specifications, you would make the correction 76.6% instead of the basic 70.6%. However, with the narrower light cone and reduced secondary magnification which goes with a short "b" distance, the 6% departure from basic is not too much.

RACK-AND-PINION FOCUSING. Most telescope builders were raised on R&P focusing and cling to this when building a Dall-Cass or Maksutov. The spring-adjustable primary is another holdover from the Newtonian which is often retained. The poor feature of the spring-mounted primary is readily controlled once you get the idea you can't adjust it willy-nilly. About 1/16 inch movement is actually all you need for collimation, and this short distance will usually not put the image beyond the focusing range. Distance "b" must be at least 4-1/4-in., as shown in top diagram on opposite page; 6 in. is better and you may have to go to 8 in. to accommodate a camera or star diagonal.

BUILT-IN ERECTOR. The lens erector is a favorite with Englishman H. E. Dall who is one of the originators of the Dall-Kirkham Cassegrain. He shows the erector in most of his Dall-Kirkham and Maksutov designs which have appeared at various times in Sky and Telescope magazine. The spacing for our standard set of optics with an erector is shown in the diagram above. Other than the nice feature of an erect image, the main advantage of the erector is that it forms a miniature picture of the primary mirror. This is the "objective stop." If you put a thin metal stop at this point you will effectively cut off all light rays except those coming from the field of view. In other words, the objective stop does the same job as the primary baffle tube. To be fully effective, the stop should be a hair smaller than the miniature picture of the objective. However, since the objective image will shift a little when you focus the telescope, it is best to make it a

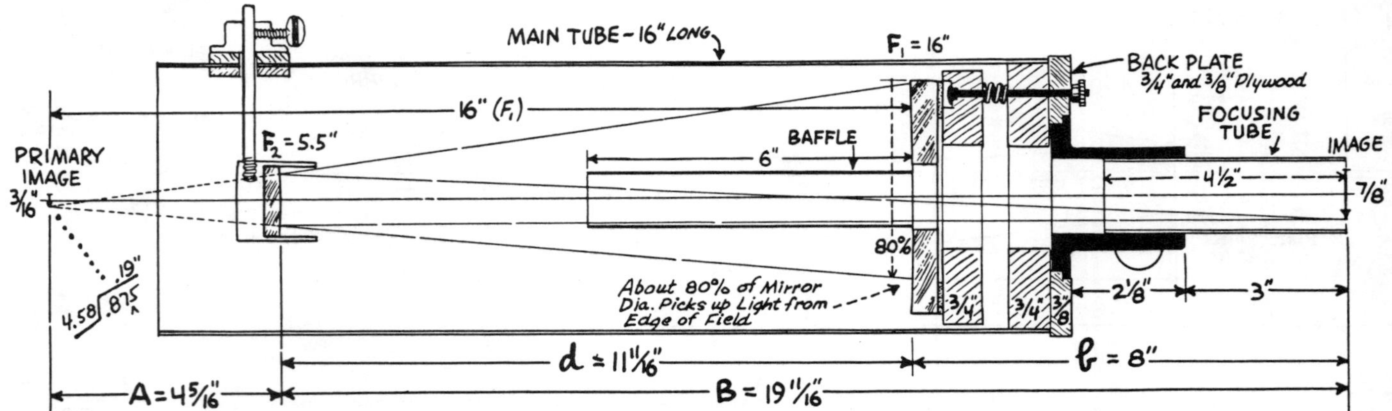

STAR DIAGONAL - *With FIXED OPTICS, A LONG "b" DISTANCE IS NEEDED, USING RACK-and-PINION FOCUSING*

SECONDARY M. = 4.58× E.F.L. = 73" f/TELESCOPE = f/17

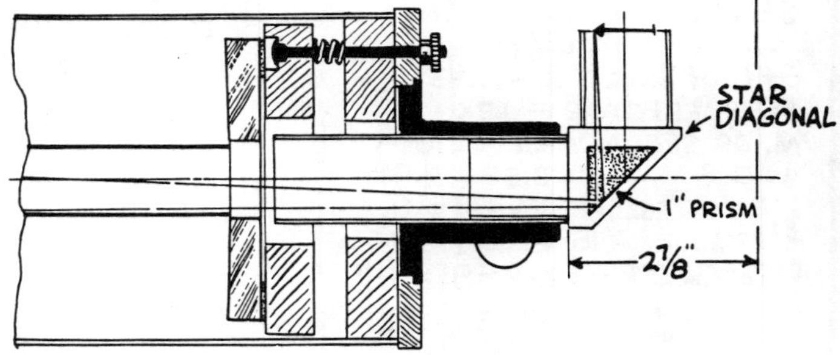

bit larger, specifically 5/32 inch for this design where the diameter of the image is 1/8 inch. With sliding primary, the extended erector tube is practically a full light baffle, making the objective stop less important but always worth fitting.

STAR DIAGONAL. A star diagonal requires an "in" focus movement of about 3 inches. The equivalent of this is easily obtained if you use internal focusing with a sliding primary. If you are using conventional rack-and-pinion focusing, the spacing for our typical Dall-Cass would be as shown in the drawing above, the main feature being the long "b" distance of 8 inches. This is line 4 in the table on a previous page, and it is well within the tolerance range.

The second diagram shows the focusing tube moved in as required to put the image at the end of the eyepiece tube.

The "in" focusing travel allowed easily takes care of a 35mm camera. If you want to use straight-ahead viewing for near objects, the needed extra "out" focus can be obtained with a short auxiliary tube, extending the image plane about 2 inches. The extension is redi-made in a 1-1/4-in. chrome sink trap extension--all you have to do is cut off about 4 in. at the flared end.

If with diagonal in place you don't have enough in-focus, you can kick the image out by manipulating the primary mirror adjusting nuts. The idea is to decrease the d-distance and this is done by unscrewing the adjusting nuts to let the mirror slip forward. One full turn on each nut should do it, and if done equally all around the collimation will remain undisturbed or at most require only a minor adjustment. Initially it is a good idea to make the spring tension somewhat strong to permit a full turn let-down and still have enough pressure for firm support.

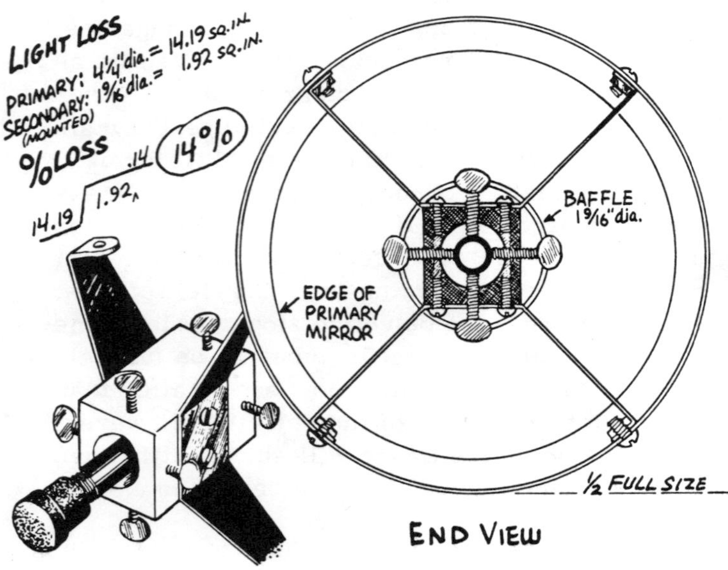

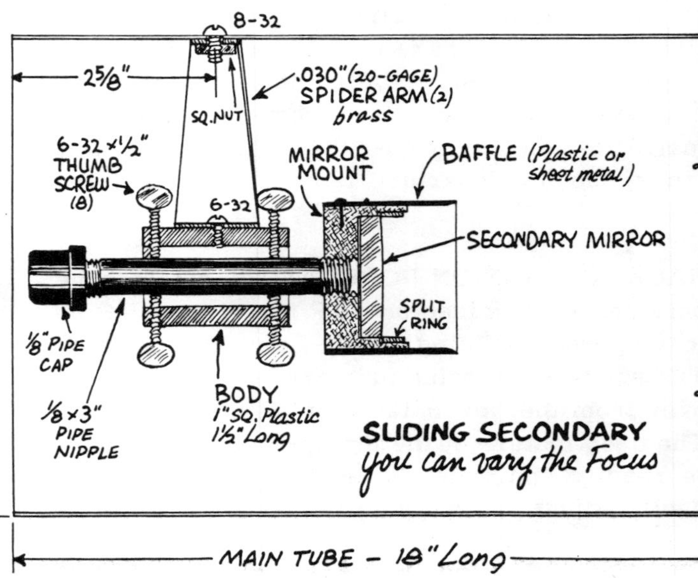

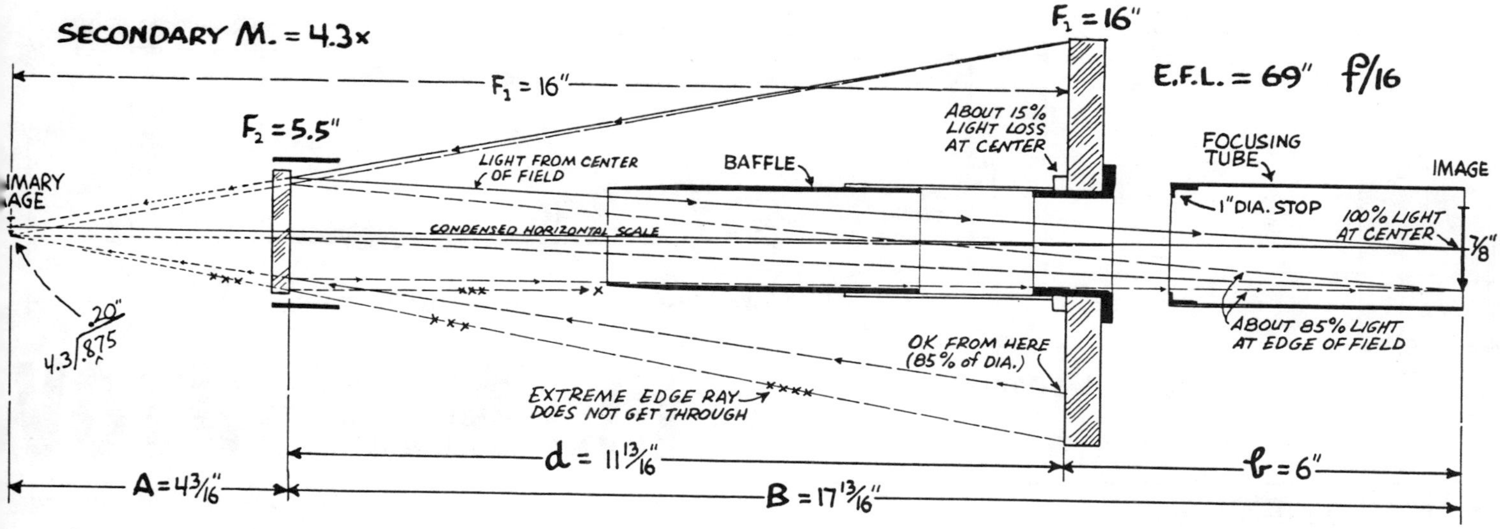

4¼-inch Dall-Kirkham with SLIDING SECONDARY

THIS IS a conventional rack-and-pinion job with the focusing range extended by a sliding secondary. The R&P is the actual focusing control; the sliding second mirror can be moved back and forth to suit certain applications of the telescope. Like if you wanted to use a star diagonal, you would find there is not enough "in" focusing movement provided. The simple solution is to move the image itself, which is done by moving the secondary about 1/8 in. closer to the primary. This will extend the image plane by about 2 inches and so put the picture within the focusing range provided.

The sliding secondary construction is elaborated to include full adjustment for centering and collimation. The set of adjusting thumbscrews nearest the primary mirror are used for centering the small secondary; the other set supplies any "tilt" needed for collimation. Four adjusting screws instead of the usual set of three make the movement less confusing. If desired, the more common three-arm spider can be used, in which case the body for the sliding mirror carrier would be round. A purchased four-arm spider with sliding mount can be used but is less convenient for centering and tilt adjustments.

The main mirror is spring-mounted for easy collimation. The main mirror is fastened to the plywood mount with clips in the usual manner. The hole in the primary is used only to support the baffle tube which sticks out an even 7 inches in front of the mirror. The rack-and-pinion focusing tube is a standard refractor part slightly modified to the dimensions given.

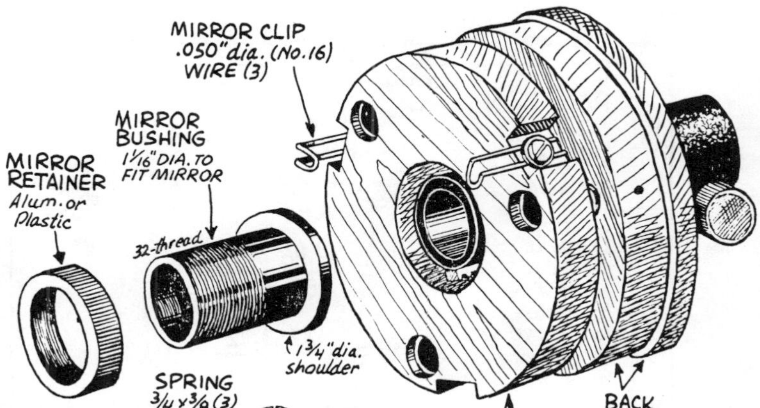

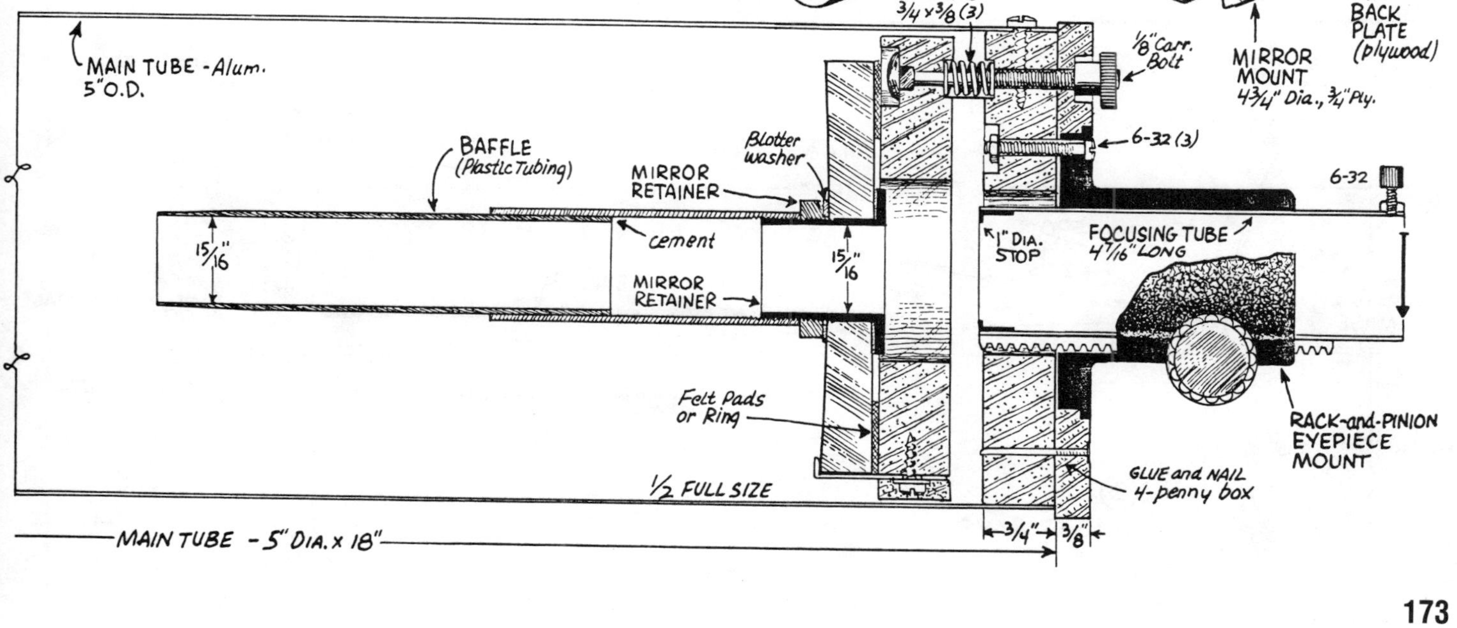

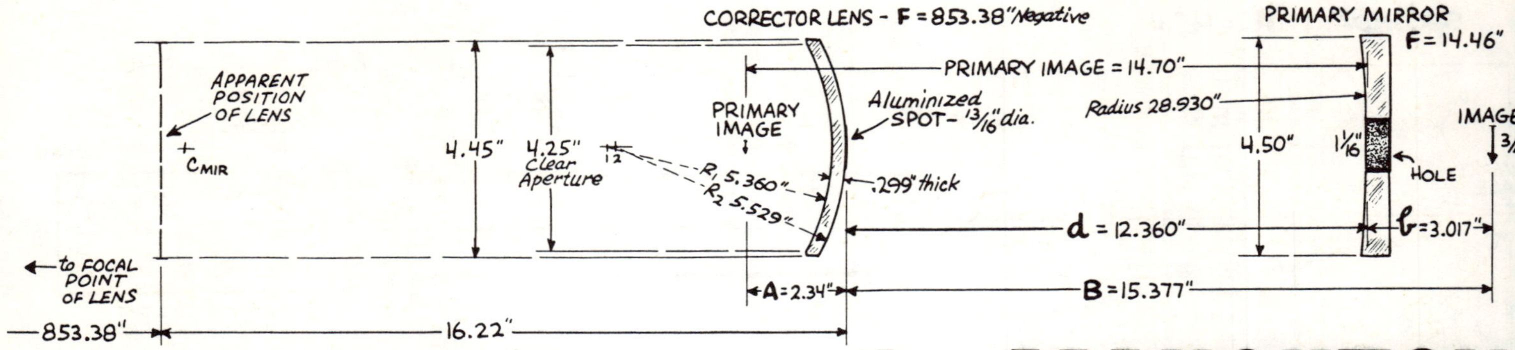

① SPECIFICATIONS and SPACING

the MAKSUTOV

The Cassegrain telescope design shown below was the invention of Bouwers in Holland and Maksutov in Russia. Hence it may be properly called Bouwers-Maksutov but it is often simply Maksutov for short. This design uses a uniform thickness meniscus corrector lens in front of a spherical primary. It is 100 inch effective focal length at f/22 but the whole instrument is only 16 inches long.

The design discussed here was originally Edmund catalog No. 1627. If you wish information about buying either the corrector blank pressing in BK-7 glass or complete finished optics, write or call our Customer Service, Technical Information.

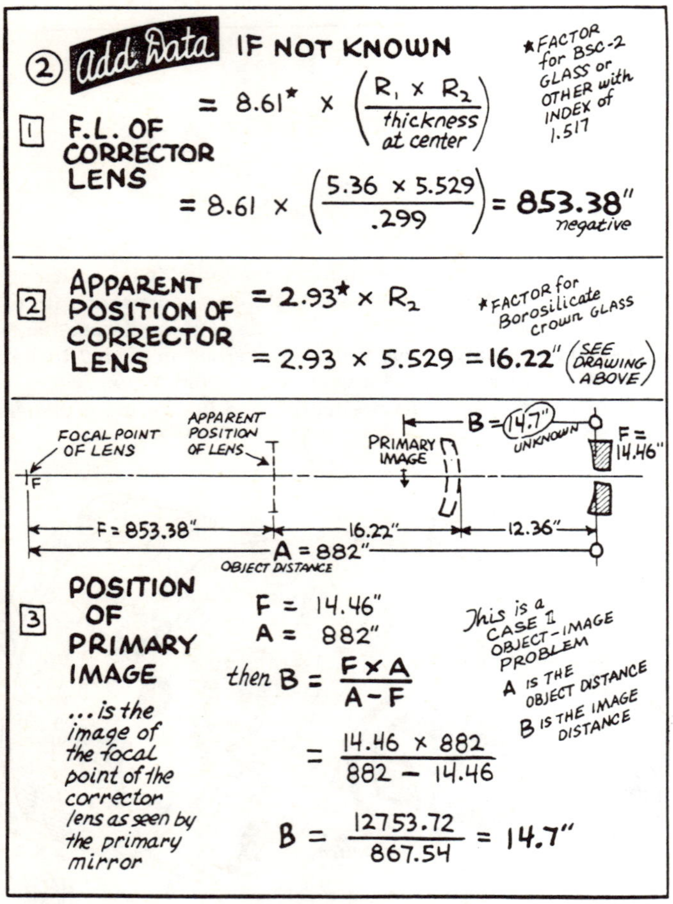

SPACING VARIATIONS. The specification sheet you get with a set of finished optics will tell you the spacing needed for a certain equivalent focal length. It is permissible to vary the spacing and if you want to do this, the table (Fig. 6) supplies the needed data; if you want a spacing not listed, the formulas at bottom of table can be used. For this or other design, additional data including f.l. of the lens, apparent position of the lens, and the position of the primary image are sometimes useful. This data can be calculated as shown in Fig. 2.

LIGHT RAYS. Fig. 3 shows how the light rays

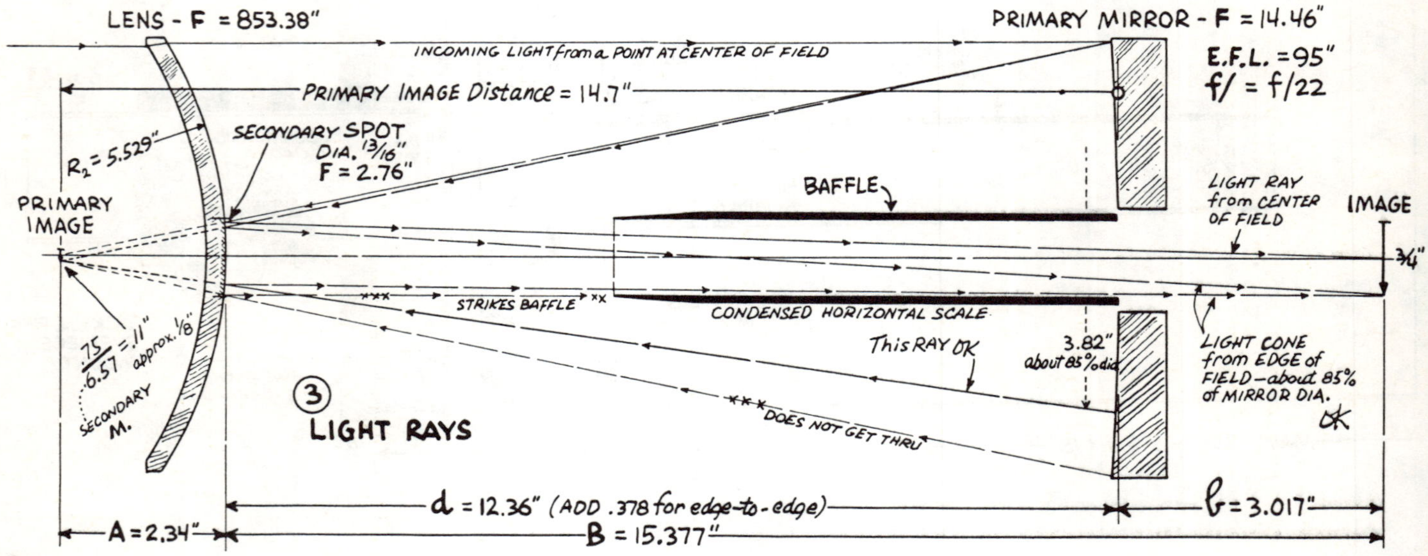

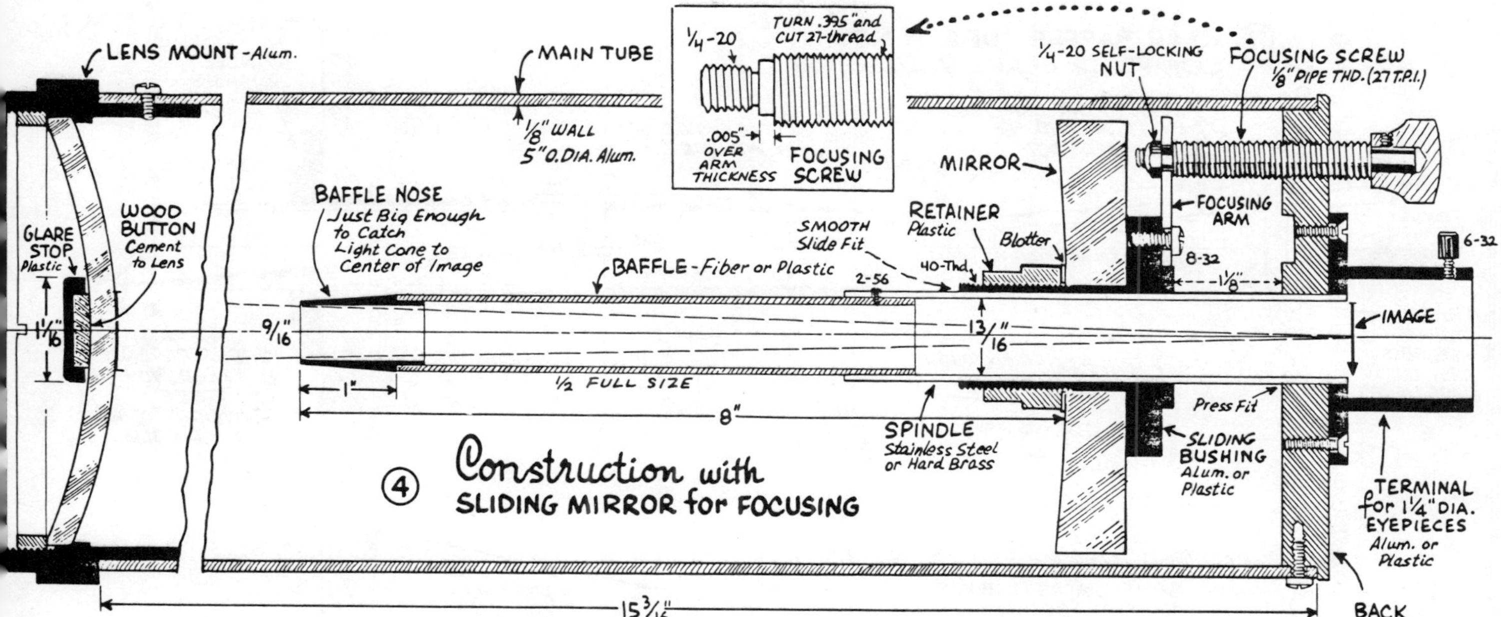

④ Construction with SLIDING MIRROR for FOCUSING

pass through a Maksutov. The secondary mirror is often just an aluminized spot on the back surface of the lens. In this design the spot has been made minimal size, the idea being to pass as much light as possible to the primary. For a high-power compound telescope, you usually consider only the light rays from the center of the field--notice the primary image is a scant 1/8-in. diameter and hardly worth considering. However, it is always best to consider the edge-of-field rays and they are shown in the diagram. In this design about 85% of the primary mirror diameter contributes to edge-of-field lighting. This is normal and entirely satisfactory. However, the 3/4 inch linear diameter of field at the image plane is not wide enough for full coverage of 35mm film; this same set of optics for photography would need a 1-1/4-in. hole through the primary. You should remember that any astronomical telescope is made specifically for visual use in viewing the night sky--it is not a camera and it is not intended for daytime use. However, most buyers and builders have come to expect this versatility. Needless to say, some buyers and builders are disappointed.

You get the full impact of this with a compound. For visual use at night, a Cass telescope can be used fairly well without any protection against glare light--you don't pick up much glare from a black sky. Inevitably however you will try it on a daytime object and you will see little or nothing but a blur of glare light. At the time when Russell W. Porter did his piece on compound telescopes for Amateur Telescope Making, baffle tubes were unknown. Porter shows his designs with no glare protection; he also comes right out and tells you not to build a compound because of its miserable performance resulting from glare light. Of course as you know baffle tubes are now in common use and give 100% glare protection

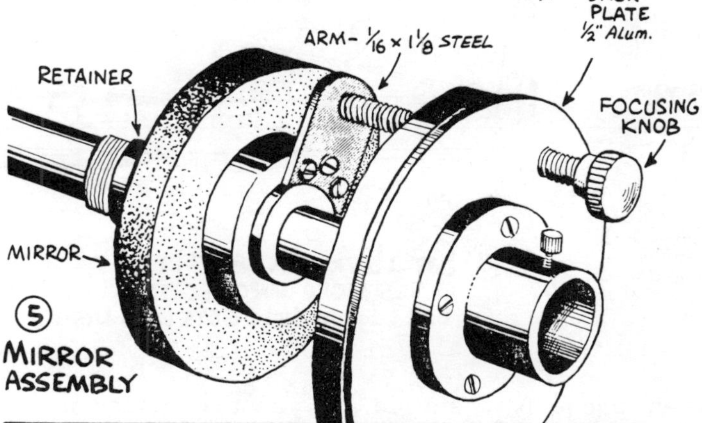

⑤ MIRROR ASSEMBLY

⑥ VARIOUS SPACINGS - *Maksutov*

LINE NO.	A	B	d	b	SEC. M.	E.F.L.	f/value
1	2.20"	10.843"	12.50"	-1.657"	4.93×	71.29"	f/16.8
2	2.24	11.889	12.46	-0.551	5.31×	76.78	f/18.1
3	2.26	12.475	12.44	+0.035	5.52×	79.82	f/18.8
4	2.28	13.110	12.42	0.690	5.75×	83.14	f/19.6
5	2.30	13.80	12.40	1.40	6.0×	86.76	f/20.4
6	2.32	14.552	12.38	2.172	6.27×	90.66	f/21.3
7	2.33	14.955	12.37	2.585	6.42×.	93.80	f/21.8
shown 8	2.34	15.377	12.36	3.017	6.57×	95.00	f/22.3
9	2.35	15.819	12.35	3.469	6.73×	97.33	f22.9
10	2.36	16.284	12.34	3.944	6.90×	99.77	f/23.4
11	2.38	17.286	12.32	4.966	7.26×	104.98	f/24.7
12	2.40	18.40	12.30	6.10	7.67×	110.91	f/26.1
13	2.42	19.645	12.28	7.365	8.12×	117.42	f/27.6
14	2.44	21.045	12.26	8.785	8.62×	124.65	f/29.3
15	2.46	22.632	12.24	10.382	9.20×	133.03	f/31.3
Formulas ⟶	$B = \frac{F_{SPOT} \times A}{F_{SPOT} - A} = \frac{2.76 \times A}{2.76 - A}$		d = PRIMARY IMAGE Minus A = 14.7−A	b = B − d	SECONDARY M. $M = \frac{B}{A}$	E.F.L. = $F_{PRI} \times M_{SEC}$ = $14.46 \times M_{SEC}$	f/value = $\frac{E.F.L.}{C.A. LENS}$ = $\frac{E.F.L.}{4.25}$

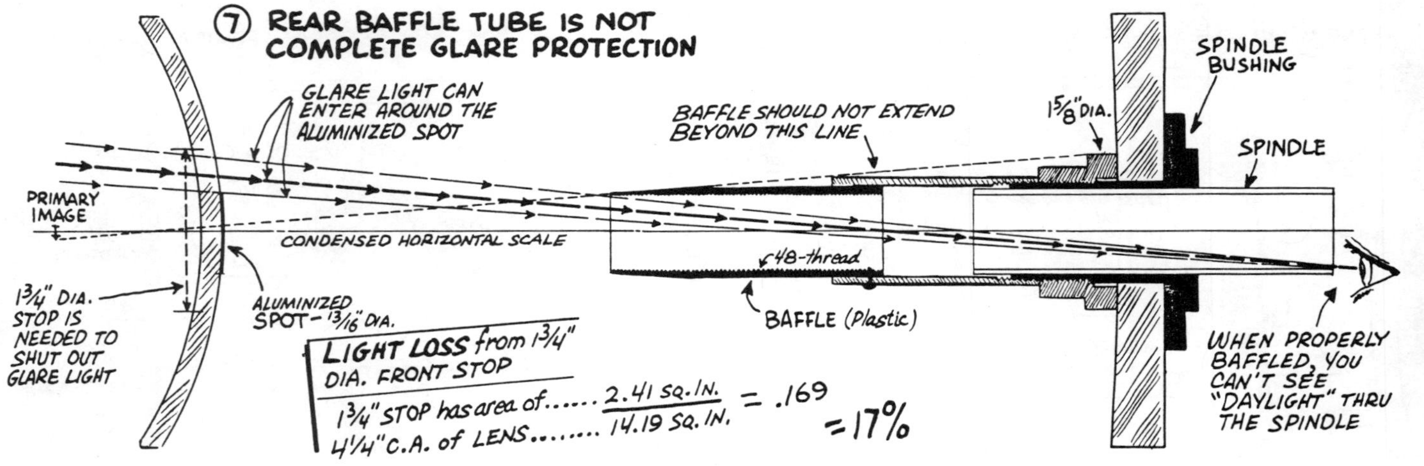

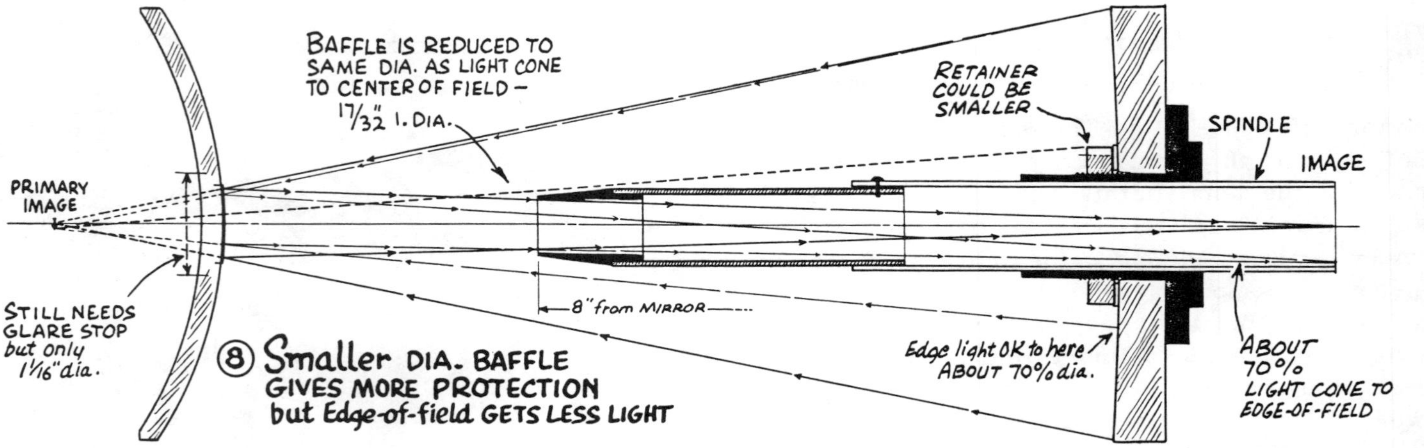

when properly installed. If you are buying a finished compound telescope of any kind, it is always wise to remove the eyepiece and look through the instrument. You should see only the secondary or its mounting; if you see "daylight" around the secondary or its mounting, you have a plain case of glare light. A little of this can be tolerated for astro use, but for the daytime scene you have to go all the way with 100% baffling of light outside the field of view.

BAFFLE TUBES. For the design shown here, it is not possible to obtain complete glare protection with a rear baffle alone, as can be seen in Fig. 7. The needed correction sounds simply awful: you paste a big 1-3/4-in disk of black paper or foil on the front surface of the lens. This is not actually as bad as it sounds because the light loss is only about 17%, as shown. You can do a bit better with a longer baffle with a nosepiece just big enough to pass the full light cone from center of field, Fig. 8. This will also provide about 70% edge-of-field lighting which is satisfactory. You still need a front stop but now it is only about 1-1/16-in. diameter. What you are doing, of course, is abandoning the original astro design to get a better scope for daytime use.

CONSTRUCTION. The Maksutov can be built in about a dozen different ways. The sliding mirror, Figs. 4 and 5, is popular but has the disadvantage of being non-adjustable for collimating. If you use this construction you must depend on accurate lathe work, and to this end a heavier main tube should be used. The sliding mirror is controlled by screw-action which should be fairly fine for comfortable focusing. A 1/8-in pipe thread is specified and is satisfactory. You can use a standard 1/8 in. straight nipple for this but both ends will have to be plugged and turned, so it is just about as practical to lathe-turn the whole thing. To tap the mating hole through the back plate, you will need a 1/8-in. pipe tap with slim shank--it must pass completely through the work to cut a full thread.

The end of the focusing screw which attaches to the focusing arm must be secure but able to turn freely. A self-locking nut is shown but it may be necessary to drill through this for a wire or pin to prevent unscrewing.

A permanent 3/4 in. wood button should be cemented to the outside of the front lens. Over this you can slip a plastic bushing when you want complete glare protection for daytime viewing. After trying both ways, it is likely you will put the larger button in place and leave it there.

Collimating a Compound

DO THE main mirror first, using a sighting ring having concentric circles on its inside surface, Fig. 2. The sighting ring is taped to the open end of the tube with the arms oriented the same as the adjusting nuts. The idea of course is to get the circles concentric around the hole in the mirror. The adjusting nuts have the same action as the familiar "center dot" method used for a Newtonian, i.e., the sighting ring reflection will move away from the nut you tighten, or toward the nut you loosen. Fig. 2 shows an example and needed adjustments.

ADJUSTING THE SECONDARY. Usually the secondary mount will have a hole or a screw on its outside surface, and you can use this for mechanical centering, using the sighting ring already mentioned and noting if the hole in the mirror mount is directly under the hole in the sighting ring.

The secondary is viewed directly from the eyepiece end of the telescope without eyepiece but with a cap having a 1/8 inch hole. Fig. 3-1 shows your view of the small mirror and its mount. If the main baffle is removed (recommended), you will also see the end of the main tube. Just inside the secondary mirror you should see a reflection of the primary, Fig. 3-2. If you can't see all of the reflection, it is likely you are blocking off too much of the secondary with a thick retaining ring; use a lighter ring or side setscrews if needed. If you are using a post mounting, any sidewise misalignment can be roughly corrected by rotating the post, Fig. 3-3. The lower adjusting nut will correct any vertical off-centering, Fig. 3-4.

Now, direct your attention to the small image of the secondary seen by double reflection. It should be in line with the secondary mounting post and concentric with the hole in main mirror, Fig. 3-5. The mirror hole is seen black and the secondary mount is also black, so the ultimate collimating view is just a black disk at the center of the main mirror reflection. It is an automatic if you have made all of the previous adjustments correctly. You can't help the centering by further manipulation of the secondary adjusting nuts because this will put the main mirror reflection off-center. You can, however, easily improve the centering of the black disk by using the main mirror adjusting nuts, consoling yourself with the thought that maybe the main mirror was not set up properly in the first place.

(1) You Can MAKE ADJUSTMENTS WHILE LOOKING THROUGH THE EYEPIECE TUBE without eyepiece

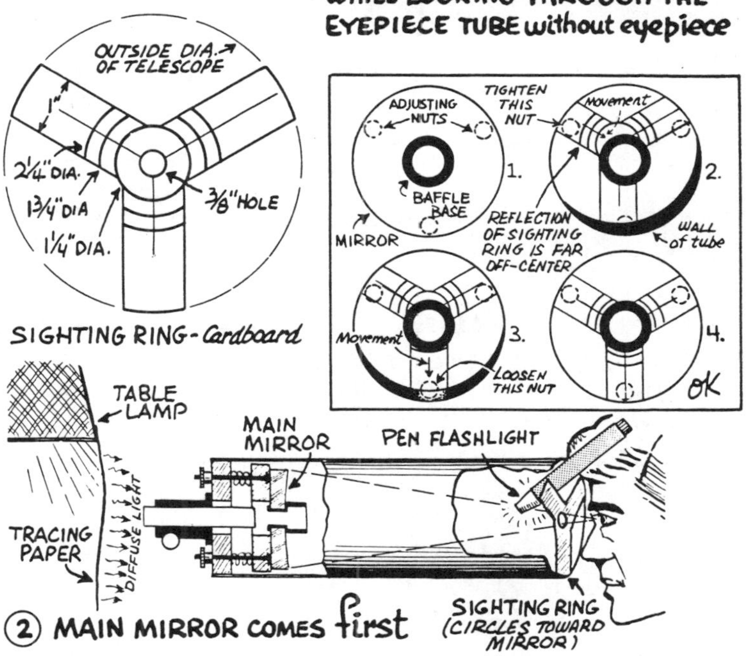

(2) MAIN MIRROR COMES first

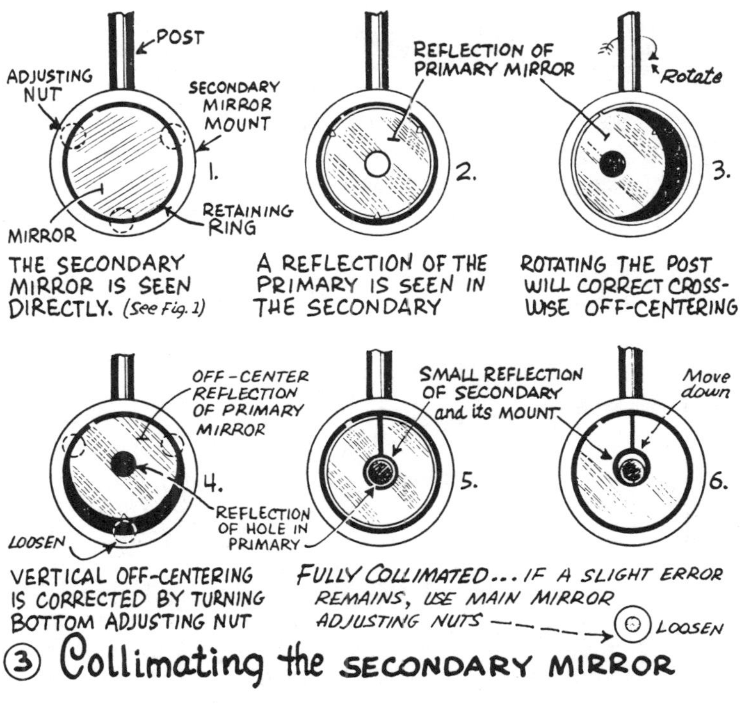

(3) Collimating the SECONDARY MIRROR

FIRST FACTS about Telescope Mirrors

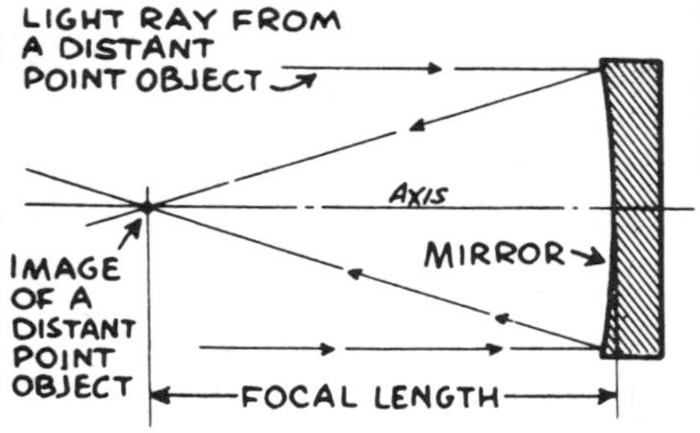

FOCAL LENGTH

DISTANCE FROM SURFACE OF A MIRROR TO THE POINT WHERE IT FORMS AN IMAGE OF DISTANT OBJECT

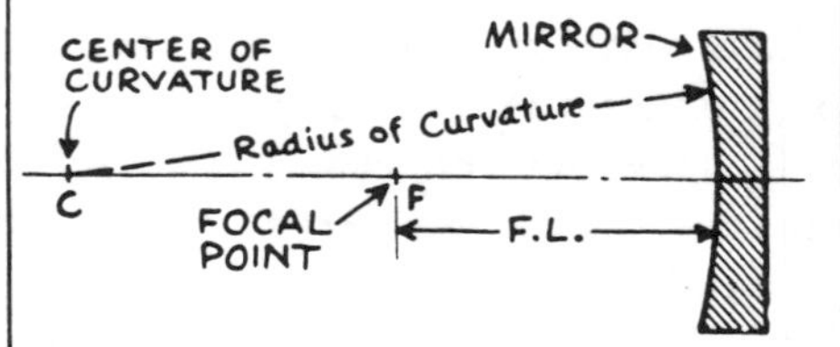

ALL TELESCOPE MIRRORS { F.L. = ½ Radius; Radius = 2 × F.L.

RADIUS OF CURVATURE

THE RADIUS OF A SPHERICAL MIRROR, OR, RADIUS OF CENTER ZONE OF A PARABOLIC MIRROR

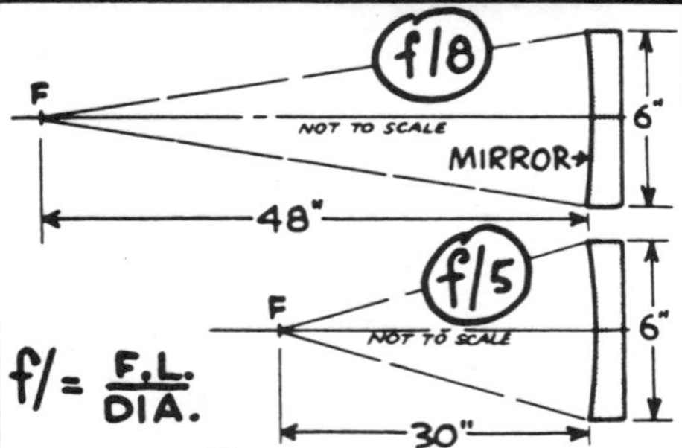

f/ VALUE

THE FOCAL LENGTH OF A LENS OR MIRROR DIVIDED BY ITS DIAMETER. f/8 IS MOST COMMON

SAGITTA FORMULA:

$$SAG = \frac{r^2}{2R}$$

(EXACT FOR PARABOLA AND NEARLY SO FOR SPHERICAL MIRROR)

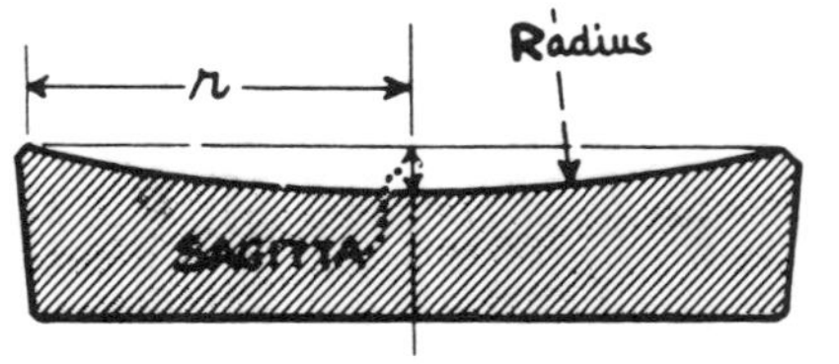

SAGITTA

THE DEPTH OF A CURVE. THE SAGITTA FORMULA IS USED TO CALCULATE ALL TELESCOPE MIRRORS

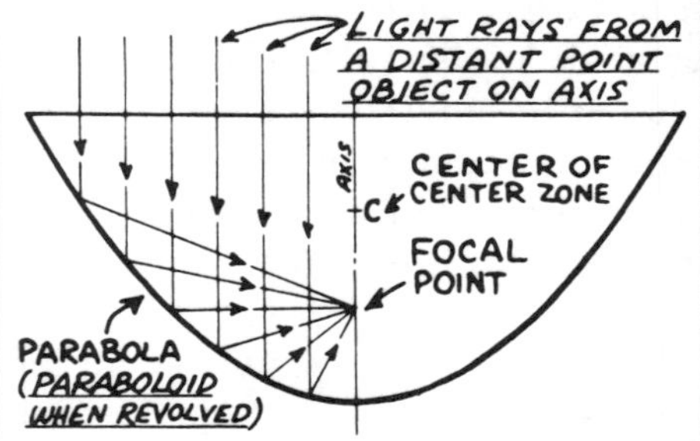

PARABOLA

A CURVE WHICH REFLECTS PARALLEL LIGHT TO A POINT. *THE PARABOLA REVOLVED FORMS A PARABOLOID*

SPHERICAL MIRROR

A MIRROR WITH A SPHERICAL SURFACE. THIS SHAPE MAKES A GOOD TELESCOPE IF f/VALUE IS f/10 OR HIGHER

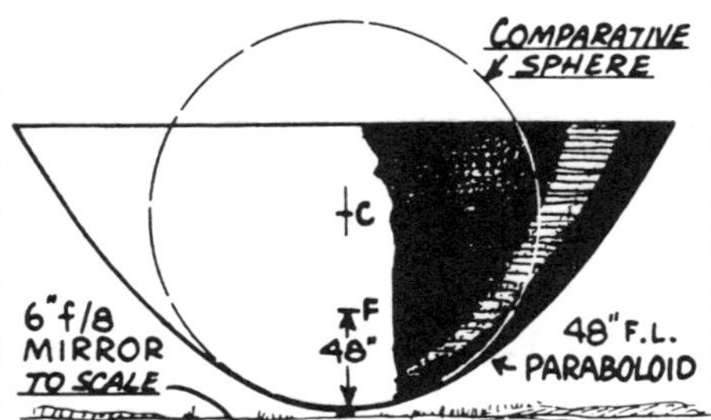

PARABOLIC MIRROR

A MIRROR WITH A SURFACE WHICH IS A REVOLUTION OF A PARABOLA. IT IS THE PERFECT SURFACE FOR TELESCOPES

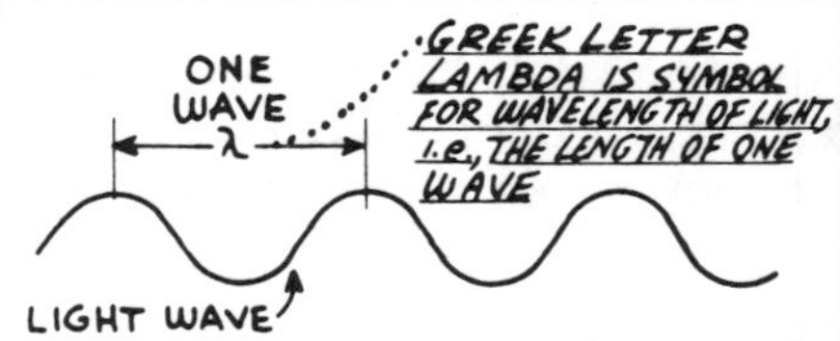

ONE WAVE (λ) = .000022" (22 MILLIONTHS-OF-AN-INCH)

¼ WAVE ($\frac{1}{4}\lambda$) = .0000055" (5½ MILLIONTHS-OF-AN-INCH)

⅛ WAVE ($\frac{1}{8}\lambda$) = .00000275" (2¾ MILLIONTHS)

WAVELENGTH OF LIGHT

. . . IS USEFUL AS A UNIT OF MEASURE TO SPECIFY THE SURFACE ACCURACY OF A LENS OR MIRROR

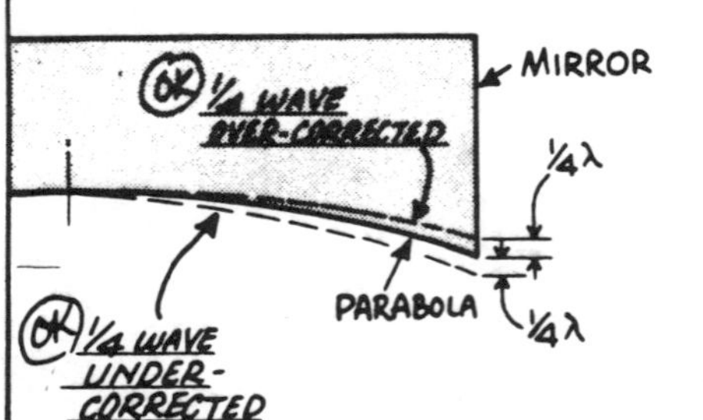

¼-WAVE MIRROR

A MIRROR WILL PERFORM CLOSE TO PERFECTION IF ITS CURVE IS SMOOTH AND NOT OVER ¼-WAVE FROM THE IDEAL PARABOLIC SHAPE

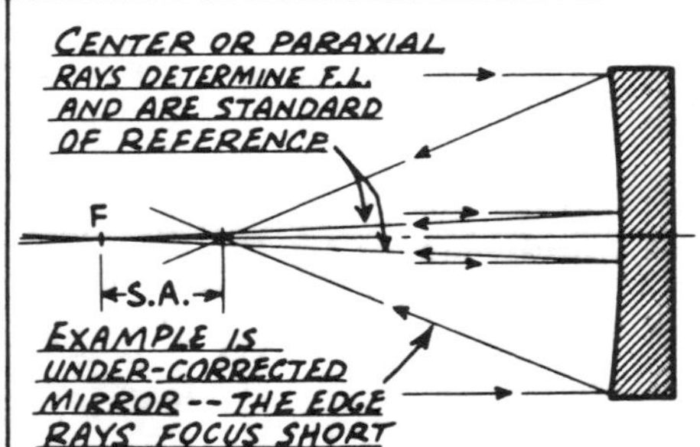

SPHERICAL ABERRATION

THE VARIATION IN FOCUS BETWEEN EDGE RAYS AND CENTER RAYS. IT IS THE ONLY ABERRATION YOU HAVE TO DEAL WITH IN GRINDING YOUR OWN MIRROR

THE SHADOW TEST

SHADOW OR KNIFE-EDGE TEST ORIGINATED BY FRENCH PHYSICIST, JEAN FOUCAULT, LETS YOU SEE AND MEASURE SHAPE OF MIRROR TO 1/12 WAVE

Mirror Grinding and Testing

TELESCOPE BUILDING is a hobby any person can enjoy regardless of manual skill or workshop equipment. The easy way, of course, is to buy your optics and parts ready-made, thereby reducing the job to a simple matter of assembly. The most satisfaction is obtained when you make some or all of the parts yourself, and the biggest thrill of all is to grind and polish your own mirror. With your own hands you can fashion a glass surface accurate to a millionth of an inch. On the practical side, you can also save money: a finished 6-inch mirror of 1/4-wave quality costs about $100 as compared to about $20 for the glass and everything needed to make your own including the cost of getting the mirror aluminized.

In precision work mirror grinding is unique in that the high degree of accuracy required can be obtained with the crudest kind of makeshift equipment. All you need is some kind of solid support to hold the work at about waist level. Then if you rub two disks of glass together, one on top of the other with abrasive grains and water between, the top disk will automatically become hollow (concave) while the bottom disk will become convex. Since you want a concave mirror, the top disk becomes the mirror, while the lower convex disk is the "tool." If you walk around the work post while rubbing the two disks together, the glass will wear uniformly all around, producing a nearly-perfect segment of a sphere for the simple reason this is the one and only curve which can remain in contact when rubbed together.

Most beginners know the rest of the story. By using finer and finer abrasive, you make the surface smoother and smoother until finally with red rouge it acquires a shining face of gemlike smoothness. In terms of ordinary accuracy, it will be a perfect spherical section, but for the super-precision required in optical work, the 25 millionths it may be in error becomes an item of considerable importance.

Up to this point, any 12-year old can do the work because the job is a routine procedure requiring only neatness and thoroughness. Youngsters being what they are, it is not strange that the most common defect is plain, ordinary lack of polish. Providing the mirror has a good polish, any shape near a sphere will form a good image.

Most of the actual work in making a first mirror of top quality comes in testing and correcting. This is more than just making a stab at parabolizing; it means that you stick with correcting technique until you acquire the know-how and skill to correct a glass surface with reasonably predictable results. This is a skill you don't acquire by mere reading. Like punching a typewriter, plastering a wall or hitting a golf ball, it takes practice. You can expect up to a hundred hours of study and practice before you become an expert glass pusher.

Mirror Blanks

STOCK SIZES

NAME-SIZE DIA.	ACTUAL DIA.	THICK	SURFACE AREA
3"	ABOUT 2 15/16"	3/8" OR 1/2"	6.78 SQ. IN.
4 1/4"	4 3/16"	3/4"	13.78
6"	5 7/8"	1"	27.11
8"	7 7/8"	1 3/8"	48.71
10"	9 7/8"	1 3/4"	76.59

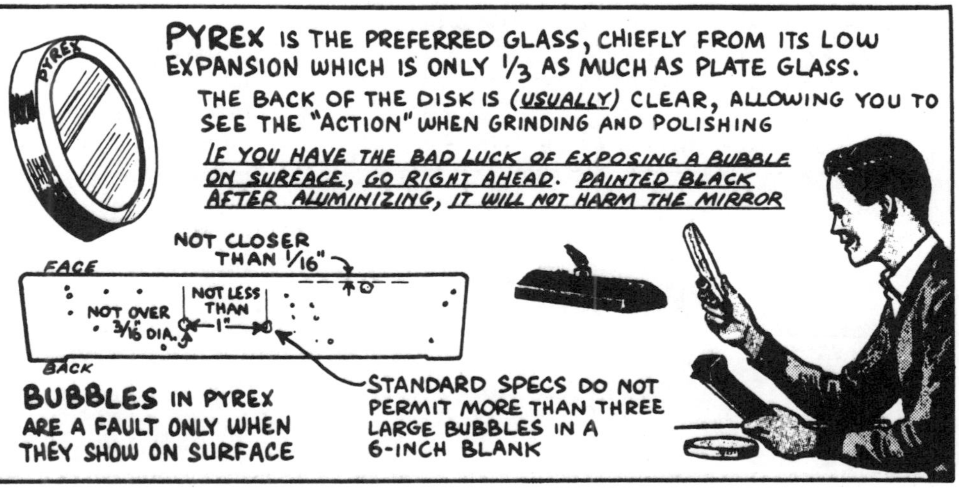

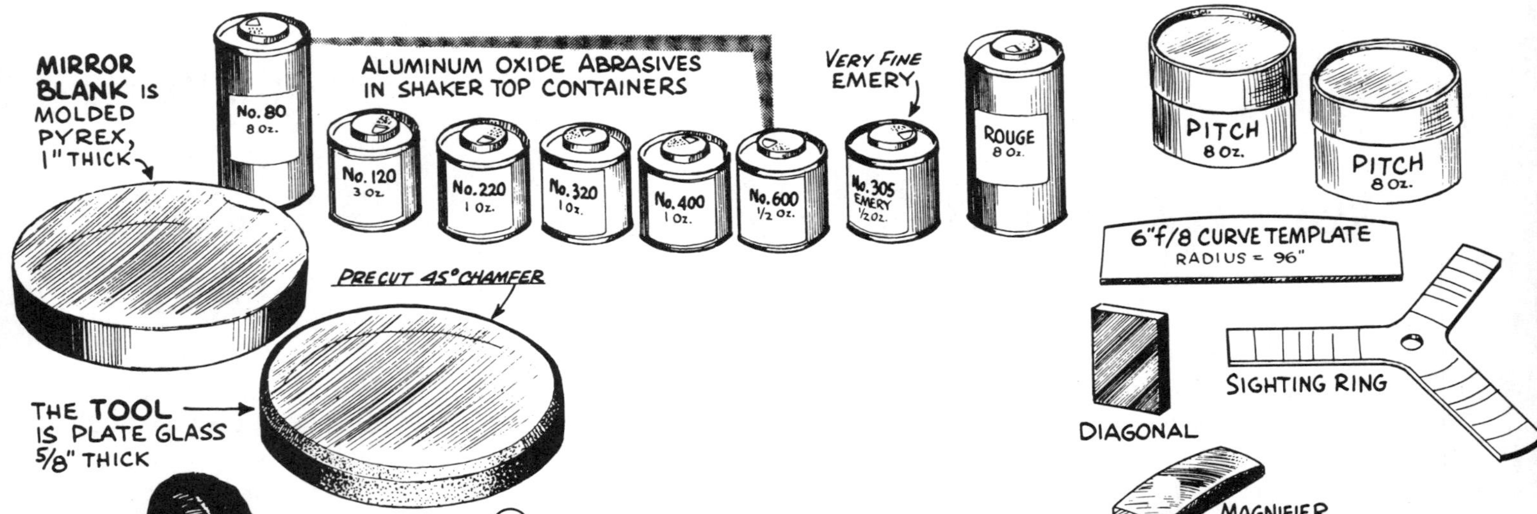

(1)

GRINDING your own MIRROR

(2)

GRINDING STAND
THIS STYLE IS ALSO USED AS PEDESTAL BASE FOR TELESCOPE

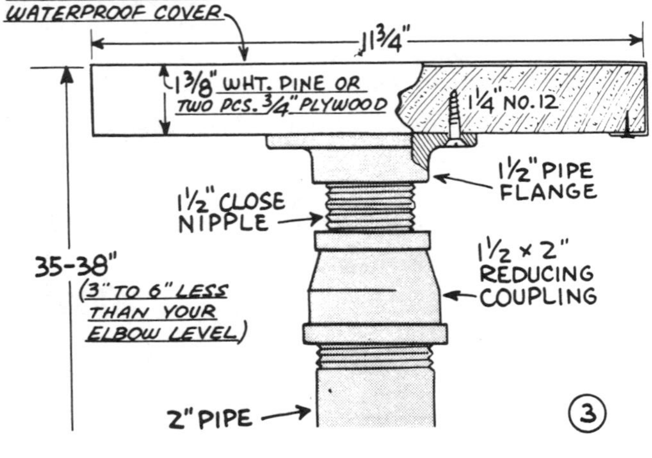

(3)

EVERYTHING needed to grind a mirror is commonly purchased in a mirror grinding kit, Fig. 1. Kits can be obtained in 4-1/4, 6, 8, 10 and 12-1/2-inch sizes, the name-size being the diameter of the mirror. The popular 6-inch size, which is a universal first choice with beginners.

GRINDING STAND. For comfort, this should be fairly high, about waist level. An old oil drum with water ballast does nicely. A sturdy box or crate loaded with a sand bag, concrete block or other weight is satisfactory. If you must make something, a happy solution is the pedestal base, Fig. 2, which you will later use to mount the telescope. Of course, you have to keep hopping over the sprawling legs, but this is altogether a minor matter. In any case you should be able to walk completely around the stand since the walk-around is one of the secrets of successful mirror grinding.

GRINDING STROKE. The back and forth stroke used in mirror grinding has its amplitude measured in mirror diameter. A full stroke indicates a total travel equal to the mirror diameter; a one-third stroke means the glass travels 2 inches, Fig. 8. The one-third stroke is used for about 75% of all grinding and polishing operations. The full stroke is used right at the beginning when you form the curve and again at the end of the polishing job when you go from sphere to parabola. In all cases, one stroke indicates the combined forward and back movement. The speed should be about 60 strokes a minute.

GREATEST WEAR. Fig. 9 diagram shows the basic

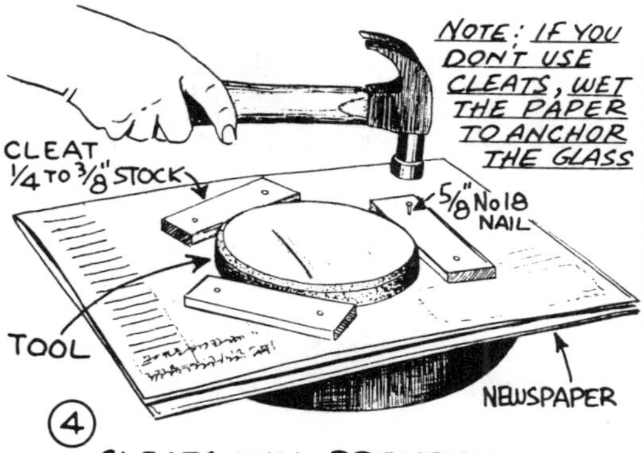

④ CLEATS WILL PREVENT SKIDDING WHEN DOING THE HEAVY, ROUGH GRINDING

⑤ ... A CHARGE OF DRY ABRASIVE

⑥ ... A SPRINKLE OF WATER

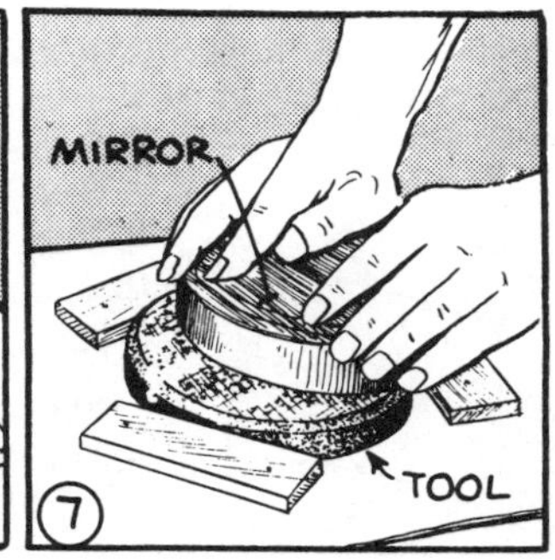

⑦ ... A FEW MINUTES OF BACK AND FORTH RUBBING

principle of mirror grinding and polishing. The top glass wears most at the center; the bottom glass wears most at the edge. The action will be rapid with a long stroke; slower but more uniform with a short stroke. Since the mirror is to have a concave surface, it becomes obvious it must be on top to form the curve. Less obvious is the case shown at right in Fig. 9 where with tool on top, the mirror curve will flatten. By using mirror or tool on top as needed, you can control the shape of the surface.

THREE MOVEMENTS. The three movements used in mirror grinding are shown separately in Fig. 10, and the combined movements which form the grinding cycle are shown in Fig. 11. There are no strict rules regarding the magnitude of the rotation and walk-around, but if you want some average figures, it can be said you will make about 15 steps around the post and the mirror will turn around twice in your hands during this period. Once you start actual work your own personal style of grinding motions will develop naturally. The mirror's edge normally gets the least grinding action. You can speed up the operation by grinding half of the time with the mirror in top position then alternating it with the tool in the top position.

FORMING THE CURVE. The commonest way of doing this is to use a full stroke, that is, 3-inch

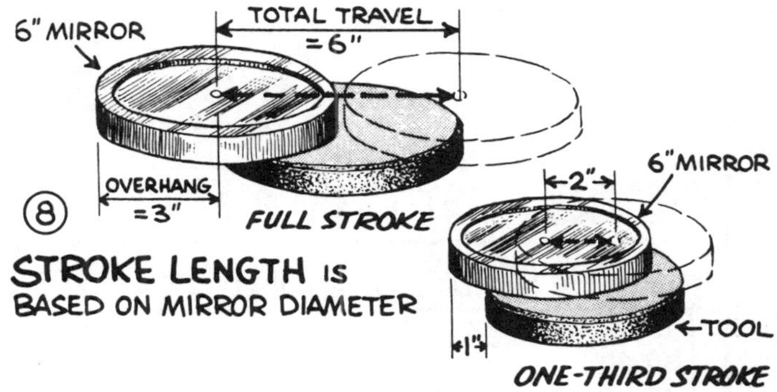

⑧ STROKE LENGTH IS BASED ON MIRROR DIAMETER

⑨ GREATEST WEAR

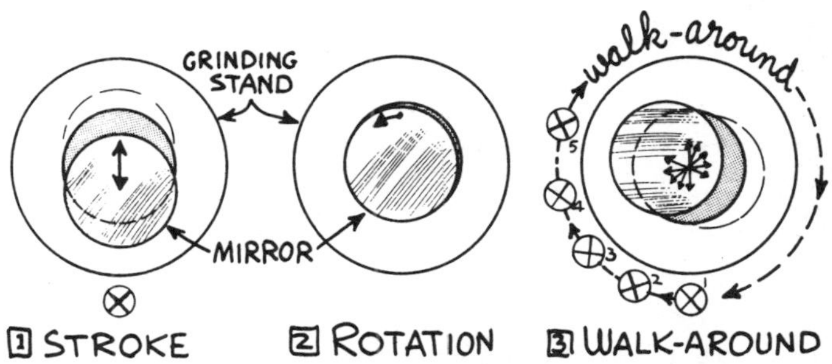

⑩ THREE MOVEMENTS OF GRINDING
THE BACK AND FORTH STROKE DOES THE ACTUAL GRINDING... *ROTATION AND WALK-AROUND DISTRIBUTE THE WEAR*

⑪ **The Grinding Cycle:**

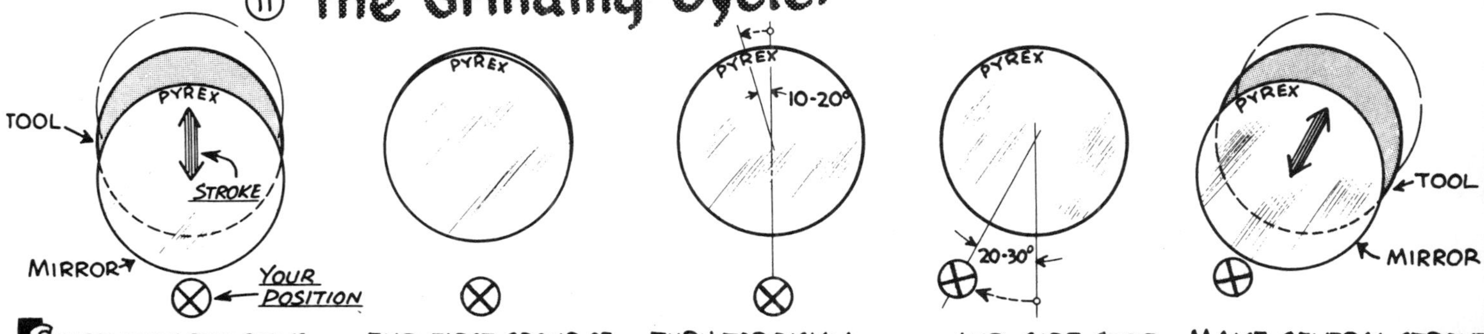

START WITH TWO DISKS CENTERED. MAKE SEVERAL STROKES IN SAME POSITION

END FIRST GROUP OF STROKES WITH DISKS APPROXIMATELY CENTERED

TURN TOP DISK A LITTLE. TAKE YOUR HANDS OFF THE GLASS

...AND SIDE STEP TO A NEW POSITION

MAKE SEVERAL STROKE IN NEW POSITION. *REPEAT THE CYCLE*

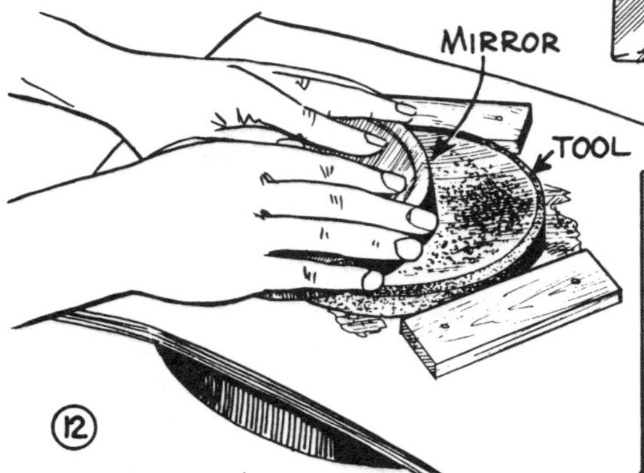

⑫ FORMING THE CURVE
IS A SIMPLE OPERATION WITH FULL STROKE, CENTER-OVER-CENTER

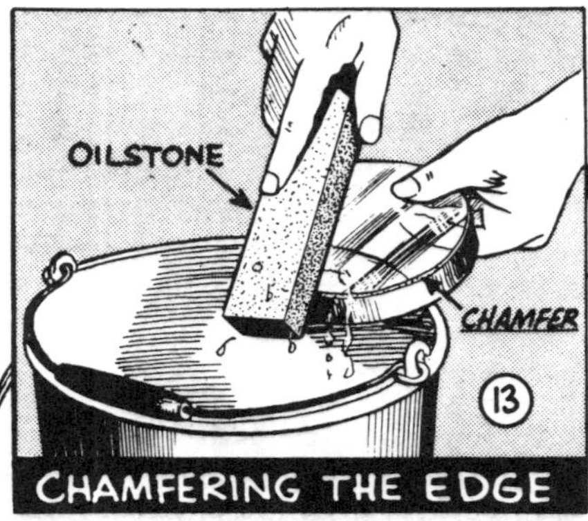

TESTING THE CURVE

TABLE 1 GRINDING SCHEDULE*

ABRASIVE		NUMBER OF WETS	AMOUNT OF ABRASIVE PER WET	TIME PER WET	TOTAL TIME
ALUMINUM OXIDE	FORMING Nº 80	20-30	FULL-SIZE	1-4 Min.	2 Hrs.
	SMOOTHING Nº 80	15-20		2-4 Min.	1¼ Hrs.
	Nº 120	12-15	½ Teaspoon	5 Min.	1 Hr.
	Nº 220	8-12		5 Min.	¾ Hr.
	Nº 320	6-10	¼ Teaspoon	5 Min.	¾ Hr.
	Nº 400	6-8	⅛ Teasp.	5-10 Min.	¾ Hr.
	Nº 600	6-8	1/16 Teasp.	5-10 Min.	¾ Hr.
EMERY	Nº 305	6-8	1/20 Teasp.	5-10 Min.	¾ Hr.
					8 Hrs.

*6" MIRROR

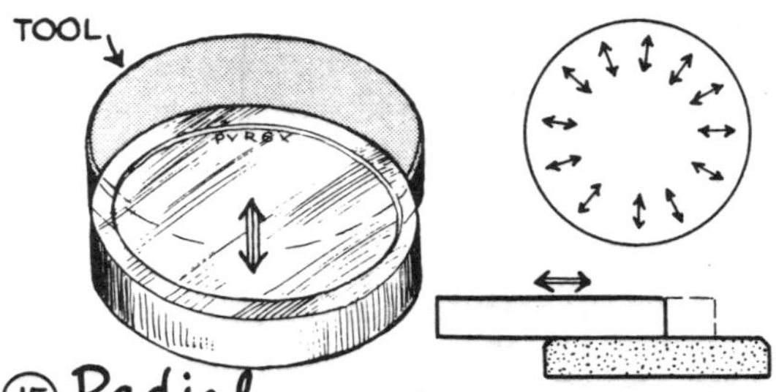

⑮ Radial

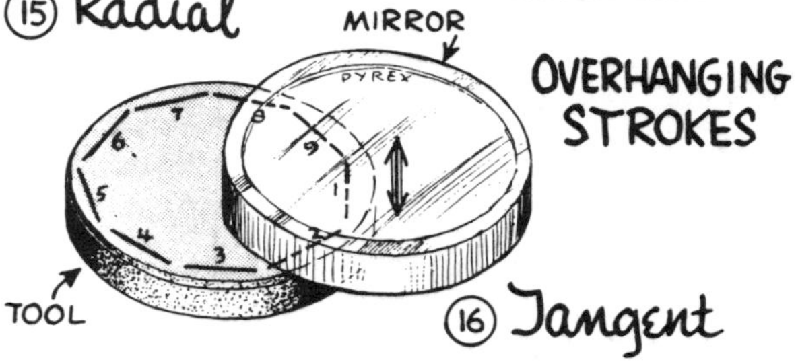

⑯ Tangent

overhang at each end of the stroke, Fig. 12. You can tell the abrasive is working by the harsh, grating growl it makes when it bites into the glass. However, you lose this quickly as most of the abrasive is pushed off over the edge of the tool. It need not be a total loss because it can be scooped up later and used again. The first few wets with No. 80 will last hardly a minute, but by the time you have gone twice around the post you will be getting more mileage out of each charge. Do a couple more turns and then dunk both mirror and tool in the water bucket to remove the sludge. Sure enough, a little hollow is beginning to appear at the center of the mirror!

As soon as you can get about three minutes of grinding time per wet, the water bottle, Fig. 6 can be discarded. Instead, you wash both mirror and tool for each new charge of abrasive, using both disks wet from the water bucket. Grinding the curve takes about two hours. The work must be done with a fair amount of pressure of from 15 to 40 lbs. The more pressure, the faster the glass will be removed, providing the stroke is not so rapid as to cause skidding. The stroke speed should be 40 to 60 pushes per minute--don't rush it. It is instructive to put the mirror on the bathroom scales and then apply your hands to register 20 lbs. This is a comfortable working pressure for rough grinding, being just a little more than the combined weight of the mirror and your hands. Work periods can be 15 or 20 minutes at a stretch, and several rounds of this at the specified pressure of 20 lbs. will get the job done in about 2 hours time. It will take an extra hour or a little more to smooth the roughly-formed curve to an accurate sphere.

CHAMFER. The edge of a Pyrex disk is gently rounded as purchased and needs no attention if you can obtain the desired surface curve without removing too much of the edge. However, if the edge starts to wear to a sharp edge liable to chip, it should be cut back with a 45-degree chamfer about 1/16 in. wide, using a fine oilstone, with water lubricant, Fig. 13. Lacking the oilstone you can use No. 220-grit abrasive made up with water to form a paste applied with a piece of metal or glass. The glass tool as purchased is already chamfered.

TESTING THE CURVE. All the time you are grinding the curve you have to watch its shape and depth. Start using a template early, Fig. 14. The cardboard template you get with the kit is satisfactory for rough testing but you may prefer to make something more accurate from sheet metal. Another useful mechanical check is to measure the sagitta, Fig. 17. If you are working a 6-inch, f/8 mirror, the ultimate goal is 48 in. focal length or 96 in. radius, but a focal length a couple inches more or less does no harm. A goodly amount of radius adjustment can be made during fine grinding. If the mirror is worked on top (the usual case), the radius will shorten; if the tool is on top, the radius will lengthen. Using a one-third stroke throughout, you can expect a maximum change of about 6 in. radius if you use one position exclusively during fine grinding.

ALTERNATE FORMING STROKES. Most mirror grinders like to see a hollow form in the center of the mirror without delay. This impatience to get on with the job has popularized various overhanging strokes, all of which are aimed at wearing the center of the mirror directly over the edge of the tool. You can use either a radial stroke, Fig. 15, or a tangent stroke, Fig. 16. The radial stroke is very much like one end of the long stroke already described. The tangent stroke is a one-third stroke with side overhang and is worked on 10 or 15 chords spaced around the edge of the mirror. Use the standard grinding cycle, Fig. 11. The first walk-around with either of these strokes is done with the center of the mirror almost directly over the edge of the tool. Following rounds are made with less and less overhang until the stroke assumes the normal center-over-center position.

SMOOTHING THE CURVE. As soon as you think the sag is about right, switch to a one-third or shorter stroke and continue with No. 80-grit until good contact is obtained. The commonest test for contact is merely a matter of watching the bubbles that form between the disks during grinding. By manipulating the top disk, you can move an air bubble from center to edge, and its changing size will show hollows and ridges if

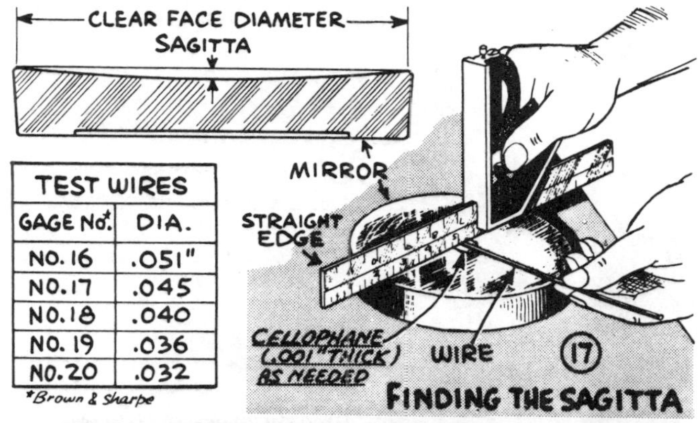

TEST WIRES	
GAGE No.*	DIA.
NO. 16	.051"
NO. 17	.045
NO. 18	.040
NO. 19	.036
NO. 20	.032

*Brown & Sharpe

(17) FINDING THE SAGITTA

TABLE 2 FOCAL LENGTH OF 6" MIRROR WHEN SAGITTA IS KNOWN

SAGITTA	CLEAR FACE DIAMETER				
	5 3/4"	5 13/16"	5 7/8"	5 15/16"	6"
.038"	54.38"	55.57"	56.77"	57.98"	59.21"
.039	52.98	54.14	55.31	56.49	57.69
.040	51.66	52.79	53.93	55.08	56.25
.041	50.40	51.50	52.61	53.74	54.88
.042	49.20	50.27	51.36	52.46	53.57
.043	48.05	49.10	50.17	51.24	52.32
.044	46.96	47.99	49.03	50.08	51.13
.045	45.92	46.92	47.94	48.96	50.00
.046	44.92	45.90	46.89	47.90	48.91
.04687					48.00
.047	43.96	44.93	45.90	46.88	47.87
.048	43.05	43.99	44.94	45.90	46.87
.049	42.17	43.09	44.02	44.97	45.92
.050	41.33	42.23	43.14	44.07	45.00
.051	40.52	41.40	42.30	43.20	44.12
.052	39.74	40.61	41.48	42.37	43.27
.053	38.99	39.84	40.70	41.57	42.45
.054	38.27	39.10	39.95	40.80	41.66
.055	37.57	38.39	39.22	40.06	40.91
.056	36.90	37.71	38.52	39.34	40.18
.057	36.25	37.04	37.84	38.65	39.47
.058	35.63	36.41	37.19	37.98	38.79
.059	35.02	35.79	36.56	37.34	38.13
.060	34.44	35.19	35.95	36.72	37.50

Example SHOWING USE OF TABLE:
YOUR MIRROR IS 6" CLEAR FACE AND SAGITTA IS .050"
From Table: F.L. = 45"

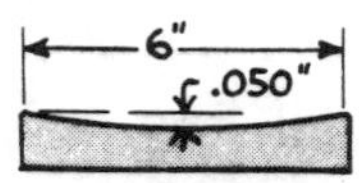

Formula FOR CASES NOT COVERED BY TABLE:

$$F.L. = \frac{r^2}{SAG} \times \frac{1}{4}$$

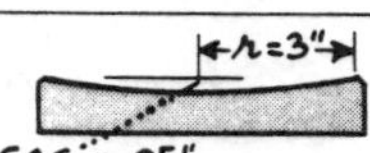

Example: $F.L. = \frac{3^2}{.05} \times \frac{1}{4} = \frac{9}{.20} = 45"$

45.
.20)9.00
80
100
100

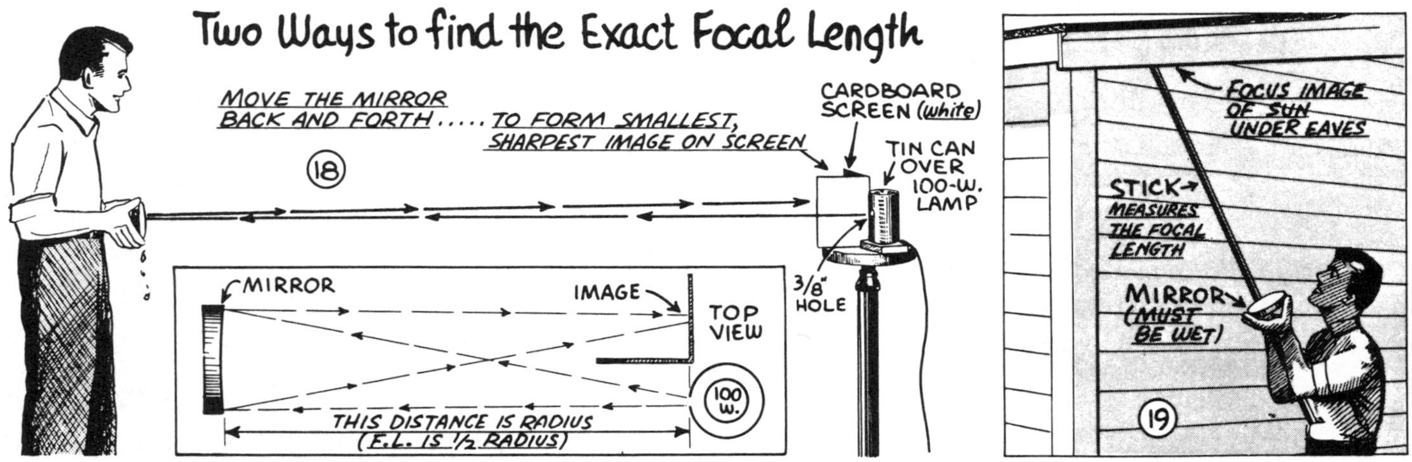

present. Another simple test is done with a wrinkled and re-smoothed strip of cellophane pressed between the glass disks--it should press equally flat all over. Also popular is the trick of drawing a line across the face of the mirror with waterproof ink and then noting if this wears evenly when grinding is resumed. Perfect contact means you have a perfect spherical shape.

FINE GRINDING. Follow the schedule in Table 1. Naturally, you have to clean up thoroughly after each grade, changing paper and water, etc., to prevent stray coarse grains from getting into the act. The one-third stroke is used throughout. A charge of abrasive should last for about five minutes or two or three turns around the barrel. When the water starts to dry, the disks will start to stick together. At this point you are probably two times or more around the barrel and can call that wet finished. Alternately, it is practical to add more water from the squeeze bottle. This prolonged grinding is especially useful for the last wet in each grade. Your main water supply comes right from the water bucket where you dunk and clean both disks before charging with fresh abrasive. A certain amount of tool-on-top grinding is recommended to assure equal wear at the edge of the mirror--use this for at least two wets per grade. If you like a handle or push block, the snap-on type shown in Fig. 21 can be fitted or removed instantly. If you get tired, a 15 lb. weight on top of the glass will substitute for hand pressure.

IMPORTANT HINT. If mirror and tool "freeze together" during grinding use wooden block against edge of mirror and drive apart with mallet.

FINDING EXACT F.L. A simple reflection test can be used, Fig. 18. The mirror must be wet with water to make it reflective; a few drops of glycerin added will keep the surface wet much longer. The room must be dark. The screen should be ample size to simplify picking up the image, which is faint and elusive in the early stages of grinding.

If sunlight is available, the test shown in Fig. 19 can be used. Stand alongside a garage or porch so that the mirror is in sunlight but your face is shaded. This test measures the focal length directly, which can be marked on a stick held in your hand as shown. The image of the sun will be a little under 1/2 in. diameter. More glycerin and less water will keep the mirror reflective for several minutes.

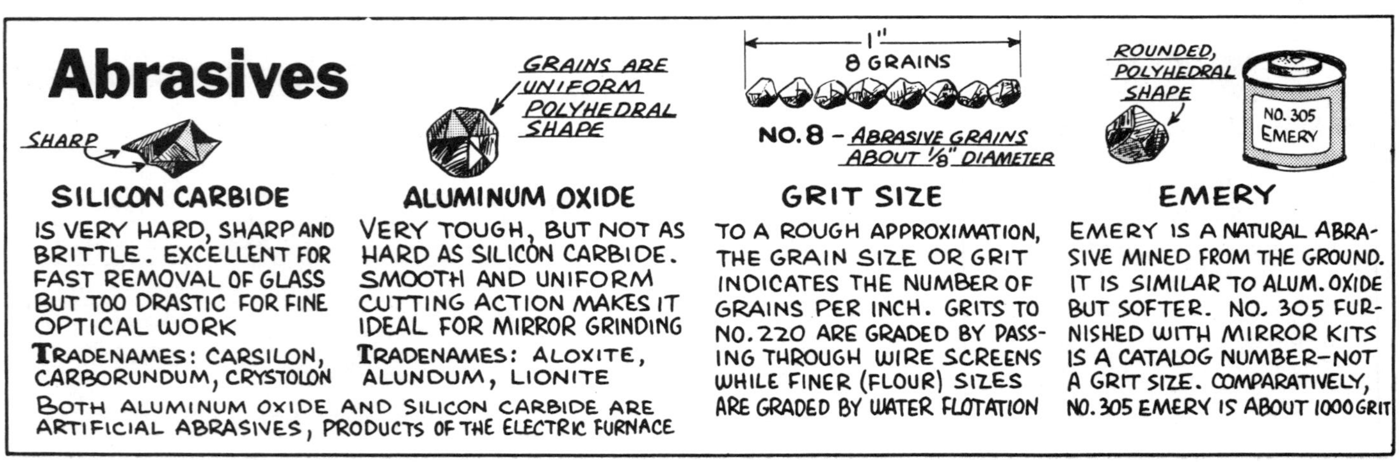

REFLECTION TEST. On grits No. 120 and finer, a reflection test is excellent for checking both the shape of the mirror and its surface smoothness. This is done with a dry mirror, Fig. 20. When the surface is coarse, you can see the lamp reflection only at a low angle, as shown. The shape of the curve is checked by turning the mirror slightly to cause the reflection to move across the mirror. Any sudden brightening indicates a ridge: any decrease in brightness means a low spot. If the reflection remains the same brightness from center to edge, the surface is spherical. The most common fault is a decrease in brightness at the edge. This is always the last to grind and polish out; hence you see the dimmer reflection from the coarser surface of the previous grade. Test after each grade and continue with each grade until the reflectivity appears uniform from center to edge of mirror.

You will pick up a secondary reflection from the back of the mirror. If annoying, this can be eliminated by spreading a thin coat of vaseline on the back of the mirror. The test lamp can be any wattage, filament or frosted. A filament lamp can be used at close range, as shown, but a frosted lamp must be viewed from 5 or 6 ft. to kill diffuse reflections.

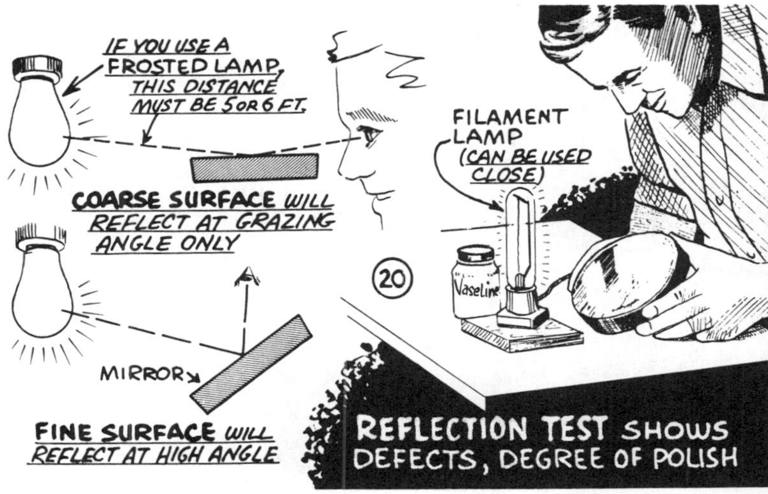

REFLECTION TEST SHOWS DEFECTS, DEGREE OF POLISH

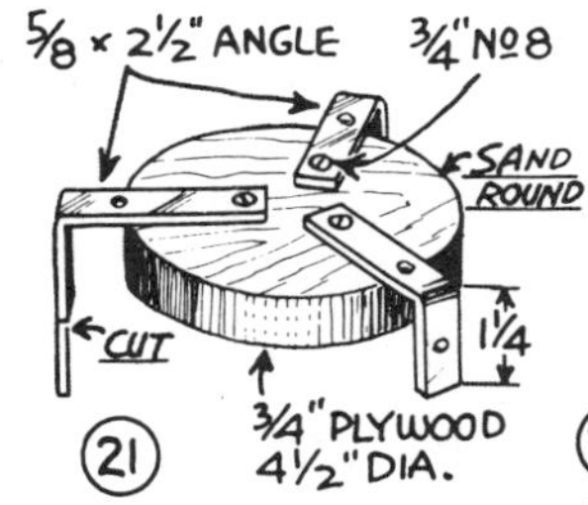

PUSH BLOCK

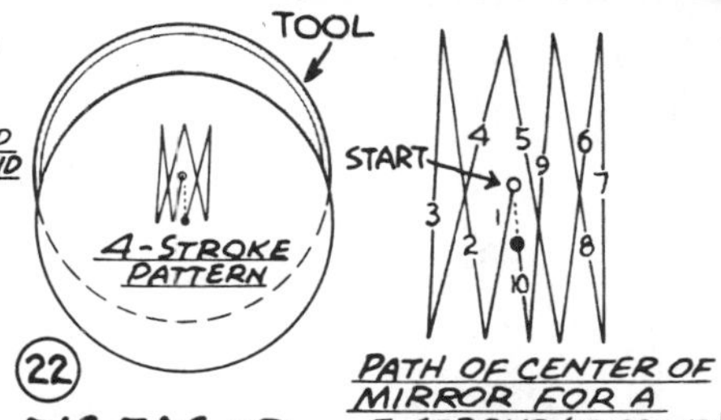

ZIG-ZAG OR W-STROKE

THE W-STROKE. Toward the end of fine grinding, start learning the zig-zag or W-stroke shown in Fig. 22. The pattern can be any number of strokes from about 4 to 10--a 5 or 6-stroke pattern is average. This is a blending stroke and produces a better surface than a straight center-over-center stroke. The side overhang should be about 1/2 in. each way.

FINAL GRINDING. The final grind with No. 305 emery can be prolonged to advantage, adding water only. Hand pressure tapers off gradually with the use of finer grits, and when you get to the finest grade it is hardly more than the weight of your hands on the glass. It is a good idea to eliminate air bubbles on this and the previous grade, and this is easily done by sliding the mirror nearly off the tool and then slowly sliding it back to center; do this after you have made about a half-turn around the barrel.

ARE YOU READY TO POLISH? Do not abandon grinding until you are sure you have a smooth surface and a good sphere. The surface is smooth if you can read a newspaper through it with the mirror several inches from the paper. The surface is both smooth and spherical if it passes the reflection test already described at a 45 degree angle. And, of course, the surface should look smooth and uniform--any kind of bloom or shading is a sure indication of insufficient grinding.

Pitch

Most of the optical pitch is obtained from certain pine and spruce trees, that from the Norway spruce being rated especially good. The resinous sap is boiled and refined to produce a semi-solid product, which is further tempered by the addition of pine rosin, turpentine and beeswax. Pitch will burn but it is not explosive in the manner of gasoline and only ordinary precautions are needed in melting it. It can be melted in a tin can directly over an electric hot plate; if you use an open flame, put the tin can in a pan of water. Heat slowly, let cool a half-minute before pouring, and pour slowly on the warm tool. The difficulty of making a pitch lap increases with its thickness--beginners take notice!

AMOUNT YOU NEED TO MAKE A LAP

DIA.	THICK	CU. IN.
4 1/4"	1/4"	4 1/2
6"	1/4"	7 1/2
6"	3/16"	5 1/2
8"	1/4"	13

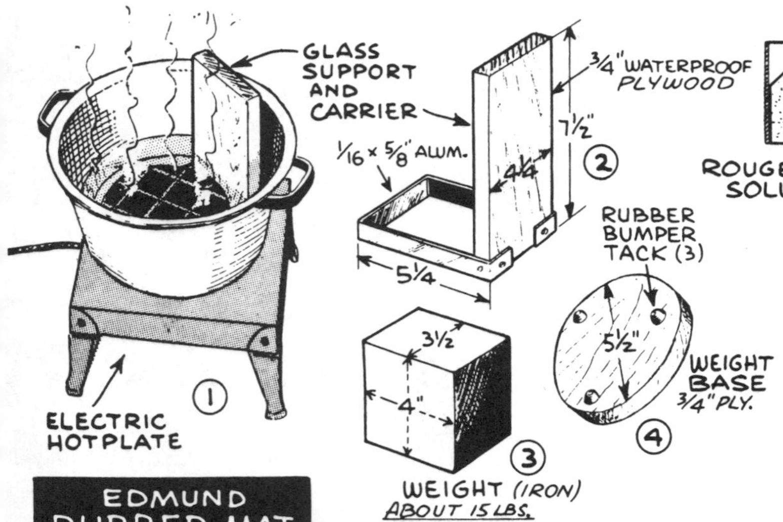

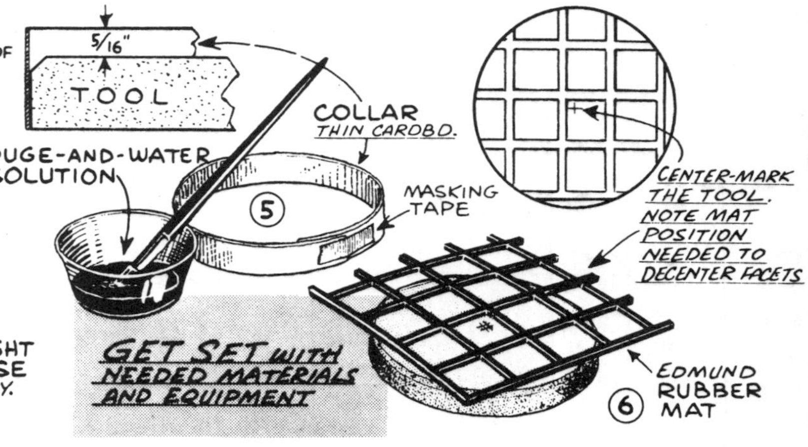

EDMUND RUBBER MAT METHOD

Making a PITCH LAP

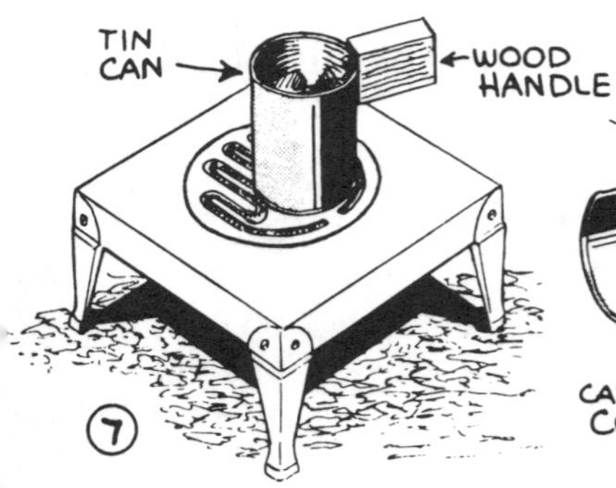

MELT THE PITCH IN A TIN CAN... IT WILL MELT IN LESS THAN 10 MINUTES

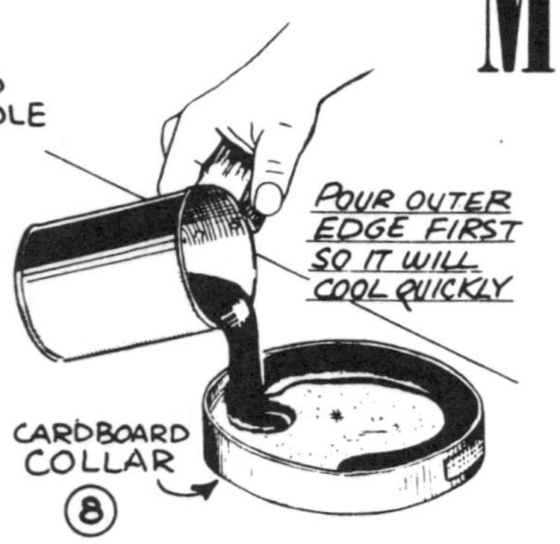

POUR THE PITCH OVER THE DRY, WARM TOOL. COLLAR IS ROUGED TO AVOID STICKING

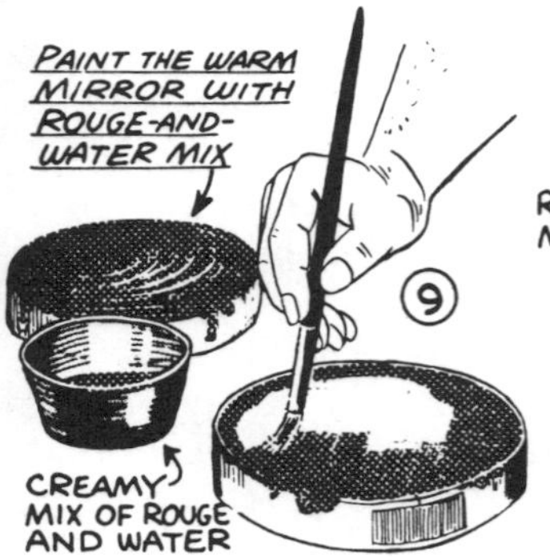

PAINT THE PITCH WITH ROUGE AND WATER. STRIP THE COLLAR AFTER PAINTING

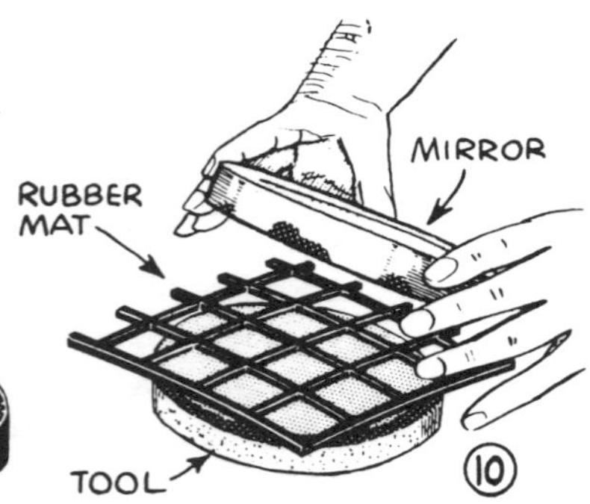

PRESS THE MAT INTO THE WARM PITCH BY PLACING THE PAINTED MIRROR OVER IT

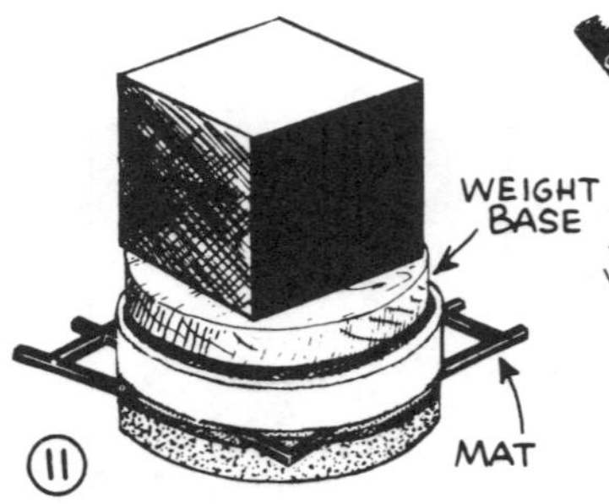

PRESS WITH WEIGHT IF PITCH GETS TOO HARD FOR EASY FORMING. *REHEAT THE WHOLE LAP IF NEEDED AS SHOWN AT TOP OF PAGE*

TRIM THE LAP WITH A WOOD CHISEL OR RAZOR BLADE. *INCLINE THE EDGE A LITTLE TO MAKE LAP ABOUT 1/16" LESS THAN MIRROR DIAMETER*

THE FASTEST and easiest way to make a pitch lap is by pressing the channels with a rubber grid while the pitch is warm. The work equipment--Figs. 1 to 4--will be used throughout the polishing operation. The rouge-and-water mixture used to prevent the hot pitch from sticking to the mat and mirror is made from one level tablespoon of rouge to four water. If you want to conserve rouge, you can substitute some or all soap flakes for the rouge. In either case, the addition of about 10% glycerin helps by retarding the drying speed. Previous to making the lap you should centermark the tool and then note how the mat must be placed to offset the facets, Fig. 6. A centered system of facets should be avoided since it tends to polish rings on the surface of the mirror. A 1/2 lb. carton of pitch will be enough for a 6-inch lap. Hit the carton several times with a hammer and then peel off the wax paper cover, working over a sheet of clean newspaper. Funnel the broken lumps into the melting pot or can.

Hot pitch adheres poorly to cold glass, so you start the job by warming the glass tool. This is done by heating a pan of water with the tool in it. The mirror can be warmed at the same time, the idea of this heating being that a cold mirror might chill the pitch too quickly in the pressing operation, Fig. 10. After heating the water a little hotter than your hand can stand, remove the water pan and start melting the pitch. When the pitch is nearly melted, remove the tool and mirror from the hot water and dry both. Paint the surface of the mirror with the rouge solution. It is also a good idea to paint the rubber mat. Put the tool on a level surface. Paint the inside of the cardboard collar and slip it over the tool. You are all set to pour the lap.

After the pitch is fully melted, remove it from the hotplate, wait about 15 seconds and then pour onto the tool to the level of the cardboard collar, Fig. 8. Wait another 15 seconds and then paint the surface of the pitch, Fig. 9. Strip off the cardboard collar. Put the rubber mat in place and press it down with the mirror, Fig. 10. The mat should sink into the pitch easily with light hand pressure. Take a peek to see how it is doing. If a full impression has not been made, you can use a weight, Fig. 11, for a few minutes or longer. After the channels are formed, remove the mat. Press down lightly with the mirror alone. This may cause the channels to close slightly and if so, put the mat back in place and repress. Repeat pressing with mat-and-mirror and mirror alone until you get a perfect impression and good contact.

If you see the pitch is too cold, put the tool with its pitch coating into the pan of hot water and put the pan on the hotplate. Heat for 3 to 8 minutes as may be needed. You can test the pitch with a screwdriver or stick and so determine the degree of softness which you think is needed to complete the pressing. If you heat the lap too much, the channels already more or less formed will start to run together. This heating needs water hotter than your hand can stand, and that's where the glass support and carrier comes in handy in removing the lap and transporting it to the bench. Then you go through the various operations again. Use plenty of rouge-and-water to prevent sticking. Trim the lap with a sharp wood chisel or razor blade.

The big 1-5/16 inch facets of the rubber mat are satisfactory for a 6-inch or larger mirror. If desired you can double-press the mat to form a second set of channels with facets about 1/2 inch square. This can be done at any time and is best delayed to the final stages of polishing where the slightly smoother action of the small facets can be used to best advantage.

USING A CHANNELING TOOL. This is identical with the rubber mat method except you use a rigid channeling tool, Fig. 13. Press all of the channels one way and then the other. Alternate with mirror alone until good contact is made.

INDIVIDUAL PITCH SQUARES. This is a long-time favorite, especially useful for steeply-curved mirrors difficult to flow-coat to uniform thickness. The wood mold is a simple job of sawing and nailing; it should be painted with rouge-and-water to prevent sticking. The pitch strips are cut into squares, Fig. 15, with a hot knife or the cutting tip of an electric soldering iron. Each facet square is heated, Fig. 16, and then pressed onto the glass tool. After all facets are in place, perfect adhesion and contact is obtained by heating the lap in water, Fig. 1, and then pressing with the mirror.

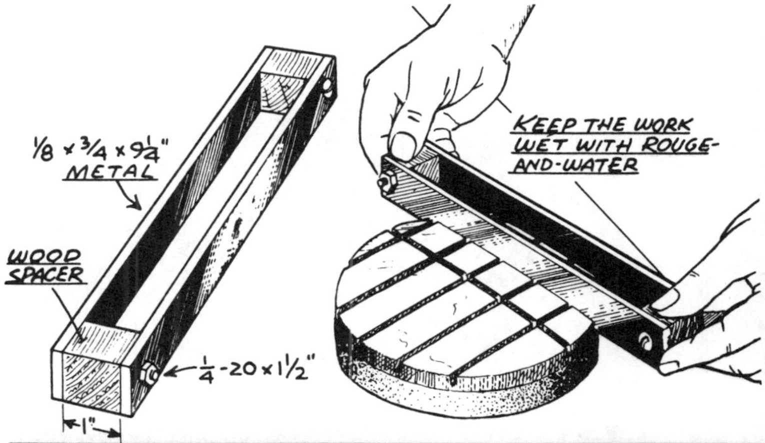

(13) USING A CHANNELING TOOL

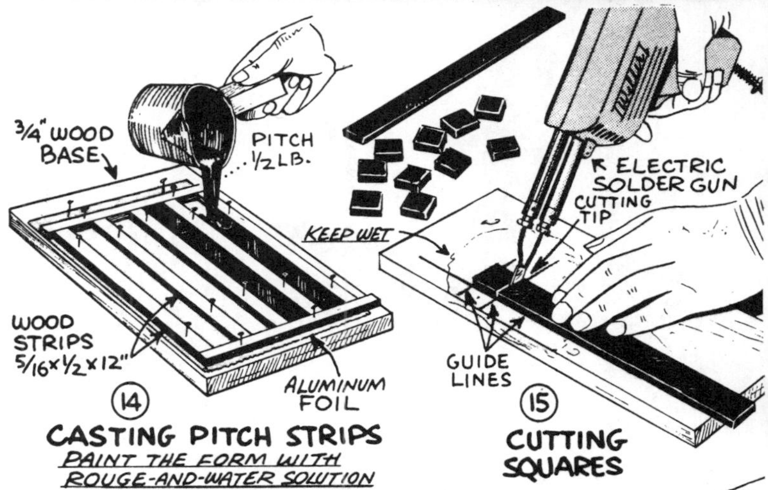

(14) CASTING PITCH STRIPS
PAINT THE FORM WITH ROUGE-AND-WATER SOLUTION

(15) CUTTING SQUARES

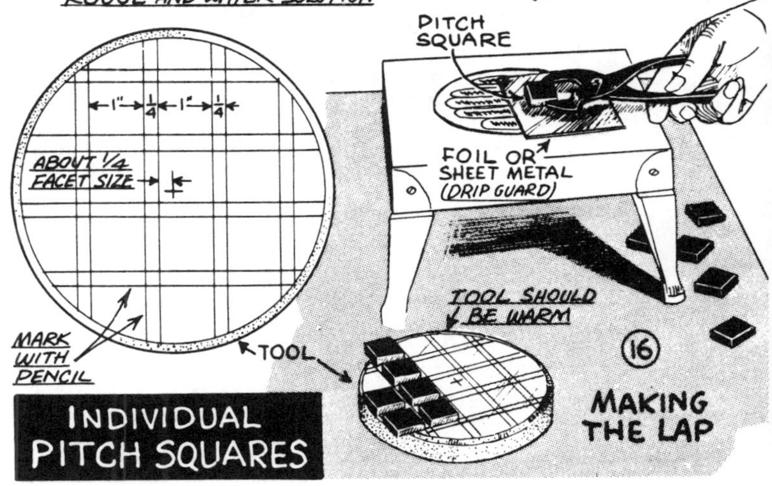

INDIVIDUAL PITCH SQUARES

(16) MAKING THE LAP

CUTTING METHOD. The direct way to make a lap is to cut the channels with a knife or saw after the pitch is cold. Soapy water is used as a lubricant. This is a messy job because the pitch tends to chip and crack in all the wrong places. If your lap is not perfect don't despair. With care and patience bubble holes and incorrectly cut facets can be repaired with hot pitch.

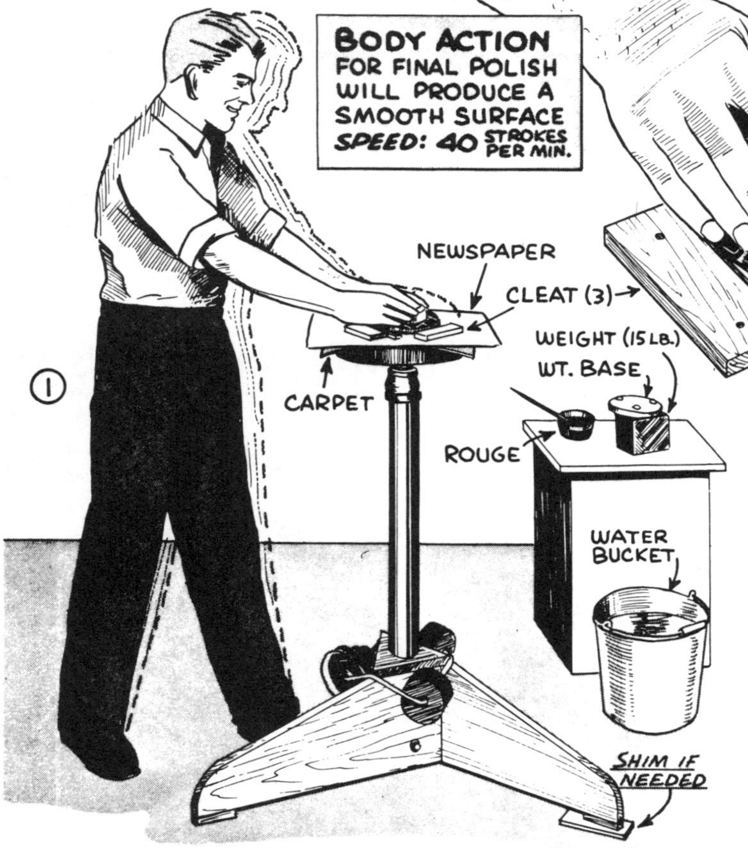

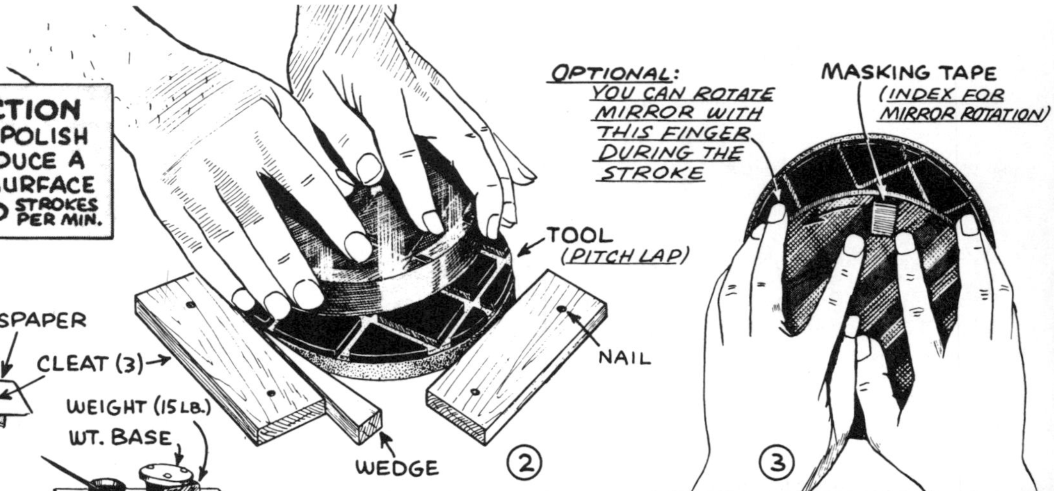

STROKE CAN BE STRAIGHT OR ZIG-ZAG - 1/3 AMPLITUD

it takes 4 to 6 hours POLISHING

POLISHING is very much like grinding except the abrasive is red rouge and the tool is now a pitch lap. The mirror is on top throughout. The job is simply to get the mirror polished "as clear as glass". This takes some 4 to 6 hours--and kid yourself not that the superficial polish which shows after the first hour is the real thing. The edge will polish out last, so don't be alarmed if a narrow halo persists after 2 or 3 hours.

THE POLISHING STROKE. This can be straight or zig-zag, with an amplitude no more than one-third, or 1 inch overhang at either end of the stroke for a 6 inch mirror. Long strokes should be avoided since the increased overhang tends to turn or flatten the edge of the mirror. Fig. 2 shows the usual hand position, with the thumbs applying modest down pressure at the center. The mirror rotation can be the stop-and-go method described for grinding, or, you can impart the turning movement with your fingers during the stroke, Fig. 3. Both the walk-around and the rotation should be as uniform as possible; stick a piece of masking tape on the mirror and note how smoothly you can make it go around.

The first two hours of polishing can be done with a normal arm action, 60 strokes per minute. For the last hour or two, you can get a smoother surface if you use a body-action stroke, Fig. 1. On a small-top grinding stand such as the one shown, the body stroke can also be done by tucking your elbows into your sides. The rouge-and-water abrasive mix should be thick at the start when contact may be poor or the lap sur-

Polishing Materials

ROUGE IS A FORM OF IRON OXIDE. IT IS A FINE, RED POWDER, MIXING FREELY WITH WATER. IT IS CHEAP AND PRODUCES A GOOD POLISH ... SLOW ACTION HELPS TO KEEP THE BEGINNER OUT OF TROUBLE.
MAIN FAULT IS THAT IT STAINS TO SOME EXTENT ... IT IS NOT "MESSY" UNLESS YOU ARE MESSY

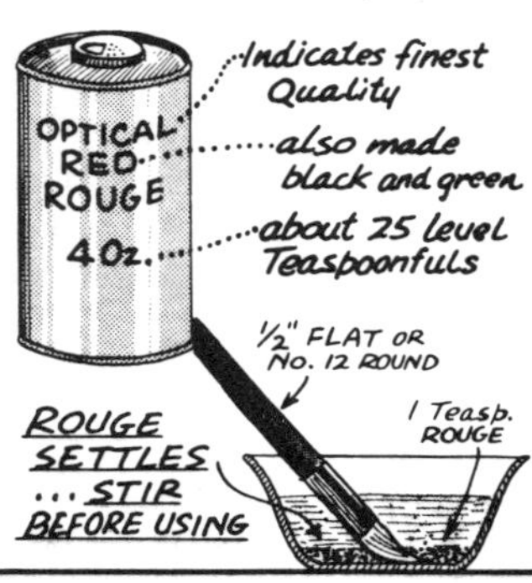

ROUGE-WATER MIXTURES	
THICK	1 PART ROUGE 4 PARTS WATER USE AS ANTI-STICK
MEDIUM	1:6 FOR INITIAL POLISHING - HOT PRESS - PARABOLIZING
THIN	1:8 GENERAL POLISHING
VERY THIN	1:10 FINAL POLISHING TO SPHERE

CERIUM OXIDE IS A PINK METALLIC OXIDE. NON-STAINING. POLISHES ABOUT TWICE AS FAST AS ROUGE. MIXES FAIRLY WELL WITH WATER AND CAN BE COMBINED WITH RED ROUGE

BARNESITE. A TRADENAME MATERIAL MADE OF SEVERAL RARE EARTHS. BROWN COLOR, NON-STAINING, EXTREMELY FINE. DOES NOT MIX READILY WITH WATER. IT CUTS TWICE AS FAST AS ROUGE

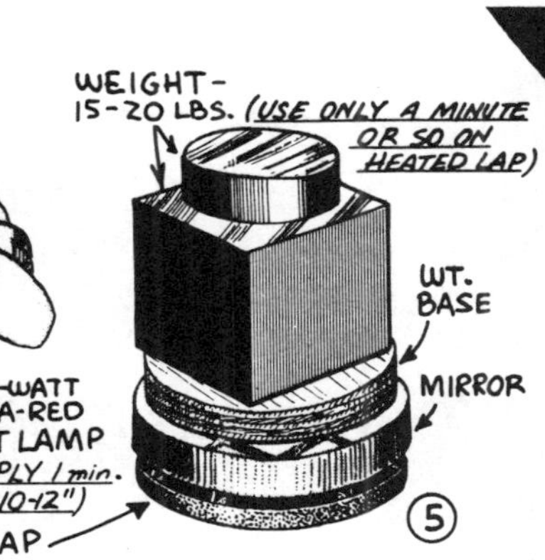

HOT PRESSING ASSURES GOOD CONTACT AND SHOULD PRECEDE ALL MAJOR WORK PERIODS. LAP CAN BE HEATED IN HOT WATER OR WITH HEAT LAMP

TRIMMING THE LAP

THE CHANNELS MUST BE KEPT OPEN - THE LAP MUST BE LESS DIAMETER THAN MIRROR

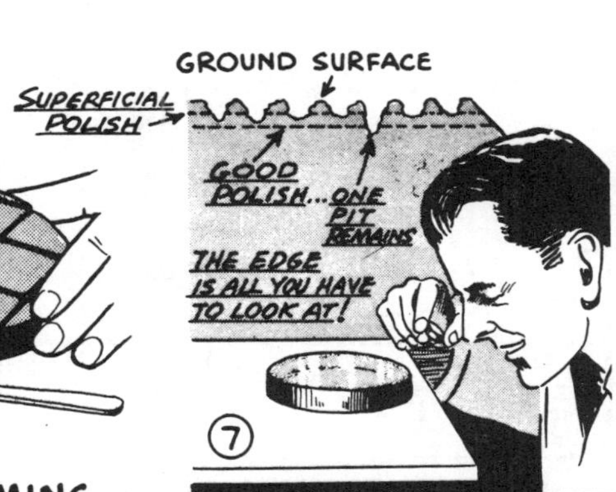

INSPECTION WILL REVEAL PITS... SUPERFICIAL POLISH IS EASY TO OBTAIN BUT A LOT OF WORK IS NEEDED TO REMOVE FINAL PITS

face sparsely charged. Each wet of rouge should be worked nearly dry since it is in this state that the rouge begins to wipe the glass with a vigorous polishing action. The drying mirror will become harder to push, and when it starts to stick and skip, it is time for a new charge of rouge. Polishing periods should be no less than 30 minutes--best to go an hour or more to get the frictional heat needed to allow the mirror to bed down snugly on its pitch pad.

HOT PRESSING. Do this with hot water heat in the manner described for lap making. Start with cold water. Heat until you see the first wisps of steam and then give it about 20 seconds more. Then proceed with the actual pressing, Fig. 5. An alternate heat treatment employs a heat lamp, Fig. 4, and this is useful for a fast press when the contact is already fairly good. An overnight cold press without weight is a perfect substitute for hot pressing; cover the work with a damp towel to retard evaporation. Various mesh materials such as onion sacking or nylon net can be hot pressed into the lap, producing numerous subfacets which facilitate making contact while giving the lap a good "bite".

TRIMMING THE LAP. After you have been polishing some 90 minutes, you will note the channels in the lap are beginning to close. Open them up with a sharp knife or chisel, Fig. 6, using a firm, bold stroke. Best results are obtained if the lap is mildly warm, as it will be immediately after polishing. An alternate here is to heat the lap and repress the rubber grid. This takes more time but in the long run it is probably the fastest method. Certainly, it is the neatest. Don't forget to trim around the edge of the lap, always keeping the lap about 1/16 inch less diameter than the mirror. It can be seen that the mirror must cover the lap completely all around to avoid forming a ridge at the edge of the lap during hot pressing. It is also easy to see that a turned edge results when the mirror plows into this ridge when polishing is resumed. Use a thinner rouge solution as polishing nears completion, and slow down the stroke for maximum smoothness.

You can stop here!

IF YOU like, you can call your polished mirror finished--9 times out of 10 the surface will be accurate enough to produce a good image. If you want proof before getting the mirror aluminized, you can assemble and test a bare glass telescope. A bare 6-inch f/8 will pick up as much light as a small, 1-1/4-inch refractor --plenty of light to look at the moon or any daytime object.

As a final touch before getting the glass aluminized, it is worthwhile to grind the face, as shown below. 20 or 30 seconds work will put a neat flat rim around your glass.

In case you are not satisfied with the imagery of the bare glass telescope, try masking to 5 in. diameter. If the only fault of the mirror is a turned edge, masking will cure it completely.

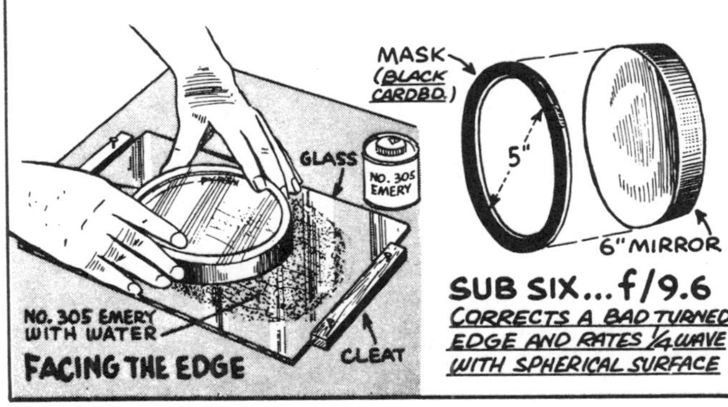

SUB SIX... f/9.6
CORRECTS A BAD TURNED EDGE AND RATES 1/4 WAVE WITH SPHERICAL SURFACE

Pre-Testing OPTICS

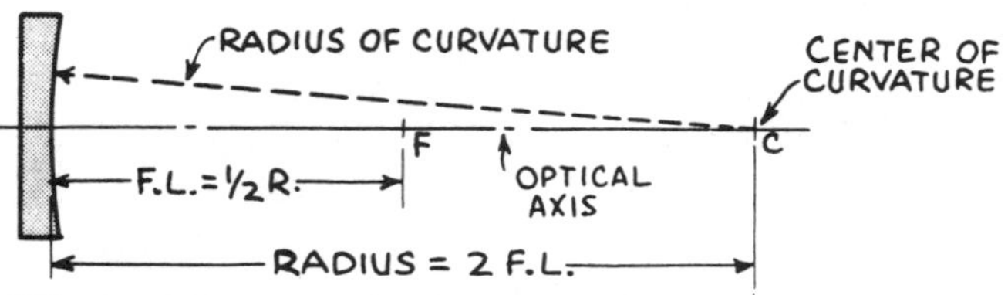

1 **GENERAL CASE-ALL MIRRORS:** THE FOCAL LENGTH OF A MIRROR = ½ THE RADIUS OF CURVATURE

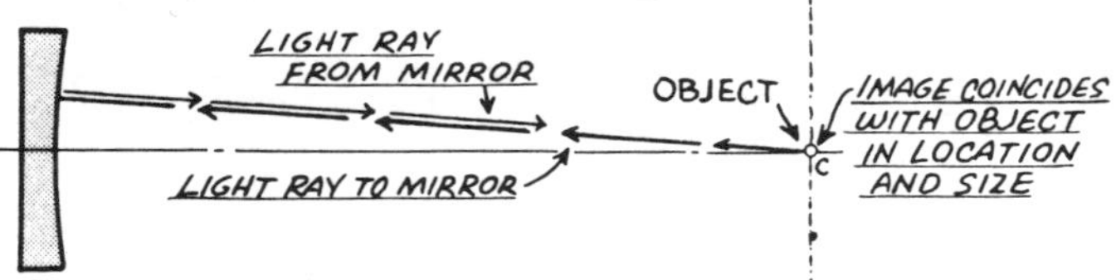

2 LIGHT RAYS FROM A SMALL OBJECT AT THE CENTER OF CURVATURE WILL BE REFLECTED OVER THE SAME PATHS TO FORM AN IMAGE COINCIDING WITH THE OBJECT

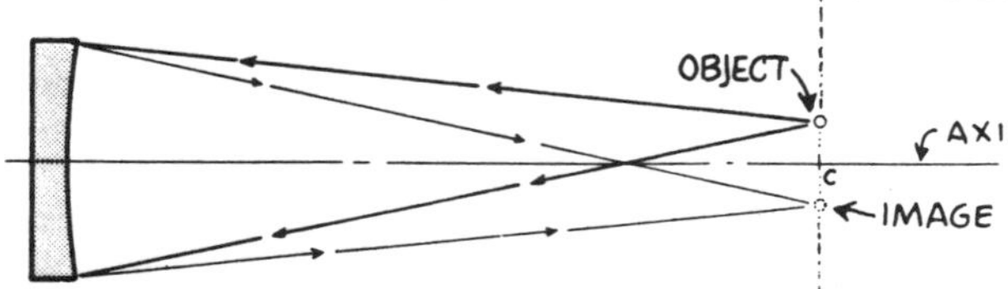

3 IF THE OBJECT IS DISPLACED TO ONE SIDE OF THE AXIS, THE IMAGE WILL FORM AT SAME DISTANCE ON OPPOSITE SIDE

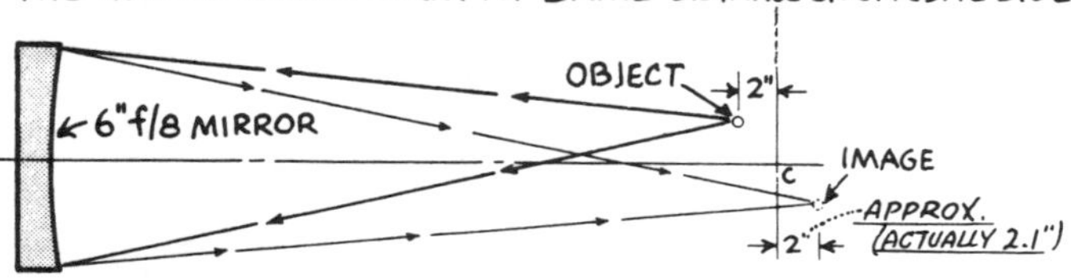

4 IF THE OBJECT IS MOVED A SHORT DISTANCE TOWARD MIRROR, THE IMAGE WILL MOVE A SHORT DISTANCE AWAY FROM MIRROR

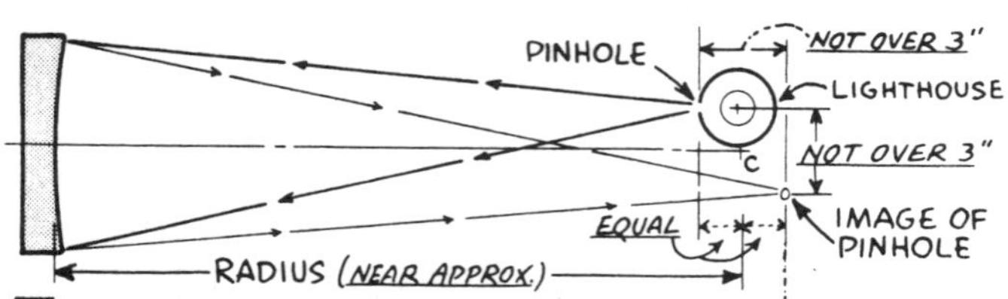

5 **USUAL SETUP FOR TESTING.** PINHOLE IMAGE AT REAR OF LIGHT IS CONVENIENTLY PLACED FOR OBSERVATION. THE SMALL DISPLACEMENTS WILL NOT AFFECT IMAGE QUALITY OR MEASUREMENTS

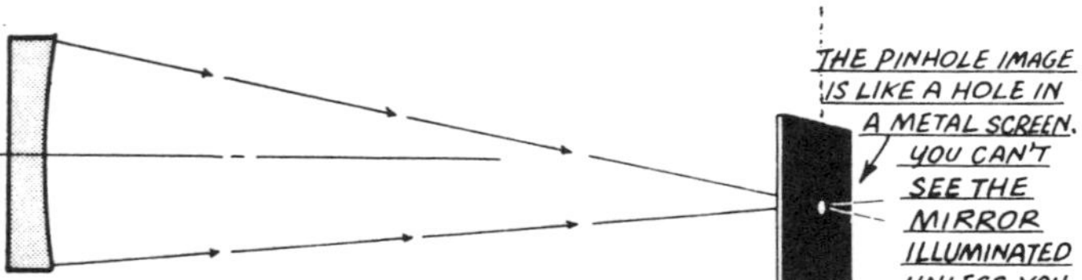

6 THE IMAGE OF ANY APERTURE IS AS REAL, OPTICALLY, AS THE APERTURE ITSELF

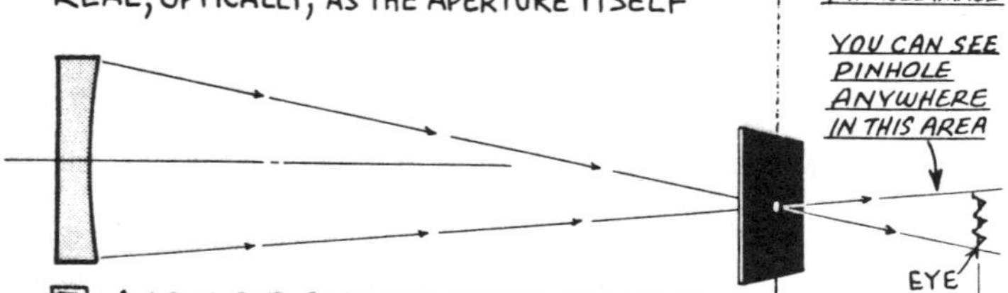

7 A LONG EYE POSITION MAKES IT EASY TO SEE THE PINHOLE IMAGE ... BUT ... (SEE NEXT COL.)

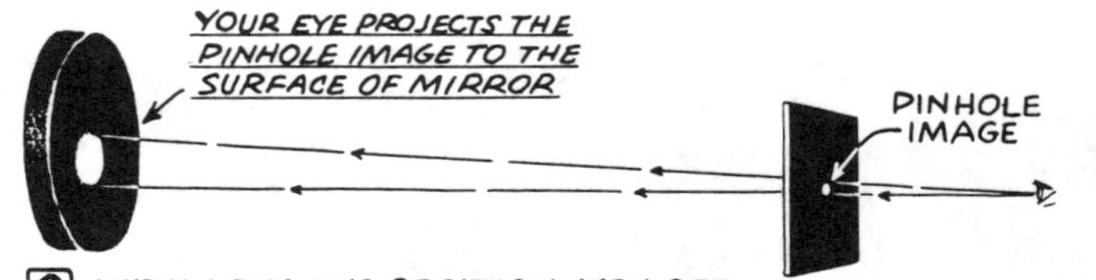

8 WITH LONG EYE POSITION, YOU SEE ONLY A SMALL PART OF THE MIRROR ILLUMINATED

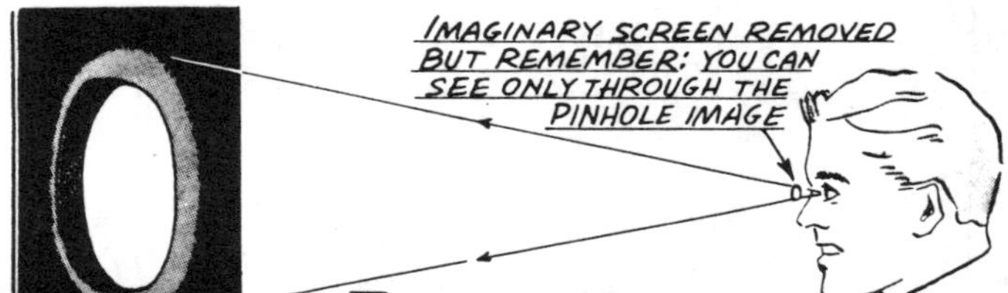

9 CLOSE EYE POSITION LETS YOU SEE ALL OF MIRROR

10 A SLIT IS EASIER TO FIND, OFFERS A BIGGER TARGET AND PROVIDES THE MOST EVEN ILLUMINATION

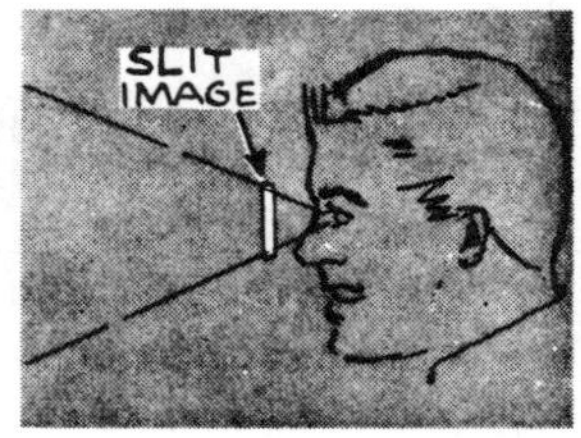

ALL TESTING SHOULD BE DONE IN SEMI-DARKNESS TO OBTAIN BEST CONTRAST. DO NOT BLACK OUT... SOME ROOM LIGHT IS DESIRABLE

Diffraction Shadows

1 **A STRAIGHT EDGE CLOSE TO YOUR EYE**

YOUR EYE CAN'T FOCUS AT SUCH SHORT RANGE. YOU SEE THE EDGE AS A DARK SHADOW BORDERED WITH A GRAY FRINGE

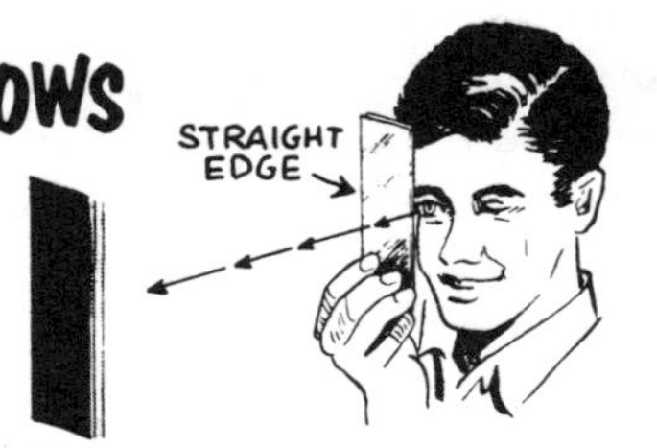

2 **TWO STRAIGHT EDGES CLOSE TOGETHER-A SLIT**

YOU SEE AN EXPANDED SLIT OPENING FILLED WITH FINE VERTICAL LINES. THE SHADING IS COARSER WITH A LONG EYE POSITION, AND IS VERY COARSE AND IRREGULAR IF THE STRAIGHT EDGES ARE ROUGH

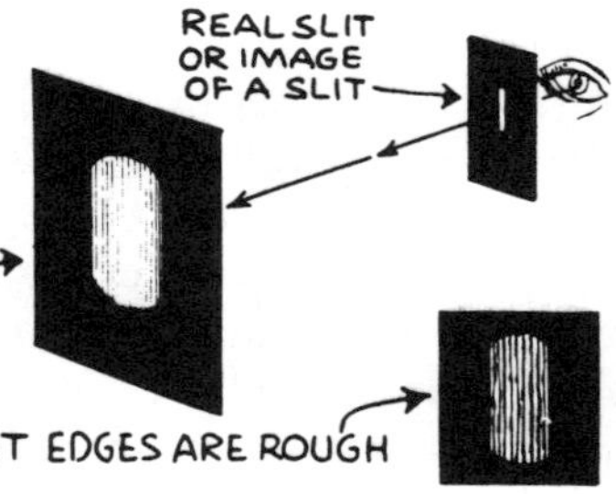

3 **KNIFE-EDGE AND SLIT COMBINED**

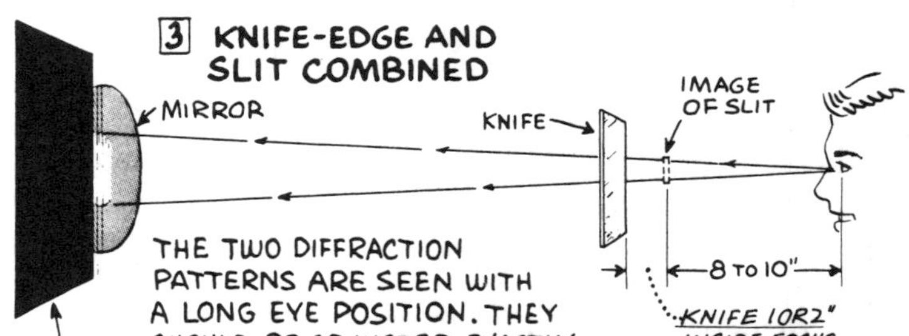

THE TWO DIFFRACTION PATTERNS ARE SEEN WITH A LONG EYE POSITION. THEY SHOULD BE ADJUSTED EXACTLY PARALLEL FOR MAXIMUM SENSITIVITY

4 **A SMALL PINHOLE**

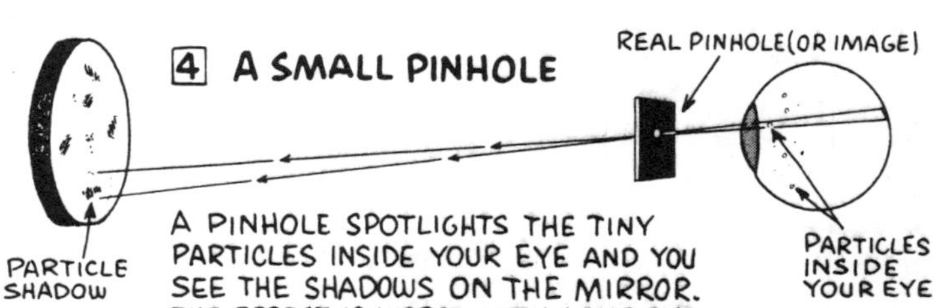

A PINHOLE SPOTLIGHTS THE TINY PARTICLES INSIDE YOUR EYE AND YOU SEE THE SHADOWS ON THE MIRROR. THE EFFECT IS WORST WITH LONG EYE POSITION ... VERY FAINT WHEN EYE IS CLOSE

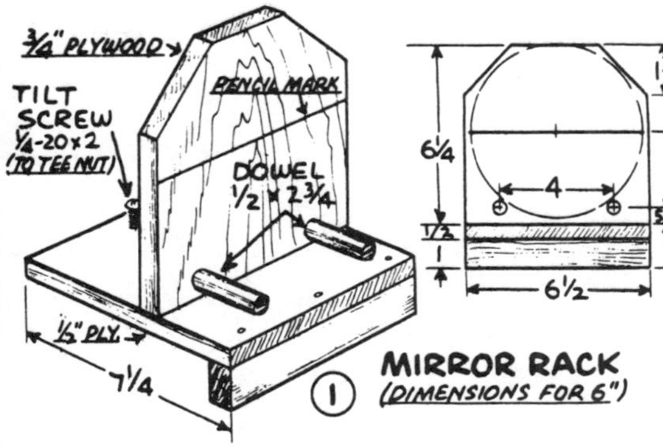

① MIRROR RACK (DIMENSIONS FOR 6")

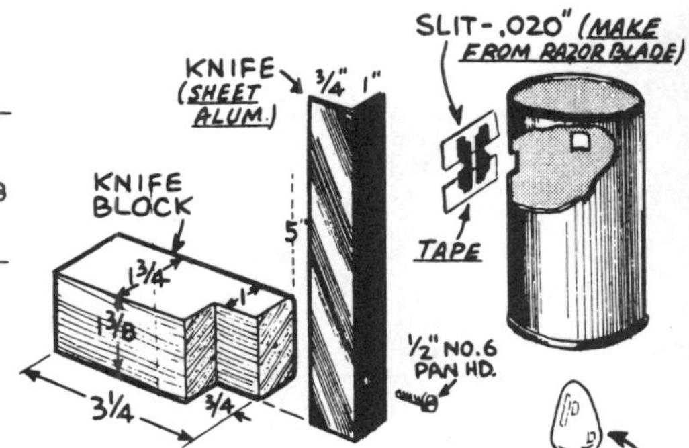

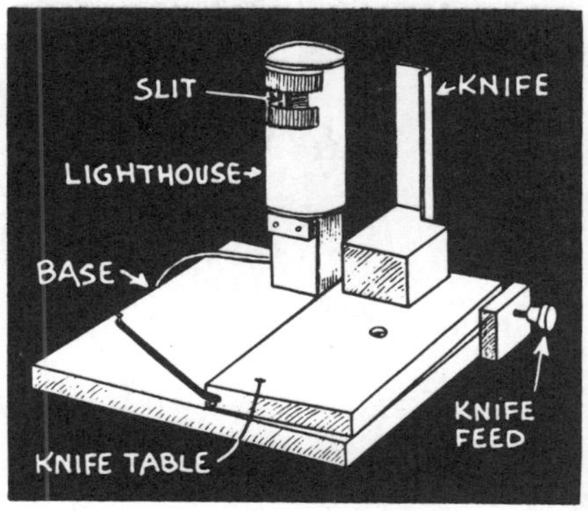

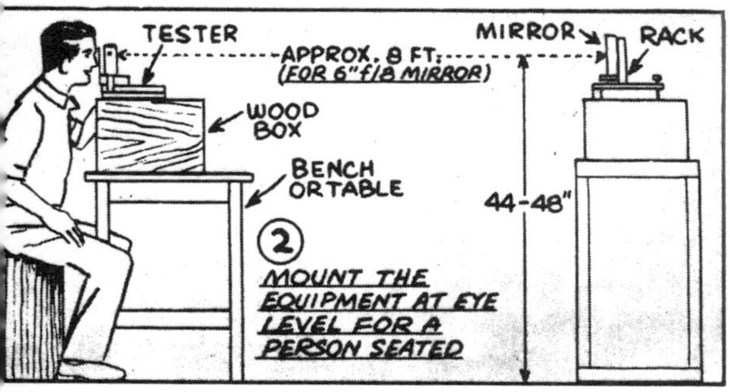

② MOUNT THE EQUIPMENT AT EYE LEVEL FOR A PERSON SEATED

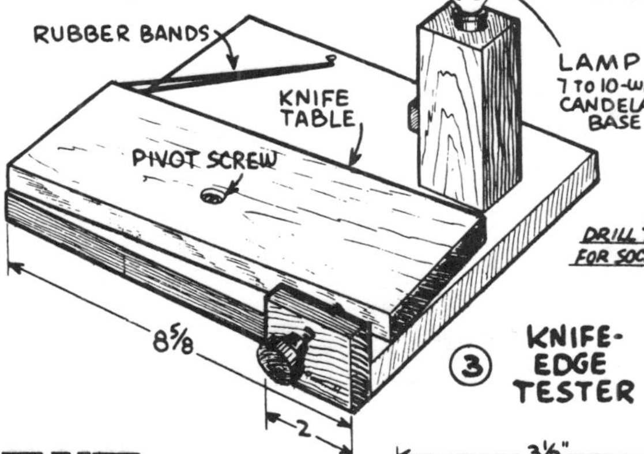

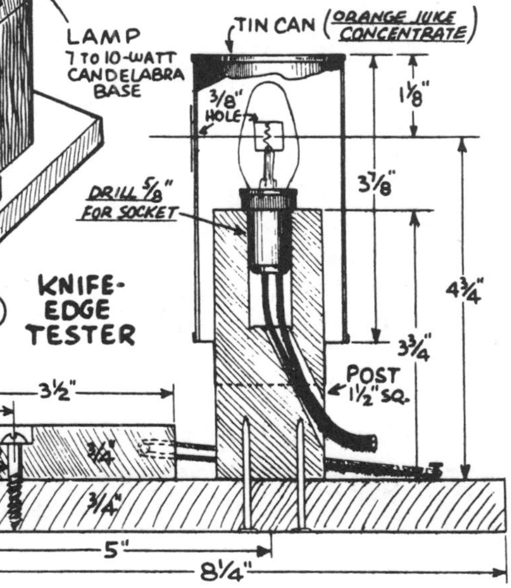

③ KNIFE-EDGE TESTER

TESTING EQUIPMENT

AFTER you have ground and polished your mirror, its surface will be a sphere or some shape very close to a sphere. The rest of the job consists of testing and correcting to an exact spherical surface, and then changing the sphere to a paraboloid. In testing the paraboloid, it is necessary to measure the exact radius of various zones, and for this delicate work a micrometer test rig is a convenience. Testing the sphere does not require zonal measurements and can be done with very simple equipment.

TESTING EQUIPMENT. Only two items are needed. One is the mirror rack, Fig. 1, and the other is the knife-edge tester itself, Fig. 3. The knife block is not fastened and can be manipulated either by hand or with the screw feed shown. The knife itself is fastened to the knife block with a single screw to permit a slight tilt adjustment when needed. The light is 10-watt, candelabra base, white or clear. If clear, tracing paper should be taped inside the can behind the slit window. The other window is used for setting up the equipment and is left open for maximum light output. A slit is recommended, and is easily made from razor blade pieces, spaced about .020" apart (five sheets of this paper). The tin can lighthouse rotates on the lighthouse post to put either window in position.

You will spend a good bit of time testing, and the idea is to be comfortable. This means the equipment should be set up at eye level for a person seated, Fig. 2. The setup should be in a

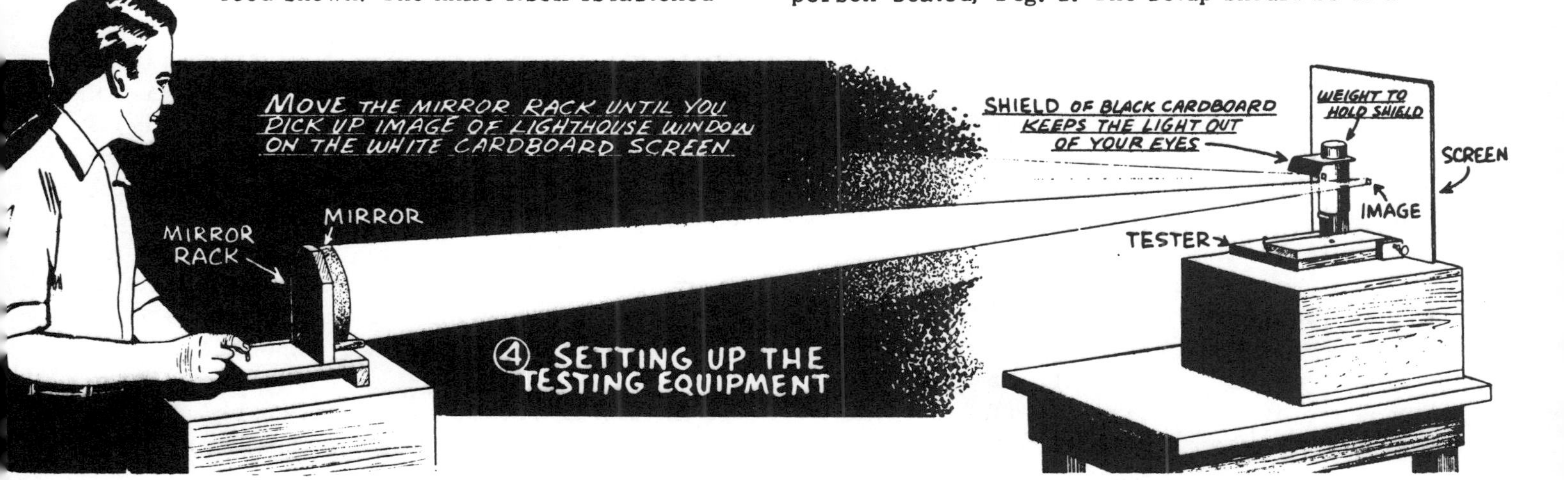

④ SETTING UP THE TESTING EQUIPMENT

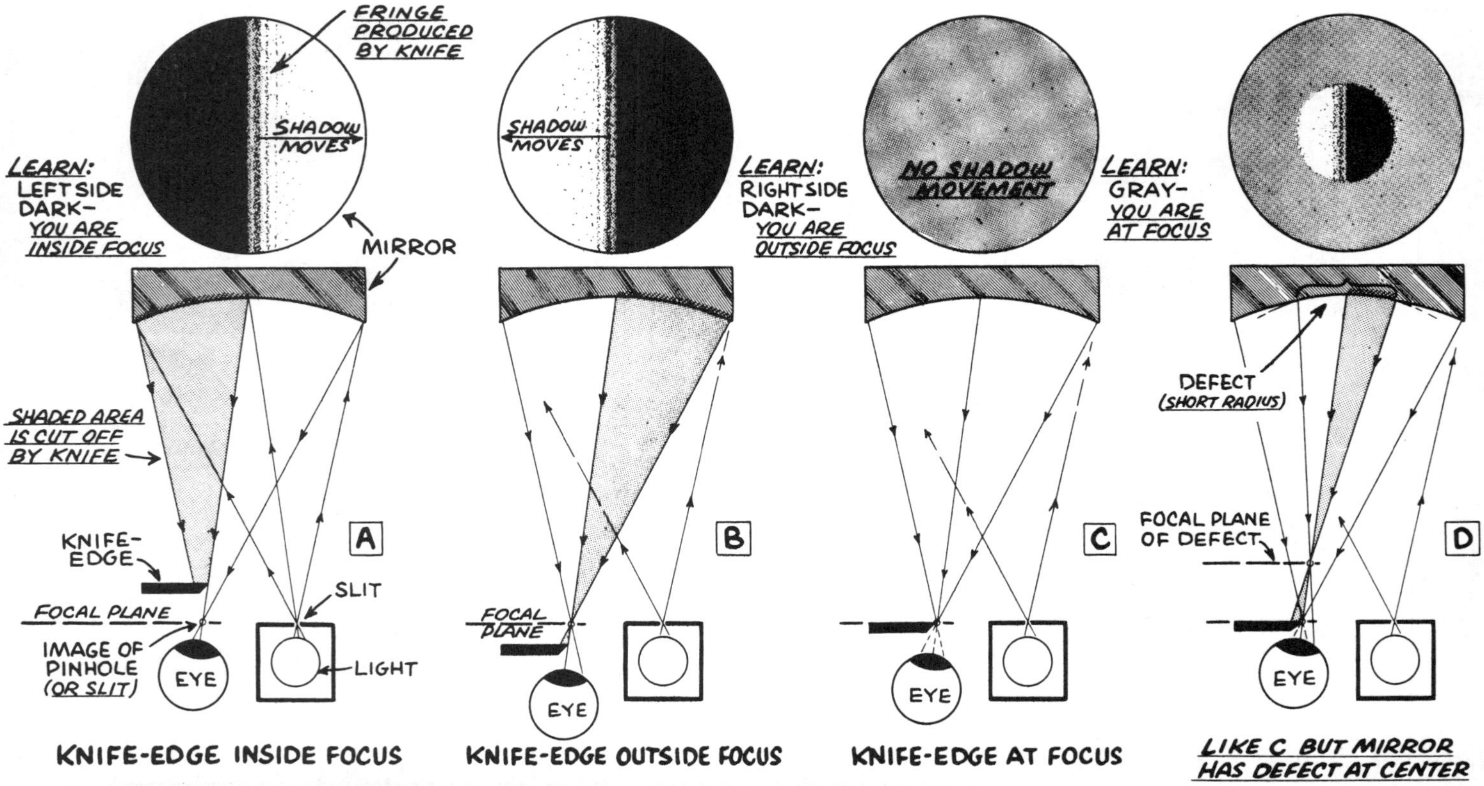

⑤ SHADOW APPEARANCE OF A SPHERICAL MIRROR

location where it can remain undisturbed until the mirror is finished. A setup near the grinding stand is preferable in order to avoid temperature changes. Vibrations and cold air drafts must be avoided.

SETTING-UP. Fig. 4 shows the equipment being adjusted for use. The room must be completely dark. Even so, it may take you a few minutes to pick up the faint reflection of the lighthouse window on the cardboard screen. Once you get the image on the screen, it is brought to a sharp focus by moving the mirror rack back and forth; final adjustments locate the image level with the window and about 3/4" from the lighthouse. Now, seat yourself behind the tester. Remove the cardboard screen. You will see the lighthouse window reflected in the mirror. Then, rotate the lighthouse until the slit image comes into view. You are ready to test.

TESTING A SPHERE. Although it is not likely your mirror is a good sphere at this stage, you can probably make it perform somewhat like a spherical mirror, Fig. 5. You will not be able to make the mirror an even gray all over, as at C, since this occurs only when the surface is a perfect sphere. The general idea in testing is that your mirror is an unlimited number of concentric mirrors or zones. You can gray any zone, and when you gray any specific zone the knife is cutting-in at the exact radius of that zone. Zones of longer radius will show dark on left side; zones of shorter radius will darken on right side. Fig. D is an example. By moving the knife forward, you could gray the center of this mirror, and the outer zone of longer radius would then show dark on the left-hand side.

Practice a little on your own mirror. Cut the knife-edge into the light beam well inside focus (toward the mirror), to form a shadow on left side of mirror. Insert the knife-edge outside focus (away from the mirror) to darken the right side. Somewhere between these two extreme knife settings, you will find an "average" knife setting where the shadows on the mirror are about evenly divided right and left. If you measure the distance from this knife position to the mirror, you will get the radius of the mirror--and, of course, half of this is the focal length. It will shrink another quarter-inch or so during correcting and parabolizing.

If your mirror shows moving clouds, you will know that the test rig is located in a current of cold air. It is possible to see through these clouds, but you will have trouble enough trying to interpret shadows without this extra complication. The trouble usually comes from a basement window and can often be cured by sealing the window with cardboard and tape.

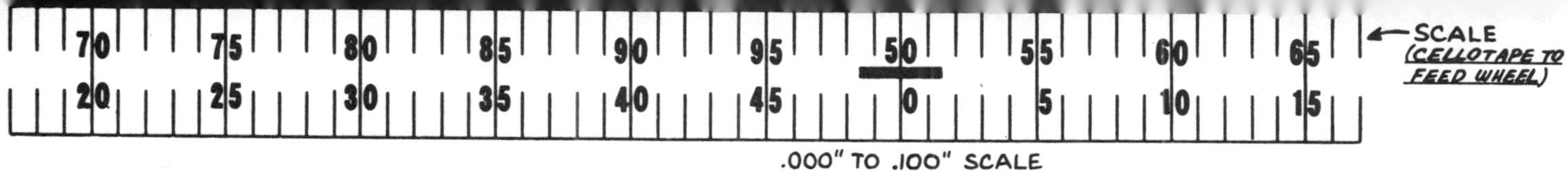

Micrometer Knife-edge Tester

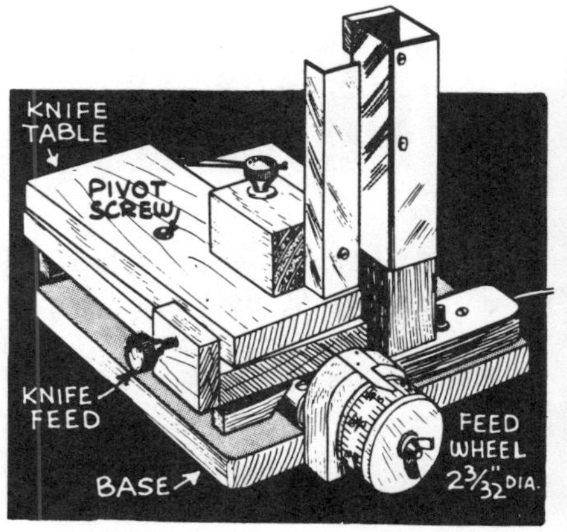

If you plan to parabolize your mirror, a tester with micrometer scale is a convenience, worth making for even a single mirror. This one is all-wood construction except the monorail bed on which the carriage slides, this part being a length of polished aluminum or steel tubing. The carriage is moved by a 1/4-20 screw, giving a feed of fifty thousandths per rev. The scale reads two revs, allowing you to read the whole correction for the average mirror. Lathe-turn the feed wheel to an exact diameter to accept the paper scale. Make the wheel a free fit on the feed screw to permit setting to a zero position.

A pivoting action is used to cut in the knife edge, controlled by the knife feed screw. A single screw holds the knife and permits tilting to a position parallel with the slit image. A fine adjustment for knife parallelism is obtained by means of the screw at right side of the carriage. The slit can be .015 to .030-inch; it is mounted on a track and can be removed when you need the big window for setting-up.

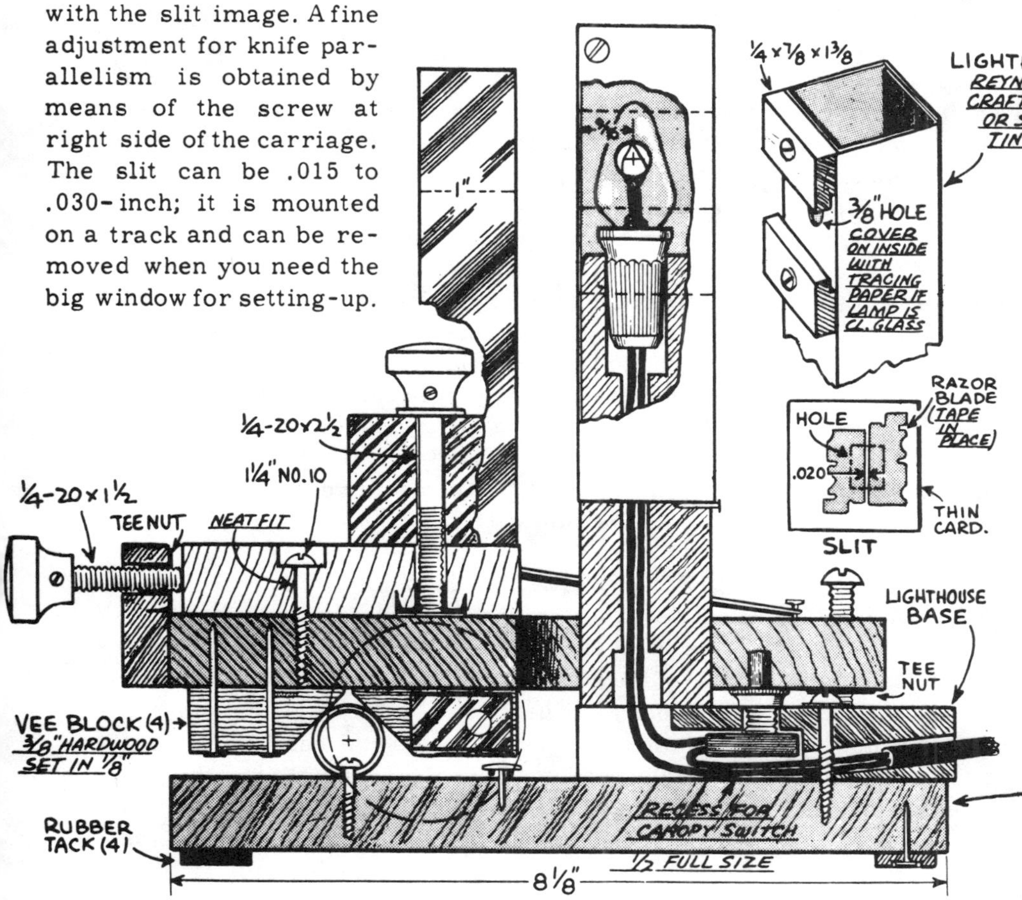

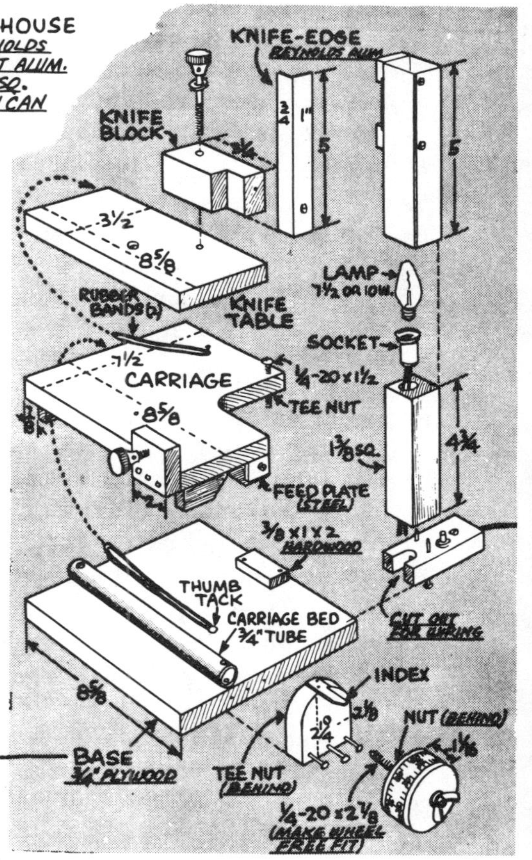

SLIT or Pinhole?

The fluid in a normal eye is fairly clear in much the same manner as the average room contains clear air. Yet a beam of sunlight will show the clear air contains dust particles...and a cone of light from a small pinhole will show the normal eye contains many tiny particles, ranging in density from transparent to opaque.

Eyes vary greatly in this respect, but few persons can use a pinhole without seeing some traces of "eye bumps" on the mirror. Eye bumps move when you move your head, and in this way are distinguished from a similar "dog biscuit" surface which results from faulty polishing. The dog biscuit stands still.

Eye bumps can be practically eliminated with a close eye position, 1/2 inch or less from the pinhole image. However, this position is difficult to obtain if you wear glasses. There is a simple solution to the whole problem: Use a slit! Then the diffraction effect from eye particles is converted to a faint vertical streaking which causes no confusion at all in interpreting the normal mirror shadows.

TESTING and CORRECTING

SHADOWS seen in knife-edge testing are easy to interpret by imagining the mirror to be side-lighted from the right, Fig. 1. Under this circumstance, a dark area on the right side means a down-slope or hole, **Fig. 3**; a light area on right side means a rising slope or hill, Fig. 4. Putting this bit of know-how to work, it is easy to make a rough sketch of the apparent section shape. The apparent shape is then corrected by suitable polishing strokes and technics. This method of working with entirely imaginary shapes works out very well. However, you may gain a better concept of the work if you keep in mind the actual glass shape; a hill, Fig. 4, is actually a flat zone of long radius; a turned edge, Fig. 5, is not an actual turning-over of the glass except possibly at the extreme edge; the parabola, Fig. 7, is not a fancy reverse curve, but simply a single smooth curve, less and less curved toward the edge.

THE BEST SHAPE. At any stage of testing, the mirror will show a variety of shadow faces, changing as the knife is moved back and forth. The general behavior of the mirror shape is that it mimics the knife, as can be seen in Fig. 8--if you move the knife toward the mirror, the mirror will bulge out toward the knife. Any of the shapes you see can be used as a basis for polishing work. Since the shape variation is caused by moving the knife, each change in shape also means a slight change in radius.

The best shape for correction is the one requiring the least work. Consider the oblate spheroid in Fig. 8. Inside average focus, you see a big hill--lots of work. Outside focus it looks like less glass, but it is very tricky working right out at the edge. The middle diagram at average focus shows a moderate hill, which you can plane down with long strokes, and there is enough glass at the rim to permit long strokes without any danger of turning the edge.

In most cases, the best apparent shape for correction is seen when the knife is at the "average" focal plane. This position is located by balancing the shadows, looking mainly at the outer 1/3 of the mirror. If the left side seems to be darker than the right, you know you are inside focus, so you pull the knife back a little. If the right side is the darker, you push the knife forward until the two sides look about the same shadow depth.

There is rarely any need to interpret a complicated shadow for the simple reason you never try to doctor a complicated shape--it's back to the rouge pot for at least 30 minutes of ordinary polishing. With anything approaching smooth, systematic stroking, you should get a fairly smooth, concentric figure. Maybe it will be an oblate spheroid or show a hill, hole or turned edge, but it will be an easily recognizable face and one you can work on.

TURNED EDGE TESTS. While you can see a hill or hole easily enough, the turned edge is not always so obvious. The one best test is the appearance of the diffraction ring at the edge of the mirror, Fig. 9A. There is always a bright ring on the right side. If this is a narrow hairline of light on right side and nearly as bright on the left side, the edge is good, although it may be turned slightly. When the diffraction ring is broad and flaring on the right side and the left side is dark

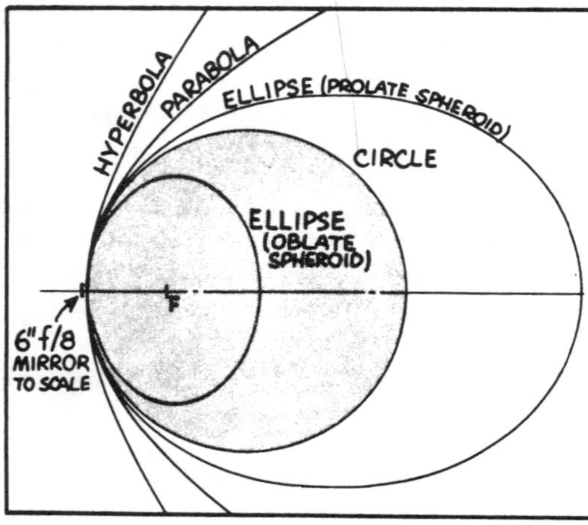

A Family of Curves

You may meet the whole family of regular curves while working a single mirror. There are many variations of the ellipse and hyperbola, but only one shape for the circle and parabola. All of the curves have about the same shape over the span of a 6-inch mirror ...differences are measured in millionths of an inch.

SPHERICAL ABERRATION (S.A.)

FIGURE	AT INFINITY FOCUS	AT CENTER OF CURVATURE
OBLATE SPHEROID	UNDER	UNDER
SPHERE	UNDER	NONE
ELLIPSOID (PROLATE SPHEROID)	UNDER	OVER
PARABOLOID	NONE	OVER
HYPERBOLOID	OVER	OVER

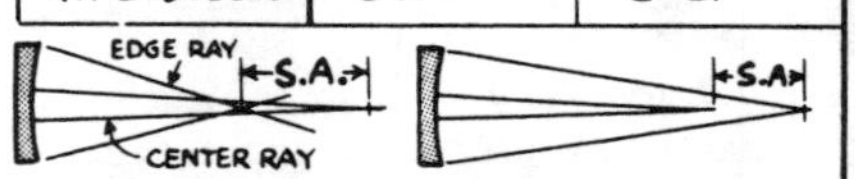

UNDER-CORRECTION *EDGE RAYS FOCUS CLOSER THAN CENTER RAYS*

OVER-CORRECTION *EDGE RAYS FOCUS LONGER THAN CENTER RAYS*

AN IMAGINARY SIDE LIGHT REVEALS THE APPARENT SHAPE OF THE MIRROR

THE LIGHT DOES NOT CAST A SHADOW. IT IS NOT STOPPED BY AN APPARENT OBSTRUCTION

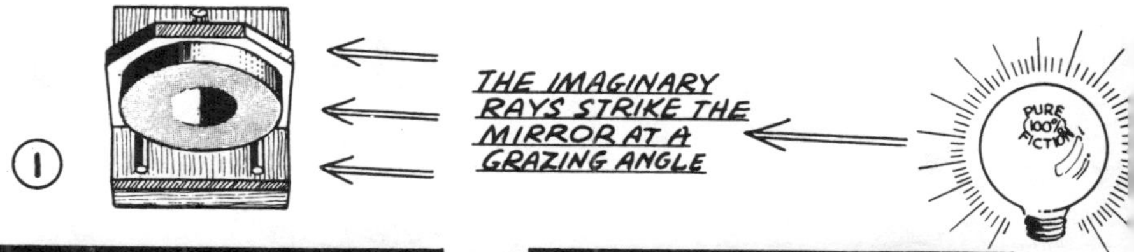

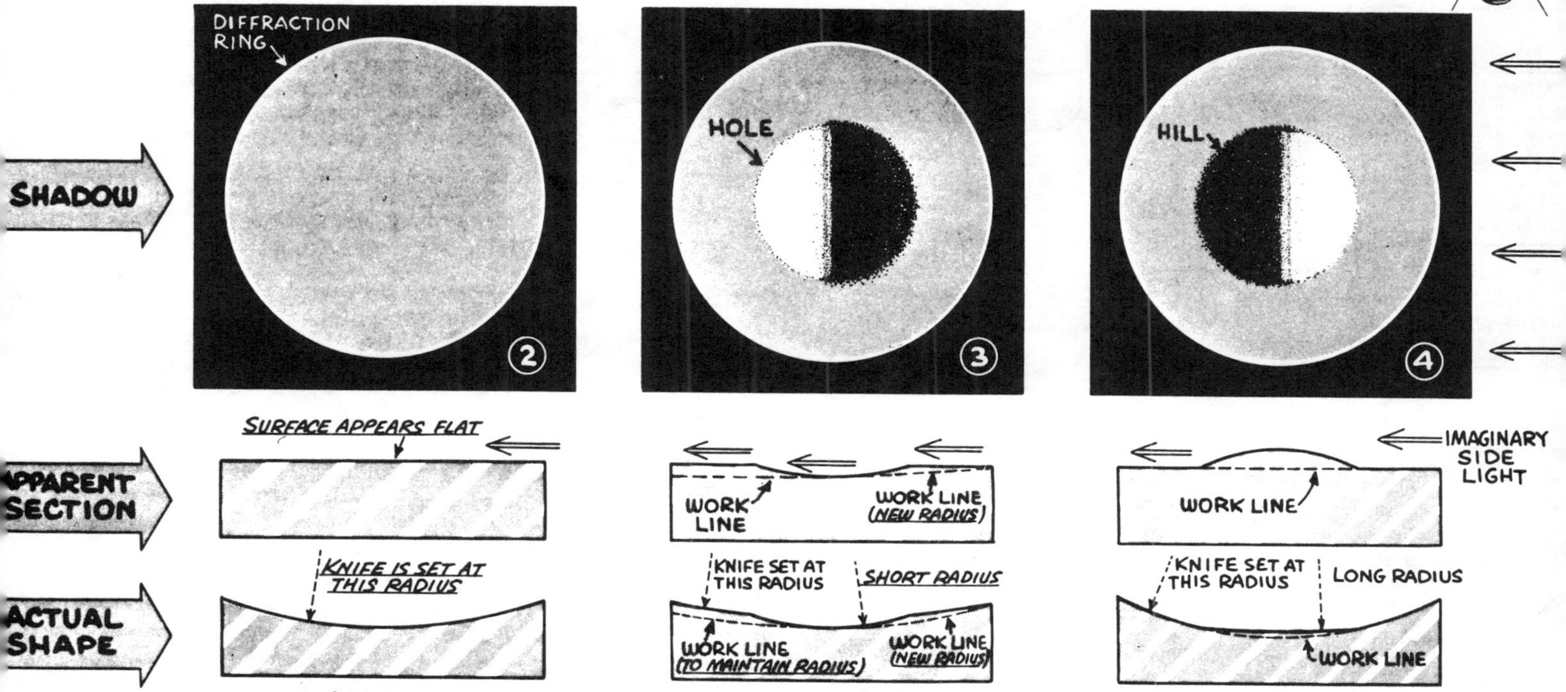

SPHERE

THE KNIFE-EDGE TEST AT CENTER OF CURVATURE IS A NULL TEST FOR THE SPHERE – THERE ARE NO SHADOWS. THE MIRROR LOOKS FLAT, GRAYS GRADUALLY TO BLACK WITH NO MOVING SHADOWS

HOLE AT CENTER

WORK LINE SHOWS GLASS TO BE REMOVED IF ORIGINAL RADIUS IS TO BE MAINTAINED. NORMALLY A SLIGHT CHANGE IN RADIUS IS PERMISSIBLE... WORK LINE IS THEN AS SHOWN IN RIGHT HALF OF DIAGRAM

HILL AT CENTER

GOING BY THE APPARENT SECTION, YOU PLANE DOWN A HILL TO A FLAT SURFACE. *BUT*... KEEP THE ACTUAL SITUATION IN MIND: A HILL IS A ZONE OF LONG RADIUS... YOU CORRECT BY DEEPENING THE CENTER

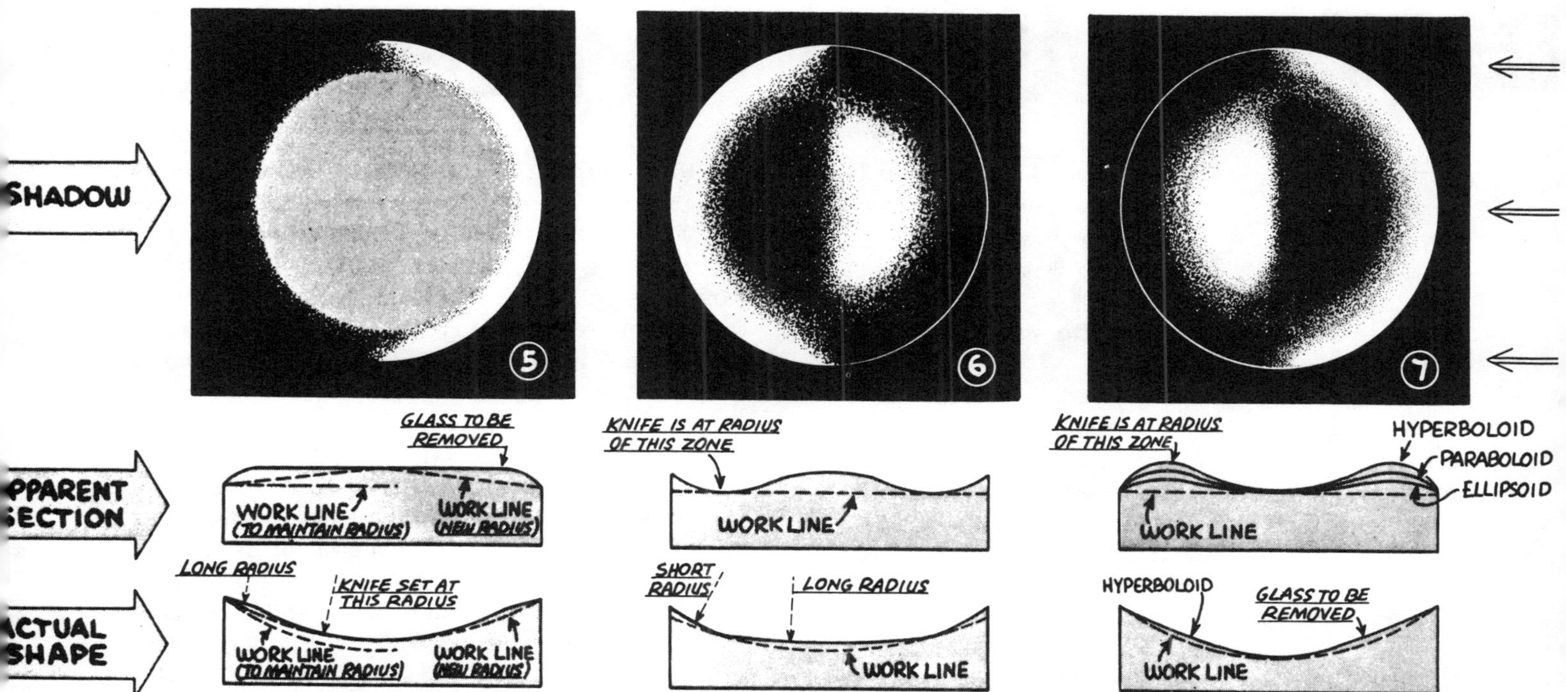

TURNED EDGE

THIS COMMON DEFECT GETS ITS NAME FROM THE TURNED-OVER APPEARANCE OF THE EDGE IN KNIFE TEST. THE ACTUAL GLASS SHAPE IS SIMPLY A FLATTENING OF THE CURVE

OBLATE SPHEROID

FAST, SHORT, STRAIGHT STROKE POLISHING ON A HARD LAP WILL USUALLY PRODUCE THIS KIND OF FIGURE. IT IS CORRECTED WITH A LONGER STROKE IN A WIDE ZIG-ZAG

ELLIPSOID, PARABOLOID, HYPERBOLOID

...ALL HAVE SAME APPARENT SHAPE AND DIFFER ONLY IN APPARENT DEPTH, THE HYPERBOLOID SHOWING THE DARKEST SHADOW. DIAGRAMS SHOW CORRECTION TO RETURN TO A SPHERE

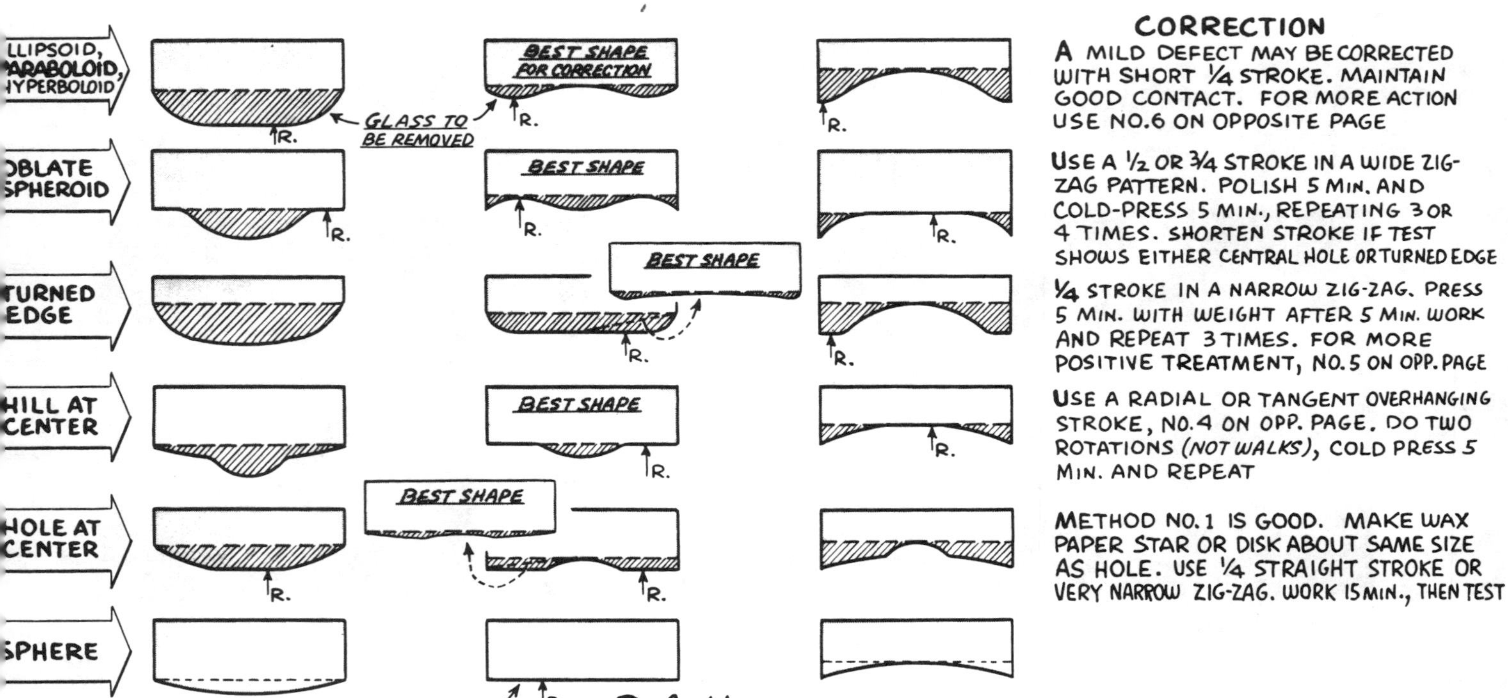

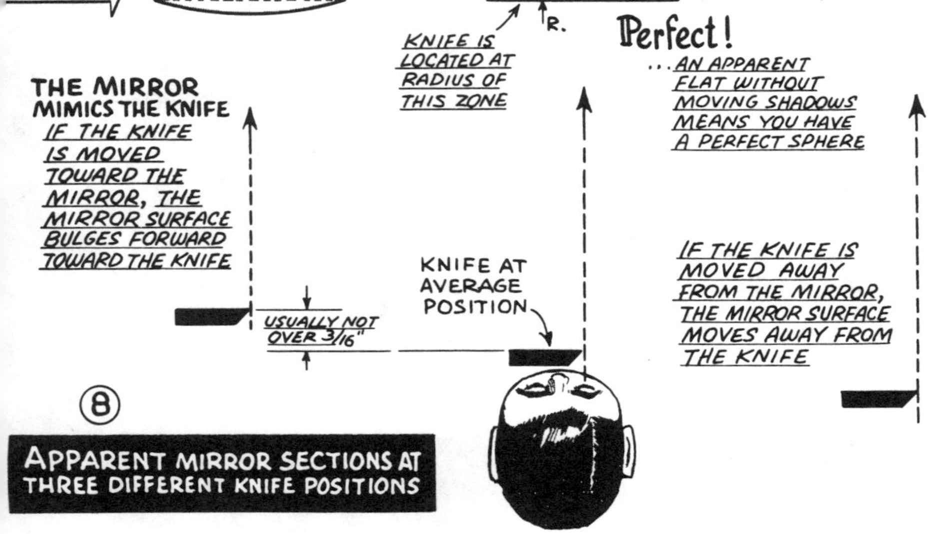

⑧ APPARENT MIRROR SECTIONS AT THREE DIFFERENT KNIFE POSITIONS

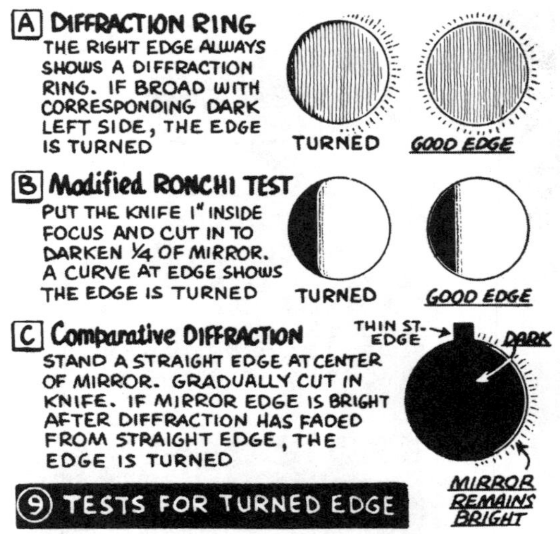

⑨ TESTS FOR TURNED EDGE

with no light at all, you have a real-for-sure turned edge. Fig. 9B shows one of the more popular tests; it works only when the light source is a slit. In applying Fig. 9C test, you must hold your head steady, eyes straight ahead, avoiding any tendency to sneak your gaze around the edge of the knife.

A turned edge is not all bad. A perfect parabola has a turned edge. A spherical mirror with a turned edge not exceeding the turned edge of a parabola is a better telescope mirror than a perfect sphere. The bad kind of turned edge is a gross fault inherited from fine grinding. If you are critical about the edge during grinding, you will not have this trouble.

CORRECTING TECHNICS. The first thing to try for most defects is simply more polishing, varying only the stroke length. For more positive action, all of the methods on opposite page are useful. They can also get you into a lot of trouble; first attempts at local retouching invariably make your mirror worse instead of better. For a hole at center, the deformed lap, No. 1 on opposite page, is usually effective. The overhanging stroke, No. 4, will reduce a central hill, while accented pressure, No. 5, is a good way to correct a turned edge. Work periods are short, ranging from 3 to 15 minutes. All work should be done with a whole number of walks or rotations, counting the mirror rotations if the mirror is on top, and the walks-around-the-post if the lap is on top. You can keep track of rotations by sticking a piece of masking tape on back of mirror; start with this between your index fingers and stop when the tape is under right index finger.

Keep a log book of your efforts. You can expect several hours of holes and hills before getting the flat, velvety moon indicating a perfect sphere. Some "dog biscuit" is permissible since you can see defects to nearly 1/100 wave, far finer than the 1/4-wave accuracy required.

Correcting Technics

1 DEPRESSING CENTER OF LAP

APPLICATION: HOLE AT CENTER

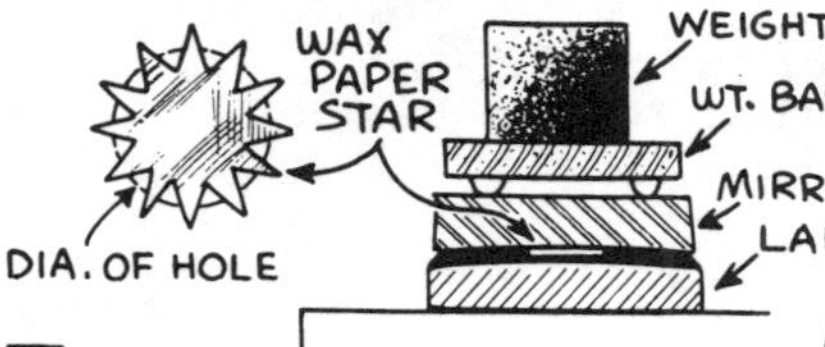

SINK WAX PAPER STAR INTO HEATED LAP. REMOVE WHEN COOL. RESUME POLISHING WITH 1/3 STROKE.

WARM LAP WITH 250-W. HEAT LAMP 1½ MIN. KEEP LAP WET WITH ROUGE

ALTERNATE: SCRATCHING, SANDING SHAVING OR FILING AT CENTER OF LAP. *NO HEAT*

THE IDEA IS TO DECREASE AREA OF CONTACT -- 15 OR 20 SCRATCHES WITH A RAZOR BLADE WILL FIX A SMALL SHALLOW HOLE

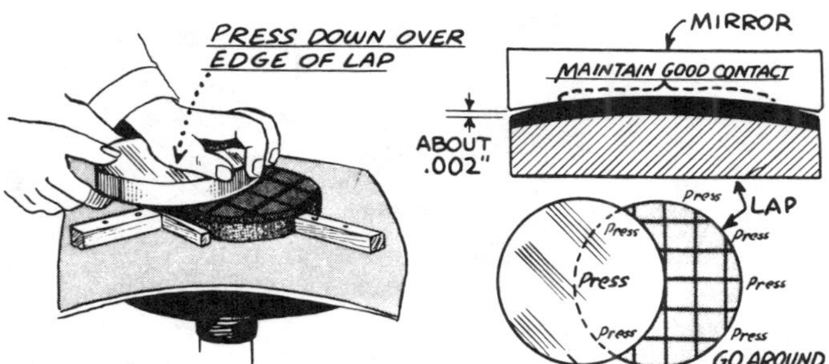

2 DEPRESSING EDGE OF LAP

APPLICATION: TURNED EDGE or HILL AT CENTER

YOU CAN DO THIS WITH EITHER HOT OR COLD PRESS, DEPENDING ON ACTION DESIRED. *GOOD INSURANCE AGAINST TURNED EDGE WHEN YOU ARE REDUCING A CENTRAL HILL*

DON'T OVERDO IT A DOMED LAP WILL FORM A HOLE

ALTERNATES:

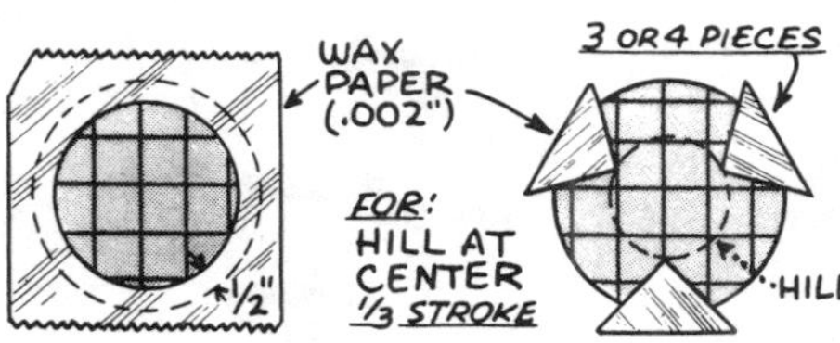

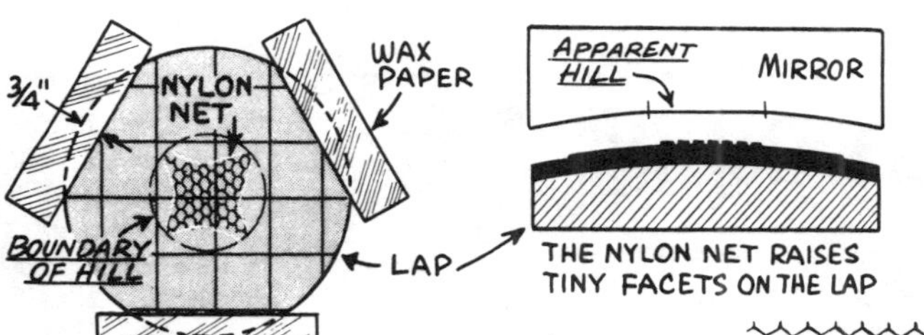

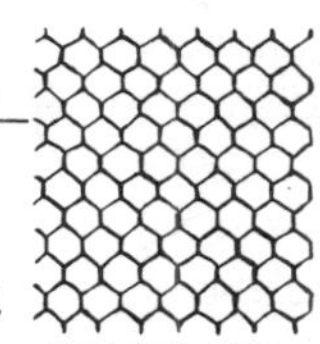

NYLON NET (.008" THICK)

3 RAISING CENTER OF LAP

APPLICATION: HILL AT CENTER

HOT-PRESS NET AND WAX PAPER... REMOVE WHEN COOL. COLD-PRESS 2 MIN... RESUME POLISHING. *CAUTION: THIS CUTS FAST*

Be Firm! DON'T WORK WITH A WARM LAP. WHENEVER YOU HOT PRESS, ALLOW 20 min. OR MORE "COOL-OFF" TIME

4 OVERHANGING MIRROR

APPLICATION: HILL AT CENTER

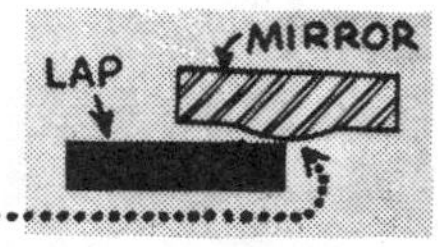

THE GENERAL IDEA IS TO WEAR THE HILL OVER THE EDGE OF THE LAP ......

THE WEIGHT OF THE OVERHANGING GLASS DOES THE WORK

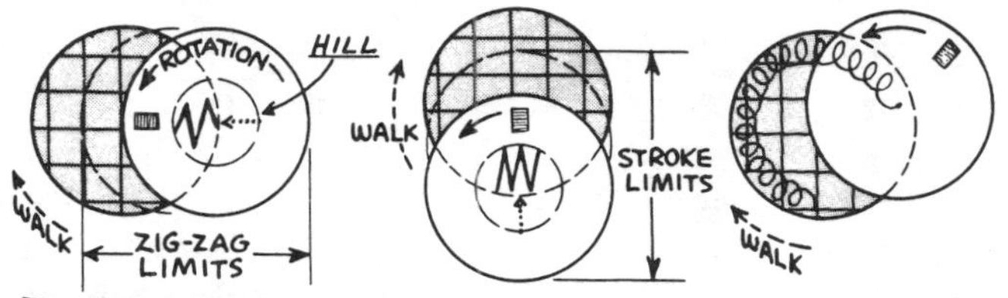

TANGENT STROKE *ZIG-ZAG 1/2 HILL DIA.*
RADIAL STROKE *STROKE 1/2 HILL DIA.*
CIRCULAR STROKE

YOU CAN USE ANY OF THE THREE STROKES SHOWN

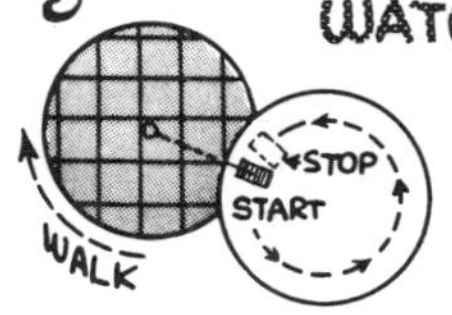

WATCH THE TAPE! ANY OVERHANGING STROKE WITH MIRROR ON TOP MUST BE PACED BY THE MIRROR ROTATION -NOT THE WALK-AROUND. *ALWAYS DO A WHOLE NUMBER OF ROTATIONS... TWO ROTATIONS WILL LEVEL A SMALL HILL*

5 ACCENTED PRESSURE

APPLICATION: TURNED EDGE

THIS TECHNIC USES THE LAP ON TOP. THE IDEA IS TO USE THE EDGE OF THE LAP AS A TOOL. THE RIGHT HAND IS THE PRESSURE HAND - THE LEFT HAND GUIDES AND ROTATES THE LAP, NO PRESSURE

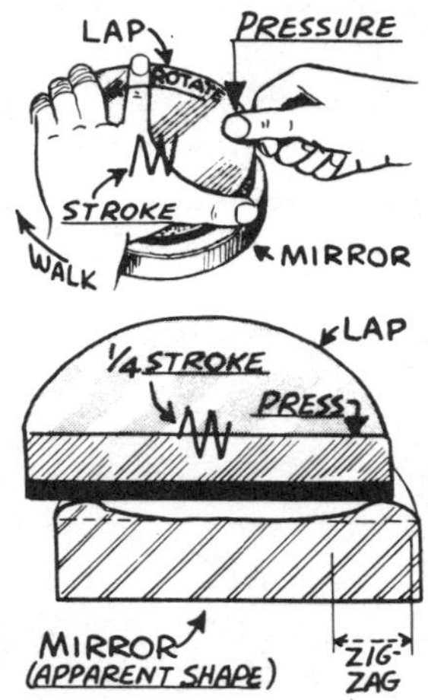

NOTE: *THIS IS AN OVERHANGING STROKE, BUT WITH LAP ON TOP, THE ACTION IS PACED BY THE WALK-AROUND*

ROTATE THE MIRROR. *DO TWO WALKS AROUND THE POST AND THEN ROTATE MIRROR A QUARTER TURN. REPEAT TWO WALKS*

6 ACCENTED PRESSURE

APPLICATION: REDUCING HYPERBOLA

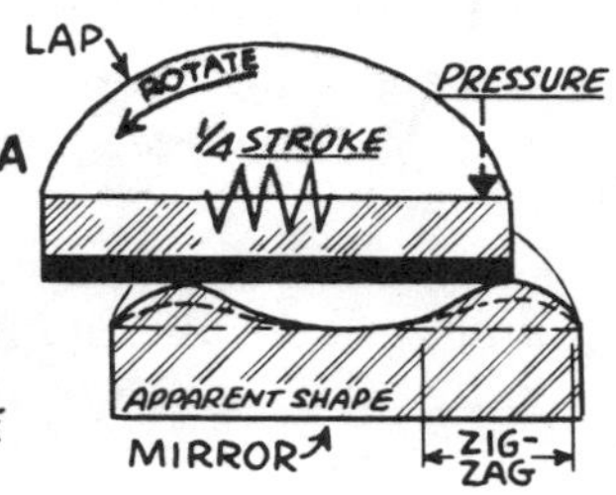

AS ABOVE EXCEPT ZIG-ZAG IS CENTERED ON APPARENT CREST. *IF THE CURVE IS SMOOTH, THE HYPERBOLA CAN BE REDUCED DIRECTLY TO PARABOLA - OTHERWISE REDUCE TO A SPHERE*

RING Defects... ARE SOMETIMES CAUSED BY LAP FAULTS. DIRT OR ABRASIVE ON A FACET WILL WEAR A DEPRESSED RING ON THE MIRROR SURFACE

REMEDY: *HOT PRESS OFTEN, USING NYLON NET OVER WHOLE LAP WHEN NEEDED TO ASSURE GOOD CONTACT AND ELIMINATE GLAZING. PATCH MISSING CORNERS OF FACETS EXCEPT AT EDGE WHERE THEY DO NO HARM. USE A ZIG-ZAG STROKE TO BLEND OUT ANY MINOR DEFECTS IN THE LAP*

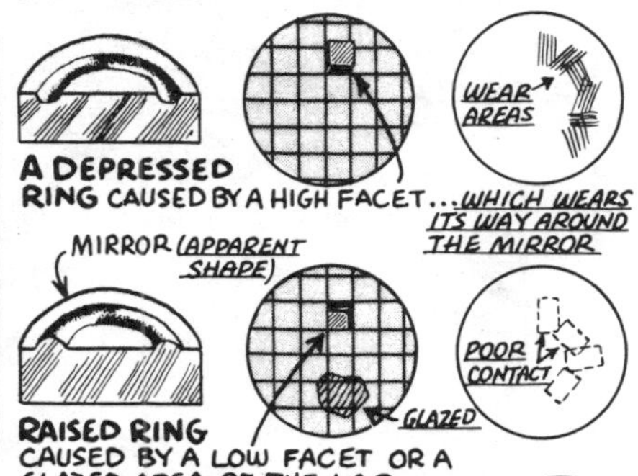

A DEPRESSED RING CAUSED BY A HIGH FACET... *WHICH WEARS ITS WAY AROUND THE MIRROR*

RAISED RING CAUSED BY A LOW FACET OR A GLAZED AREA OF THE LAP

DOG BISCUIT LAP FAULTS PERSISTING FOR 10 OR 15 Seconds CONTRIBUTE TO A "DOG BISCUIT" MIRROR →

Figuring the PARABOLOID

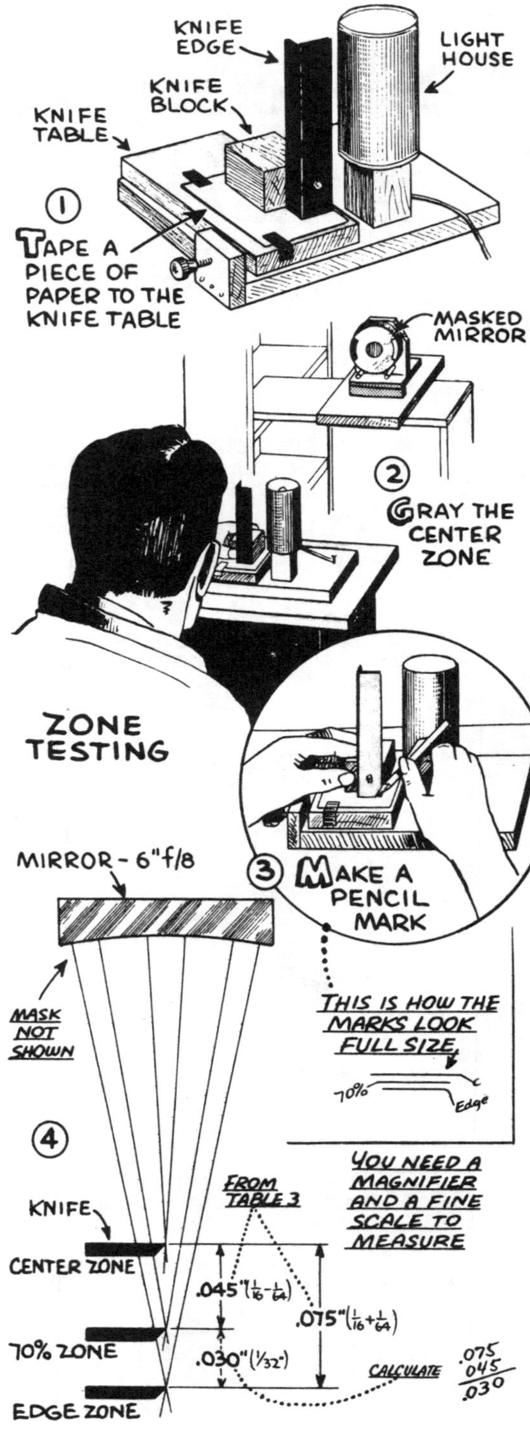

AS YOU may already know, the paraboloid is a defective surface when looking at a near object, suffering from a considerable amount of spherical aberration. If the near object is located at two focal lengths, as it is in the knife-edge test, the spherical aberration is easily calculated or can be obtained directly from Table 3. This particular amount of spherical aberration is known as the Mirror Correction. When your mirror shows the exact amount of "correction" specified in the table, the surface is a paraboloid, and it will have no S.A. at all when used for its intended purpose of looking at distant objects. Any departure from the mirror correction given is, of course, real spherical aberration and will affect the performance of your telescope.

ZONE TESTING. Testing involves measuring the difference in radii of three different zones of the mirror, or, as already explained, you measure the amount of spherical aberration. The zone to be tested is isolated with a thin cardboard mask. If you are using a simple tester, the procedure is as shown in Figs. 1, 2 and 3. After marking the position of the center zone, you move the knife back until a position is found where both edge zone openings in the mask show equally gray as the knife is cut into the light beam. Another mark is made on the paper. The operation is repeated for the 70% zone, and the result is a set of three marks, Fig. 4. They are close together and you will need a magnifier and fine scale or a direct-reading scale magnifier to measure exactly. If the marks measure within the values given in Table 3, you have a parabola or a near-parabola.

If you are using a micrometer tester, the center zone is tested first, after which the micrometer scale is set at zero, Fig. 6. The edge and 70% zones are then direct readings from the scale.

After each zonal measurement, you should remove the mask and get acquainted with the full-mirror shadow at that particular knife setting, Fig. 7. Of greatest interest is the shadow with

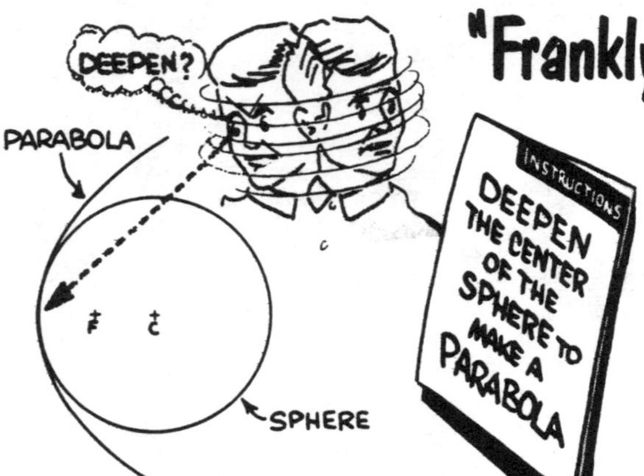

"Frankly, I Don't Get It!"

The parabola is commonly shown in relation to a circle of the same radius, as shown at left. What actually happens in making a mirror is that you form a smaller parabola--it is about 1/16" shorter radius and is more sharply curved at the vertex than the original sphere.

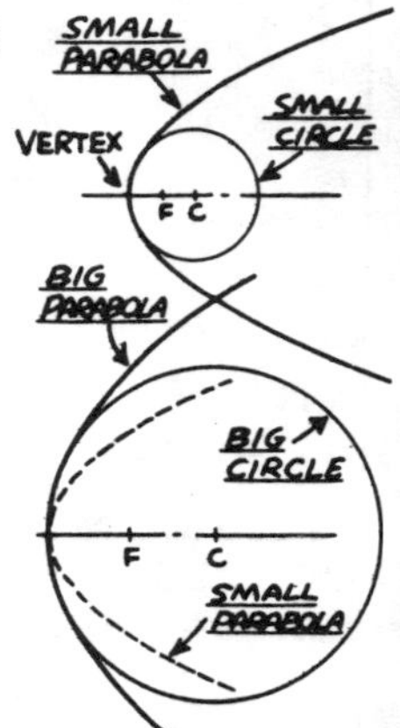

LIKE THE CIRCLE, THE PARABOLA HAS ONLY ONE SHAPE AND VARIES ONLY IN SIZE (RADIUS) THE RADIUS OF A PARABOLA IS MEASURED AT THE VERTEX

WHEN CIRCLE AND PARABOLA HAVE THE SAME RADIUS, THE PARABOLA IS LESS CURVED -- IT IS OUTSIDE THE SPHERE. HOWEVER, A SMALL PARABOLA FITS INSIDE A BIGGER CIRCLE

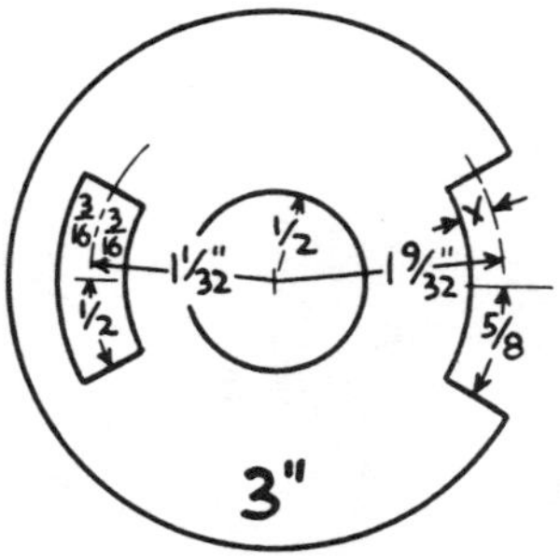

CLEAR FACE	X
3"	7/32"
2 15/16	3/16
2 7/8	5/32

<AVERAGE

MASKS

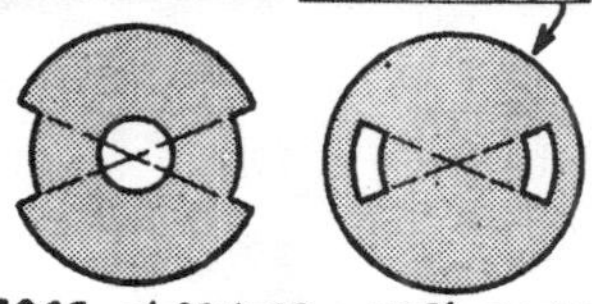

MAKE SET OF TWO FROM THIN WHITE CARDBOARD

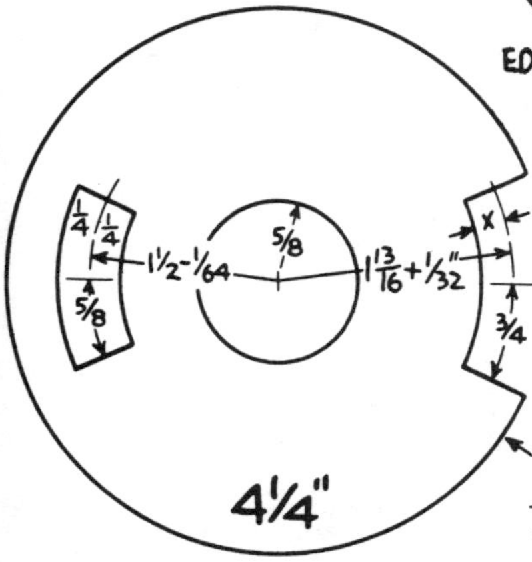

DIMENSION X ACCORDING TO CLEAR FACE OF MIRROR

DIAMETER SAME AS MAX. PHYSICAL DIA. OF YOUR MIRROR

CLEAR FACE	X
4 1/4"	9/32"
4 3/16	1/4
4 1/8	7/32
4 1/16	3/16

CLEAR FACE	X
6"	5/16"
5 15/16	9/32
5 7/8	1/4
5 13/16	7/32

FORMULA: CL. FACE RADIUS TIMES .707

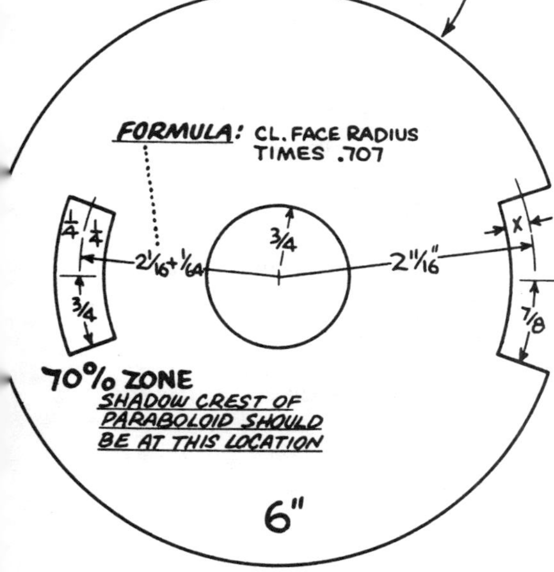

70% ZONE

SHADOW CREST OF PARABOLOID SHOULD BE AT THIS LOCATION

CLEAR FACE	X
8"	3/8"
7 15/16	11/32
7 7/8	5/16
7 13/16	9/32

70% ZONE DIVIDES YOUR MIRROR INTO TWO PARTS OF EQUAL AREA

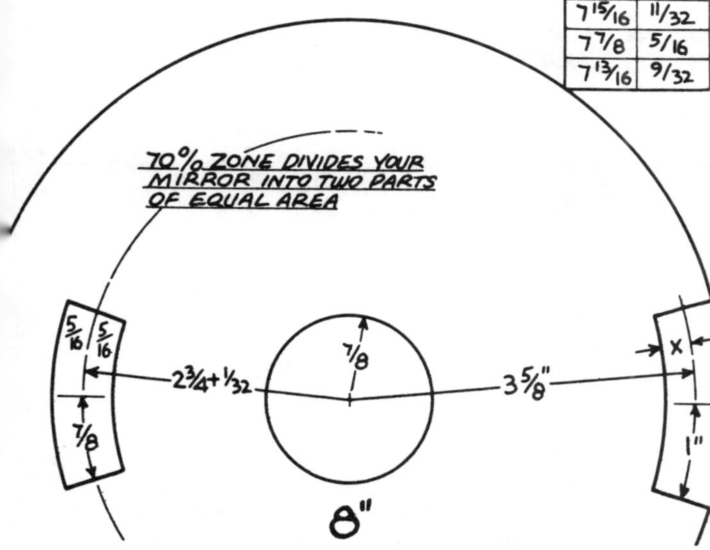

TABLE 3 — MIRROR CORRECTION [1]

NAME-SIZE DIA. [2]	NAME-SIZE f/	F.L.	R.	EDGE ZONE MIN.	EDGE ZONE IDEAL	EDGE ZONE MAX.	70% ZONE [3] MIN.	70% ZONE IDEAL	70% ZONE MAX.	WAVE RATING OF SPHERE [4]
3"	f/6	18"	36"	.028"	**.046"**	.064"	.018"	**.030"**	.042"	2/3 WAVE
	f/7	21	42	.014	**.039**	.064	.010	**.026**	.042	2/5
	f/8	24	48	.002	**.034**	.066	.001	**.023**	.045	1/4
	f/9	27	54	-.011	**.030**	.071	-.007	**.020**	.047	1/5
	f/10	30	60	-.024	**.027**	.078	-.016	**.018**	.052	1/8
AV. CL. FACE:	f/11	33	66	-.037	**.025**	.087	-.024	**.016**	.056	1/10
2 15/16"	f/12	36	72	-.052	**.023**	.098	-.034	**.015**	.064	1/12
4 1/4"	f/7	29 3/4	59 1/2	.032	**.057**	.082	.020	**.037**	.054	1/2
	f/8	34	68	.016	**.050**	.084	.010	**.032**	.054	1/3
	f/9	38 1/4	76 1/2	.002	**.044**	.086	.001	**.029**	.057	1/4
	f/10	42 1/2	85	-.013	**.040**	.093	-.008	**.025**	.058	1/5
	f/10.6	45	90	-.022	**.038**	.098	-.014	**.024**	.062	1/6
AV. CL. FACE:	f/11	46 3/4	93 1/2	-.028	**.037**	.102	-.018	**.023**	.064	1/7
4 3/16"	f/12	51	102	-.043	**.033**	.109	-.021	**.021**	.069	1/9
6"	f/5	30	60	.106	**.120**	.134	.063	**.072**	.081	2 1/8
	f/6	36	72	.080	**.100**	.120	.048	**.060**	.072	1 1/4
	f/7	42	84	.059	**.086**	.113	.035	**.051**	.067	4/5
	f/7.3	44	88	.051	**.082**	.113	.031	**.049**	.067	2/3
	f/7.7	46	92	.044	**.078**	.112	.027	**.047**	.067	9/16
	f/8	48	96	.039	**.075**	.111	.023	**.045**	.067	1/2
	f/8.3	50	100	.034	**.072**	.110	.020	**.043**	.066	1/2
	f/8.7	52	104	.027	**.069**	.111	.016	**.041**	.066	7/16
	f/9	54	108	.021	**.067**	.113	.012	**.040**	.068	1/3
	f/10	60	120	.004	**.060**	.116	.003	**.036**	.069	1/4
AV. CL. FACE:	f/11	66	132	-.014	**.055**	.124	-.008	**.033**	.074	1/5
5 7/8"	f/12	72	144	-.030	**.050**	.130	-.018	**.030**	.078	1/6
8"	f/4	32	64	.196	**.205**	.214	.116	**.121**	.126	5 3/4
	f/5	40	80	.149	**.164**	.179	.088	**.097**	.106	2 2/3
	f/6	48	96	.117	**.137**	.157	.069	**.081**	.093	1 2/3
	f/7	56	112	.089	**.117**	.145	.052	**.069**	.086	1
	f/8	64	128	.066	**.103**	.140	.038	**.060**	.082	2/3
	f/9	72	144	.044	**.091**	.138	.026	**.054**	.082	1/2
AV. CL. FACE:	f/10	80	160	.025	**.082**	.139	.014	**.048**	.082	1/3
7 7/8"	f/11	88	176	.005	**.075**	.145	.003	**.044**	.085	1/4

[1] THE MIRROR CORRECTION IS THE DIFFERENCE IN FOCUS BETWEEN CENTER ZONE AND EDGE ZONE (OR CENTER AND 70% ZONE) WHEN TESTED WITH A FIXED LIGHT SOURCE, USING THE MASKS SHOWN. THE TOLERANCE IS 1/4 WAVE UNDER AND OVER-CORRECTED

[2] THE ACTUAL CLEAR FACE IS A LITTLE LESS THAN NAME-SIZE

[3] DOUBLE THE VALUES AT 70% ZONE WILL GIVE THE FULL CORRECTION AND TOLERANCE AT EXTREME EDGE OF AVERAGE-SIZE MIRROR

[4] THIS COLUMN GIVES WAVE-RATING OF MIRROR IF GROUND TO A SPHERICAL SHAPE. MIRRORS RATED 1/4 WAVE OR LESS WILL BE SATISFACTORY, PROVIDING THE SURFACE IS A SMOOTH CURVE

Example:

YOUR MIRROR IS 6" f/8

FROM TABLE, READ:

CORRECTION BETWEEN EDGE AND CENTER ZONES:

IDEAL (*PERF. PARABOLA*) = **.075"**
MIN. (*1/4 WAVE UNDER*) = **.039**
MAX. (*1/4 WAVE OVER*) = **.111**

CORRECTION BETWEEN 70% ZONE AND CENTER ZONE:

IDEAL (*PERF. PARABOLA*) = **.045"**
MIN. (*1/4 WAVE UNDER*) = **.023**
MAX. (*1/4 WAVE OVER*) = **.067**

ANY CORRECTION BETWEEN MINIMUM-MAXIMUM VALUES IS A GOOD MIRROR, PROVIDING ALWAYS THE SURFACE IS A SMOOTH CURVE

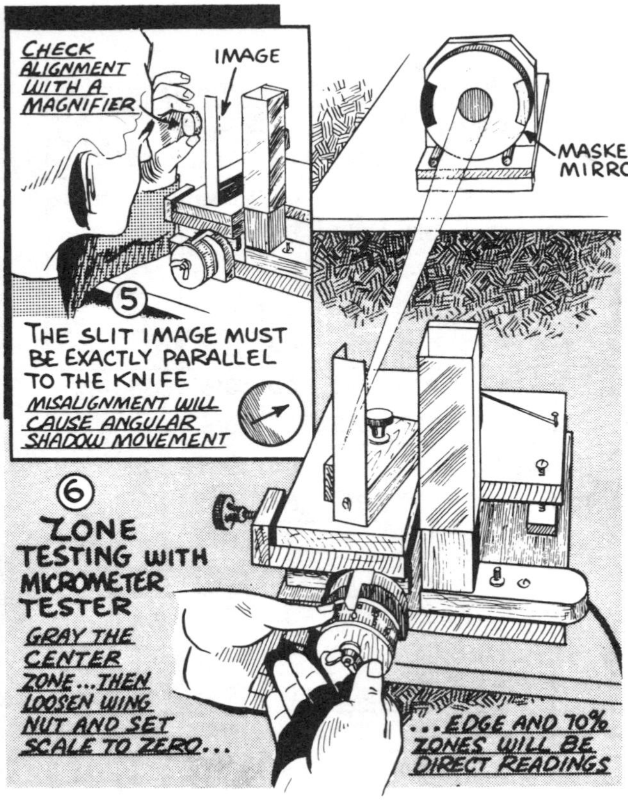

7 THE THREE COMMON FACES OF THE PARABOLOID

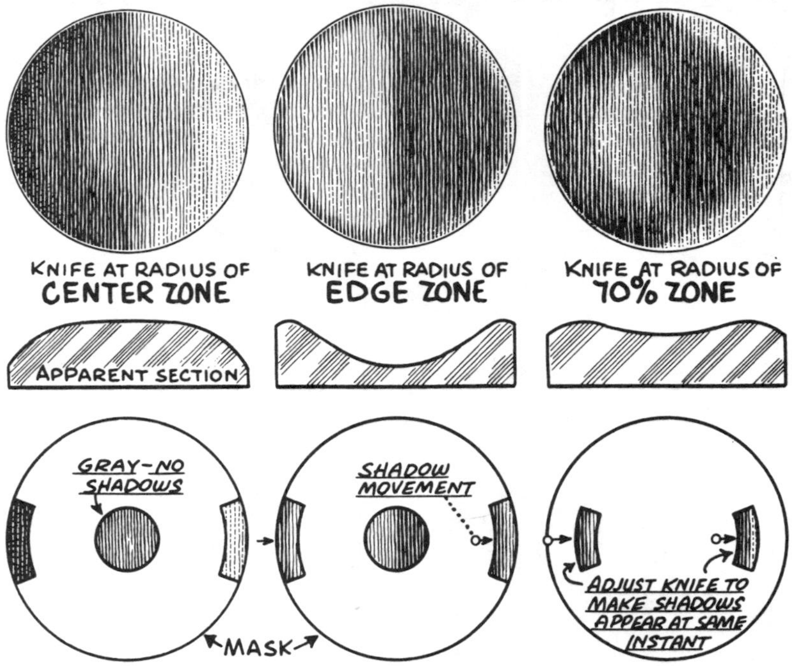

knife graying the 70% zone; this is the face of the mirror you analyze and study for proof of the paraboloid. It is a lightly-shaded figure when first tested, barely visible, but becoming stronger and more contrasty as you approach full correction. The first shadow to appear is at the left edge, followed immediately by a second shadow which originates well inside the right edge. Both shadows advance to the right as the knife is pushed more into the light beam, and it is this advancing shadow which you try to equalize in zone testing. Instead of judging equal grayness when using the mask, you will find it easier to watch for the first wisp of shadows in the mask openings--when they appear at the same instant, you have the knife in the proper location.

Taking zone measurements is a delicate operation. Beginners are often mystified when successive measurements at the same zone vary as much as .030 inch. Did something slip? Not at all--it's just a case where you have to sharpen your eye and be super critical. With practice, you can reduce your observing error to about .015 inch, and if you take the average of three or four readings, you will be in error no more than about .010 inch. The center is especially difficult because the light beam is very narrow. This zone has no moving shadow at all--it simply goes gray gradually all over and if you can detect the least shadow movement from either side, you are not in the proper position.

8 PARABOLIZING WITH LONG STROKES

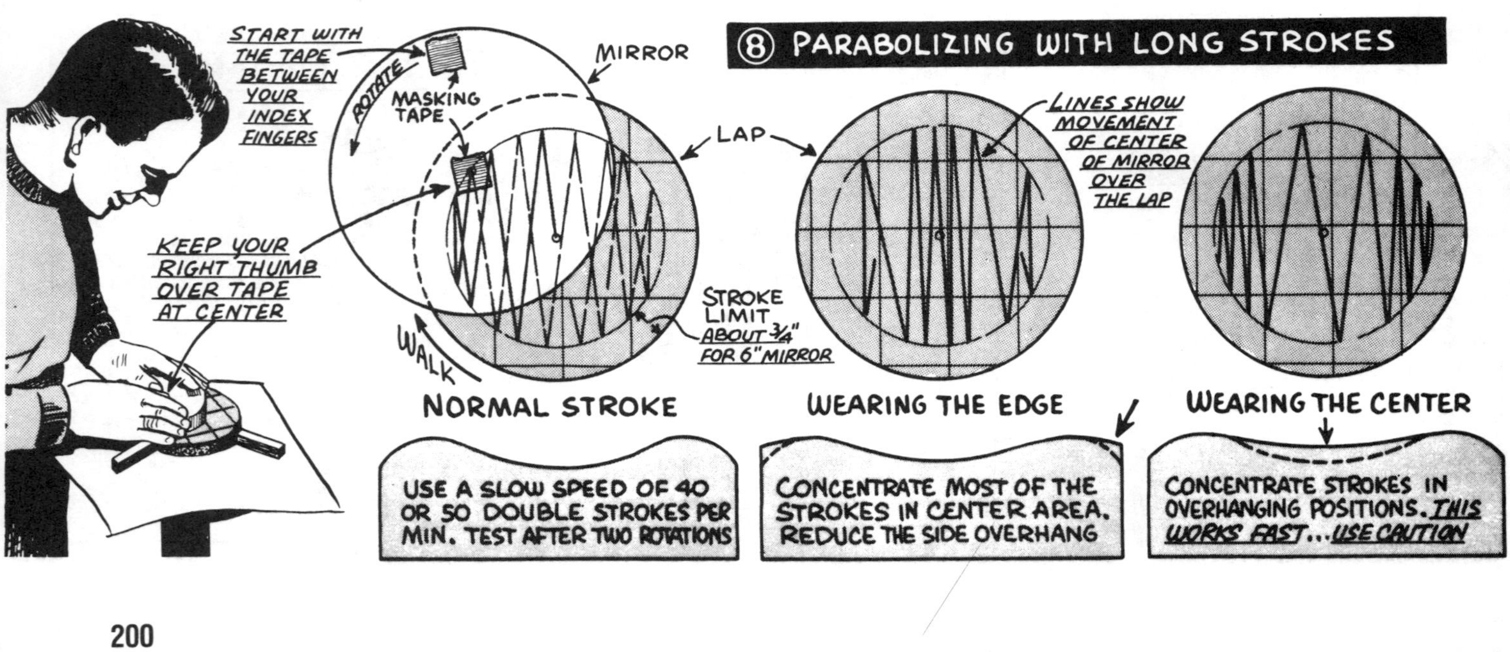

LONG STROKE PARABOLIZING. The actual work of parabolizing from a sphere or near-sphere is done with a stroke of maximum length and sidewise zig-zag, Fig. 8. Under good conditions you can parabolize a 6-inch f/8 in as little as five minutes. However, it is best to stretch the work over a longer period of about 20 minutes, using a slightly shorter stroke and less side overhang.

Long stroke parabolizing works beautifully when the lap is of the proper temper, which is medium soft. The actual situation is that your much-used lap is now very thin and hard as a rock. Unless the lap deforms under the long stroke, the paraboloid will not develop -- you will only make a hole at the center. The best way out is to make a new lap, adding about a half-teaspoonful of turpentine to the heated pitch and stirring gently for at least three minutes. Alternately, you can hot press and then start work while there is a trace of heat in the lap. This requires fine judgement but it is worth trying.

JOB PROCEDURE. Hot press for contact. Dunking the lap 5 min. in warm water (about 110 deg.) is satisfactory. Press with weight, about 10 lbs. for 5 min. and then let the mirror remain on the lap for another 15 min. without weight. In other words: be sure you make good contact; be sure the lap has lost its heat. Use a fairly heavy rouge mix, about 1:5. Do two rotations (not walks), using a piece of masking tape on back of mirror as an index marker. Another piece of tape at center of mirror will be found useful; put your right thumb over this and then think in terms of pushing your thumb over the lap to the limits shown in Fig. 8.

Place mirror on test rack and let stand at least 15 min. before testing in order to normalize. However, it is instructive to zone test immediately for comparison with a measurement at the same zone taken when the mirror has normalized. The idea here is that any kind of polishing generates more or less heat in the mirror. Then, when you put the mirror on the test rack, it is in the process of cooling-down to room temperature; a zone measurement of a warm mirror is not reliable. The change can also be noted in the appearance of the shadow.

The zone measurements together with the shadow appearance will dictate the work procedure. What you strive for is "normal wear" in parabolizing, as shown alongside Fig. 12. The initial period of parabolizing with the long stroke (two rotations) should about half parabolize your mirror. In other words, you should get zone

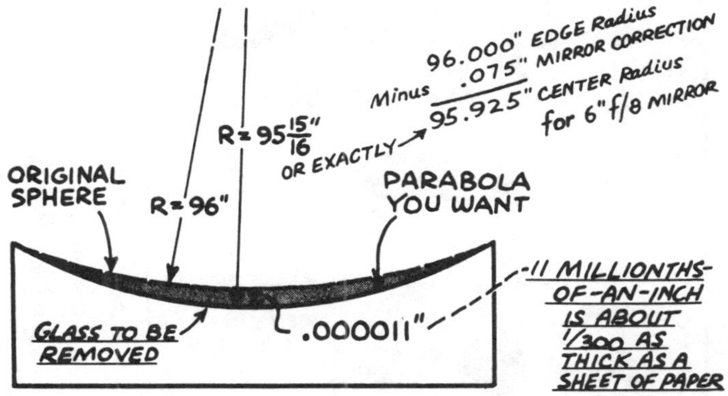

(9) KEEP THE REAL GLASS SHAPE IN MIND

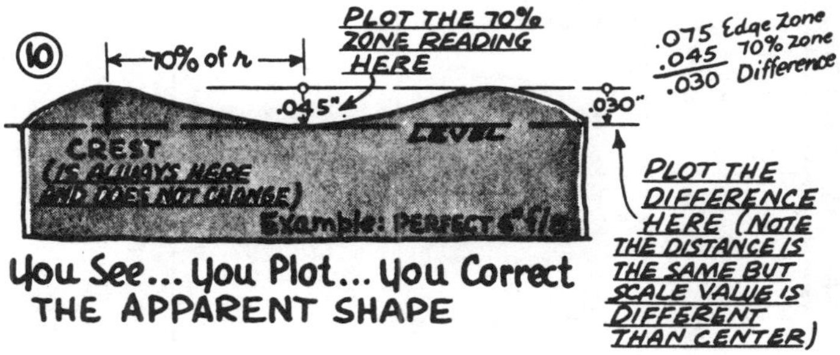

You See... You Plot... You Correct THE APPARENT SHAPE

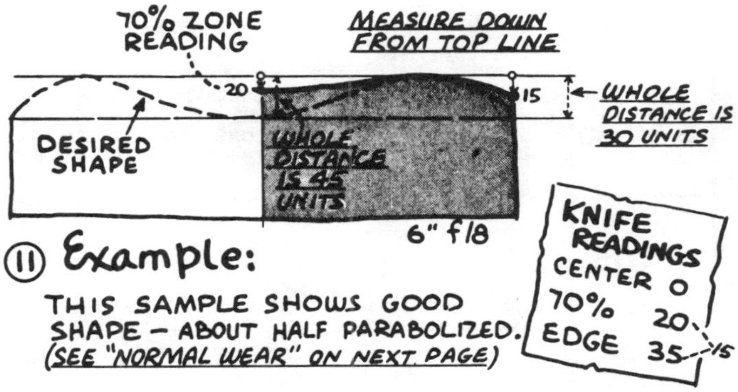

(11) Example:

THIS SAMPLE SHOWS GOOD SHAPE – ABOUT HALF PARABOLIZED. (SEE "NORMAL WEAR" ON NEXT PAGE)

measurements of about 0-20-35 thousandths-of-an-inch for center, 70% and edge zones respectively, using the micrometer tester. Of course, the chances of this happening on a first attempt are practically zero--something always happens! If you are not too far from Normal Wear, you can continue parabolizing; if you are off a mile, you simply return to the sphere and try again. You can expect to do this many times on a first mirror, but every time should make you a little more skilful. When you do get the parabola, it is liable to be rather sudden--after all, the sphere itself is only a half-wave from the perfect shape.

SHAPE DIAGRAMS. A sketch diagram of the mirror shape can be be constructed from the zone measurements. Graphs of this kind are set off from a straight baseline. If you zero the diagram at either center or edge zone knife reading, it will reveal a plain curve showing the physical shape,

⑫ SHAPE EXAMPLES - 6" f/8

READING CEN. 70% EDGE	APPARENT SHAPE	REMARKS and CORRECTION
DIFFERENCE 0 0 30 30	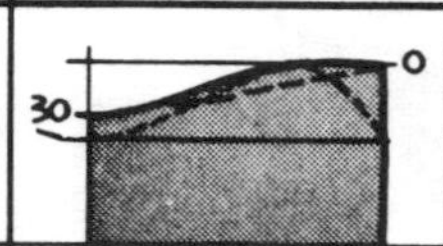	ALL HOLE. RETURN TO SPHERE AND TRY AGAIN
75 0 0 75	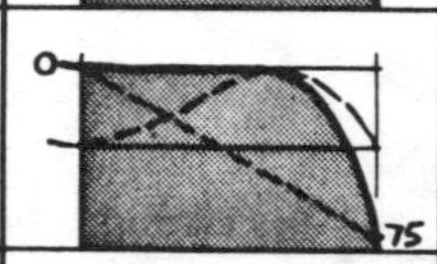	ALL TURNED EDGE. RETURN TO SPHERE
15 0 20 35	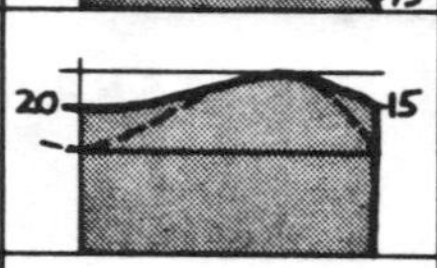	NORMAL WEAR... KEEP GOING JUST AS BEFORE
5 0 25 30	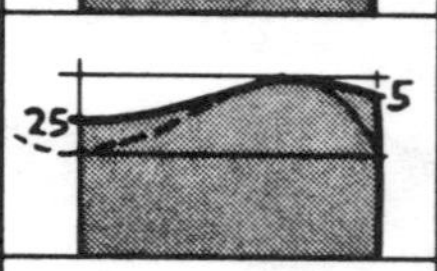	HOLE STARTING AT CENTER. NEEDS MORE WEAR AT EDGE
30 0 10 40	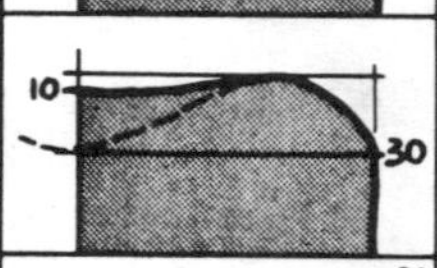	TOO SHALLOW AT CENTER. NEEDS WEAR AT CENTER
-20 0 60 40	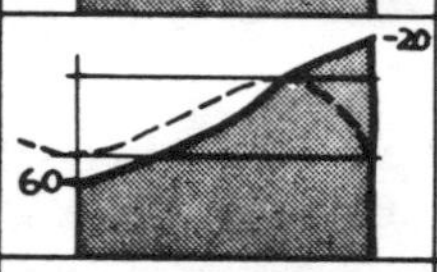	DEEP HOLE AND RAISED EDGE. GO BACK TO SPHERE
50 0 30 80	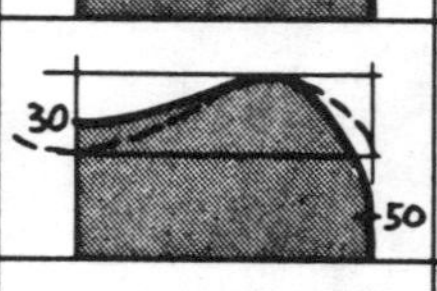	TURNED EDGE. REDUCE AT 70% ZONE WITH LAP ON TOP
30 0 90 120	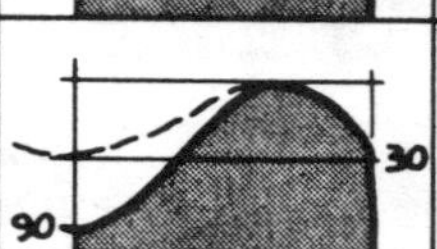	HOLE AT CENTER. REDUCE AT 70% ZONE WITH LAP ON TOP
35 0 55 90		OVER-CORRECTED. OK AS IS OR WEAR BACK TO PARABOLA

YOU ZERO AT CENTER. ALL READINGS IN THOUSANDTHS-OF-AN-INCH

NORMAL WEAR

0	0	0
SPHERE		
0	3	5
0	6	10
0	9	15
0	12	20
0	15	25
0	18	30
0	21	35
0	24	40
¼ WAVE LIMIT*		
0	27	45
0	30	50
0	33	55
0	36	60
0	39	65
0	42	70
0	45	75
PARABOLA*		
0	48	80
0	51	85
0	54	90
0	57	95
0	60	100
0	63	105
0	66	110
¼ WAVE LIMIT*		
0	69	115
0	72	120
0	75	125
0	78	130
0	81	135
0	84	140
0	87	145
0	90	150
0	93	155
0	96	160
0	99	165
0	102	170
0	105	175

*for 6" f/8

VALUES LISTED ARE EXACT for ANY 6-INCH MIRROR, and NEARLY EXACT for OTHERS with the MASK SPECIFIED

⑬ Wave-rating Your MIRROR

A LITTLE DETECTIVE WORK ON TABLE 3 WILL REVEAL THAT THE ¼ WAVE TOLERANCE FOR A 6" f/8 MIRROR AT THE EDGE ZONE IS .036". HENCE, ONE WAVE = 4 × .036 = .144"

.075
.039
.036

TO RATE YOUR MIRROR, DETERMINE HOW MUCH IT DEPARTS FROM THE IDEAL .075" CORRECTION. SAY YOUR READING IS .060"... YOU ARE .015" UNDER-CORRECTED. OMIT THE DECIMAL MARK AND WRITE THE FIGURE AS A FRACTION:

$$\frac{15}{144} = \frac{1}{10} \text{ WAVE}$$

AVERAGE $\frac{1}{11}$ WAVE

DO THE SAME AT 70% ZONE WHERE 1-WAVE = .088" SAY YOU ARE .007" OFF EITHER OVER OR UNDER

$$\frac{7}{88} = \frac{1}{12} \text{ WAVE}$$

like Fig. 9. By making a diagram zeroed at the crests, you can obtain the more familiar doughnut, which is the shape you see and analyze when testing. Fig. 10 shows this construction. The zone measurements are set off, downwards from a straight line. A little juggling is needed in applying the zone measurements to the diagram. You zero the diagram at the crest; the depth at the center of diagram is the 70% zone reading; the depth at the edge is the edge zone reading minus the 70% zone reading.

The perfect "doughnut" shadow of a parabola has the same depth at edge and center. Fig. 10 shows the perfect shape--edge and center are level. The way the readings are applied, the diagram will be 45 units deep at the center and 30 units deep at the edge for a perfect 6-inch f/8 parabola. Edge and center being the same depth on the diagram, as already explained, the scale of the diagram at these points will be different.

Fig. 11 shows a sample diagram. You have made zone measurements with a micrometer knife-edge tester and have obtained readings of 0, ·20 and 35 units (thousandths-of-an-inch) for center, 70% and edge zones, respectively. These distances are set off on the diagram in the manner described. At the center, 20 units is shy of half the whole ideal distance of 45 units; at the edge, 15 units is exactly half of the desired 30 units. With the crest of the shadow at zero, you can now sketch in the approximate doughnut shape. In this particular example, the job is going along just fine, the mirror being about half way to full parabolization.

Other shape diagrams are shown in Fig. 12, covering most of the actual work situations. Since you always zero at the crest, you will get no direct information about this part of the mirror from the diagram. However, the diagram shows plainly if glass is to be removed at center, at edge, or both. Meanwhile, the readings alone will give you a pretty fair idea of the mirror shape and depth, while the visible shadow itself when tested at the 70% zone will show the same kind of pattern as your shape diagram.

RATING YOUR MIRROR. When you get a pretty fair set of zone measurements and the 70% zone shadow looks smooth without too much dog biscuit, you can say your mirror is done. You will be interested in giving it a wave-rating, and this is easily done by the method shown in Fig. 13. Common practice is to try for at least 1/8 wave, just to be sure you have an honest 1/4-wave mirror. You have to be an expert with a lot of practice before you can read shadow positions

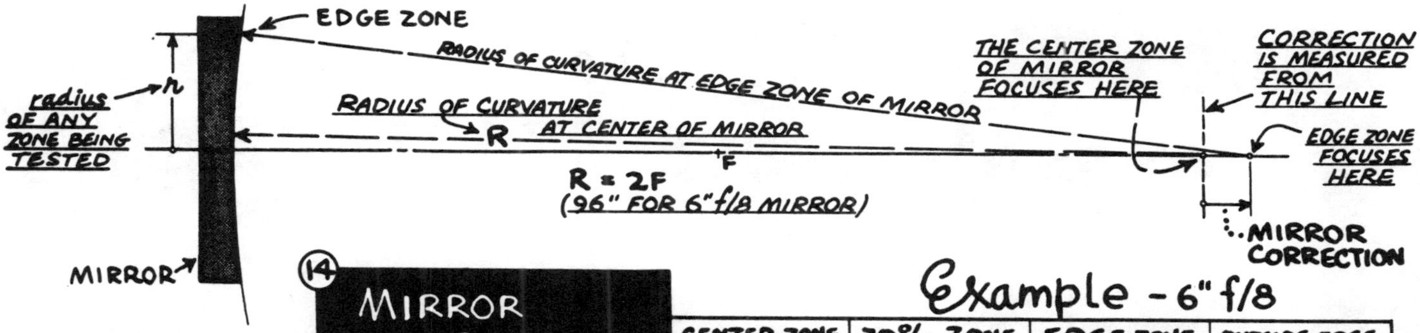

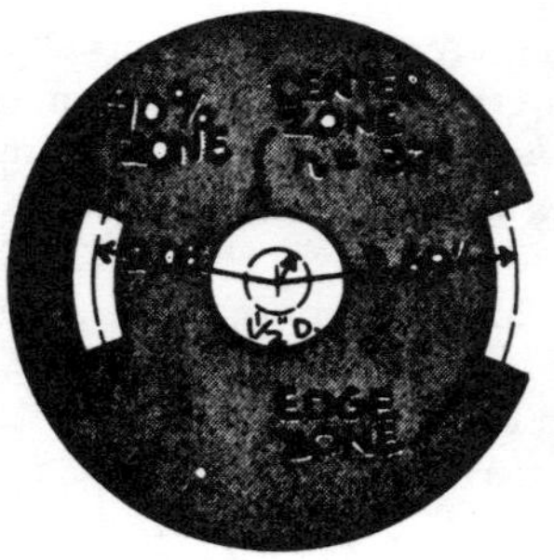

TEST ZONES - 6" MIRROR

14

Example - 6" f/8

MIRROR CORRECTION	CENTER ZONE $r = .37''$	70% ZONE $r = 2.08''$	EDGE ZONE $r = 2.69''$	OUTSIDE EDGE $r = 3''$
Formula: CORRECTION $= \frac{r^2}{R}$	Corr. $= \frac{.37^2}{96}$ $= \frac{.14}{96} = .001''$	Corr. $= \frac{2.08^2}{96}$ $= \frac{4.33}{96}$ $= .045''$	Corr. $= \frac{2.69^2}{96}$ $= \frac{7.24}{96}$ $= .075''$	Corr. $= \frac{3^2}{96}$ $= \frac{9}{96}$ $= .094''$ (96) 9.000; 864; 360; 384)
THE FORMULA ABOVE IS FOR ANY TEST RIG WHERE THE LIGHT IS STATIONARY. WHEN LIGHT AND KNIFE MOVE TOGETHER: CORRECTION $= \frac{r^2}{2R}$	THIS IS USUALLY NEGLECTED... CENTER ZONE IS SIMPLY ZERO			NOTE: IT IS NOT POSSIBLE TO TEST AT EXTREME EDGE
	THESE CALCULATIONS FOR VARIOUS MIRRORS ARE WORKED OUT IN TABLE 3			

much closer than .010 inch. So, anytime you get within .020 of the required corrections--over or under--it is just as well to call the job done, which indeed it may be.

MIRROR CORRECTION. At times you may work a mirror not covered by Table 3 specifications, or, you may wish to test at zones other than those specified. For such cases, you can calculate the proper "correction" for any zone (radius) of the mirror with the familiar formula: r^2 / R. The lower case r is the radius of the mirror or any zone of the mirror, while the big R is the radius of curvature, which is two times the focal length.

Fig. 14 gives the equation and also works out as an example the correction for a 6-inch, f/8 mirror. The basic form of the equation calls for a division by 2R (2 times the radius of curvature), and you use the equation in this form if you are using a test rig where pinhole and knife edge move together as a unit. More commonly, the pinhole is in a fixed position, so that the knife must move twice as far as when the "correction" is split between knife movement and pinhole movement. In this case, the "correction" is the more familiar r^2 / R formula. You apply this to any zone of any mirror; four test zones are often used for mirrors over 6-inch diameter. These zones can be located as desired, the sole idea being to get a series of zones representative of the whole mirror. The "correction" at center zone mask opening is very small, being in the neighborhood of .001 inch, and this item is usually neglected--the center is plain zero. Note that while you can calculate the correction for the extreme edge of the mirror, you can't test the extreme edge because you need a strip of mirror about 1/2 inch wide for testing, and the center of this strip is your test radius. As shown in Fig. 14, the correction for a 6-inch f/8 mirror is .094 inches, this being the calculated value for the extreme edge of a mirror exactly 6 in. diameter or 3 in. radius. But the radius of the outer test zone is only 2-11/16 in., with the result the actual working correction for the mirror is the considerably smaller .075 inch.

Many beginners over-correct their mirrors by using the full .094 inch correction when testing the edge zone. This is well inside the quarter-wave limit, as given in Table 3, but it means the mirror has gone beyond the parabola and is now a hyperboloid.

ALUMINIZING SERVICE. Very few amateurs attempt mirror silvering. Today, the work is done with a vapor vacuum process, and the metal coating is aluminum instead of the traditional silver. Aluminum is slightly less reflective than silver but it stays bright indefinitely and soon out shines the silver which tends to blacken with age. A good job of aluminizing will last 10 to 12 years or even longer. The common failure is minor pinholing, which develops slowly and does not greatly impair the performance of the mirror. Aluminizing and over-coating a 6-inch mirror costs about $10 (1976). Contact the firm of your choice and get a price before sending your mirror. Various firms doing this kind of work are listed in the index.

telescope gadgets You Can Build

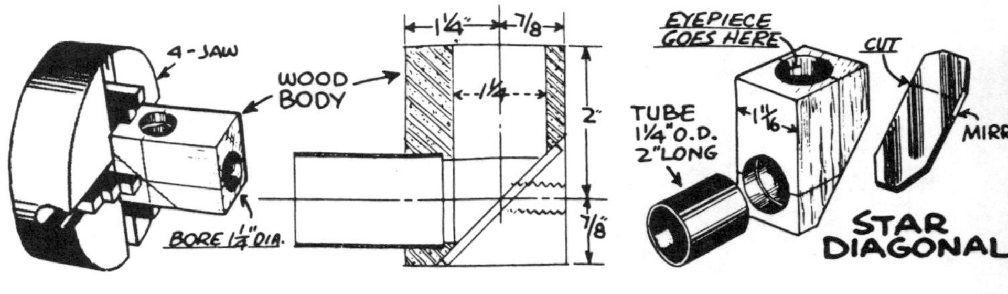

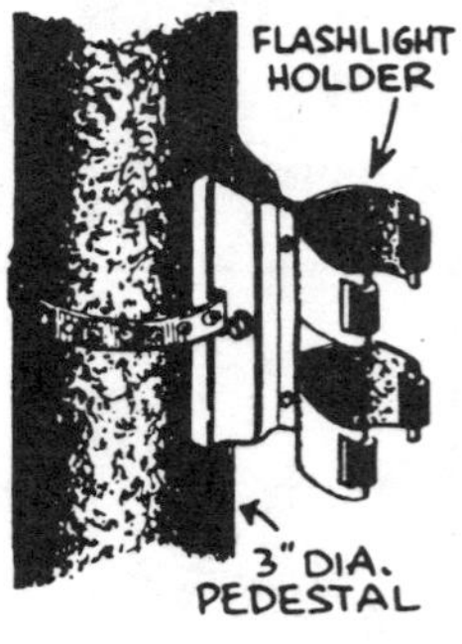

IF YOU have a pedestal base mount, a flashlight holder is just the thing for mounting a battery case on the pedestal. You can buy a holder for automobile steering post for about $1.00 at most hardware stores. It has an adjustable metal strap that fits posts from 1-1/2″ to 3″ diameter. Mount the holder on the south side of the pedestal where it will not be in the way of the telescope. You can also use the same mounting to hold a shelf.

STAR DIAGONAL. If you have a refractor, a star diagonal is almost a necessity for looking at stars near the zenith. It is easy to make one from a first surface mirror and a block of wood. A layout is first made on the side of the wood block, after which the block is mounted in a 4-jaw chuck for boring in the lathe. The diagram gives dimensions to suit standard 1-1/4" diameter eyepieces; a smaller size can be designed in the same manner, and the smaller holes may permit simple drilling on the drill press. Band sawing is satisfactory for the 45-degree angle cut which is made last. The mirror is simply taped in place with adhesive or masking tape. The unit shown requires 3-1/4" of "in" focusing travel.

ERECT DIAGONAL. You can get a normal erect image with any reflecting telescope by using a penta roof prism in place of the usual mirror diagonal. A homely but satisfactory mounting is made by sawing a piece of plastic or hardwood to fit the prism, taping or cementing the prism in place, as shown. This is a fairly big prism and is best suited for a 6-inch reflector where it obstructs about 8% of the light. However, it works fairly well on a 4-1/4" reflector despite the 15% obstruction.

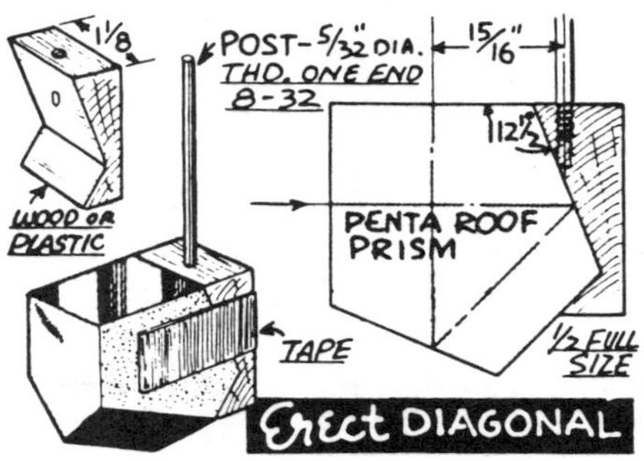

You can build your telescope around a prism diagonal like this, or you can use it merely as an accessory item. In any case it takes about 1-3/4 inch "in" focusing travel. This can be obtained in any reflector but may require adjustment of the main mirror. Alternately, you can get "in" by using a long eyepiece, such as a 2 inch f.l. symmetrical, which of itself will "reach in" to the image.

PIGGY BACK ERECTOR. When you already have a battery of eyepieces, the practical way to make an erecting unit is to use the eyepieces you have, mounting one on top of another, piggy-back fashion. A good combination for this system is Edmund 28mm RKE eyepiece and Edmund 3/4" symmetrical, arranged as shown. An adapter is needed to slip over the RKE eyepiece which becomes the erecting member of the optical system. The free end of the eyepiece tube will take any standard astro eyepiece, and you may like to experiment with other combinations. The system shown gives a nice clear image although the apparent field of 30 degrees is on the small side. About 1 inch "out" focus travel from the position of the 28mm RKE used alone will put the combination in focus.

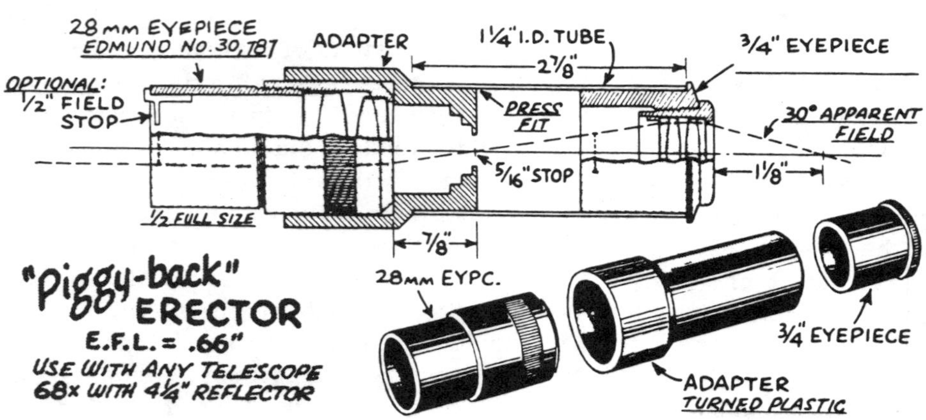

ILLUMINATED EYEPIECE. This is a real joy to use in any finder or tracking scope. The ideal illumination is a grain of wheat lamp with rheostat control. There are several styles of this miniature lamp; the one used is the brass "button" type. A socket is available for this, but since you have to make the base block anyhow, it is perhaps just as easy to make the whole socket-base from a plastic knob, as shown. If the power supply is mounted on the column of pedestal base, the preferable power line is a flexible braided two-wire cord; if the run is short, as it will be if the battery case is mounted on the telescope, the original wire leads from the battery holder will be long enough. The reticle itself can be a simple crossline scribed on clear plastic with a sharp needle, using the very lightest pressure. For tracking as required for star photography, a grid of crosslines is preferable--you can use any intersection. Be sure lamp contact points are scraped clean.

ZOOM BARLOW. Actually this is only a half-zoom because it does not keep the image in focus at all settings. What it does is set the degree of magnification, after which you focus with rack-and-pinion in the usual way. It is fun to use and makes all powers instantly available. The mechanism is a common one in which rotation without

vancing the end member (the eye-
ece) causes the Barlow lens to move
side the sleeve. High power is ob-
ned by screwing the eyepiece clock-
se, causing the Barlow cell to ad-
nce. With either a 4-1/4 or 6 inch
flector, the overall magnification
creases about 6X per turn of the
epiece. The top M is about 100X.

N DIAGONAL. A Herschel wedge in
mbination with a No. 4 sun filter
the perfectly safe way to see the
n spots with a refractor. All of the
at is directed against the back sur-
e of the box, which is covered with
bestos paper. The metal cover
ps over the wood body and is held
the eyepiece tube which should be
removable press fit, allowing dis-
sembly for cleaning. Like the more
niliar star diagonal, the sun diagon-
needs about 3 inches of "in" focus-
travel.

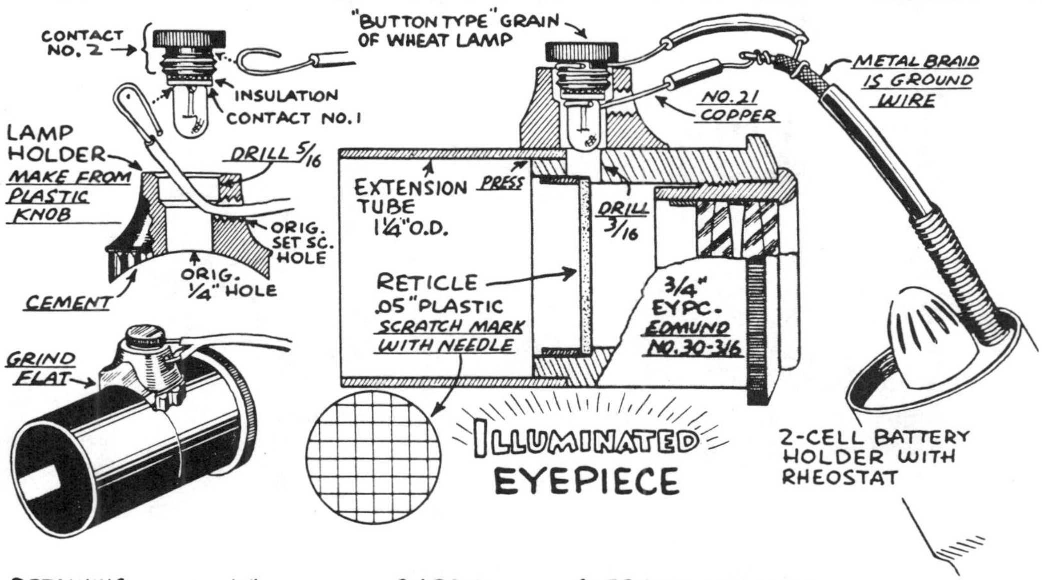

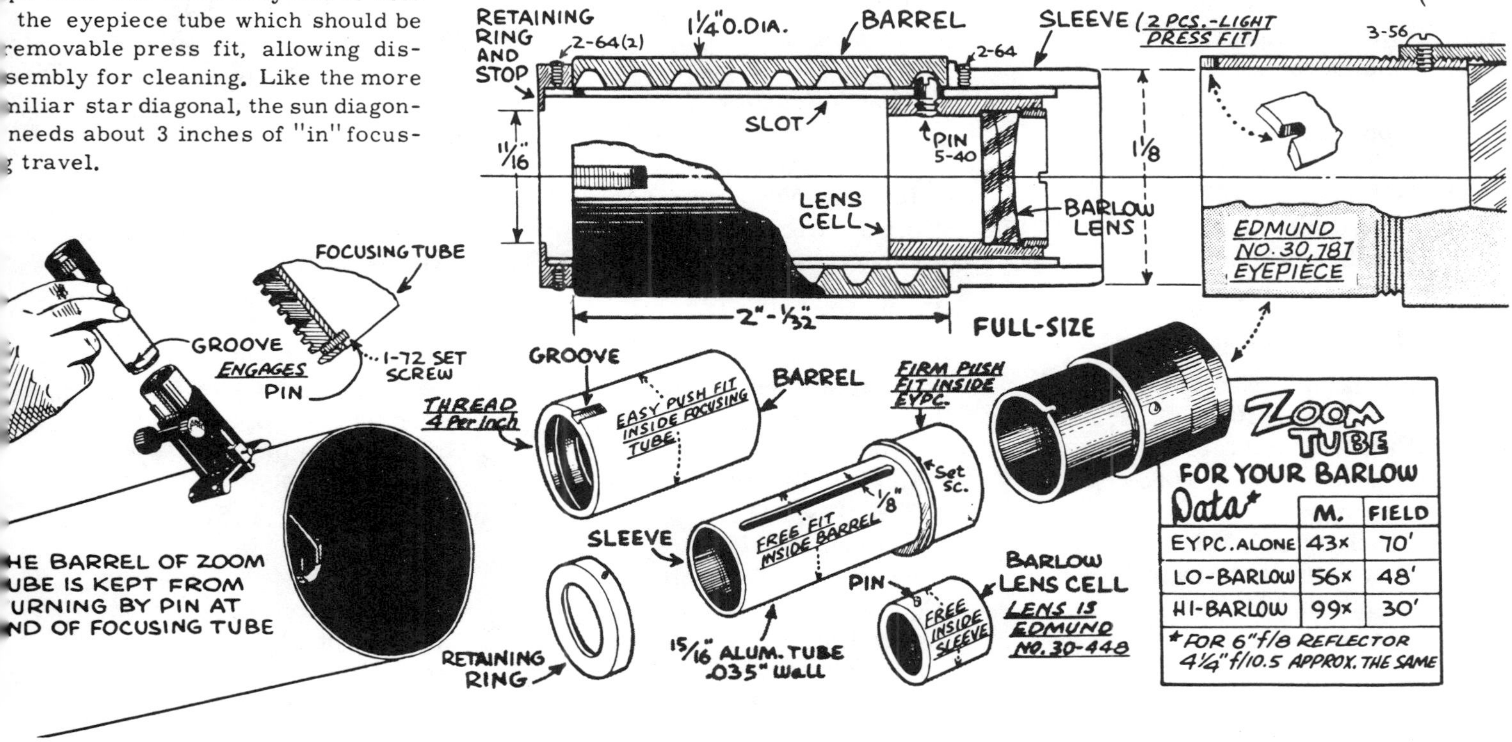

Data*	M.	FIELD
EYPC. ALONE	43x	70'
LO-BARLOW	56x	48'
HI-BARLOW	99x	30'

*FOR 6" f/8 REFLECTOR 4¼" f/10.5 APPROX. THE SAME

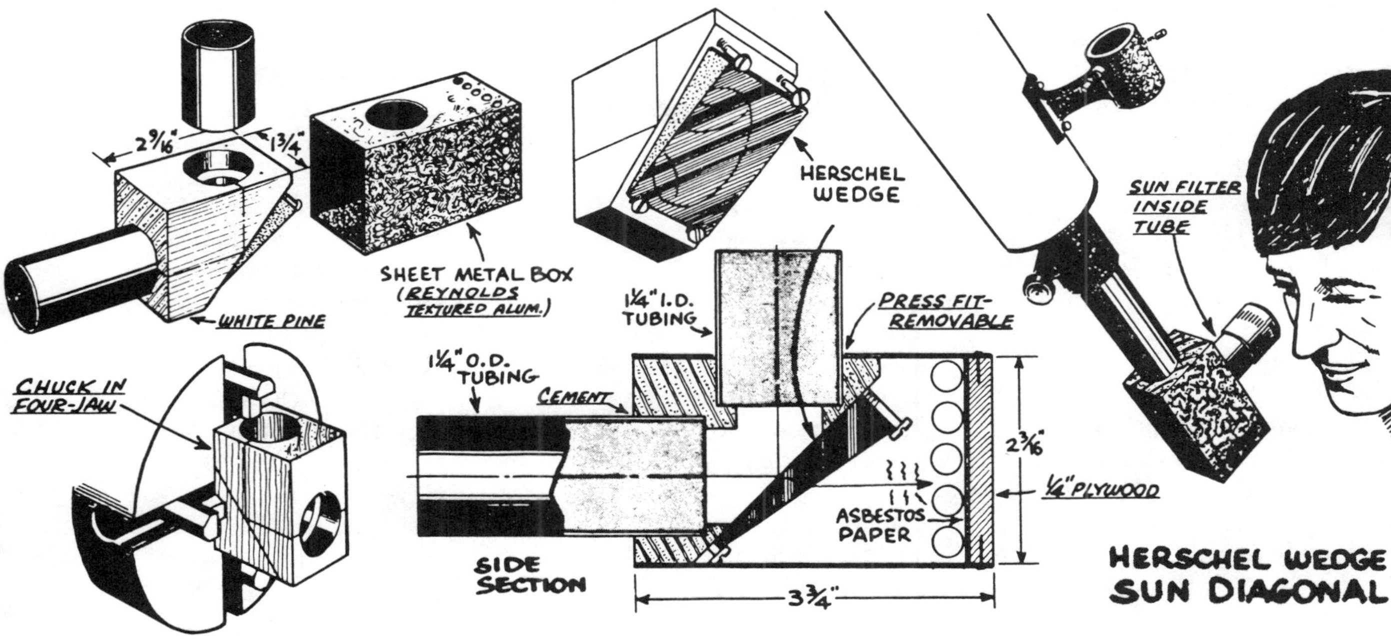

Index